Environmental Geology laboratory

2nd edition

Written and illustrated by

Tom Freeman

University of Missouri-Columbia

Oil invading wetlands and marshes of the Mississippi delta, flyway for countless migratory waterfowl (May 23, 2010)
inapcache.boston.com/universal/site_graphics/blogs/bigpicture/oil_05_24/o16_23539669.jpg
(Cover story is on pages 216-217. Related eruption of methane story is on page 256)

WILEY

www.wiley.com/college/freeman

You, too, can visit these Google Earth™ sites

I have replaced all air-photo stereograms that appeared in the previous edition of *Environmental Geology Laboratory* with images from Google Earth™.

If you wish to experiment with any of the sites from which I extracted these images, do the following:

(a) Download the newest version of free Google Earth™ from the Web (if you haven't already done so), and launch it.

(b) From the main menu, pull down the *View* menu and open the *Sidebar*.

(c) On the *Sidebar*, in the *Fly to* labeling space, keystroke the site's latitude and longitude that appear along with the location map, graphic scale, and the image (e.g., page 22).

> *Tip:* In keystroking the latitude and longitude, you need not include the symbols for degrees, minutes, and seconds. So (e.g., page 22), N. 43° 46' 39", W. 111° 57' 54" becomes N 43 46 39 W 111 57 54.

(d) Click on the magnifying glass icon to *Begin search*.

Voilá. You're at the site, which—via *View*, *Show Navigation*, *Always*—you can zoom in and out, rotate, tilt, and, with *Sun* (which began with version 4.3), slide the time of day control to create the shadow you prefer.

ISBN 978-0-470-13632-4

Printed in the United States of America.

10 9 8 7 6 5 4 3 2 1

CONTENTS

Tools supplied by instructor

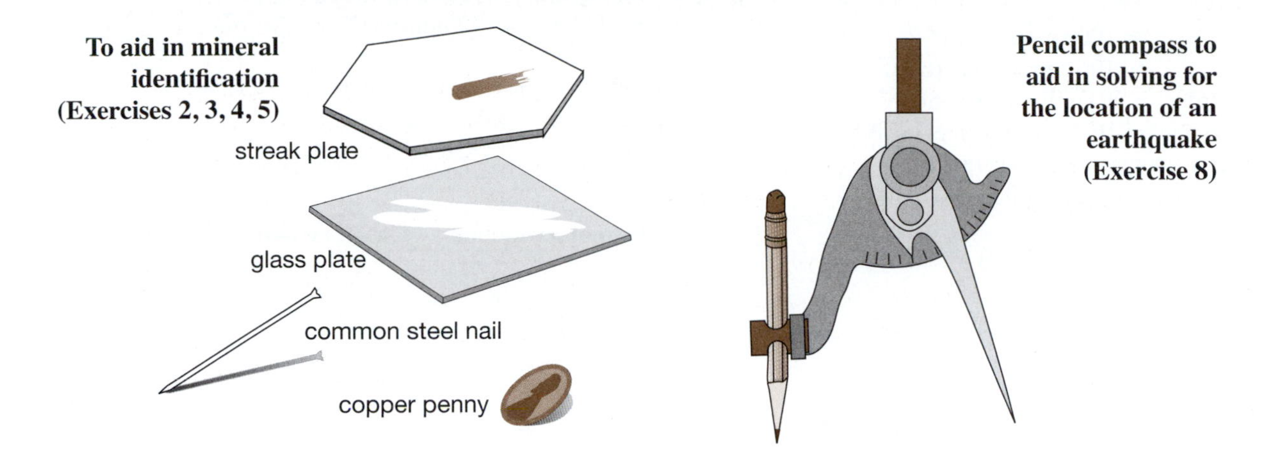

Two additional tools are included here

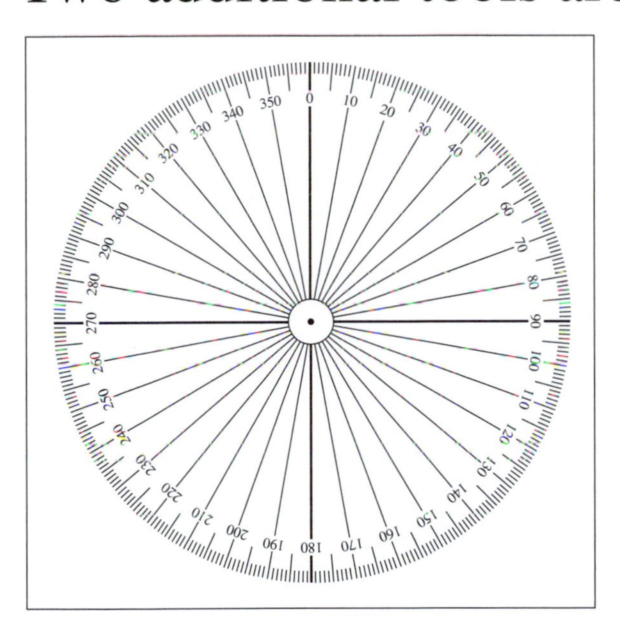

Protractor and ruler for measuring map directions and distances (Exercises 1)
Also for measuring slope angle (Exercise 11)

Inches and millimeters
(drawn to scale)

Additional on the Web

• Measurements and their conversions • Temperature conversions
• International system of units

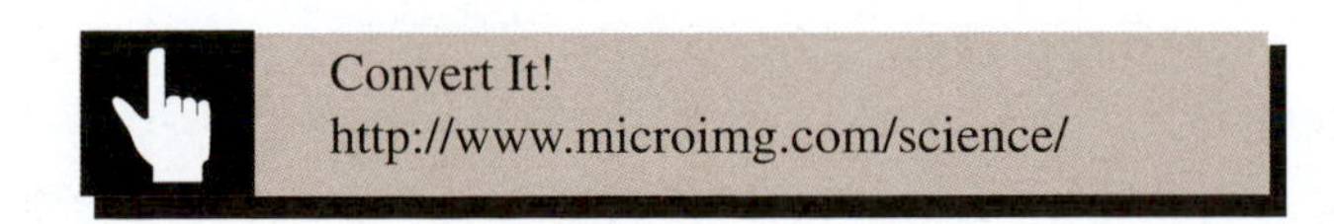

Measurements and their conversions

Lengths

Metric measure

1 kilometer (km) = 1000 meters (m)
1 meter (m) = 100 centimeters (cm)
1 centimeter (cm) = 10 millimeters (mm)
1 millimeter (mm) = 1000 micrometers (μm), formerly known as microns (μ)
1 micrometer (μm) = 0.001 millimeter (mm)

Nonmetric measure

1 mile (mi) = 5280 feet (ft) = 1760 yards (yd)
1 yard (yd) = 3 feet (ft)
1 fathom (fath) = 6 feet (ft)

Conversions

1 kilometer (km) = 0.6214 mile (mi)
1 meter (m) = 1.094 yards (yd) = 3.281 feet (ft)
1 centimeter (cm) = 0.3937 inch (in)
1 millimeter (mm) = 0.0394 inch (in)
1 mile (mi) = 1.609 kilometers (km)
1 yard (yd) = 0.9144 meter (m)
1 inch (in) = 2.54 centimeters (cm)
1 inch (in) = 25.4 millimeters (mm)
1 fathom (fath) = 1.8288 meters (m)

Area

Metric measure

1 hectare (ha) = 10,000 square meters (m^2)

Nonmetric measure

1 square mile (mi^2) = 640 acres (ac)
1 acre (ac) = 4840 square yards (yd^2)
1 square foot (ft^2) = 144 square inches (in^2)

Conversions

1 square kilometer (km^2) = 0.386 square mile (mi^2)
1 hectare (ha) = 2.471 acres (ac)
1 acre (ac) = 0.4047 hectare (ha)

Volume

Metric measure

1 cubic meter (m^3) = 1,000,000 cubic centimeters (cm^3)
1 liter (1 L) = 1000 milliliters (ml)
1 centiliter (1 cL) = 10 milliliters (ml)
1 milliliter (1 mL) = 1 cubic centimeter (cm^3)

Nonmetric measure

1 cubic yard (yd^3) = 27 cubic feet (ft^3)
1 cubic foot (ft^3) = 7.481 U.S. gallons (gal)
1 barrel oil (bbl) = 42 U.S. gallons (gal)

Conversions

1 cubic meter = 264.2 U.S. gallons (gal)
= 35.314 cubic feet (ft^3)
1 liter (1 L) = 1.057 quarts (qt)
1 cubic centimeter (cm^3) = 0.0610 cubic inch (in^3)
1 gallon (gal) = 3.784 liters (l)

Mass

Metric measure

1000 kilograms (kg) = 1 metric ton (a tonne) (m.t)
1 kilogram (kg) = 1000 grams

Nonmetric measure

1 short ton (sh.t) = 2000 pounds (lb)
1 long ton (l.t) = 2240 pounds (lb)
1 pound (Troy) = 12 ounces (Troy) (Tr. oz)

Conversions

1 kilogram (kg) = 2.205 pounds (lb)
1 gram (g) = 0.03527 ounce (oz)
1 pound (lb) = 0.4536 kilogram (kg)

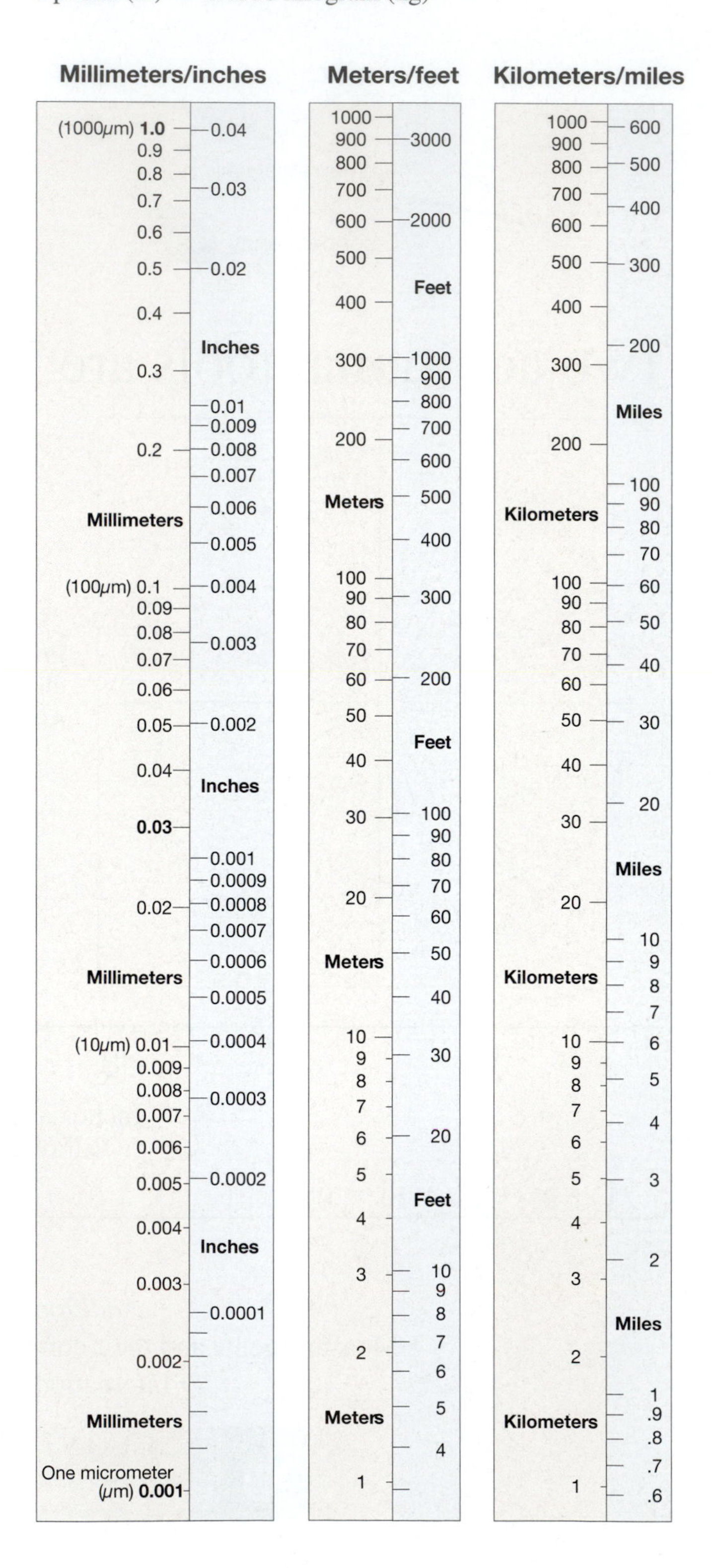

Degrees and conversions of temperature.

Comparing
A Celsius (Centigrade) degree and a Fahrenheit degree.
1 C° = 1.8 F°
1 F° = 0.5556 C°
Conversions
°F = 9/5 °C + 32
To convert °C to °F: Multiply °C by 1.8 and add 32 degrees.
°C = 5/9 (°F – 32)
To convert °F to °C: Subtract 32 degrees from °F and divide by 1.8.
Note:
°F is read degrees Fahrenheit.
F° is read Fahrenheit degree.
°C is read degrees Centigrade or degrees Celsius.
C° is read Centigrade degree or Celsius degree.

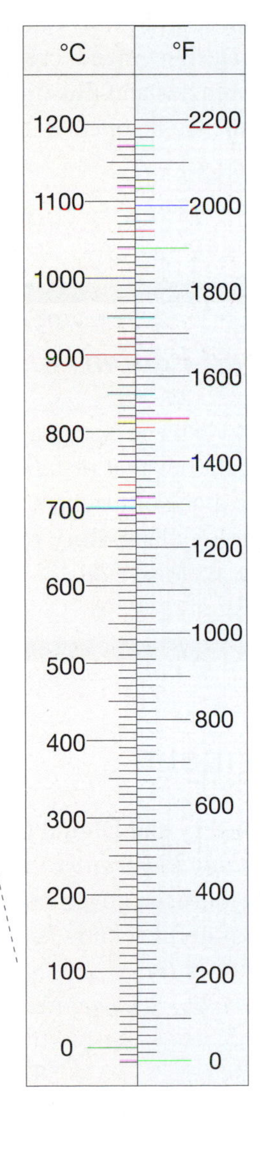

International system of units (ISU).

The International System of Units (SI) is the official system of symbols, numbers, base-10 numerals, powers of 10, and prefixes in the current metric system.

Symbol	Number	Numeral	Power of 10	Prefix
T	one trillion	1,000,000,000,000	10^{12}	tera-
G	one billion	1,000,000,000	10^{9}	giga-
M	one million	1,000,000	10^{6}	mega-
k	one thousand	1,000	10^{3}	kilo-
h	one hundred	100	10^{2}	hecto-
da	ten	10	10^{1}	deka-
	one	1	10^{0}	
d	one-tenth	0.1	10^{-1}	deci-
c	one-hundredth	0.01	10^{-2}	centi-
m	one-thousandth	0.001	10^{-3}	milli-
μ	one-millionth	0.000,001	10^{-6}	micro-
n	one-billionth	0.000,000,001	10^{-9}	nano-
p	one-trillionth	0.000,000,000,001	10^{-12}	pico-

Examples
1 meter (1 m) = 0.001 kilometer (0.001 km), 10 decimeters (10 dm), 100 centimeters (100 cm), or 1000 millimeters (1000 mm).

1 kilometer (1 km) = 1000 meters (1000 m).

1 micrometer (1 μm) (formerly known as micron: μ) = 0.000,001 meter (0.000001 m) or 0.001 millimeters (0.001 mm).

1 kilogram (kg) = 1000 grams (1000 g).

1 gram (1 g) = 0.001 kilogram (0.001 kg).

1 metric ton (1 m.t) = 1000 kilograms (1000 kg).

1 liter (1 L) = 1000 milliliters (1000 ml).

1 milliliter (1 mL) = 0.001 liter (0.001 l).

Convert It!
http://www.microimg.com/science/

Preface

organization

"Inquiry-Based Format" is a quote from a number of teachers who use my manuals. That comment stems from my embedding questions and activities within the text so as to make my manuals *interactive*. Some say *conversational*. Simple in concept, difficult in doing—especially given the fact that many students come with little or no background in earth science. To lessen students' frustrations in handling activities, I encourage them to work in small groups in an effort to pool their backgrounds. It's been said, "College students learn more from fellow students than from teachers."

Exercises differ in length. Some exercises might prove too long for the time available. If so, the organization of lettered topics within exercises allows for easy picking and choosing by the instructor. Some exercises work well as homework activities.

for instructors

For instructors using *Environmental Geology Laboratory 2ed*, I provide the *Painless Instructor's Guide* that briefly covers the following topics:

Format features
Guide to tools and supplies
Philosophy for managing your lab
Suggestions for each exercise
The Complete Answer Key

I also provide instructors with a CD with all line-art in both jpg and eps formats. The jpg images can be placed in PowerPoint and any page-layout file (e.g., InDesign), sized, cropped, and used for making transparencies and illustrating quizzes. The eps images can also be opened in Adobe Illustrator and edited.

New in this Second Edition

- Google Earth™ images of Digital Globe® photography have replaced air-photo stereograms
- Color photos of rocks and minerals are organized in galleries that match tables of classification
- Corrections and clarifications have been made on the advice of teachers using the first edition
- The exercise *Climate Change* is new to entry-level geology laboratory manuals
- Cover photo of Gulf oil spill has been worked into the exercise *Coastal Processes and Problems*

grateful acknowledgment

In addition to Google Earth™ imagery and Digital Globe® photography, *Environmental Geology Laboratory* 2nd edition draws extensively from scientific journals, government reports, and the popular media. It is with grateful acknowledgment that I recognize the following: *Earth, GSA Today, Journal of Geological Education, Scientific American, Science, Nature, and the U.S. Geological Survey.*

Tom Freeman
FreemanT@missouri.edu

1 Maps

Topics

A. What are two contributions made by Herodotus to the field of earth science? What was John Snow's approach to discovering the source of cholera in London's 1854 epidemic?
B. What is the Global Positioning System? What theory of Alfred Wegener has the Global Positioning System documented?
C. What is the Global Reference System? What is the purpose of latitude and longitude?
D. How does a fractional map scale differ from a graphic map scale?
E. What is the Land Survey System?
F. What are topographic (aka 'quadrangle') maps?
G. What are the 'three norths,' and what is the basis for each? How does the method of measuring map direction using *azimuth* differ from that using *bearing*?
H. What are the two most common map scales used in measuring distances on a map?
I. What are contour lines, and what do they reveal about land portrayed on a contour map? What are the nine rules of contours? How is a contour map constructed from elevation data? How is a topographic profile constructed along a straight line on a contour map?

A. Maps—their history and importance

A brief history of maps

The earliest map was probably drawn in the dirt by an aborigine sharing the whereabouts of distant water. Then, perhaps there came maps of territory-minded peoples in efforts to delineate land ownership. Then later there came navigation charts and the geography of trade routes, etc.

Fast-forward to the Greek Herodotus, who might be called the earliest *paleogeographer* (**paleogeography**: the charting of ancient lands and seas) because in 450 B.C. he discovered fossil seashells in limestones near Cairo, Egypt (Fig. 1.1) and correctly inferred that the Mediterranean Sea once spread farther southward than it does today.

Incidentally, Herodotus also coined the term **delta** for that part of the Nile valley characterized by the curious branching of the river near its entry into the Mediterranean.

Q1.1 What do you suppose prompted Herodotus to coin the term *delta* for the area marked by branches of the Nile? *Hint:* Draw an equilateral triangle that encloses all five branches of the Nile in Figure 1.1.

Figure 1.1 The Pyramids of Giza were built of quarried blocks of fossiliferous Eocene-age limestone. The nearby Sphinx was carved from this same limestone *in situ*.

A brief history of maps (cont.)—Fast-forward again, this time to Sir Francis Bacon, noted British essayist. It was a map of the world that prompted Bacon in 1620 to exclaim, "I say!" (Or some such.) "It appears as though Africa and South America were once united and have since drifted apart." Alfred Wegener and others eventually compiled abundant evidence in support of 'continental drift,' but it was a cursory view of maps by Bacon and others before him that first suggested the notion.

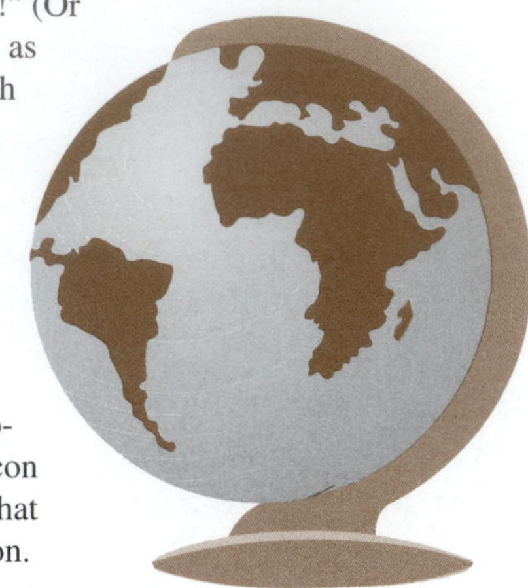

The importance of maps

There is no better way to marshall spatial and temporal information—*and to manage and interpret that information*—than on a map. Ways in which maps have been used to the benefit of humankind could fill volumes. Maps are essential for...

- commerce
- social sciences
- industry
- history
- recreation
- ecology
- military
- earth science

There is hardly an imaginable societal activity that doesn't rely to some extent on maps. And now we have satellite photography enhanced by Google Earth.

Q1.2 Pool your and your classmates' experiences and make a list of ways in which members of your group have used maps, beginning with familiar highway maps.

A map launched the field of epidemiology

The time: Septembr 1854. The place: London, England. The epidemic: More than 500 people died during the first week of an outbreak of dreaded **cholera**. In a classic example of medical detective work, physician John Snow—known as the *Father of Epidemiology*—solved the mystery of the source of cholera.

Cholera—a disease of the small intestine, with severe diarrhea, vomiting, and rapid dehydration. Mortality rate reaches 50% if untreated.

It wasn't until 1886 that the discovery of the bacterium *Vibrio cholerae* confirmed that cholera is transmitted by water. At the time of the 1854 epidemic in London, people wondered whether cholera was transmitted by air, by water, by people, or what.

Cholera broke out in central London on August 31, 1854. Dr. Snow, who had investigated earlier epidemics, suspected that water from a community pump was the culprit.

In the 1800s, London community wells were shallow and subject to contamination by sewage—the breeding sites of cholera.

Between September 4 and September 7 Snow obtained from the General Register Office the names and addresses of cholera victims during that brief span of time. He then methodically plotted those addresses on a map of central London (Fig. 1.2), thereby documenting the undeniable correlation between (a) *fatalities* and (b) *proximity* to a particular community well. These two variables—plus the importance of time—have become watchwords of the field of epidemiology: **people**, **place**, **time**.

On the evening of September 7 Snow presented his map to the Board of Guardians of St. James's Parish. The Board ordered that the handle of the indicted pump be removed immediately. London's 1854 cholera epidemic soon abated.

Q1.3 Examine the map in Figure 1.2 (facing page) and name the community pump (whose name is taken from the street on which it still stands) whose location (*place*) best correlates with numbers of fatalities (*people*).

There's more—Epidemiology is not only the science of answering the question, *why*? It is also the science of answering the question, *why not*? Two 'why-not' anomalies became evident on John Snow's map: Near ground-zero (i.e., the indicted pump in Figure 1.2) there stood (1) a workhouse, with 535 inmates, with *only 5 fatalities*, and (2) a brewery, with 70 employees living in the vicinity, with *no fatalities*.

Q1.4 Why were fatalities few-to-nonexistent among inmates in the workhouse and employees of the brewery? The answer to the first part is that the workhouse had its own water well within its walls. You might be able to imagine the answer to the second part; i.e., why were there no fatalities among the 70 employees of the brewery? *Hint*: There was a particular employee benefit provided by brewery management.

Figure 1.2 John Snow's map of fatalities resulting from London's 1854 cholera epidemic. Thirteen community pump-wells are shown, and black bars mark addresses of fatalities. Two why-not anomalies—the workhouse and the brewery—are highlighted.

B. Global Positioning System (GPS)

Alfred Wegener's hope for his theory of continental drift: "This must be left to the **geodesists**. I have no doubt that in the not too distant future we will be successful in making a precise measurement of the drift of North America relative to Europe." *Alfred Wegener, 1929*

> *Geodesy—the field of measuring the shape of the Earth and distances among points on the Earth.*

Fast-forward one more time…to the late 1970s and the launching by the Department of Defense of an array of 21 geodetic NAVSTAR satellites some 20,000 km overhead. These satellites circle Earth twice each day, transmitting signals on a pair of microwave carrier frequencies. By determining the ranges to a minimum of three satellites from a combination of *signal delays and satellite orbit information,* a single land-based receiver can determine its own location. This is the **Global Positioning System (GPS)**.

GPS surveying—A satellite can determine the *distance* to a receiver, but not the *direction* to that receiver. That is, a satellite's signal is not a vector (i.e., a line designating both magnitude and direction) pointing to a receiver. If it were, a single satellite could determine the *location* of a receiver, correct? So how does GPS surveying work?

The distance between a receiver and a satellite can be envisioned as describing a circle on the surface of the Earth, with the receiver anywhere on that circle (Fig. 1.3). So, as you can surmise from Figure 1.3, the intersection of three circles from three satellites is required to determine the location of a receiver. Additional satellites (a) enhance accuracy and (b) provide solutions for elevation as well.

Q1.5 (Ref: Figure 1.3) To what extent can only two satellites constrain the location of a receiver? *Hint:* Mentally 'remove' Bird #2 and its sweep from Figure 1.3, and study the sweeps of Birds #1 and #3.

A simple handheld receiver (Fig. 1.4) can determine the operator's location with a precision of 100 m.

Figure 1.3 The distance to an earth-bound receiver is determined by each of three satellites. Each distance describes a circle of points, any one of which might mark the location of the receiver. The intersection of three circles marks the receiver's location.

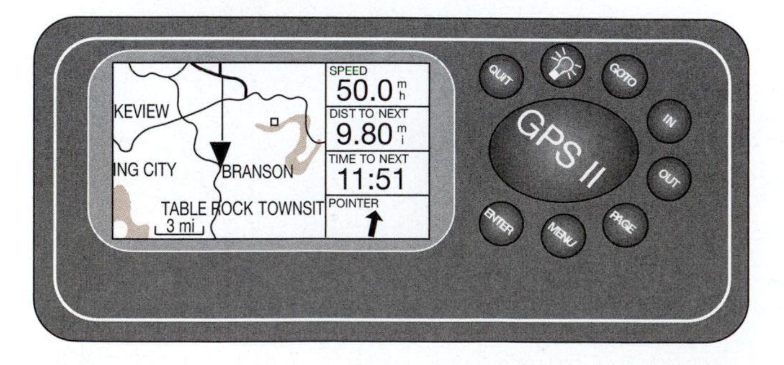

Figure 1.4 This hand-held GPS receiver was manufactured by Garmin. This type of receiver is now commonplace in vehicles of all types, where they are useful in indicating location and, when moving, speed and direction. Four AA batteries power the processing of information received from satellites.

UNAVCO, Inc.'s brochure, with illustrated examples of geological applications of GPS, is at… http://www.unavco.org/research_science/brochure/brochure.html

Seeing **geologic processes**—High-precision GPS geodetics can achieve precision of location on the order of 2 to 5 mm (Table 1.1). As illustrated in Figure 1.5, Wegener's vision of continental drift can now be documented. With GPS technology, we can see India crunching into Asia at the rate of a few centimeters per year; and, we can observe the lofty Himalayan Mountains to the north being squeezed upward a few millimeters per year.

Table 1.1
GPS Applications and Methods

Precision	Method	Science	Scale
2–5 mm	High-precision geodetic (dual-frequency)	Plate motions, glacial rebound, volcanoes	10–1000's km
2–5 mm	High-precision geodetic single-frequency	Volcanoes, fault zones, tide gauges, buildings/structures	< 10 km
1–10 cm	Real-time kinematic, rapid static (use carrier phase)	High-precision faults, volcanoes, buildings/structures	< 10 km
10 cm–1 m	Differential code, high-end receivers (requires base station)	Precdise geologic mapping, ice motions, topography	< 100 km
1–10 m	Differential code, handheld receivers (requires base station)	Moderate precision mapping, seismic exploration surveying, navigation	< 100 km
100 m	Single GPS receiver (precision limited by selective availability)	Coarse locations of geological, biological, archaeological, etc. sample points, maps, navigation	100 m to 1000 km

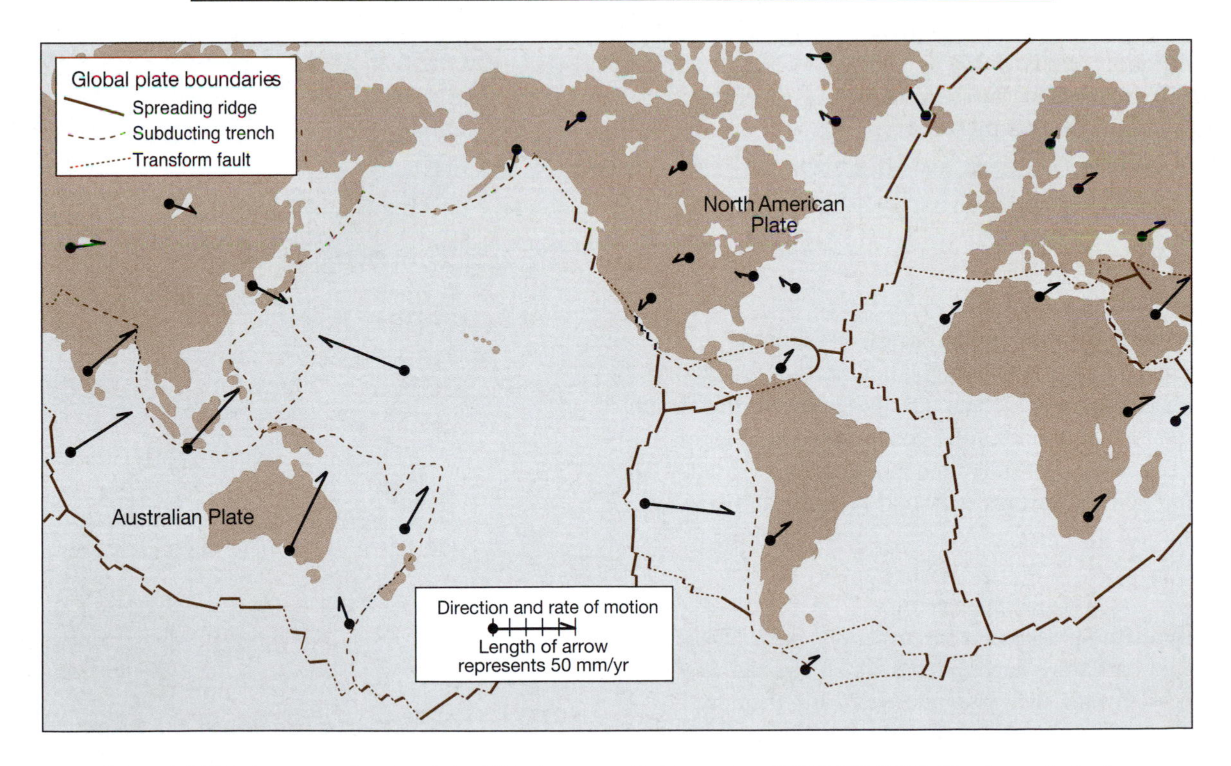

Figure 1.5 This is a map of some of the GPS sites compiled by project UNAVCO 2000. The dot at the butt of each arrow marks the location of a GPS receiver. The orientation of each arrow indicates direction of plate motion, and its length indicates the rate of motion relative to the arrow in the legend.

Q1.6 Referring to the legend in Figure 1.5, state the general direction and the approximate rate of motion of the North American Plate. Do the same for the Australian Plate.

C. Global Reference System (GRS)

Land-use planning has given new importance to a great variety of maps, all of which utilize the *Global Reference System* of **latitude** and **longitude** as a means of locating points on Earth's surface (Fig. 1.6). Latitude is the angle measured both northward and southward from the **equator**, which is assigned a value of zero. Longitude is the angle measured both westward and eastward from a meridian running through Greenwich, England (the **prime meridian**), which is assigned a value of zero.

Q1.7 What is the latitude at (A) the north pole? (B) The south pole?

Q1.8. What do you imagine is the maximum longitude value on Earth? *Hint*: It is on the opposite side of the Earth from the prime meridian.

Q1.9. A young man drafted by a team in the National Basketball Association was quoted by the press as having said, "We're going to turn this team around 360 degrees." What's wrong with this picture?

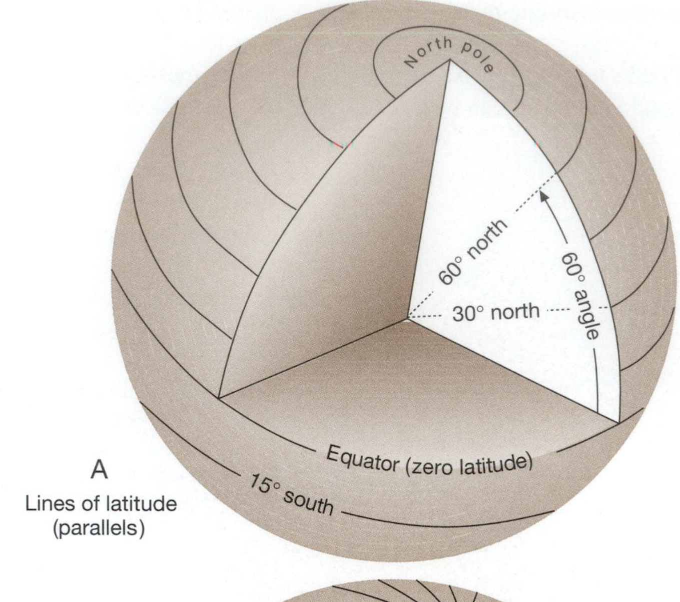

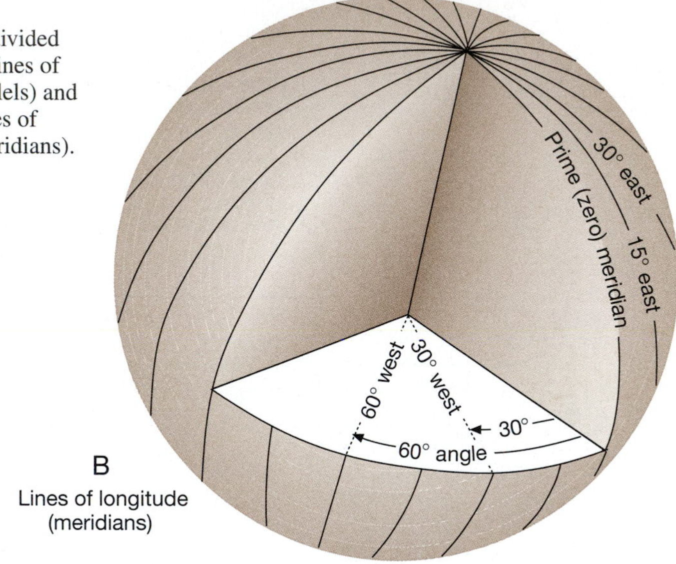

Figure 1.6 Earth is divided (A) north–south by lines of latitude (called parallels) and (B) east–west by lines of longitude (called meridians).

Locating with latitude and longitude—The most fundamental method of expressing the location of any spot on Earth's surface is some number of **degrees** north or south of the equator (= latitude), followed by some number of **degrees** east or west of the prime meridian (= longitude). The location of Willow Springs, Missouri is shown in Figure 1.7.

(Optional teaser)

Fun with latitude: *Seen in a Sunday supplement:* There is one point on Earth's surface from which one can walk one mile southward, then one mile westward, then one mile northward and be back at the starting point. *Question*: Where is this starting point? *Answer*: The north pole. Did you find that one easy? If so, try the following: There is an *infinite* number of points on Earth's surface from which one can walk one mile southward, then one mile westward, then one mile northward and be back at the starting point. *Question*: Where are these countless starting points?

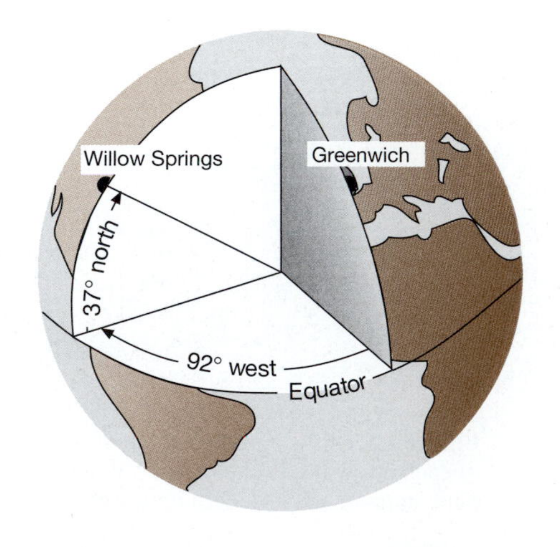

Figure 1.7 Location of Willow Springs, Missouri: 37° north latitude, 92° west longitude.

D. Map scales

The scale of a map shows the relationship between the length of a line on that map and the distance represented by that line on the ground. This relationship can be expressed by a **fractional scale**, which is either a *ratio* or a *representative fraction* (RF), designating the length of the line on the map equal to 1. For example, a scale of 1:24,000 (or 1/24,000 RF) is taken to mean that the length of a measurement along the edge of an English or metric rule represents 24,000 times that length in actual distance on Earth's surface. To illustrate…

Q1.10 Given a scale of 1:24,000: (A) What is the ground distance in *feet* represented by three inches on the map? (B) What is the ground distance in *meters* represented by 5 cm on the map? (C) How about the ground distance in *kilometers* represented by 5 cm on the map? *Hint:* Multiples of units of measures—and their conversions to other systems of measures—appear on pages *i* and *ii* at the front of this manual.

A relatively large fraction, for example, 1:24,000, is called a *large-scale map*; a relatively small fraction, for example, 1:250,000, is called a *small-scale map*. A large-scale map represents a smaller land area in greater detail than does a small-scale map on paper of the same size.

Another kind of scale is the **graphic scale**, the application of which is demonstrated in Figure 1.8 in two simple steps.

A graphic scale has two advantages over a fractional scale.

- First, a graphic scale can be applied visually, rather than through mathematical computation.

- Second, suppose you have a map on which both fractional and graphic scales are indicated. You would like an enlargement of your map, so you take it to a copy center and ask to have it enlarged by whatever percent you choose.

Q1.11 What happens to the usefulness of (A) the fractional scale, and (B) the graphic scale, as you enlarge your map?

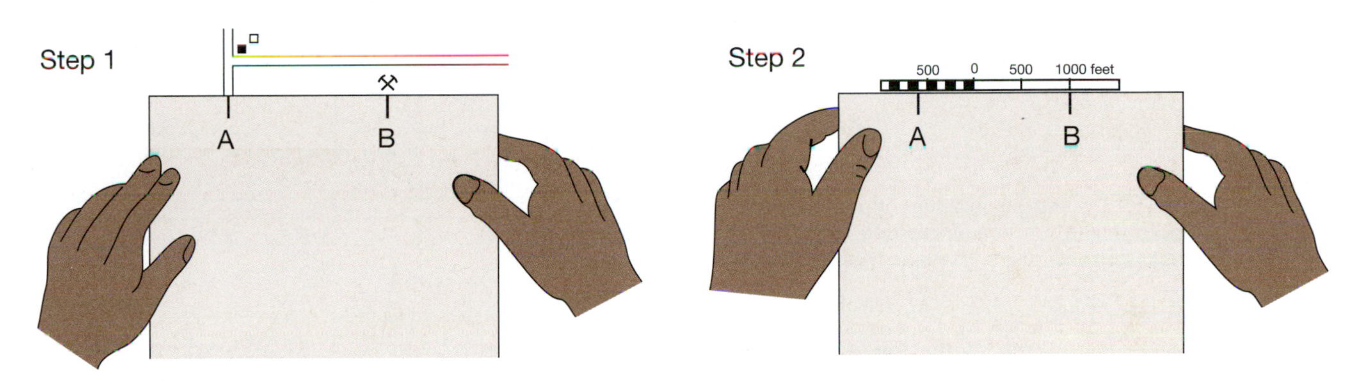

Figure 1.8 A simple way in which to measure the ground distance between two points on a map using a graphic scale:

Step 1: Place two marks labeled A and B on a sheet of scrap paper, indicating the separation between two points on the map, e.g., the north-south road and the quarry.

Step 2: Bring the paper with the two marks to the graphic scale on the map and read the ground distance directly: *1,600 feet*.

E. Land survey system

Townships—In 1785 the Continental Congress, directed by Thomas Jefferson, established a rectangular system of public land survey that has since been adopted by some 30 states. The original thirteen states and five others, plus Texas and Hawaii, use a variety of different systems (Fig. 1.10 on the facing page).

The area of these 30 states is subdivided by regional **base lines** (running east-west) and **principal meridians** (running north-south) that serve as reference lines for **townships** (Fig. 1.9). A township is bounded on its east and on its west by north-south **range lines**, and on its north and on its south by east-west **township lines**.

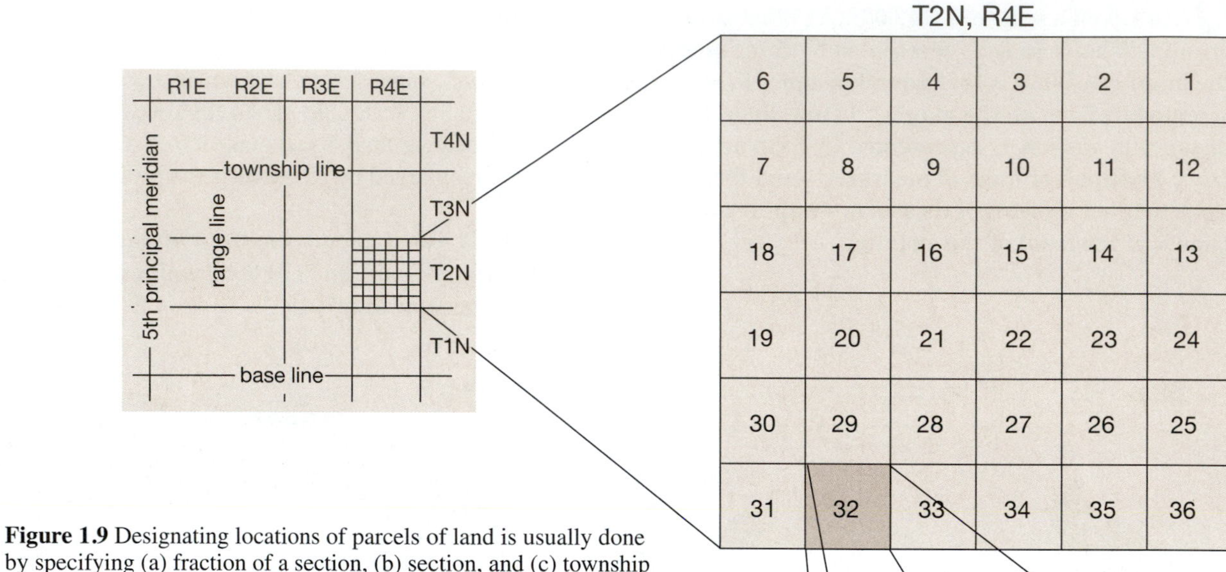

Figure 1.9 Designating locations of parcels of land is usually done by specifying (a) fraction of a section, (b) section, and (c) township (by township and range). Examples A through D follow below.

A: NW 1/4, Sec. 32, T2N, R4E
B: SW 1/4, NE 1/4, Sec. 32, T2N, R4E
C: W 1/2, SW 1/4, Sec. 32, T2N, R4E
D: S 1/2, SE 1/4, Sec. 32, T2N, R4E

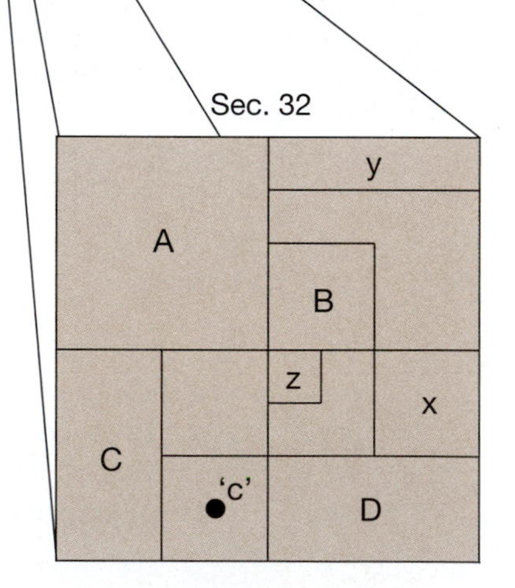

Notice that the order of fractions of a section is as though the words *of the* were inserted between fractions. A point, rather than a parcel, can be described as the center (c) of some fraction of a section.

Sections—Ideally, a township consists of a square block of 36 **sections**. Each section is one square mile (= 640 acres) in area (Fig. 1.9). Sections are delineated by section lines, and each bears its number in its center. (Notice the curious way in which sections are numbered.) When it is necessary to describe the location of a parcel of land within a section, the section can be subdivided into quarters—and further into halves of quarters—and further into quarters of quarters.

Q1.12 Referring to Figure 1.9, give the locations for (A) parcel x, (B) parcel y, (C) parcel z, and (D) the point indicated by the letter c.

Q1.13 What are the mileage dimensions of a parcel of land that is three townships wide (east-west) and two townships high (north-south)? (State the parcel's width and height in miles.)

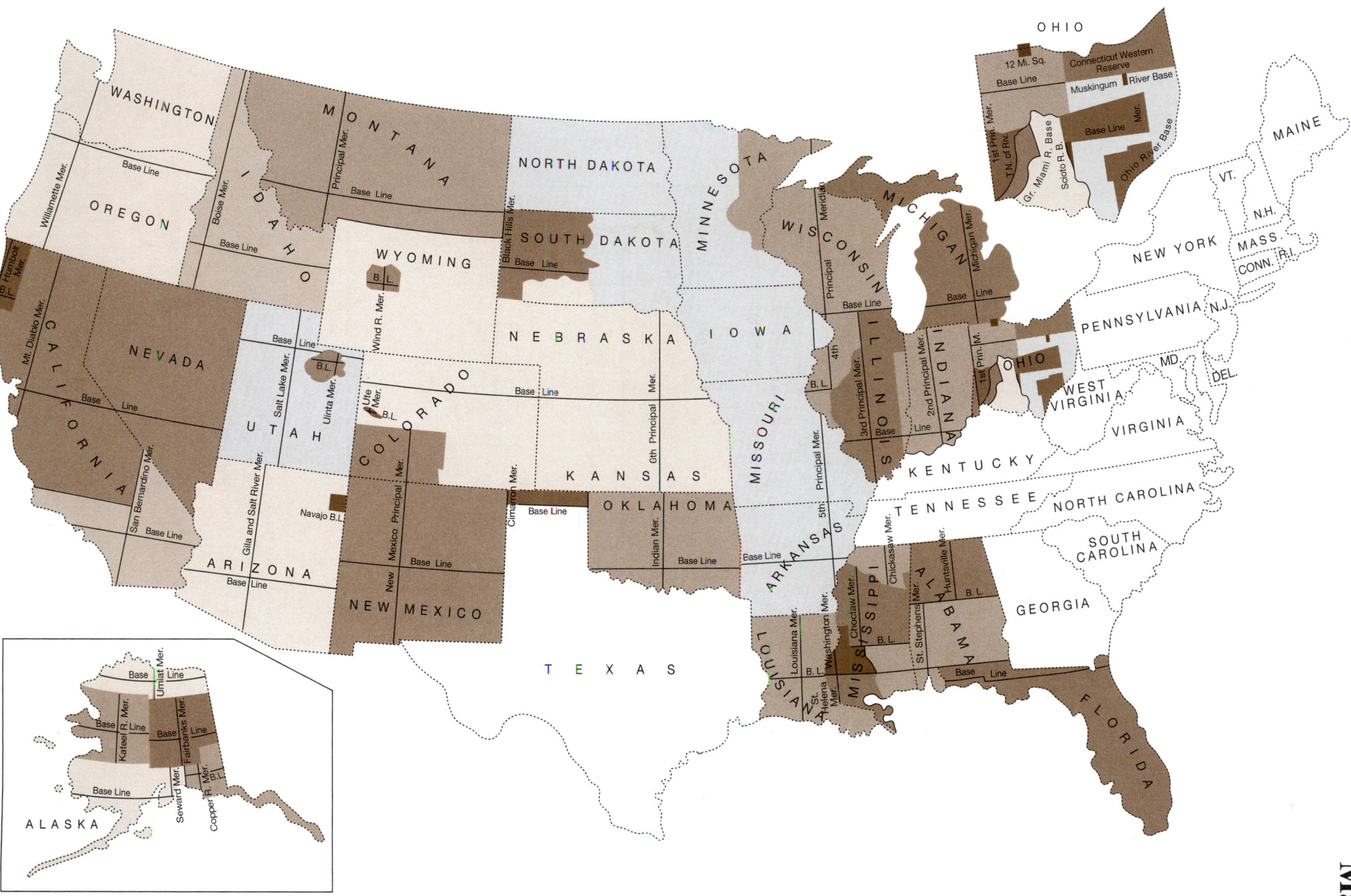

Figure 1.10 The Public Land Survey System. *(From U.S. Department of Interior, Bureau of Land Management.)*

F. Topographic (or 'quadrangle') maps

The U.S. Geological Survey was established by Congress in 1879 to consolidate four earlier organizations that had been engaged in the mapping of public lands. In 1882 a general plan was adopted for the Survey to produce a series of topographic (or 'quadrangle') maps, examples of which appear in this manual.

Quadrangle maps—The name comes from the fact that a quad is 'square' in that its latitude and longitude dimensions are equal. That is, most quads are square in their coverage of the global reference system.

Latitude and longitude dimensions of most quads are less than one degree, so they are measured in fractions of a degree called **minutes** (') (Fig. 1.11). One minute is 1/60 of a degree, and one minute is further divisible into 60 **seconds** ("). So, angular measurements (degrees, minutes, and seconds) are fractionally analogous to the temporal divisions hours, minutes, and seconds.

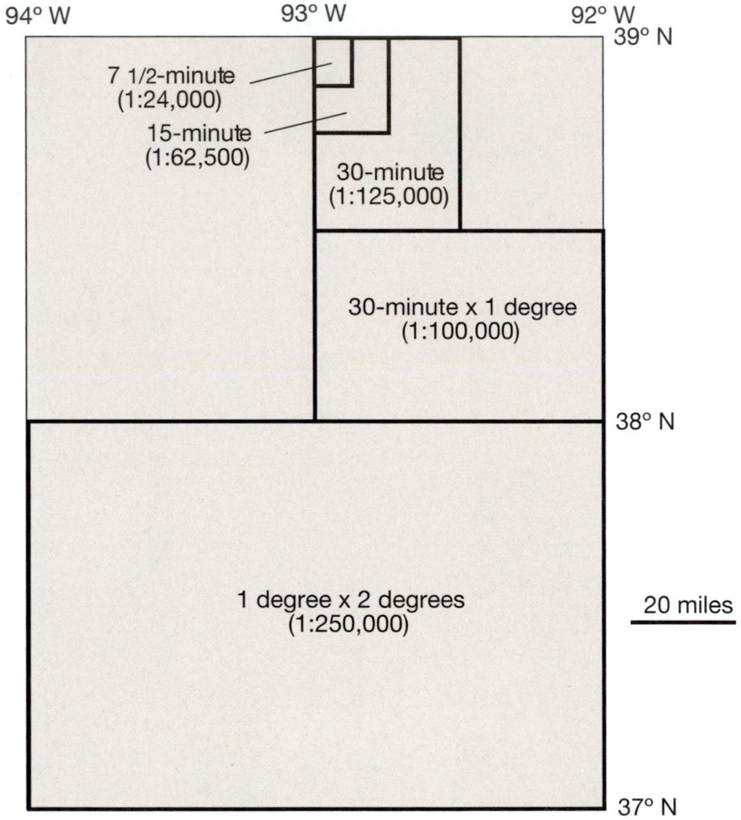

94° W
93° W
92° W
39° N
7 1/2-minute (1:24,000)
15-minute (1:62,500)
30-minute (1:125,000)
30-minute x 1 degree (1:100,000)
38° N
1 degree x 2 degrees (1:250,000)
20 miles
37° N

Figure 1.11 Quadrangle maps come in five different scales. Notice that latitude and longitude dimensions generally increase by a factor of two. The reason that the fractional scales are not also exact multiples of two is because the sheets of paper on which maps of different scales are printed differ in size.

Q1.14 As shown in Figure 1.11, 7 1/2-, 15-, and 30-minute quads each cover an amount of latitude equal to that of longitude, yet not one of these quads appears to be square. Instead, all three are *rectangles* elongated north-south. Why is this? *Hint*: Notice the longitude and latitude scales at the margins of this diagram. Why is the distance between 92° W and 93° W shorter than the distance between 38° N and 39° N? *Hint*: Study Figure 1.6B on page 6.

The primary Web gateway to the USGS Geography Division Server is at…http://mapping.usgs.gov/

G. The three norths

Included in the information within the margins of quadrangle maps is a diagram indicating the angle between true north and magnetic north. And, on newer quads, there is an indication of the orientation of grid north as well (Fig. 1.12).

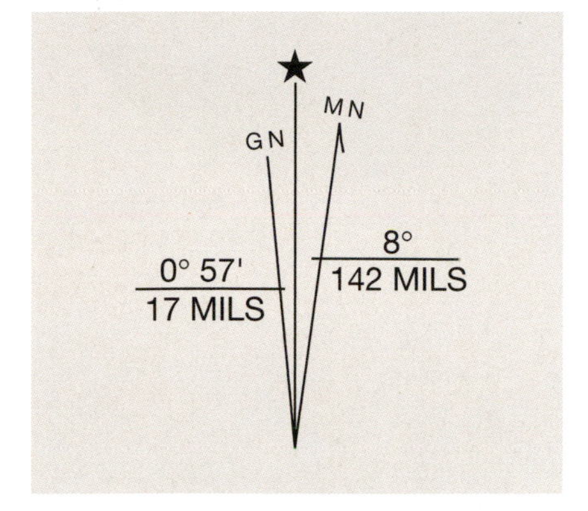

Figure 1.12 Map margin information graphically shows directions of true north (star), grid north (GN), and magnetic north (MN). In this case, grid north is west of true north by 0 degrees, 57 minutes, which is equal to 17 mils. (There are 6,400 mils in a circle.) Magnetic north is east of true north by 8 degrees, which is equal to 142 mils. Notice that the angles indicated by the lines to GN and MN are exaggerated. They're intended only to show whether GN and MN are east or west of true north.

True north is the direction to Earth's north rotational pole. Meridians converge on the rotational poles (Fig. 1.6B on page 6), so a true-north arrow can be viewed as a tiny segment of an imaginary meridian.

Grid north is more complicated. **The Universal Transverse Mercator System (UTM)**, which was developed by our military in 1947, employs a reference system of rectangular map zones (which are divided into square map areas) that are superimposed on meridians. To repeat—meridians converge on the poles. A visual metaphor: The 'orange slices' of Earth's surface bounded by meridians taper toward the poles, so there is an angle between a meridian and the side of a UTM square map area. On quads indicating grid north there are imaginary north-south lines indicated by labeled tick-marks along the north and south margins of the map. The angle between these imaginary lines and the meridians that forms the east and west margins of the quad is the angle labeled *grid north*. The size of the angle between true north and grid north depends both on the latitude of the rectangular zone and on the position of the square map area within the zone. Whew!

Q1.15 Judging from Figure 1.13, how does the angle between true north and grid north at 120° west longitude differ as a function of latitude; i.e., does the angle *increase* with increasing latitude, or does it *decrease* with increasing latitude?

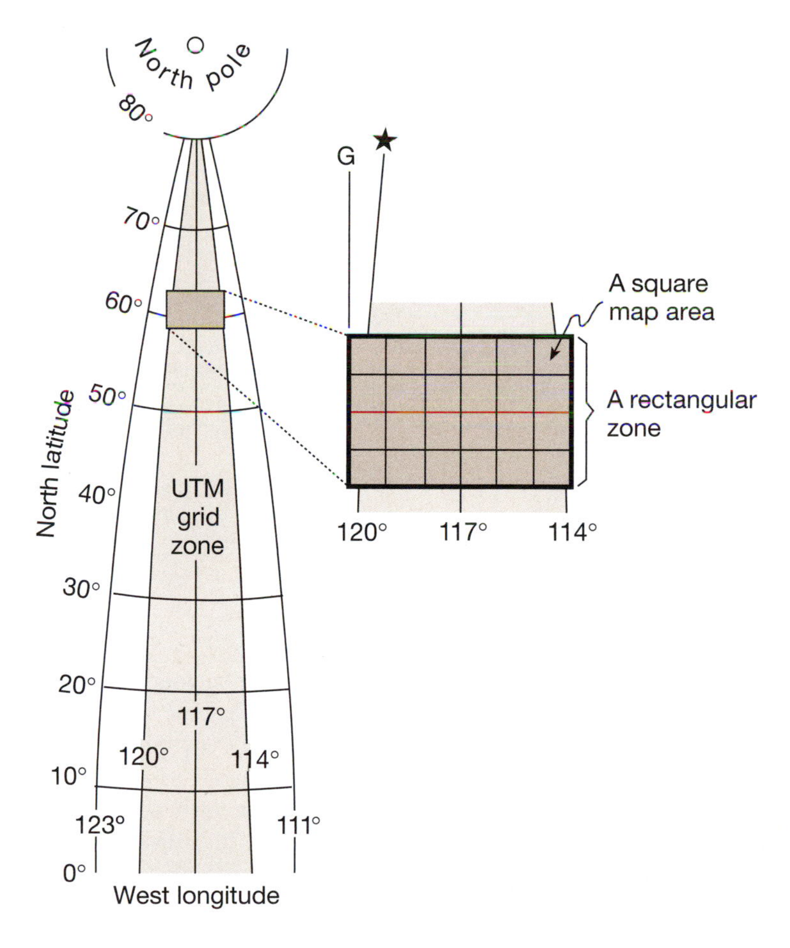

Figure 1.13 The Universal Transverse Mercator reference system of square map areas is superimposed on meridians. Here, segments of parallels are shown from the equator to the north pole between 111° and 123° west longitude. Segments of four meridians are shown extending from the equator to 80° north latitude. As you can see, the converging meridians can form angles with the sides of the square map areas. At some longitudes—here at 117° west—true north and grid north are the same along some map margins.

Magnetic north—Earth's magnetic field is believed to be in part a consequence of Earth's rotation, but Earth's magnetic north pole is not exactly coincident with Earth's north rotational pole. In fact, the two are separated by some 1,290 kilometers (Fig. 1.14).

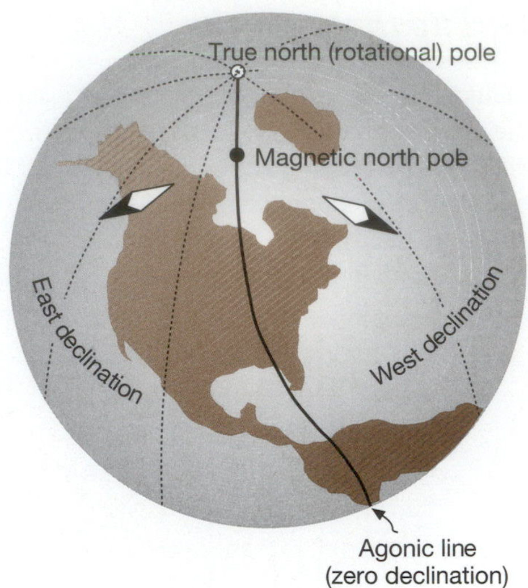

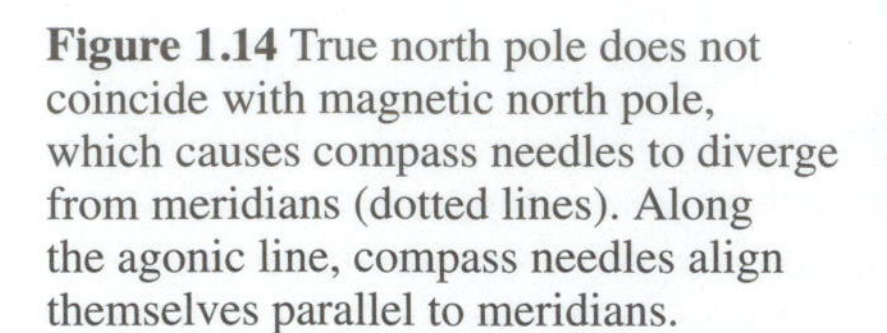

Figure 1.14 True north pole does not coincide with magnetic north pole, which causes compass needles to diverge from meridians (dotted lines). Along the agonic line, compass needles align themselves parallel to meridians.

Another wrinkle is the fact that the orientation of Earth's magnetic field is continually changing. Figure 1.15 shows the amount of magnetic declination (along any *solid* line) for the year 1990, as well as the rate at which the orientation of the magnetic field is changing (along any *dashed* line).

Sample problem: Referring to the map for 1990 in Figure 1.15, compute the magnetic declination for the year 2000 at Seattle, Washington.

Solution:

- 10 (years between 1990 and 2000)
- 10 years multiplied by 6 (minutes W of north change per year) = 60 minutes W = 1 degree W
- algebraically add 1 degree W to 20 degrees E = 19 degrees E

Q1.16 Again referring to Figure 1.15, what will be the magnetic declination at Ann Arbor, Michigan, in the year 2010?

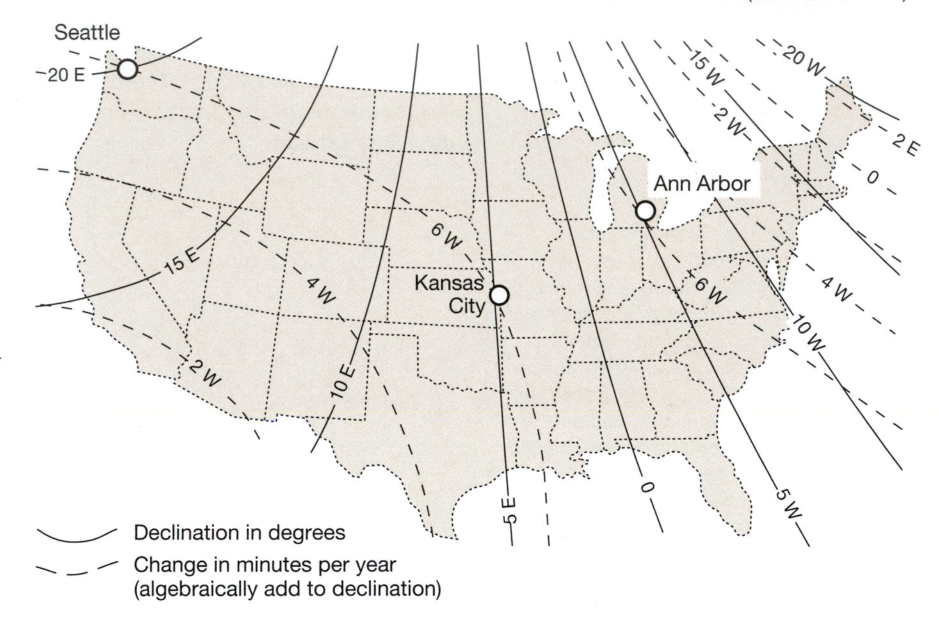

Figure 1.15 This is an isogonic map showing magnetic declinations for the year 1990; and, rates of change in the orientation of the magnetic field in minutes per year.

Based on a true story—The year is 1990, and you have recently moved into an older home in the suburbs of Kansas City. It is clear from the plat that front and back property lines run due east–west and side property lines run due north–south (Fig. 1.16). Iron stakes provide locations of front corners of lots, but only vague understandings of the locations of back corners guide homeowners in activities such as mowing. (Back stakes can no longer be found, and there are no fences.)

You decide to build a pen for your dog so that it will be least disruptive to your backyard activities, so you envision a site in the extreme northwest corner of your lot.

As you begin to prepare the site, the neighbor on your west side approaches you and asserts, "Hey, you're a bit on me, aren't you, Sport?" Then, armed with an aging scouting compass, your neighbor deftly demonstrates that the property line runs through the middle of your prospective site. But you have had a course in geology, so you offer your neighbor a short course in the nuances of *correctly* using a magnetic compass.

Q1.17 What is the main principle on which you structure your short course?

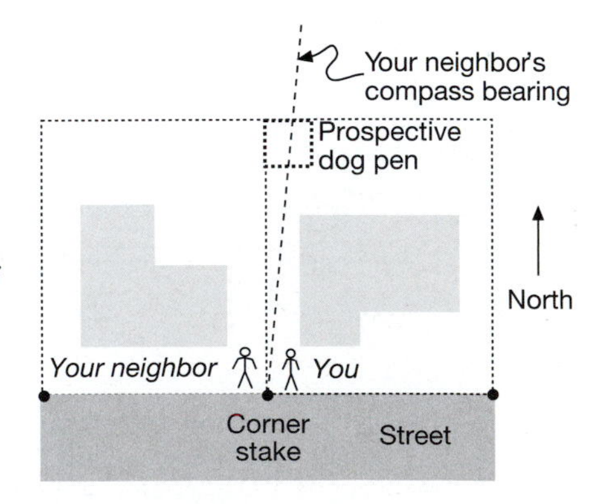

Figure 1.16 Your neighbor's compass bearing appears to bisect the site of your prospective dog pen.

Measuring map direction—There are two common ways to express map direction—**azimuth** and **bearing**.

(1) **Azimuth** is like the minute-hand on a clock, only instead of a clockwise sweep of 60 minutes, azimuth is a clockwise sweep of 360 degrees (Fig. 1.17).

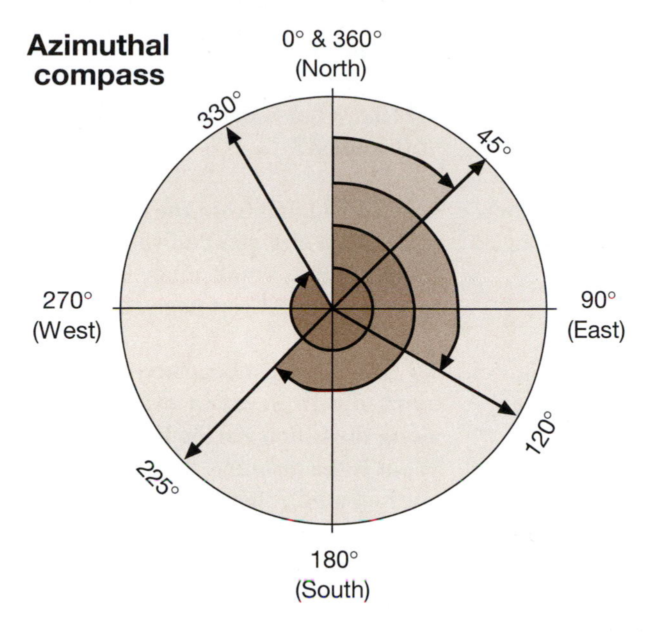

Figure 1.17 Four different directions (indicated by arrows), each expressed as azimuth.

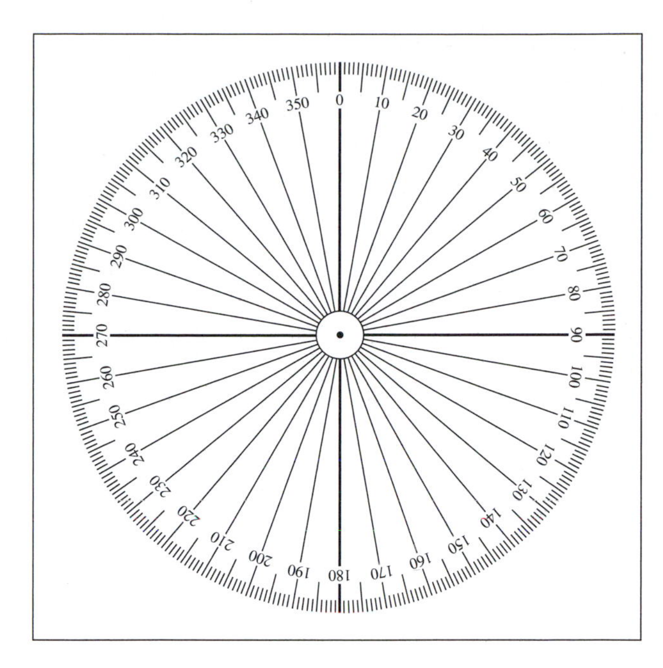

Need a protractor? Make a transparency of the one above with a copy machine.

(2) **Bearing** is a carryover from the ancient compass rose with its four quadrants. Bearing is an angle measured eastward or westward from either north or south, whichever is closer. The method employs a circle divided into four 0°–90° quadrants: northeast (NE), northwest (NW), southeast (SE), and southwest (SW) (Fig. 1.18A).

A bearing can be expressed by stating (first) the *pole*—north or south—from which the angle is measured; (second) the *magnitude* of the angle measured (between 0° and 90°); and (third) the *direction*—east or west—toward which the angle is measured. Four examples are shown in Figure 1.18B.

Bearing is more traditional. Azimuth is simpler. 'This is America,' so you can use whichever method you choose.

Speaking of America…and speaking of angles…aficionados of snowboarding learned all about '540s' (one and one-half 360° airborne turns) and '720s' (two 360° airborne turns) during the 2002 winter Olympics in Salt Lake City as Americans swept gold, silver, and bronze in that event.

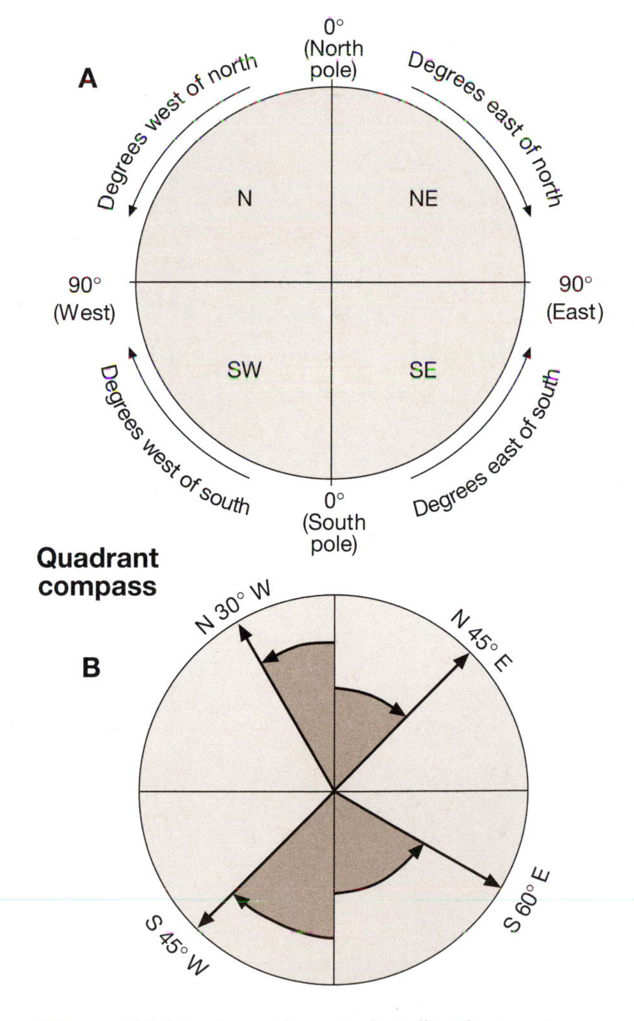

Figure 1.18 This shows the same four directions as those in Figure 1.17, but with each expressed as a bearing.

H. Working with a real topographic quadrangle map—Menan Buttes quadrangle, Idaho

(Conventional map symbols appear on pages 16 and 17)

Note: In addition to elevations being printed with brown ink astride contour lines, elevations are also printed at scattered locations in black. The initials BM preceding an elevation value marks the location of a U.S. Geological Survey field *benchmark*.

Menan Buttes quadrangle, Idaho (facing page). Questions 1.18 through 1.26 apply to this quadrangle.

Q1.18 (A) What is the latitude indicated at the southwest corner of this quad? (B) What is the longitude? (The latitude value lies along an imaginary east–west extension of the southern boundary of the map; the longitude value lies along an imaginary north-south extension of the western boundary.) (C) In what two counties does North Menan Butte occur?

Q1.19 (A) What is the magnetic declination—that is, the difference (in degrees) between true north (star) and magnetic north (MN arrow)? (B) What is the difference (in degrees and minutes) between true north (star) and grid north (GN)—in this case, the line to the left of the true north line?

Q1.20 (A) Butte Market Lake Canal is a tributary of Snake River. In what section do they join? (B) What is the elevation of the triangulation marker in section 3?

Q1.21 What is the fractional scale of this quad?

Q1.22 Judging from the fractional scale, what size quad do you suppose Menan Buttes Quadrangle is? (See again Figure 1.11 on page 10.)

Q1.23 There's a benchmark (BM) marked with elevation 4826 feet along the Union Pacific Railroad. What is the distance, to the nearest tenth of a mile, between that benchmark and the X marking the approximate center of South Menan Butte?

Q1.24 How far—to the nearest tenth of a kilometer—is the well in section 11 from the southwest corner of section 10?

Q1.25 The southwest corner of section 10 is approximately what direction—expressed as a *bearing*—from the X that marks the approximate center of North Menan Butte?

Q1.26 See the stippled areas in section 23, southeast of South Menan Butte. What does this stippling designate? (See the second of Surface Features of Topographic map symbols on page 16.)

USGS topographic mapping home page is at http://mac.usgs.gov/isb/pubs/booklets/topo/topo.html

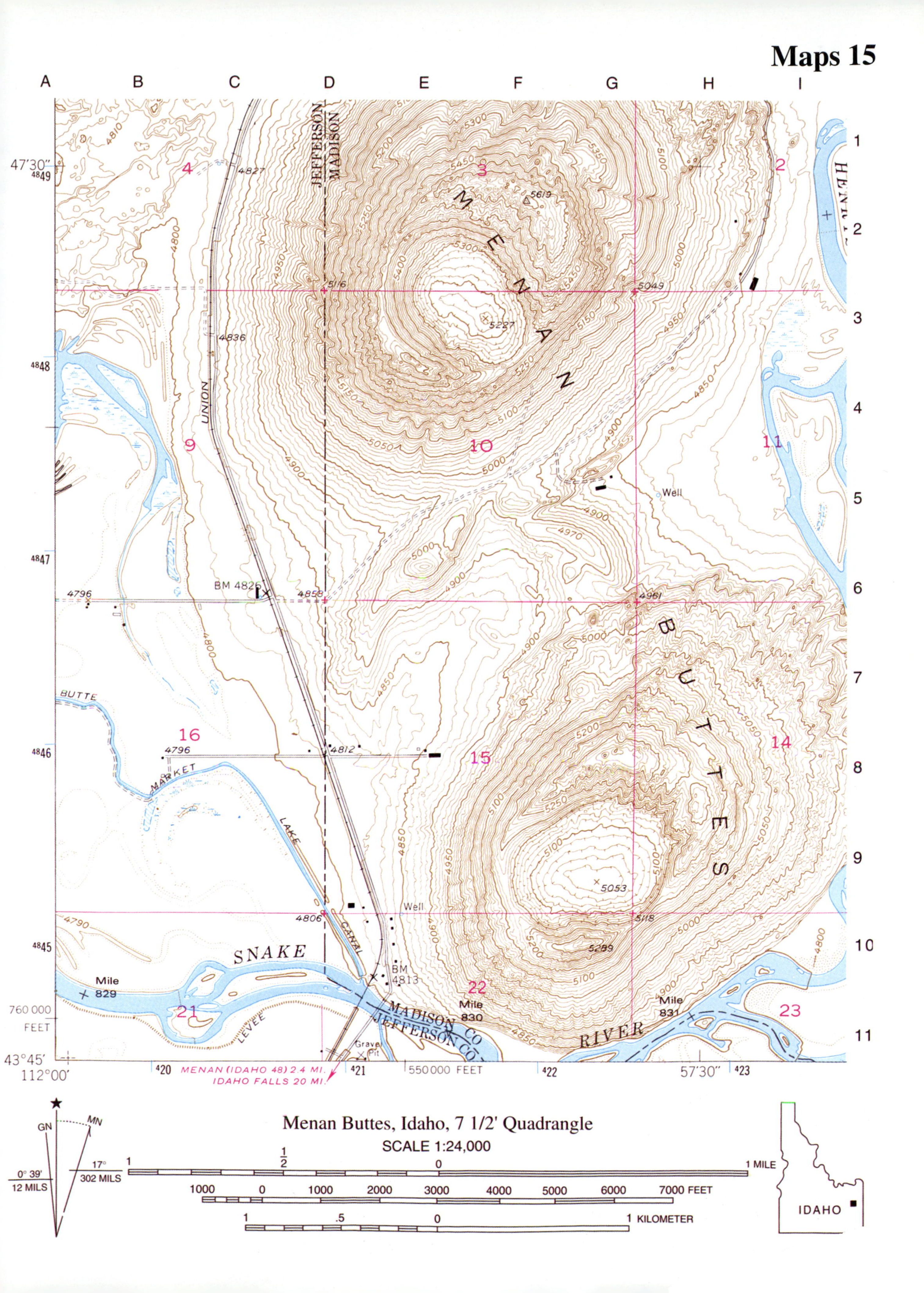

Menan Buttes, Idaho, 7 1/2' Quadrangle

Topographic map symbols

CONTROL DATA AND MONUMENTS

Aerial photograph roll and frame number*	3-20
Horizontal control	
Third order or better, permanent mark	Neace Neace
With third order or better elevation	BM 45.1 Pike BM 45.1
Checked spot elevation	19.5
Coincident with section corner	Cactus Cactus
Unmonumented*	
Vertical control	
Third order or better, with tablet	BM 16.3
Third order or better, recoverable mark	120.0
Bench mark at found section corner	BM 18.6
Spot elevation	5.3
Boundary monument	
With tablet	BM 21.6 BM 71
Without tablet	171.3
With number and elevation	67 301.1
U.S. mineral or location monument	

CONTOURS

Topographic	
Intermediate	
Index	
Supplementary	
Depression	
Cut; fill	
Bathymetric	
Intermediate	
Index	
Primary	
Index Primary	
Supplementary	

BOUNDARIES

National	
State or territorial	
County or equivalent	
Civil township or equivalent	
Incorporated city or equivalent	
Park, reservation, or monument	
Small park	

*Provisional Edition maps only

...nal Edition maps were established to expedite completion of the ...ge scale topographic quadrangles of the conterminous United ...ain essentially the same level of information as the standard ...s can be easily recongnized by the title "Provisional ...nd corner.

LAND SURVEY SYSTEMS

U.S. Public Land Survey System	
Township or range line	
Location doubtful	
Section line	
Location doubtful	
Found section corner; found closing corner	
Witness corner; meander corner	WC MC
Other land surveys	
Township or range line	
Section line	
Land grant or mining claim; monument	
Fence line	

SURFACE FEATURES

Levee	Levee
Sand or mud area, dunes, or shifting sand	Sand
Intricate surface area	Strip mine
Gravel beach or glacial moraine	Gravel
Tailings pond	Tailings Pond

MINES AND CAVES

Quarry or open pit mine	
Gravel, sand, clay, or borrow pit	
Mine tunnel or cave entrance	
Prospect; mine shaft	
Mine dump	Mine dump
Tailings	Tailings

VEGETATION

Woods	
Scrub	
Orchard	
Vineyard	
Mangrove	Mangrove

GLACIERS AND PERMANENT SNOWFIELDS

Contours and limits	
Form lines	

MARINE SHORELINE

Topographic maps	
Approximate mean high water	
Indefinite or unsurveyed	
Topographic-bathymetric maps	
Mean high water	
Apparent (edge of vegetation)	

Topographic map symbols

COASTAL FEATURES

- Foreshore flat
- Rock or coral reef
- Rock bare or awash
- Group of rocks bare or awash
- Exposed wreck
- Depth curve; sounding
- Breakwater, pier, jetty, or wharf
- Seawall

BATHYMETRIC FEATURES

- Area exposed at mean low tide; sounding datum
- Channel
- Offshore oil or gas: well; platform
- Sunken rock

RIVERS, LAKES, AND CANALS

- Intermittent stream
- Intermittent river
- Disappearing stream
- Perennial stream
- Perennial river
- Small falls; small rapids
- Large falls; large rapids
- Masonry dam
- Dam with lock
- Dam carrying road
- Perennial lake; Intermittent lake or pond
- Dry lake
- Narrow wash
- Wide wash
- Canal, flume, or aqueduct with lock
- Elevated aqueduct, flume, or conduit
- Aqueduct tunnel
- Well or spring; spring or seep

SUBMERGED AREAS AND BOGS

- Marsh or swamp
- Submerged marsh or swamp
- Wooded marsh or swamp
- Submerged wooded marsh or swamp
- Rice field
- Land subject to inundation

BUILDINGS AND RELATED FEATURES

- Building
- School; church
- Built-up Area
- Racetrack
- Airport
- Landing strip
- Well (other than water); windmill
- Tanks
- Covered reservoir
- Gaging station
- Landmark object (feature as labeled)
- Campground; picnic area
- Cemetery: small; large

ROADS AND RELATED FEATURES

Roads on Provisional edition maps are not classified as primary, secondary, or light duty. They are all symbolized as light duty roads.

- Primary highway
- Secondary highway
- Light duty road
- Unimproved road
- Trail
- Dual highway
- Dual highway with median strip
- Road under construction
- Underpass; overpass
- Bridge
- Drawbridge
- Tunnel

RAILROADS AND RELATED FEATURES

- Standard gauge single track; station
- Standard gauge multiple track
- Abandoned
- Under construction
- Narrow gauge single track
- Narrow gauge multiple track
- Railroad in street
- Juxtaposition
- Roundhouse and turntable

TRANSMISSION LINES AND PIPELINES

- Power transmission line: pole; tower
- Telephone line
- Aboveground oil or gas pipeline
- Underground oil or gas pipeline

I. Contour maps

A contour line is a line that connects points of equal elevation on a map, so contour maps illustrate the **topography** (two-dimensional shape) of a land surface. Figure 1.19 is a simplified presentation of the meaning of contours.

Q1.27 (A) What can you say about the elevation, in feet, of the higher of the two hills in Figure 1.19? (Be as specific as you can.) (B) How about the lower hill?

Incidentally, although the metric system has grown in use within our scientific community, we are heavily committed to *feet* as the unit of measurement for contours. With thousands of topographic maps already published with elevations in feet, it would be an enormous expense to change. Even so, metric maps are beginning to be produced by government agencies such as the U.S. Geological Survey.

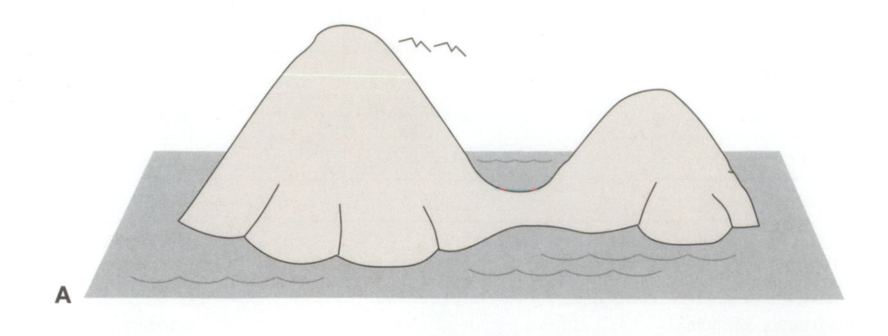

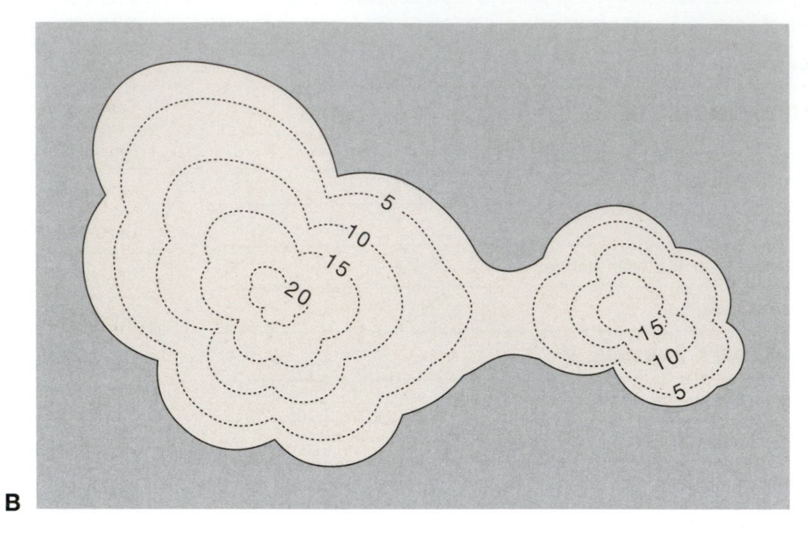

Figure 1.19 The zero-foot contour line is drawn at average sea level. Contours drawn at intervals of 5 feet continue upward to the tops of the two hills.

Contours in stereogram—Figure 1.20 is a stereographic contour map that might help you envision just how contours portray a land surface. Viewing Figure 1.20 with a stereoscope should enable you to discern highlands from lowlands and to understand how contours represent **physiography** (i.e., the shape of the land). If you don't have a stereoscope handy, try the following:

Figure 1.20 A stereographic contour map of Johnson Shut-Ins State Park, Missouri.

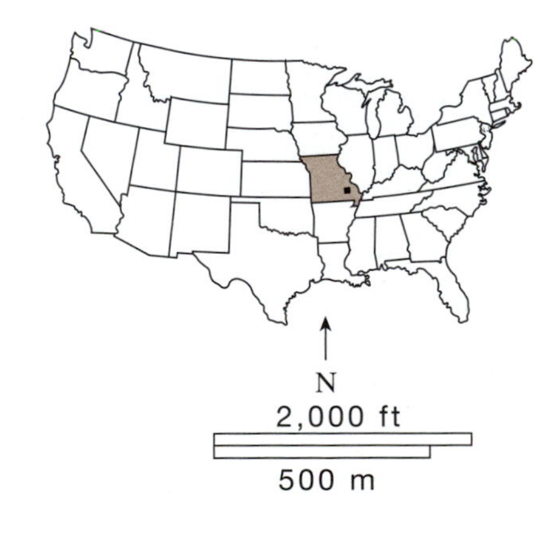

Stereoscopic vision with the unaided eye—Have you ever noticed someone daydreaming in such a manner that their eyes appear to 'glaze over,' like those of a mannequin in a store front window? Try fixing your eyes in this manner—as though staring at infinity—and then bring Figure 1.20 to a foot or so in front of your eyes while keeping your eyes fixed on infinity. The trick is to focus *beyond* the page. Voilá, stereographic vision.

Q1.28 In the stereographic contour map in Figure 1.20, (A) which of the points—A, B, or C—appears to be the highest? (B) Which appears to be the lowest?

Q1.29 In Figure 1.20 there is a narrow gorge with a stream. In which direction does the stream appear to be flowing, southward or northward?

Nine rules of contours are illustrated in Figure 1.21. Two additional rules are:

(1) Contours do not cross one another.
(2) Individual contours do not branch. Contours appear to branch in cases of vertical cliffs, but the apparent branches are actually separate contour lines superposed on one another.

Contour interval—The succession of elevation values represented by contour lines is commonly graduated in some multiple of five or ten feet. The choice of contour interval depends on (a) the scale of the map, and (b) the magnitude of **relief**—which is the difference in elevation between the highest point and the lowest point within the map area. If the contour interval is too small for a particular area, contour lines will be overly crowded. If too large, there will be too few contour lines to adequately portray the topography.

Another thing: By convention, every fifth contour line is drawn heavier in line weight than others. This difference makes it a bit easier to visually follow an individual contour across a map. And, it imparts a more obvious fabric to the map.

Q1.30 What is the contour interval shown in Figure 1.21B?

Q1.31 In Figure 1.21B, what is the relief between the lowest and highest contour lines?

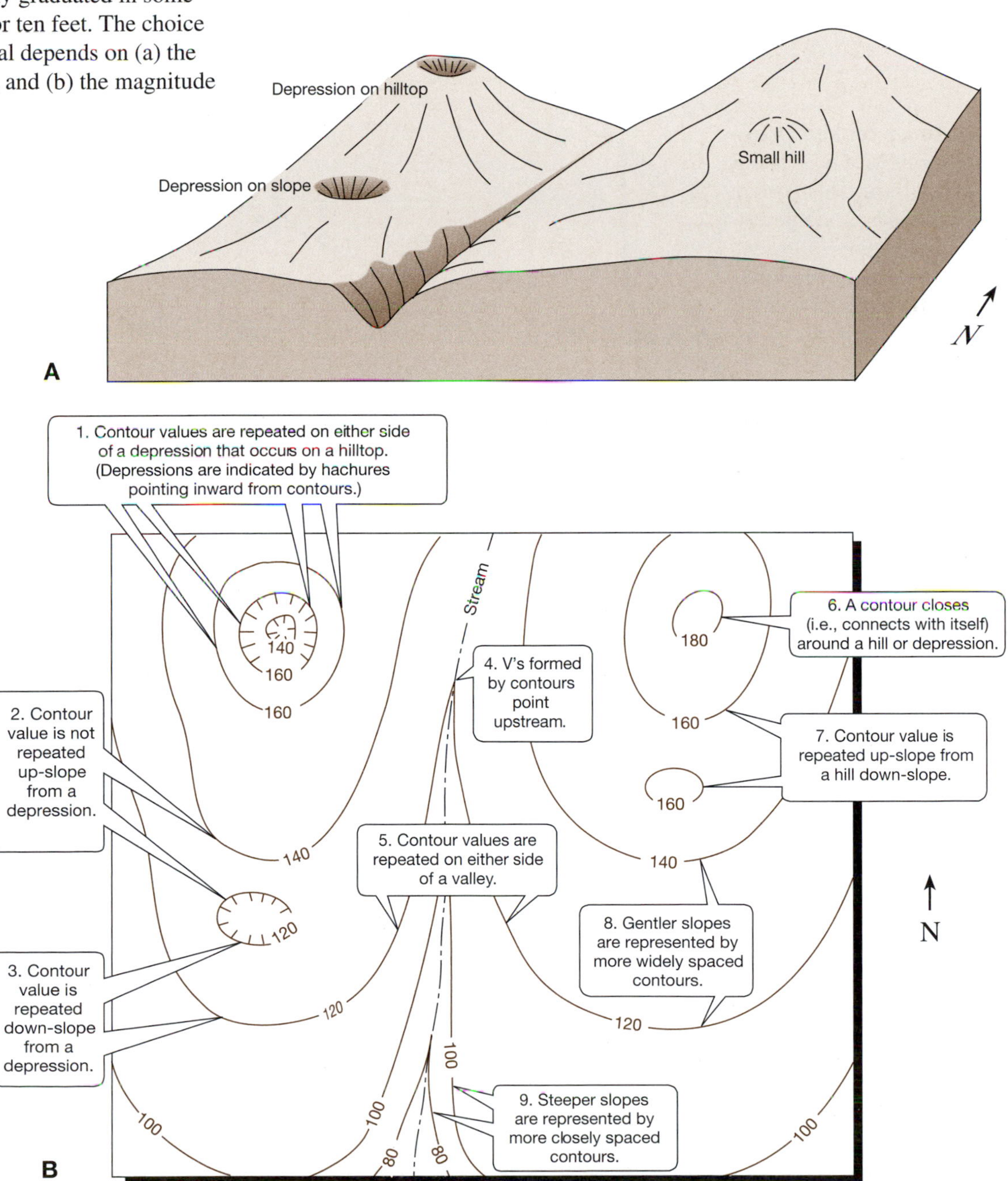

Figure 1.21 The rules of contours. **A**. An oblique sketch of a land surface. **B**. A contour map illustrating the land surface shown in A.

Contours on a topographic quadrangle map—Refer once again to the Menan Buttes quadrangle map on page 15. These pre-historic buttes are examples of volcanic cinder cones, which develop through the process of hot dry ash being thrown aloft from volcanic vents before falling to earth like a snowfall. This process has resulted in the burial of entire cities—for example that of the Italian city of Pompeii by Mount Vesuvius volcano in 79 A.D. Volcanic ash from Island also shut down air traffic for days over Europe in the spring of 2010.

Q1.32 What is the contour interval on the Menan Buttes topographic quadrangle (page 15)?

Q1.33 At the crest of south Menan Butte there are contour lines with tick-marks. (A) What do these tick-marks indicate? *Hint:* See the feature in the northwest quarter of Figure 1.21B on page 19. (B) What is the approximate height, or the approximate depth, of the feature bounded by these tick-marks on south Menan Butte?

Q1.34 Again, with reference to south Menan Butte. In the elevation range of 4900–5100 feet, where is the slope steeper—on the northeast side of the Butte or on the southwest side of the Butte?

Q1.35 Given the origin of cinder cones described in the first paragraph on this page, what do you suppose accounts for the asymmetry of south Menan Butte (i.e., one slope's being steeper than the other)?

Constructing a contour map—Drawing a contour line is a trial-and-error art form. The most important thing to bear in mind is that you must keep points of lesser values on one side of your line, while keeping points of greater values on the other side (Fig. 1.22)—a bit like slalom racing in that you must go to the right of some flags and to the left of others. But in contouring, rather than going downhill (as in skiing) you are proceeding along the side of a hill at a constant elevation.

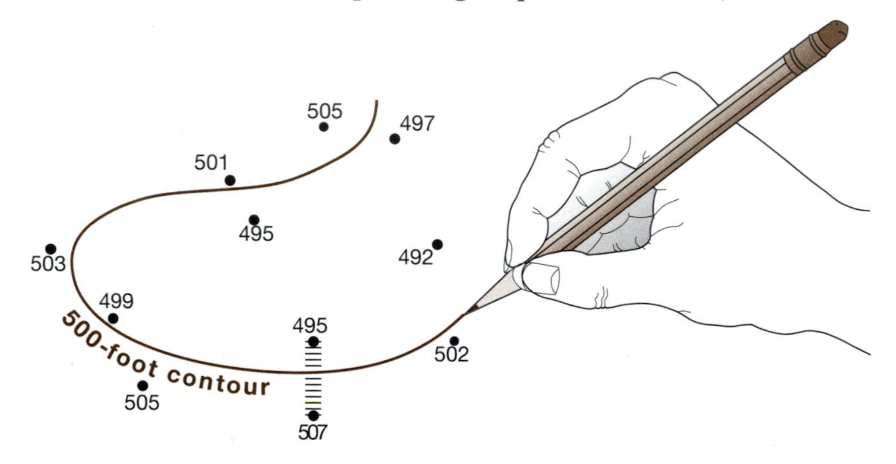

Figure 1.22 A contour line segregates elevation points. Notice that in this case of drawing a 500-foot contour line, values inside the contour are less than 500 feet, and values outside the contour are greater than 500 feet. Also, the distance between a contour line and an elevation point is proportional to their difference in elevation.

Also, the placement of a contour between points should be proportionate to the difference between the contour value and that of each of the two points. Confusing? See in Figure 1.22 that the 500-foot contour is drawn five one-foot increments from the 495-foot point and seven one-foot increments from the 507-foot point. And, notice how close the 500-foot contour is to the 499-foot point.

For an added touch of realism—draw each contour line so that it forms a 'V' where it crosses a stream (with the apex of the 'V' pointing upstream), and a smooth curve on ridges between streams (Fig. 1.23).

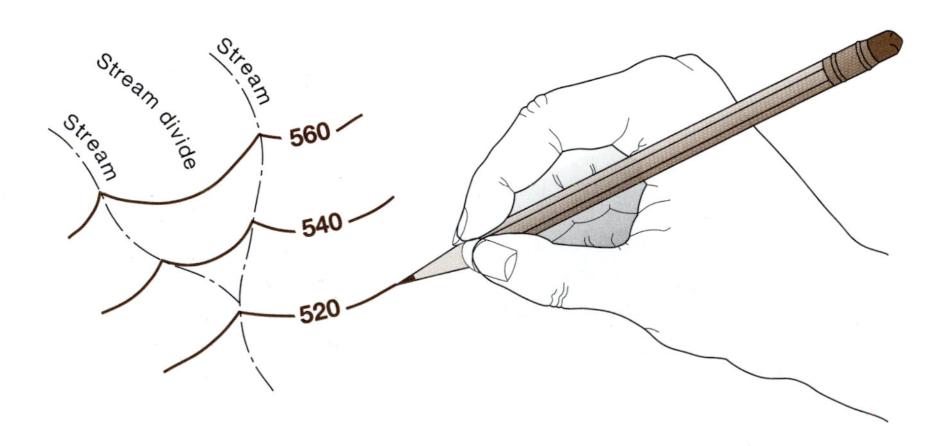

Figure 1.23 Contour lines typically form abrupt angles (like the apex of a 'V') where they cross streams, and smooth curves where they traverse stream divides.

Q1.36 Now go to *'Constructing a contour map'* on Answer Page 24 and follow instructions for completing Figure 1.28.

Constructing a topographic profile—Drawing a topographic profile—*a side view of a landscape*—along profile line A-B in Figure 1.24.

The base for a topographic profile consists of two graphic elements:

(1) A vertical scale appropriate to the relief (i.e., the range in elevations). In Figure 1.24 the vertical scale is 1,000 feet per inch.

(2) Horizontal guidelines on a graduated profile grid as per the contour interval. In Figure 1.24 the contour interval is 100 feet.

Step 1 (Fig. 1.24)—Where a contour line intersects the profile line A-B on the map, project the value of that contour straight down to the guideline of that same value on your profile grid. You can orient a straight-edge perpendicular to the profile line to use as a guide in placing points on your profile.

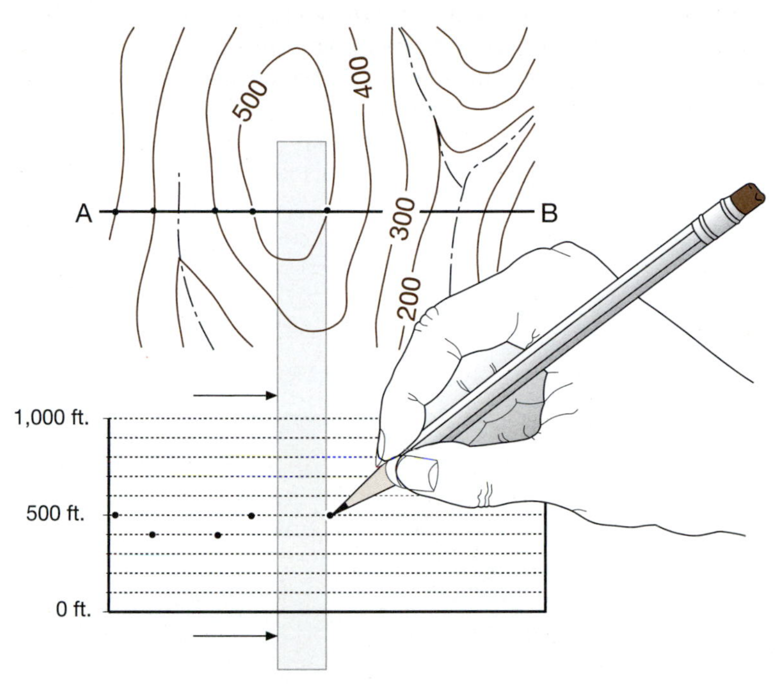

Figure 1.24 Points where the line of profile (A–B) intersects contour lines should be projected straight down onto a profile graph. (Here a transparent ruler serves as a guide for plotting points directly beneath the line of profile.)

Step 2 (Fig. 1.25)—Connect the dots on your profile graph. Voilá! You have constructed a topographic profile.

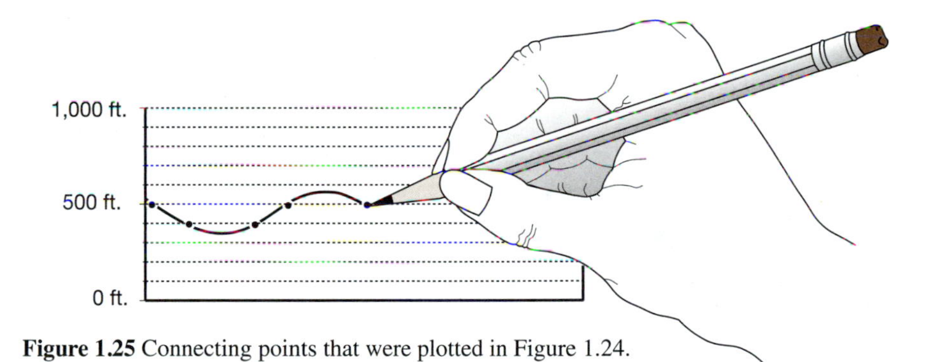

Figure 1.25 Connecting points that were plotted in Figure 1.24.

Important (Fig. 1.26): For maximum authenticity, where your profile crosses a valley or a hilltop, you should extend your profile toward—but not all the way to—the next contour value. After all, it's not likely that the bottom of a valley or the top of a hill would be precisely at a contour-interval value—thereby indicating that the valley or the hilltop is perfectly flat.

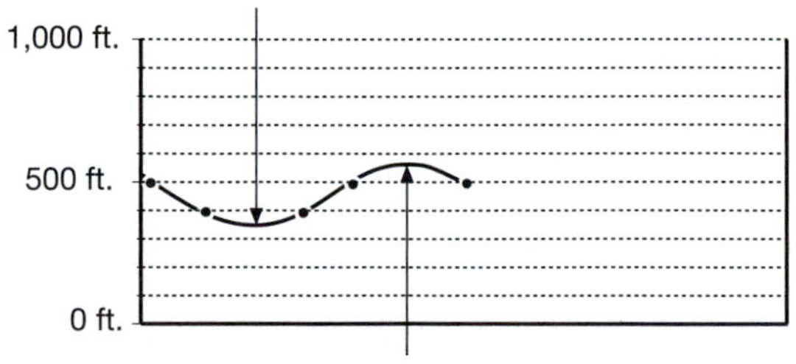

1,000 ft.
500 ft.
0 ft.

Figure 1.26 The bottoms of valleys typically extend a bit below the lowest elevation points, and the crests of hilltops typically extend a bit above the highest elevation points.

Q1.37 Go to *Constructing a topographic profile* (Fig. 1.29) across Big Island, Hawaii, on Answer Page 25 and follow instructions.

Q1.38 Refer once again to Menan Buttes, Idaho quadrangle. On the southeast slope of south Menan Butte contours are relatively straight, whereas on the northeast and southwest slopes the contours are wavy. It is difficult to decipher from the quadrangle map exactly what the wavy shapes of the contours represent. But on the satellite image below (Fig. 1.27) the meaning of the wavy contour lines is more apparent. What kind of landscape feature do you think is reflected by the wavy contours? *Hint:* One common term will do.

Volcanic cinder cones
Menan Buttes, Idaho

Menan Buttes, Idaho
7 1/2' quadrangle
N. 43° 46' 39", W. 111° 57' 54"

Figure 1.27 Menan Buttes, Idaho.

(Student's name) (Day) (Hour)

(Lab instructor's name)

ANSWER PAGE

1.1

1.2

1.3

1.4

1.5

1.6

1.7 (A) (B)

1.8

1.9

1.10 (A) (B) (C)

1.11 (A)

(B)

1.12 (A)

(B)

(C)

(D)

1.13

1.14

1.15

1.16

1.17

1.18 (A) (B)

(C)

1.19 (A) (B)

1.20 (A) (B)

1.21

1.22 ______

1.23 ______

1.24 ______

1.25 ______

1.26 ______

1.27 (A) ______

______

(B) ______

______

1.28 (A) (B) ______

1.29 ______

1.30 ______

1.31 ______

1.32 ______

1.33 (A) (B) ______

1.34 ______

1.35 ______

______

______

1.38 ______

Constructing a contour map

On Figure 1.28 (below) three stream courses appear as dashed lines. Elevation points (black dots with elevations shown in feet) have been surveyed along stream courses and along stream divides. The 480-foot contour line has been provided. Draw all other contours, applying a 20-foot contour interval. Label each of your contours with its elevation.

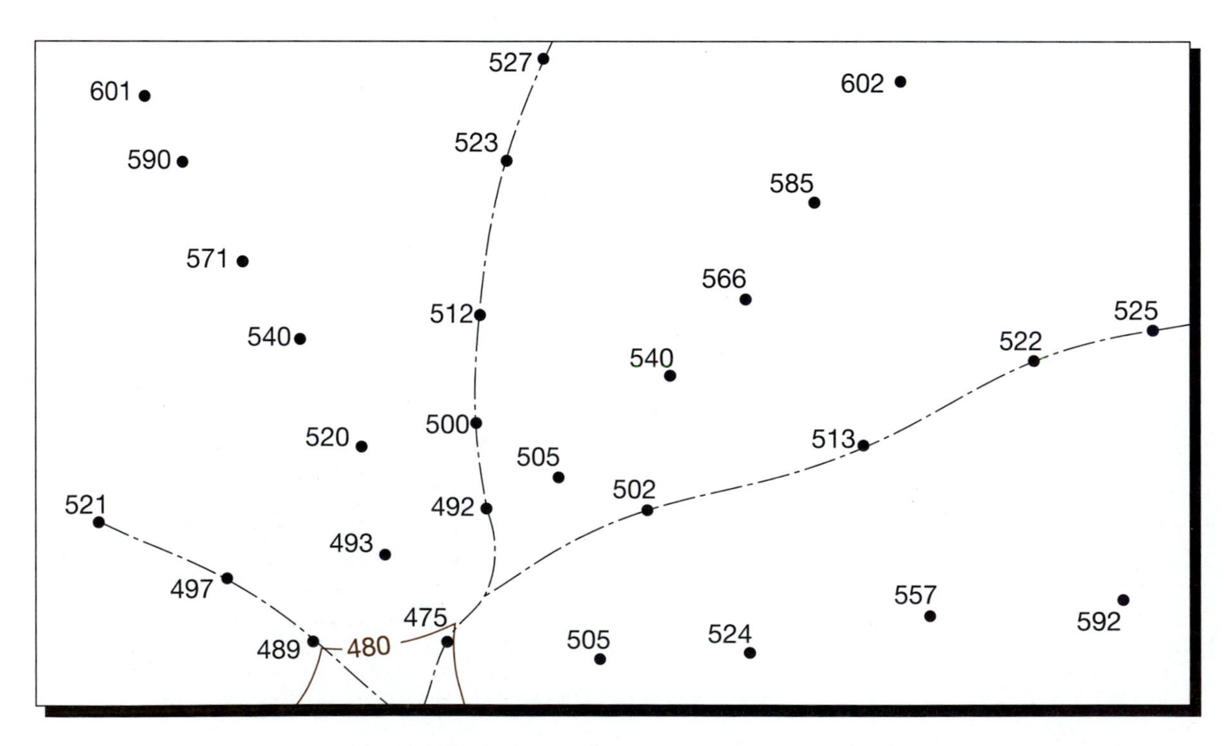

Figure 1.28 Base map for drawing a contour map. Stream courses and one contour line (at 480 feet) have been provided.

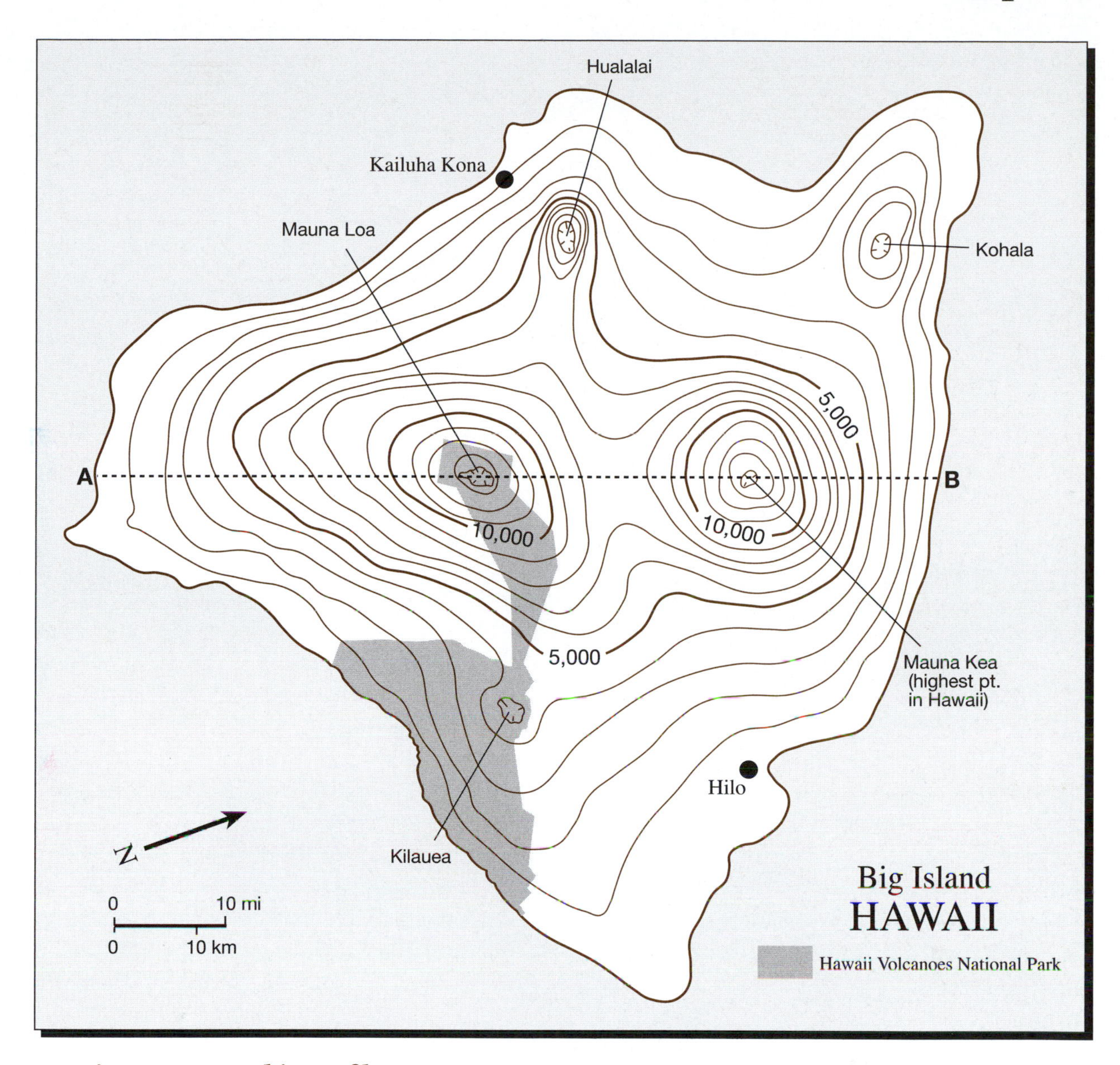

Constructing a topographic profile

Figure 1.29 Big Island Hawaii. Notice the orientation of the north arrow.

First—Tear out this page and fold it to bring the profile grid (below) up along line A-B of the above map.

Then—Where line A-B intersects a contour line, project that point straight down to a line of equal value on the profile grid. Place a dot at that spot on the grid. (The first four points, beginning at the left, are already plotted.)

Finally—Connect all of the projected dots to complete your topographic profile.

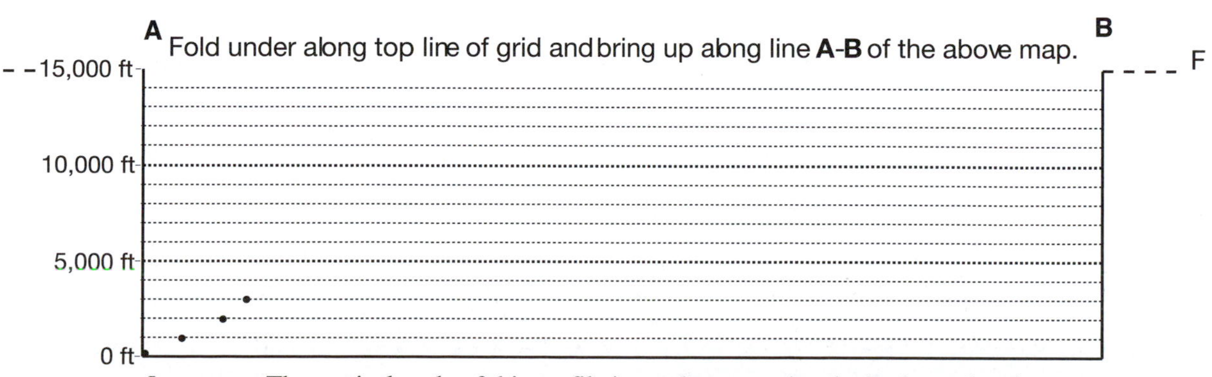

Important: The vertical scale of this profile is much greater than its horizontal scale. Therefore, true relief is exaggerated, and slopes appear to be much steeper than they are.

Intentionally Blank

2 Minerals

Topics

A. How do atoms differ from ions? What is meant by electron donor and electron recipient? What are the eight most abundant elements in Earth's crust, and what are their charges?
B. What are the two parameters that define a particular mineral species? Why is it that halite has both cubic crystal structure and cubic cleavage?
C. What is the fundamental building block of silicate minerals? Why is it called a complex ion? What two features of volcanoes are influenced by the abundance of silicate anions? What do crocidolite and chrysotile have to do with the asbestos problem? What expertise might a mineralogist provide in considering the practicality in removing asbestos from a building?
D. How are the shapes of crystals influenced by the space constraints under which they develop? What are the physical properties of minerals that aid in their identification? What are the two reasons why muscovite cleaves so easily? What is the difference between a cleavage direction and a cleavage face? What are the two parameters that define a particular mineral species?
E. What are some of the common economic uses of minerals?

A. The stuff of minerals

Minerals consist of **chemical elements**, *the most fundamental substances into which matter can be subdivided by chemical means*. Perhaps you learned in an early chemistry course that two or more elements can unite to form a compound. Well, most minerals are compounds.

The name of every element is abbreviated in chemical formulas with its chemical symbol. In some cases the symbol is taken from its name in English, for example, O for oxygen. In other cases, the symbol is taken from Latin and Greek roots, for example, Fe from Latin *ferrum*, meaning iron.

An **atom** is *the smallest particle that possesses the properties of a particular chemical element*. Atoms consist of a nucleus containing **protons** (each with one positive electrical charge) and **neutrons** (each of which is electrically neutral), plus **electrons** (each with one negative electrical charge) revolving in orbits about the nucleus (Fig. 2.1).

Q2.1 What does the net electrical charge on the carbon-12 atom shown in Figure 2.1 appear to be? *Hint:* Algebraically add the number of positively charged protons to the number of negatively charged electrons.

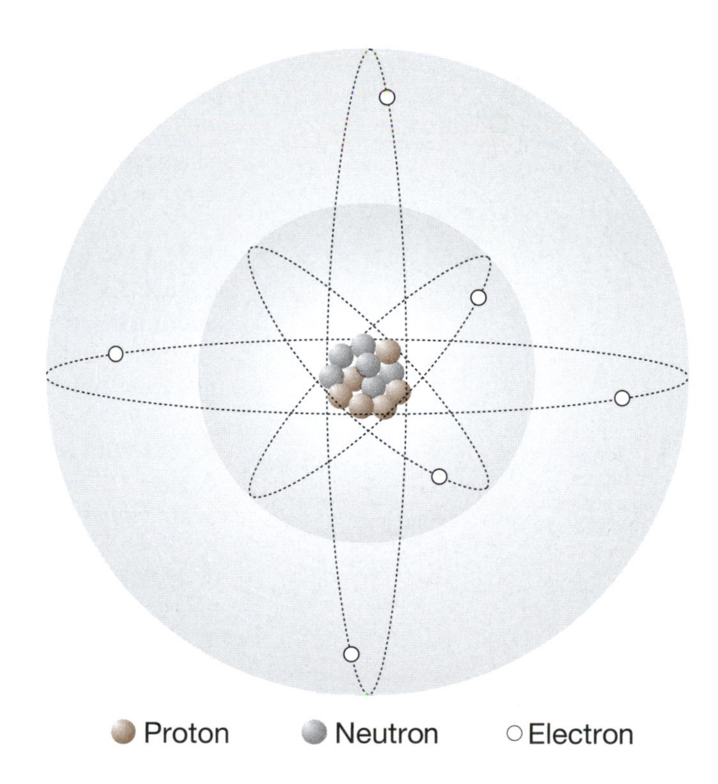

Figure 2.1 A carbon-12 (^{12}C) atom. Electrons revolve about the nucleus within an inner shell and an outer shell.

Ions—An atom is electrically neutral because its number of protons (+) is equal to its number of electrons (-). But an **ion** is another matter. Ions are electrically charged.

Only a certain maximum number of electrons can occur within each successive shell of an atom. Shell 1 (nearest the nucleus) can contain up to 2 electrons; shell 2, up to 8; shell 3, up to 18; shell 4, up to 32. An atom with its outer shell at or near capacity is relatively *stable*, whereas an atom in which its outer shell has few electrons is relatively *unstable*. When two such atoms come in contact, the atom with few electrons in its outer shell becomes an **electron donor**, and the atom at or near capacity in its outer shell becomes an **electron recipient**. For example, in the case of lithium fluoride (Fig. 2.2), lithium loses one electron, and so becomes a positive-one *cation*, whereas fluorine gains one electron, and so becomes a negative-one anion. The electrostatic force between the cation and the anion bonds them together. This *ionic bonding* is the most common type of bonding in minerals.

Q2.2 (A) What do you suppose is the ratio (proportion) of lithium-to-fluorine in lithium fluoride? (B) How about hydrogen-to-oxygen in water (H_2O)?

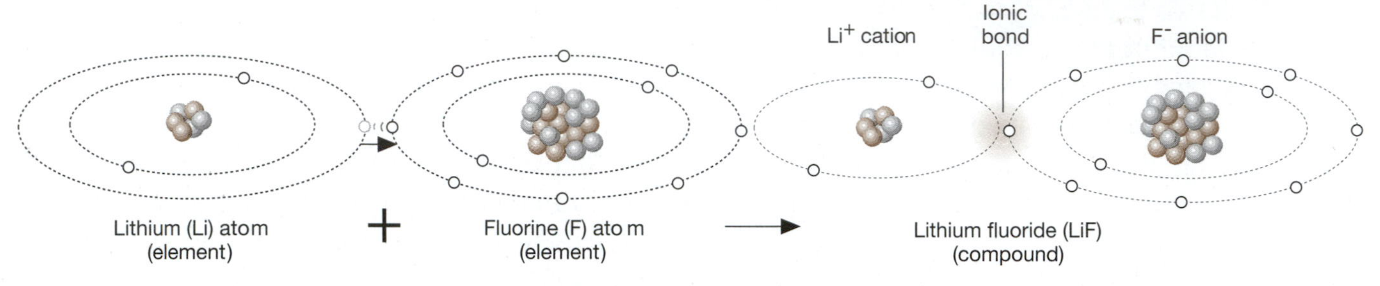

Figure 2.2 An atom of lithium combines with an atom of fluorine to form lithium fluoride. The transfer of one electron from lithium to fluorine creates two ions, a Li^+ cation and a F^- anion. The electrostatic force of the ionic bond holds the two ions together.

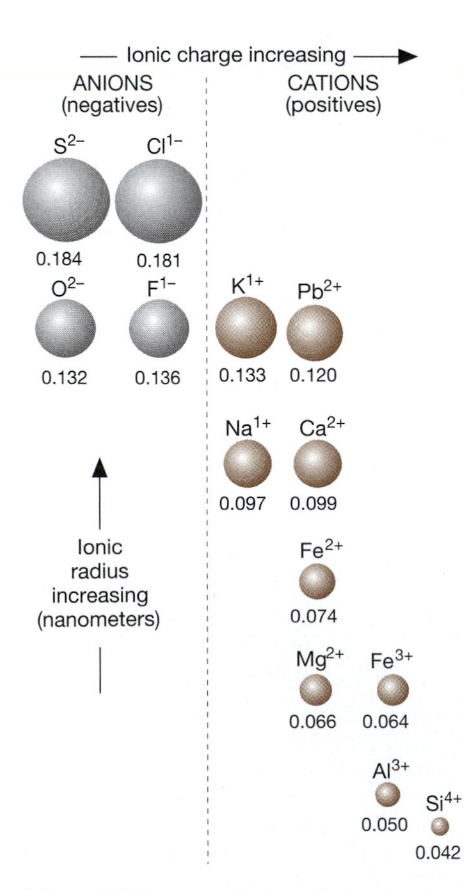

Figure 2.3 These are the ions that are most important in Earth's crust.

The manner in which elements behave in nature—i.e., with what other elements they combine, and in what proportions—is affected by (a) *ionic size* (radius in nanometers) and (b) *ionic charge* (+ vs. –). So, if ions of two elements are similar in size and charge, they can *substitute* for each other in chemical compounds.

Q2.3 (A) Judging from details in Figure 2.3, for what element can sodium most readily substitute? (See chemical symbols in Table 2.1.) (B) How about lead?

Q2.4 With reference to the ions in Figure 2.3, how do *anions* compare with *cations* as concerns ionic radii?

Q2.5 (Refer to the cations in Figure 2.2.) What is the relationship, if any, between ionic radius and charge?

Most abundant elements in Earth's crust—Estimates of the proportions of elements (by weight) in Earth's crust are shown in Table 2.1. The eight most abundant elements comprise some 98.5% of Earth's crust, with all other elements accounting for the meager balance.

Table 2.1
Most common elements in Earth's crust

Element	Symbol	% by weight
Oxygen	O^{2-}	46.6
Silicon	Si^{4+}	27.7
Aluminum	Al^{3+}	8.1
Iron	$Fe^{2+,3+}$	5.0
Calcium	Ca^{2+}	3.6
Sodium	Na^{1+}	2.8
Potassium	K^{1+}	2.6
Magnesium	Mg^{2+}	2.1

Additional common element

Element	Symbol
Barium	Ba^{2+}
Carbon	C^{4+}
Copper	Cu^{2+}
Lead	Pb^{2+}
Zinc	Zn^{2+}
Chlorine	Cl^{1-}
Fluorine	F^{1-}
Hydrogen	H^{1+}
Sulfur	S^{2-}

Q2.6 Oxygen and silicon comprise the bulk of Earth's crust. What is the sum of their two percentages by weight?

B. Crystal structure

A **mineral** is a naturally formed, solid, chemical substance having (a) a specific *composition* and (b) a characteristic *crystal structure*. These two parameters define any particular mineral species. Composition sounds simple enough, i.e., the elements present and their relative proportions to one another, but how about crystal structure?

Perhaps as a child you assembled Tinkertoys or Legos and learned that pieces fit together in only certain ways (without bending and warping your creations). **Crystallography**—*the study of crystal structures*—deals with much the same principles of geometry.

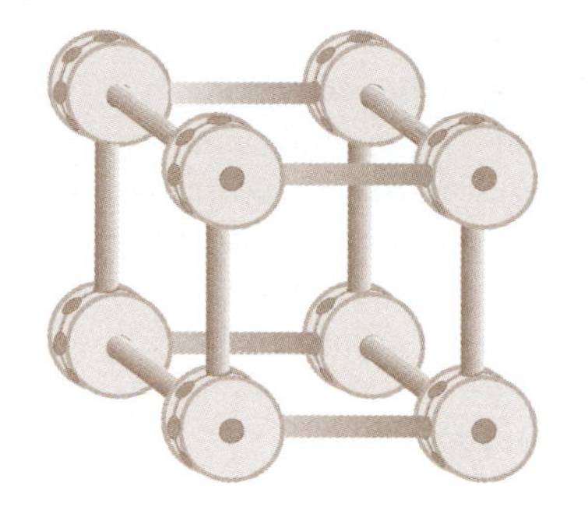

A textbook example—Both the chemical composition and the crystal structure can easily be visualized in the familiar mineral **halite (common table salt)**, which consists of sodium and chlorine arranged in an ion ratio of 1:1, which accounts for its chemical formula, NaCl, and its boxlike crystal structure (Fig. 2.4). Four chlorine ions and four sodium ions define a cube, which is manifest both in the shape of halite crystals and in the manner in which halite *cleaves* or breaks.

Ionic bonding of sodium and chlorine is analogous to that of lithium and fluorine illustrated in Figure 2.2.

Q2.7 Consider an atom of sodium brought in contact with an atom of chlorine (Fig. 2.5). (A) Which will become the electron donor, sodium or chlorine? (B) Which will become the electron recipient? *Hint:* See 2nd paragraph, top of facing page 28.

Q2.8 (A) After electron transfer, what will be the charge on the sodium ion? (B) What will be the charge on the chlorine ion?

Q2.9 (A) How many electron shells will the sodium ion have? (B) How many electron shells will the chlorine ion have?

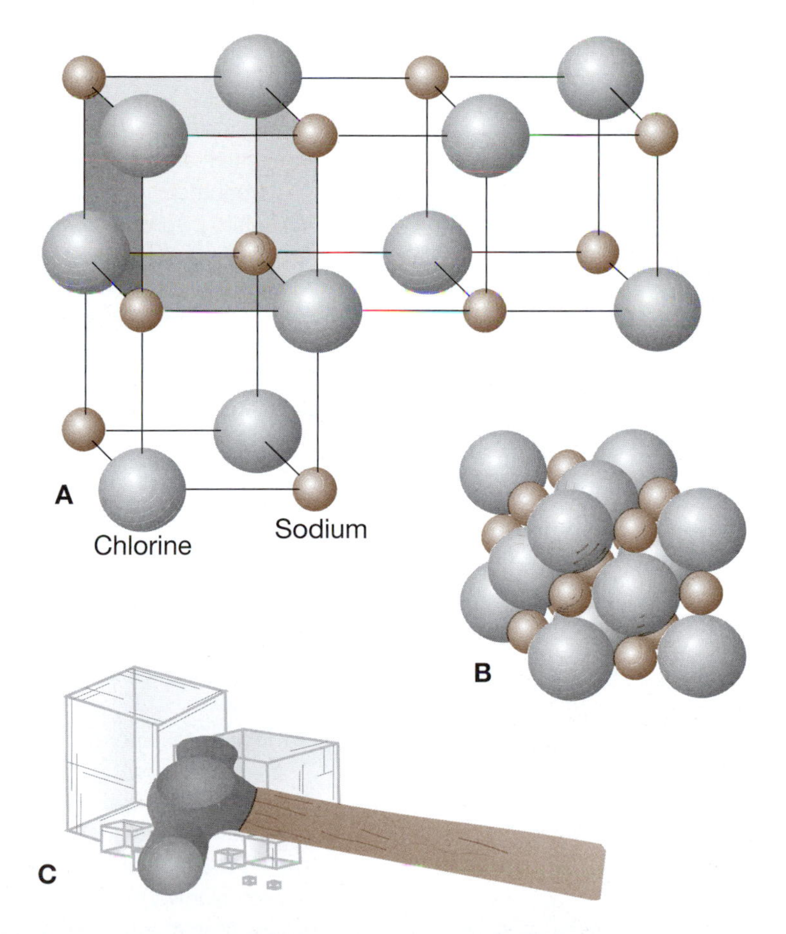

Figure 2.4 The crystal structure of halite owes its cubic character to a 1:1 mix of sodium and chlorine. **A.** An expanded model shows the geometry of four unit cubes. **B.** Actual packing of ions. C. Halite crystals and halite cleavage fragments are not necessarily cubes, but their bounding surfaces are *parallel* to those of unit cubes within the crystal.

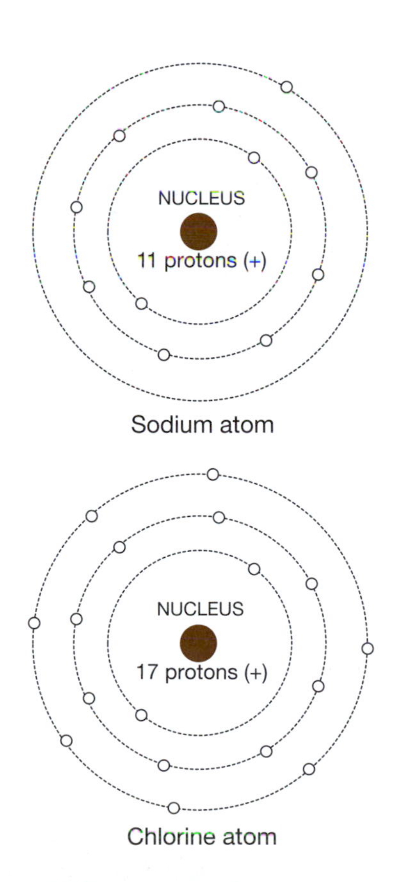

Figure 2.5 Atoms of sodium and chlorine each have three electron shells. Each nucleus is generalized as a solid sphere, with numbers of protons shown.

C. Silicate minerals

Oxygen and silicon comprise approximately 75% of all elements within Earth's crust. Together, they combine with one or more of the other six elements in Table 2.1 to form **silicate minerals**—a group within the *chemical classification* of minerals (Table 2.2). Silicate minerals are igneous in origin, that is, they crystallize from molten rock (*magma* underground, *lava* above ground).

Silicates have as their fundamental building block the **silica tetrahedron** (Gr. tetra, four) (Fig. 2.6). Oxygen (2^-) and silicon (4^+) have such an affinity for each other that they commonly unite within molten rock to form a **complex ion**—*complex*, because it consists of more than one element (oxygen plus silicon in this case)—*ion*, because it possesses an electrical charge. Chemically speaking, the negatively charged silicon-oxygen molecule is a **silicate anion**.

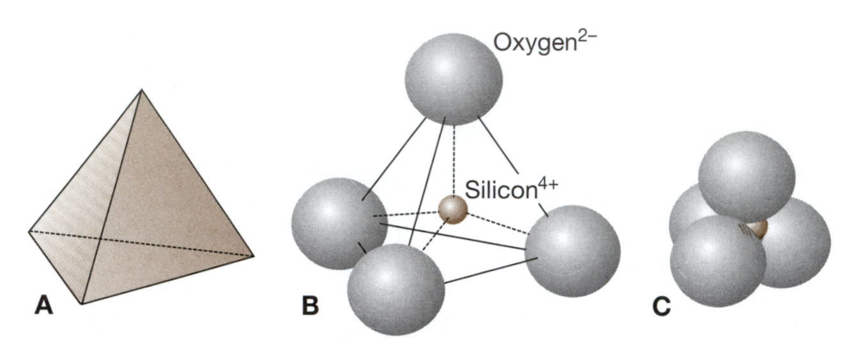

Figure 2.6A An object bounded by four equilateral triangles is a tetrahedron. **B** A silica tetrahedron is shown expanded to reveal the central position of the silicon ion. **C** This is a more realistic packing of the four oxygen anions and the one silicon cation.

Q2.10 What is the electrical charge on a silicate anion?

Q2.11 With which of the six other most common elements in Earth's crust (Table 2.1) might a silicate anion combine so as to balance electrical charges? *Recall:* Anions (-) and cations (+) combine with one another.

Various crystal structures within silicate minerals result from the degree to which silicate anions are packed together—which in turn reflects the abundance of silicate anions relative to that of various cations. In situations where silicate anions are sparse within molten rock, a single silicate anion might become entirely surrounded by cations in a mineral (Fig. 2.8A). But where silicate anions are abundant, they join into linkages with other silicate anions, with groups of linked anions separated by various cations (Fig. 2.8B–E). So the richer the magma or lava is in silicate anions, the greater the abundance of linkages.

We will see on facing page 31 that the environmental issue of asbestos is largely a matter of silicate crystal structures. And, Exercise 3, Igneous Rocks, deals with how the abundance of silicate anions within lava affects the shape of volcanic cones and the severity of eruptions (Fig. 2.7).

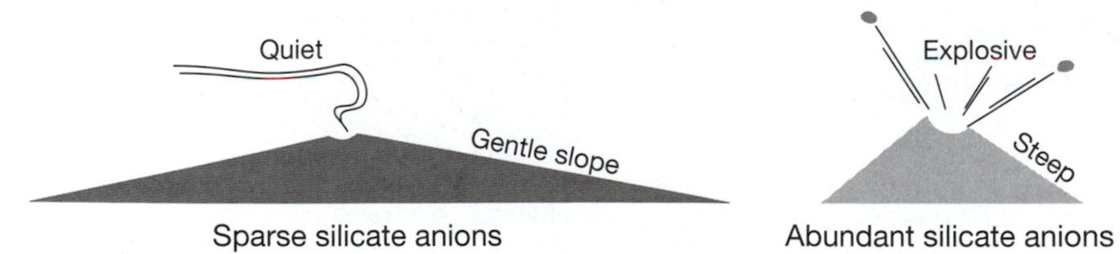

Figure 2.7 The silicate anion content of lava affects volcanic activity.

Table 2.2 Chemical classification of minerals that appear in Table 2.3 on pages 36–37

NATIVE ELEMENTS
- Graphite (carbon)
- Sulfur

SULFIDES
- Galena (lead sulfide)
- Sphalerite (zinc sulfide)
- Pyrite (iron sulfide)
- Chalcopyrite (copper iron sulfide)

HALIDES

A. Chlorides
- Halite (sodium chloride)

B. Fluorides
- Fluorite (calcium fluoride)

OXIDES
- Magnetite (iron oxide)
- Hematite (iron oxide)
- Limonite (hydrous iron oxide)
- Bauxite (hydrous aluminum oxide)

OXYGEN SALTS

A. Carbonates
- Calcite (calcium carbonate)
- Dolomite (calcium magnesium carbonate)
- Malachite (hydrous copper carbonate)
- Azurite (hydrous copper carbonate)

B. Silicates
- Quartz (silicon dioxide)
- Biotite (potassium, iron, aluminum silicate)
- Chlorite (magnesium aluminum silicate)
- Pyroxene (calcium magnesium silicate)
- Hornblende (calcium magnesium iron aluminum silicate)
- Serpentine (magnesium silicate)
- Talc (magnesium silicate)
- Muscovite (potassium aluminum silicate)
- Orthoclase feldspar (potassium aluminum silicate)
- Plagioclase feldspar (sodium calcium aluminum silicate)
- Kaolinite (aluminum silicate)
- Garnet (aluminum silicate)
- Olivine (iron magnesium silicate)

C. Sulfates
- Barite (barium sulfate)
- Gypsum (hydrous calcium sulfate)

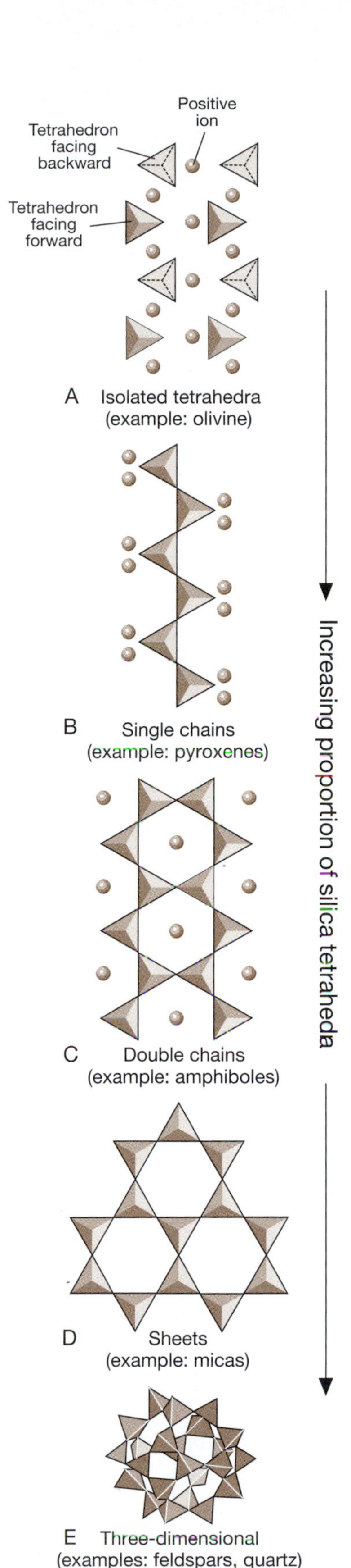

Figure 2.8 Crystal structures of silicate minerals are built upon the basic building block—the silicate anion (indicated here by tetrahedra). Cations (indicated by spheres) are shown only in A, B, and C.

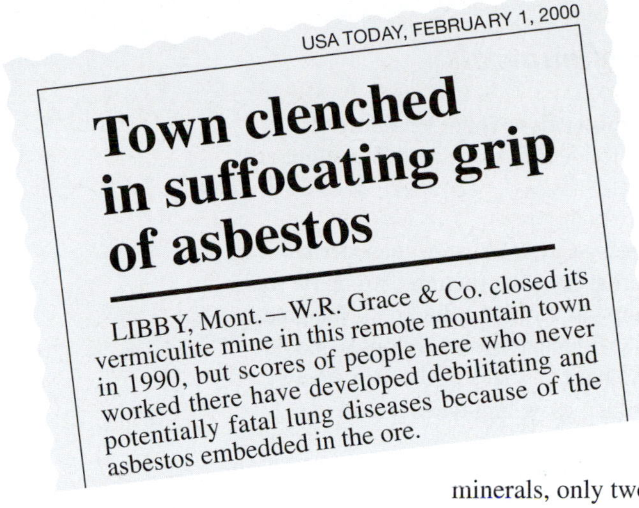

USA TODAY, FEBRUARY 1, 2000

Town clenched in suffocating grip of asbestos

LIBBY, Mont.—W.R. Grace & Co. closed its vermiculite mine in this remote mountain town in 1990, but scores of people here who never worked there have developed debilitating and potentially fatal lung diseases because of the asbestos embedded in the ore.

The problem with asbestos—Some 200,000 people are suing companies to recover damages from alleged asbestos exposure. The problem with asbestos is its delicate fibrous structure, fragments of which can create dust that penetrates lung tissue and causes lung cancer and asbestosis—a hardening of the lungs.

In 1970 the U.S. Occupational Safety and Health Administration (OSHA) began regulating asbestos in response to reports of lung disease coming from miners and construction workers handling asbestos. In the late 1970s, the Environmental Protection Agency (EPA) designated asbestos a Class A human carcinogen. Since that time, tens of billions of dollars have been spent on the removal of asbestos from public buildings—including schools. But were these horrendous costs justified? The disturbing answer arising from a chorus of earth scientists is *no!*

Asbestos, the mineral—Asbestos (Gr. *asbestos*, unaffected by fire) is a generic name for a half-dozen fibrous silicate minerals, only two of which—*crocidolite* and *chrysotile*—have been of commercial importance in America.

Crocidolite, $NaFe(SiO_3)_2$, is a double-chain amphibole with a microscopic appearance of needle-like fibers (Fig. 2.9A). Crocidolite accounts for 5% of commercial asbestos. A 1990 survey of 33 people who, in 1953, worked in a factory where crocidolite was used in the manufacture of cigarette filters, found that 19 had died of asbestos-related illnesses. *Crocidolite has been implicated as a carcinogen.*

Chrysotile, $H_4Mg_3Si_2O_9$, is a double-chain hydrous silicate member of the serpentine group masquerading as a sheet structure (Fig. 2.9B). Chrysotile accounts for 95% of commercial asbestos. A study in Thetford, Quebec evaluated people who had been exposed to high levels of chrysotile in their workplace and found that no deaths could be ascribed to that variety of asbestos. *Chrysotile has not been implicated as a carcinogen.* Unfortunately, *all* asbestos is treated equally.

Figure 2.9 **(A)** Crocidolite has the crystal habit of tiny needles, whereas **(B)** chrysotile has the crystal habit of coarser sheet-like fibers.

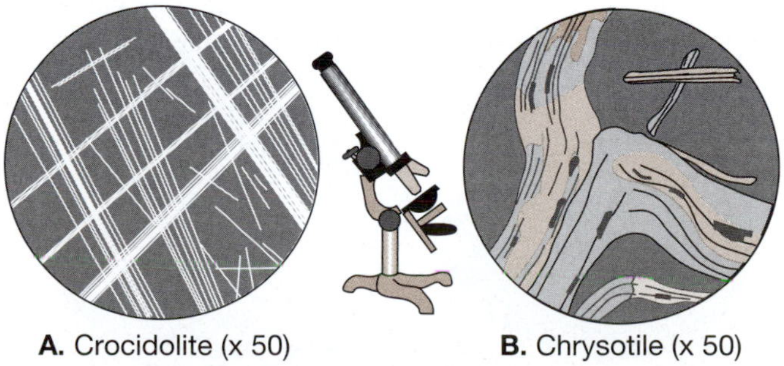

A. Crocidolite (x 50) **B.** Chrysotile (x 50)

Q2.12 What role might a mineral microscopist play in determining whether or not asbestos should be removed from a building. *Hint:* Keep it simple.

D. Identification of minerals

Physical properties of minerals—There's a litany of physical properties used to identify minerals. Only eight are discussed here.

(1)

Crystal form (i.e., shape)—Crystal form is the external manifestation of an internal crystal structure. A mineral that grows free of obstruction—as within molten rock (magma or lava) or within a water-filled cavity—develops a geometric shape bounded by crystal faces (Fig 2.10). But where two crystals abut each other, crowding takes its toll.

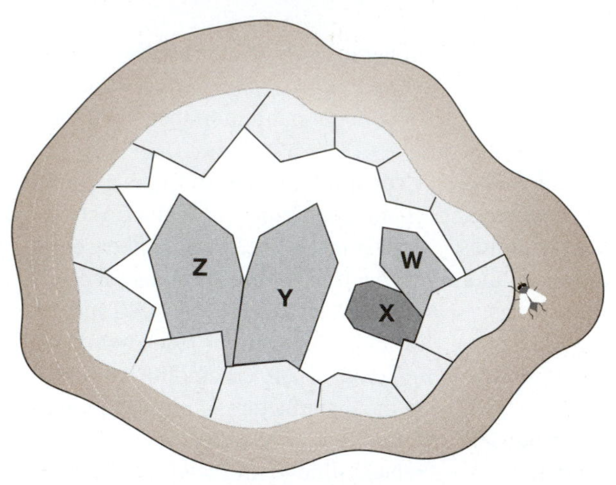

Figure 2.10 This is a slice through a *geode*—a cantaloupe-like rock-rind with crystals protruding into its central cavity. These four crystals (W, X, Y, Z) grew within a water-filled cavity, so they tended to develop crystal faces insofar as open space would allow. *Note:* All four crystals are in the same plane, that is, no one crystal is standing in front or behind its neighbor.

Q2.13 (A) Judging from physical relationships in Figure 2.10, which crystal grew first, W or X? (B) Which crystal grew first, Y or Z? (Caution.)

A similar story about shapes appears in common mollusks (Fig. 2.11). Clams, which move about, are free to grow unobstructed, and so follow the plan dictated by genetics. In contrast, oysters, although exhibiting some degree of plan, are irregular in shape because they grow crowded together in oyster reefs, and so impinge on their neighbors. Any two oysters are noticeably different in shape.

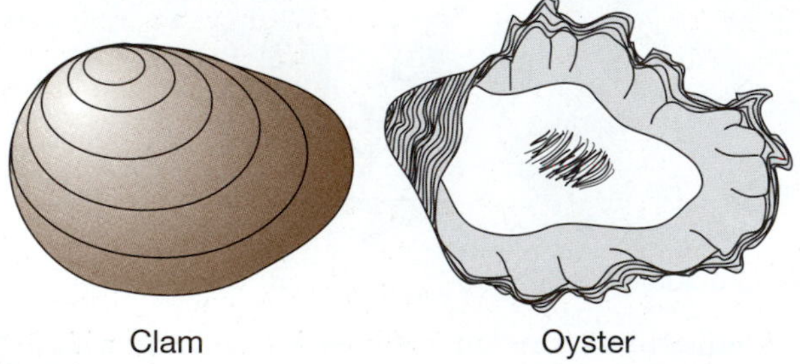

Figure 2.11 Clams appear more regular in plan than do oysters.

With few exceptions, the mineral specimens provided for study in your laboratory are not likely to be bounded by crystal faces. The reason: Crystal forms are rare. Suppliers of rocks and minerals collect specimens in bulk samples with sledge hammers, not with dental tools. Specimens with crystal faces are available, but they are costly, and most schools lack the resources necessary to purchase them. An exception: *quartz* 'points' collected from within rock cavities in the Ouachita Mountains of Arkansas and sold by suppliers for less than a dollar each.

Quartz

(2)

Cleavage—The strength of chemical bonding differs among different elements within minerals. When a mineral breaks, it tends to part between elements where bonding is weaker, much as wood splits along its grain. If breakage produces a smooth plane, the mineral is said to possess **cleavage**. (Recall halite on page 29.) Cleavage is said to have a *direction* (the orientation of a plane) because it is repetitive within a mineral, because crystal structure is pervasive.

Micas—e.g., *muscovite* and *biotite*—exhibit remarkable sheet-like cleavage. Two reasons: (1) Micas are sheet silicates, with silica tetrahedra in layers (Fig. 2.12). (2) In muscovite, layers of potassium cations are sandwiched between layers of silicon anions, and the bonds between potassium cations and silicon anions are weak. So muscovite cleaves along potassium layers.

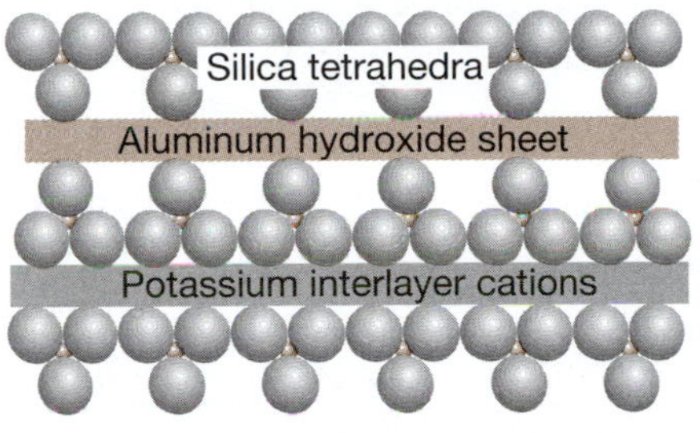

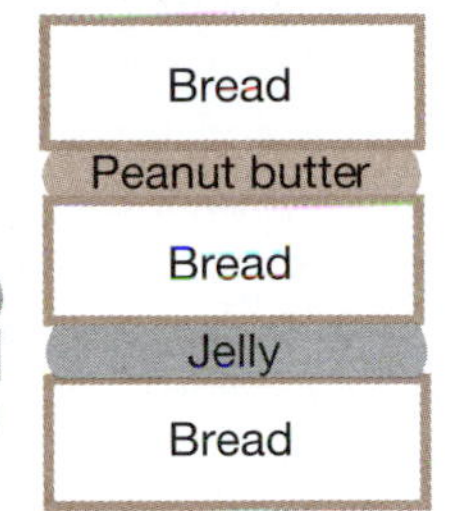

Figure 2.12 A double-decker sandwich of peanut butter and jelly illustrates the cation layers within muscovite. The sandwich would part at the layer of jelly more easily than at a layer of peanut butter, right? So it is with muscovite mica, which cleaves along layers of potassium because the bond between potassium cations and silicon anions is much weaker than that between aluminum hydroxide cations and silicon anions.

Figure 2.13 (A–F). Examples of cleavage. A tiny meat cleaver cleaves minerals along planes of weaker ionic bonding. Each such plane defines a cleavage *direction*.

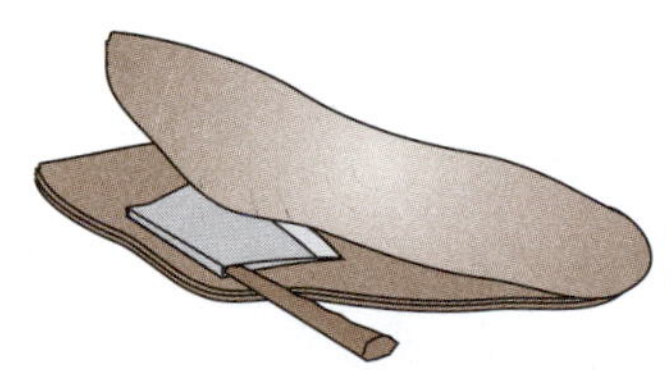

A. Biotite and muscovite micas exhibit one excellent direction of cleavage, allowing these minerals to be 'peeled' like an onion.

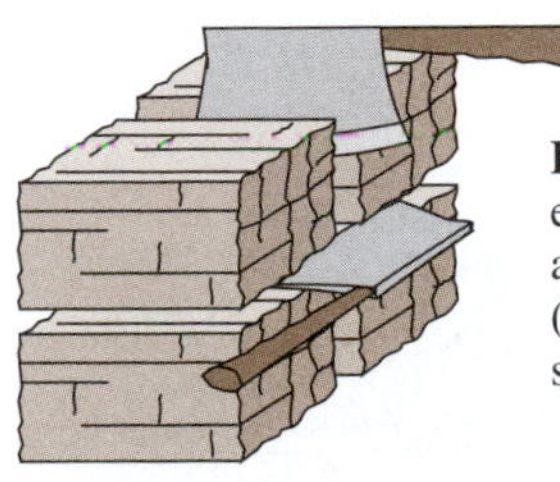

B. Orthoclase and plagioclase feldspars exhibit two good directions of cleavage at approximately 90° from each other. (The irregular, darkest surface in this sketch is a fracture, not cleavage.)

The question arises: How can one distinguish between crystal faces and cleavage surfaces? Within transparent and translucent minerals, cleavage surfaces are visibly *repetitive* and *penetrative*, like the pervasive grain of splintered wood, whereas crystal faces are nonrepetitive surface forms. (Obviously, in opaque minerals this is not so evident.)

C. Hornblende (an amphibole) has two good directions of cleavage at 124° from each other. (The irregular, darkest surface is a fracture, not cleavage.)

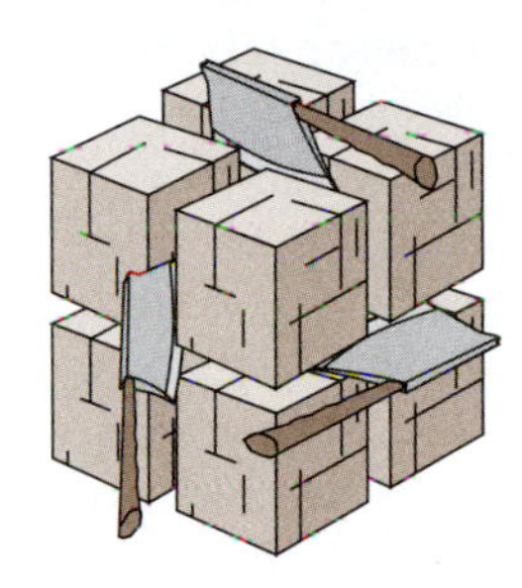

D. Galena has three good directions of cleavage that form two 90° angles in two perpendicular planes. This type of cleavage, called cubic, produces a form with six sides. (Again, recall halite.)

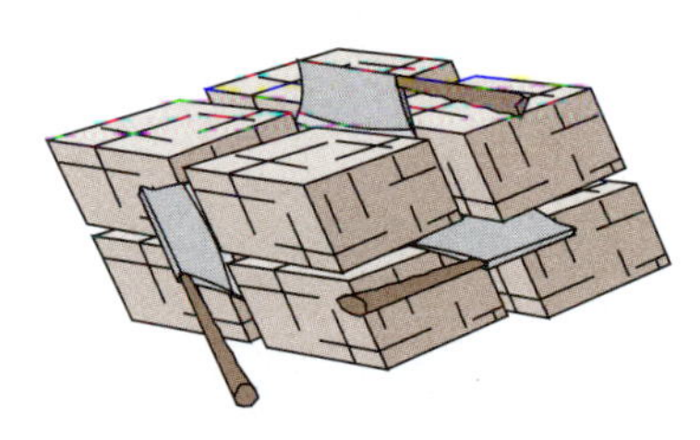

E. Calcite exhibits three good directions of cleavage that form angles of 105° in one plane and 75° in another. This rhombohedral (or rhombic) cleavage produces a form with six sides that 'leans' toward one of its corners.

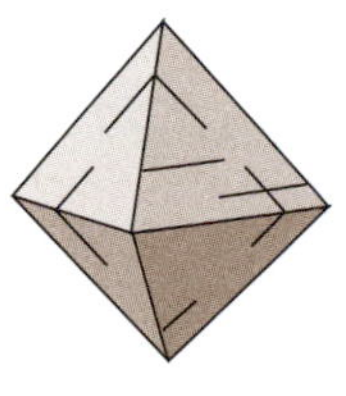

F. Fluorite has four good directions of cleavage. This type of cleavage, called octahedral, produces a form with eight sides.

Q2.14 Table 2.3 (page 37, 6th mineral from the top) lists sphalerite (a common zinc ore mineral) as having 'six good cleavage *directions*.' What is the maximum number of cleavage *faces* theoretically possible in a specimen of sphalerite?

(3)

Fracture—In some minerals the chemical bonding is so uniform in all directions that cleavage is hardly apparent when the mineral is broken. Instead, the mineral is said to fracture (break in an irregular manner). There are two common types of fracture: (a) **Conchoidal** fracture produces smooth, curved surfaces, like those of broken glass. (Conchoidal comes from Gr. *konche*, a shell —with its concavities and convexities). (b) **Uneven** fracture produces rough surfaces, like the darker surfaces in Figures 2.13B and 13C (previous page). A mineral can exhibit cleavage in one direction and fracture in another (same figures).

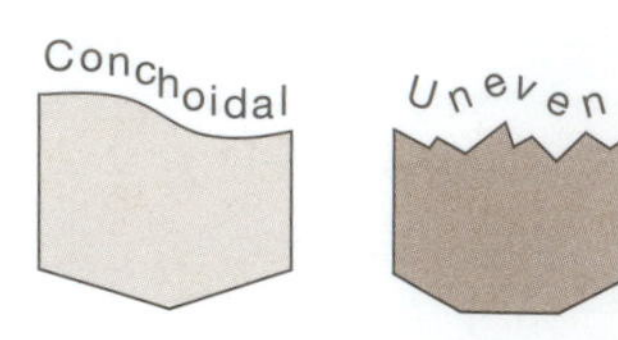

(4)

Luster—Luster is the manner in which a mineral reflects light. **Glassy** luster is characteristic of many minerals, such as *quartz*. **Metallic** luster is characteristic of metallic sulfides, such as *galena* (lead sulfide) and *pyrite* (iron sulfide). As with hardness, weathering can obscure a mineral's true luster by producing a dull earthy appearance, so look for a freshly broken surface when assessing luster. Some minerals do exhibit dull luster even on fresh surfaces. Incidentally, the brilliance of gemstones is enhanced by faceting and polishing. Gemstones are commonly dull in their natural setting.

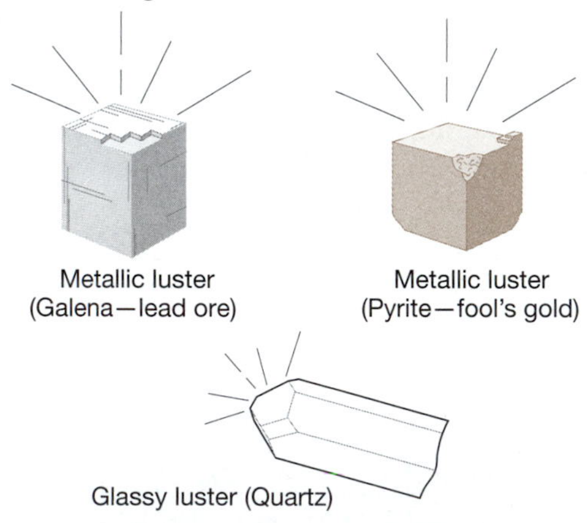

(5)

Color…obvious but variable—The color of a mineral is commonly one of its least diagnostic properties because it can vary within a mineral species. For example, *fluorite* can be almost any color, as can *quartz*. Colors of metallic minerals (e.g., *galena* and *pyrite*) are less variable, and so are more reliable aids in identification.

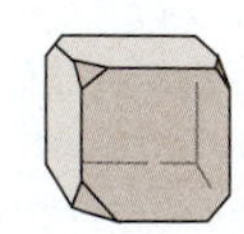
Fluorite
(honey-colored)

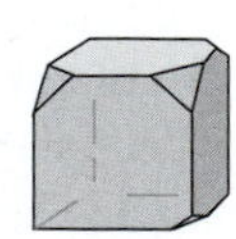
Fluorite
(violet-blue)

(6)

Streak—More reliable than the color of a mineral is the color of its streak. Streak, which is powdered mineral, is produced by rubbing the mineral on a *streak plate*, which is a piece of unglazed porcelain. The hardness of the streak plate is 7, so streak cannot be produced from a mineral that is harder than 7.

A light-colored streak of a darker mineral

(7)

Specific gravity (SG)—The specific gravity of a mineral is a number that expresses the ratio of the weight of the mineral to the weight of an equal volume of water. So the SG of water—the arbitrary standard—is 1.0. **Density**, which is weight per unit volume of a substance (e.g., grams per cubic centimeter), is measured differently from SG, but the higher the SG, the greater the density, so—at the elementary level—they can be viewed as practically synonymous. In geology, SG is usually applied to *minerals*, and density is usually applied to *rocks*.

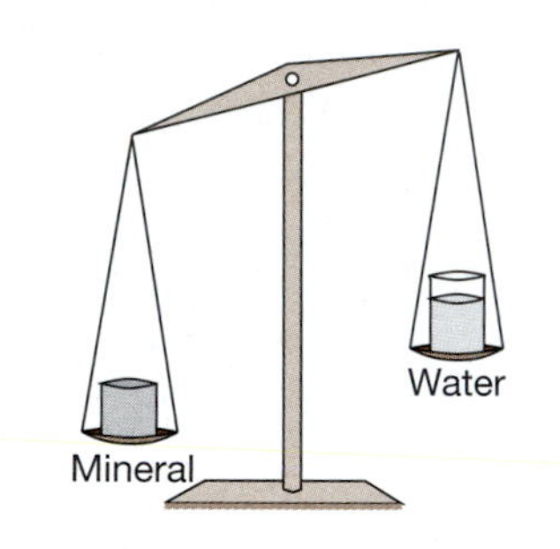

The SG of metallic minerals is noticeably high, averaging around 5.0 (e.g., *pyrite*, *magnetite*, and *hematite*). *Galena* weighs in at a whopping 7.4! An exceptionally heavy nonmetallic mineral is *barite* (barium sulfate). Barite's SG of 4.5 accounts for its utility in oil-well drilling mud, the weight of which is intended to prevent blowouts when the drill encounters high-pressure fluids. We heard a lot about drilling mud in discussions of technology surrounding the summer 2010 blow-out in the Gulf of Mexico. Most other nonmetallic minerals, for example, *quartz* and *feldspar*, exhibit the narrow range of 2.6–2.7 SG.

(8, 9, 10)

Feel (as to the touch)—example: the soapy feel of *talc*.
Taste—example: *halite* (salt).
Magnetic quality—*Magnetite* is the only strongly magnetic mineral. A few other iron-bearing minerals (e.g., *hematite*) can be weakly magnetic.

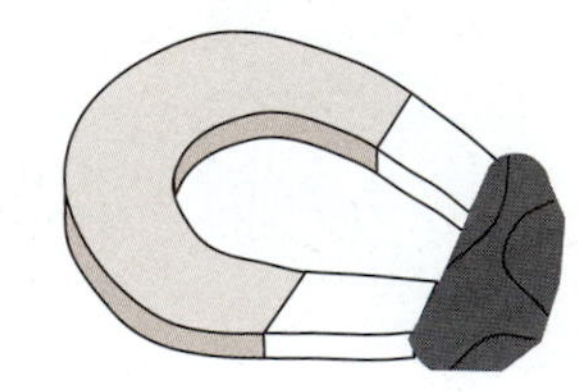

(11)

Hardness—The hardness of a mineral is its resistance to abrasion, which is a measure of the strength of bonding among its chemical elements. The degree of hardness can be estimated by the comparative ease, or difficulty, with which one mineral can be scratched by another mineral whose hardness is known.

Q2.15 Name a couple of commercial items that are useful because of their hardnesses.

In the 19th century the German mineralogist Friedrich Mohs assigned arbitrary numbers (1 through 10) to ten well-known minerals (Fig. 2.14).

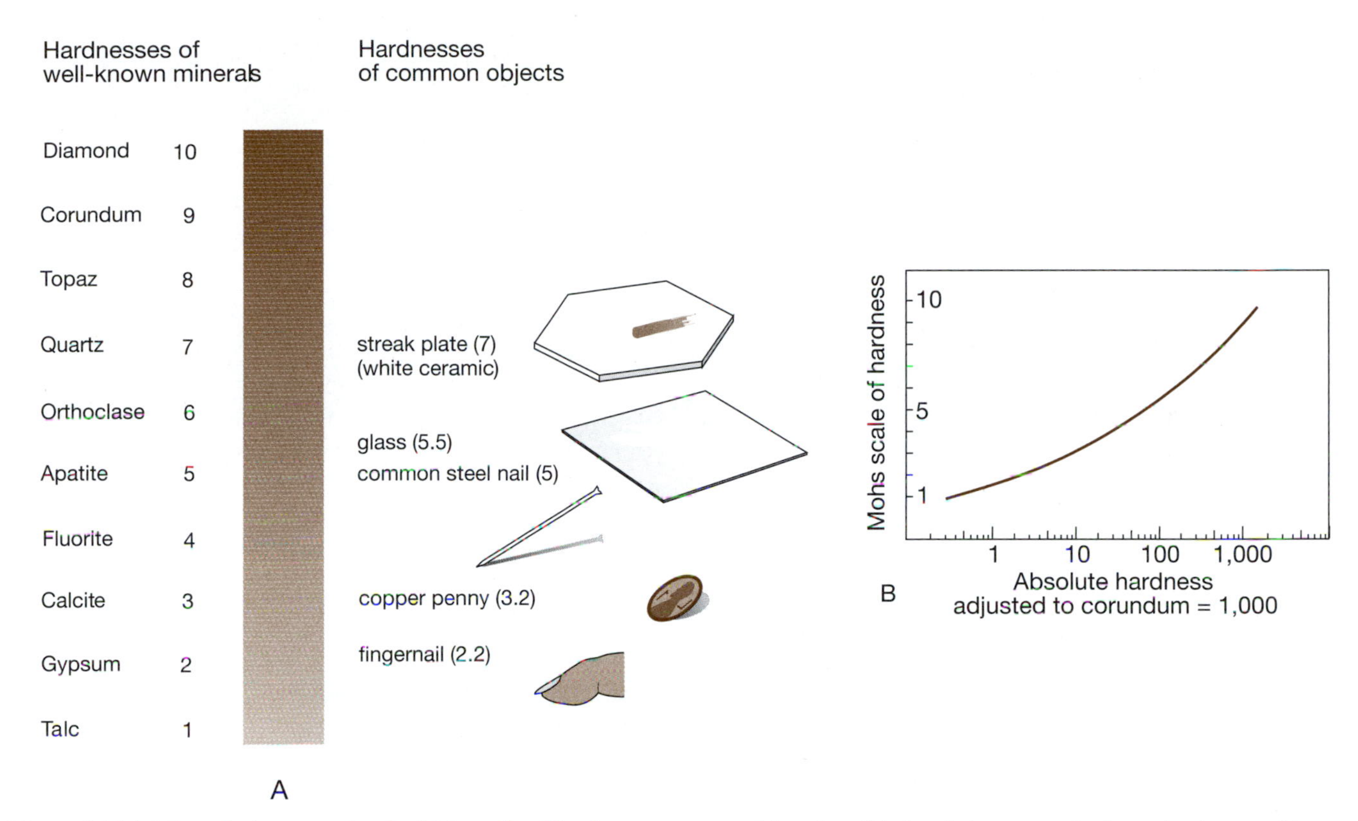

Figure 2.14 A Minerals that comprise the Mohs scale of hardness are arranged in order of their relative—*not absolute*—hardnesses. For example, diamond is not 2 times as hard as apatite, it's actually about 16 or 17 times as hard as apatite; so Mohs' scale can be described as crudely exponential in character, as is illustrated in **B**.

The hardness of a mineral can be assessed by testing it against common objects—e.g., fingernail, copper penny, steel nail, glass plate, streak plate. A harder object scratches a softer object. ('Brings to mind the childhood game rock-paper-scissors.) But a softer object might leave a *smear* on a harder object that could be mistaken for a scratch on the harder object. Unlike a scratch, however, a smear can be rubbed off. Also, some specimens might be weathered and exhibit an earthy coating. The hardness of such a coating is likely to be less than that of the fresh mineral, so you should use only a clean surface when testing hardness.

When testing the hardness of a mineral against that of glass, it's safest to lay the glass plate flat on the table and try to scratch it with a corner of the mineral. A mineral having a hardness of 7 will scratch the glass easily; one with a hardness of 6, with difficulty. When testing the hardness of a mineral against that of a nail, try scratching the mineral with the pointed end of the nail. If no scratch is made, find a sharp edge on the mineral and rotate the nail against it. A mineral having a hardness of 7 will scratch the nail easily; whereas one with hardness 6 will scratch the nail only slightly, and the edge of the mineral will likely be crushed.

Recognizing mineral species

Table 2.3 (on following pages 36–37) lists 32 common minerals, along with their physical properties and chemical compositions. (Photographs of these minerals follow Table 2.3 on pages 38–41.) Ideally, if one can accurately assess the physical properties of a specimen, its name can be keyed-out in Table 2.4 on page 44. In practice, however, mineral identification is not so mechanical. Learning to identify minerals is a bit like learning to spell (e.g., *'i before e except after c'*). There are useful rules in spelling, but there's no substitute for word recognition.

Table 2.3 Mineral properties

Minerals with metallic luster

(in order of decreasing hardness)

tool (hardness)	hardness	common color(s)	streak color	other properties (SG where diagnostic)	name and composition
streak plate (7.0)					
	6–6.5	pale brass yellow	greenish black, brownish black	cubic crystals; conchoidal to uneven fracture; SG 5.0	*PYRITE* FeS_2
	5.5–6.5	iron-black	black	magnetic (clings to magnet); rare crystal faces; SG 5.1; indistinct cleavage; metallic to submetallic (dull) luster	*MAGNETITE* Fe_3O_4
	5.5–6.5 (softer where weathered)	silver to iron-black	reddish brown	can exhibit glittering tiny crystal faces; common traces of rust; SG 5.2	*HEMATITE* Fe_2O_3
glass plate (5.5)					
steel nail (5.0)					
	2.5–4	brass yellow; commonly with iridescent tarnish	greenish black, brownish black	can exhibit pyramid-shaped crystals; more commonly scaly; SG 4.2	*CHALCOPYRITE* $CuFeS_2$
copper penny (2.2)					
	2.5	silver gray	gray to black	heft and appearance of lead; cubic crystals; 3 cleavage directions at 90° produce cubic fragments; SG 7.4	*GALENA* *PbS*
fingernail (2.2)					
	1	pencil-lead gray	gray to black	cleavage of tiny crystals indicated by luster; greasy to the touch; stains textured objects; SG 2.3	*GRAPHITE* *C*

Minerals with nonmetallic luster (continued next page)

(in order of decreasing hardness)

tool (hardness)	hardness	common color(s)	streak color	other properties (SG where diagnostic)	name and composition
	6.5–7.5	dark red to black	no streak (harder than streak plate)	commonly 12-sided crystals (dodecahedra); conchoidal fracture	*GARNET* $Al_2(SiO_4)_3$ *+ other metallic elements*
	7	clear, milky white, purple, rose, smoky	ditto	commonly 6-sided crystals; conchoidal fracture	*QUARTZ* SiO_2
	6.5–7	green	ditto	crystals with slightly conchoidal fracture; common granular appearance	*OLIVINE* $(Fe, Mg)_2SiO_4$
streak plate (7.0)					
	6–6.5	white to pink	no streak (too close to hardness of s. plate)	2 cleavage directions at 90°; irregular stringers as in wood grain	*ORTHOCLASE FELDSPAR* $KAlSi_3O_8$
	6–6.5	white to dark gray	ditto	2 cleavage directions at near 90° ; fine straight parallel striations	*PLAGIOCLASE FELDSPAR* $(Na,Ca)AlSi_3O_8$
	6	dark green to black or brown	white to pale green	common satin sheen; 2 cleavage directions at 124°	*AMPHIBOLE GP. (e.g., HORNBLENDE) NaCaMgFeAlSiO*
	6	dark green to black	white to pale green	luster approaches a sheen; 2 cleavage directions at near 90°	*PYROXENE GP. (e.g., AUGITE)* $CaMgSi_2O_6$ *plus Fe*

Table 2.3 Mineral properties (cont.)

Minerals with nonmetallic luster (continued)

(in order of decreasing hardness)

tool (hardness)	hardness	common color(s)	streak color	other properties (SG where diagnostic)	name and composition
	5.5–6.5	red, brown, yellow	red, brown, yellow	rusty earthy appearance; SG 5.2	*HEMATITE* Fe_2O_3
glass plate (5.5)					
	2–5.5	brownish yellow, ocher-yellow	yellowish brown	earthy appearance; SG 2.6–4.4	*LIMONITE* $Fe_2O_3 \bullet 2H_2O$
steel nail (5.0)					
	4	honey-colored, violet-blue, yellow, green	white, or none	glassy; possibly translucent cubic crystals; 4 cleavage directions produce octahedrons	*FLUORITE* CaF_2
	2.5–4	bright green	pale green	crystal faces rare; glassy to earthy	*MALACHITE* $CuCO_3 \bullet Cu(OH)_2$
	2.5–4	bright blue	pale blue	crystal faces rare; glassy to earthy	*AZURITE* $2CuCO_3 \bullet Cu(OH)_2$
	2.5–4	yellowish brown	brownish to light yellow to red	resinous luster; odor of rotten eggs when scratched; 6 good cleavage directions	*SPHALERITE* ZnS
	2.5–4	White, gray, pink	no streak (or white)	glassy; rhombic crystals; some crystals are curved; 3 cleavage directions produce rhombic fragments	*DOLOMITE* $CaMg(CO_3)_2$
	2.5–4	light to blackish green	white	splintered and/or layered; soapy to the touch	*SERPENTINE* $H_4Mg_3Si_2O_9$
copper penny (2.2)					
	3	clear, white, gray, yellow	no streak (or white)	glassy; hexagonal crystals; 3 cleavage directions produce rhombic fragments	*CALCITE* $CaCO_3$
	2.5–2.5	clear to white to blue	white	heavy; 2 good cleavage directions; bladed appearance; SG 4.5	*BARITE* $BaSO_4$
	2.5–3	brown to black	light tan to brown to black	1 perfect (flaky) cleavage direction; translucent in thin sheets	*BIOTITE* $K(Mg, Fe)_3AlSi_3O_{10}(OH)_2$
	2–3	clear to yellow	colorless to white	1 perfect (flaky) cleavage, transparent in thin sheets	*MUSCOVITE* $KAl_2(AlSi_3O_{10})(OH)_2$
	2.5	white, tan, red, brown, black	tan to brownish red	distinct pebble-like character; earthy	*BAUXITE* $Al_2O_3 \bullet 2H_2O$
	2.5	clear to dark gray	no streak (or white)	glassy; cubic crystals; 3 cleavage directions at 90° produce cubic fragments; salty taste	*HALITE* $NaCl$
	1.4–2.5	yellow to red	pale yellow	bright to dull colors; brittle; conchoidal fracture	*SULFUR* S
	2 –2.5	green to very dark green	pale green	irregular cleavage produces layered fabric	*CHLORITE* $H_4Mg_2Al_2SiO_9$
fingernail (2.2)					
	2 –2.5	white, yellow, pink	white	chalky; sticky when wet	*KAOLINITE* $H_4Al_2Si_2O_9$
	2	clear to yellow	white	glassy transparent plates or satin-white rods; 1 very good and 2 good cleavage directions produce parallelogram fragments in plan view	*GYPSUM* $CaSO_4 \bullet 2H_2O$
	1 –1.5	silvery or greenish white	white	pearly luster; soapy to the touch; flakes easily	*TALC* $Mg_3Si_4O_{10}(OH)_2$

Minerals

Minerals with metallic luster

Magnetite
Iron-black;
magnetic (clings to magnet); rare crystal faces; indistinct cleavage; metallic to submetallic (dull) luster

Pyrite
Pale brass yellow;
cubic crystals;
conchoidal to uneven fracture

Hematite
Silver to dark gray;
can exhibit glittering tiny crystal faces;
common traces of rust

Chalcopyrite
(here with white to gray quartz)
Brass yellow with iridescent tarnish;
can exhibit tiny pyramid-shaped crystals, but scaly appearance is more common

Galena
Silver gray;
heft and appearance of lead;
cubic crystals;
3 cleavage directions at 90° produce cubic fragments

Graphite
Dark gray like pencil lead;
cleavage of tiny crystals indicated by luster;
greasy to the touch; stains textured objects

Minerals with nonmetallic luster

Quartz
Clear, milky white, purple, rose, smoky;
hexagonal crystal faces common;
conchoidal fracture

Garnet
Dark red to black;
commonly 12-sided crystals;
conchoidal fracture

Olivine
Yellowish green;
crystals with slightly conchoidal
fracture; common granular appearance

Orthoclase feldspar
White to pink;
2 cleavage directions at 90°;
irregular stringers as in wood grain

Plagioclase feldspar
White to dark gray;
2 cleavage directions at near 90°;
fine straight parallel striations

Amphibole (ex., hornblende)
Dark green to black or brown;
common satin sheen;
2 cleavage directions at 124°
(like a low-pitched roof)

Pyroxene (e.g., augite)
Dark green to black;
luster approaches a sheen;
2 cleavage directions at near 90°

Minerals with nonmetallic luster (continued)

Hematite
Red, brown, yellow;
rusty earthy appearance

Limonite
Brownish yellow, ocher-yellow;
earthy appearance

Fluorite
Honey-colored; violet-blue; yellow; green;
glassy; possibly translucent; cubic crystals;
4 cleavage directions produce octahedrons

Malachite (green)
Azurite (blue)
Bright colors;
crystal faces rare;
glassy to earthy

Sphalerite
Yellowish brown;
resinous luster; odor of rotten eggs
when scratched; 6 good
cleavage directions

Dolomite
(here with crystals of brassy chalcopyrite)
White, gray, pink;
glassy; rhombic crystals; some crystals
are curved; 3 cleavage directions produce
rhombic fragments

Asbestos serpentine
Light green to blackish green;
splintered and/or layered;
soapy to the touch

Calcite
Clear, white, gray, yellow;
glassy; hexagonal crystals; 3 cleavage
directions produce rhombic fragments

Barite
Clear to white to blue;
heavy; 2 good cleavage directions;
bladed appearance

Minerals with nonmetallic luster (continued)

Biotite
Brown to black;
1 perfect (flaky) cleavage direction;
translucent in thin sheets

Muscovite
Colorless to white;
1 perfect (flaky) cleavage direction;
transparent in thin sheets

Bauxite
White, tan, red, brown, black;
pebbly appearance;
earthy texture

Halite
Clear to dark gray;
glassy; cubic crystals;
3 cleavage directions at 90°
produce cubic fragments; salty taste

Sulfur
Yellow to red;
bright to dull colors;
brittle; conchoidal fracture

Chlorite
Green to very dark green;
irregular cleavage produces
layered fabric

Kaolinite
White, yellow, pink;
chalky; sticky when wet

Gypsum
Clear to yellow;
glassy transparent plates or satin-white rods;
1 very good and 2 good cleavage directions
produce parallelogram fragments in plan view;
easily scratched with fingernail

Talc
Silvery or greenish white;
pearly luster; soapy to the touch;
flakes easily

E. Uses of some industrial minerals

Pool the life experiences of a number of your classmates and see if you can answer the following questions having to do with practical uses of minerals in this exercise.

Galena [PbS]—the principal source of lead

Q2.16 Name one industrial use for lead.

Magnetite [Fe_3O_4], hematite [Fe_2O_3], and limonite [$Fe_2O_3 \cdot 2H_2O$]—sources of iron

Q2.17 Name one industrial use for iron.

Bauxite [$Al_2O_3 \cdot 2H_2O$]—the singular ore of aluminum

Q2.18 Name one industrial use for aluminum, other than that in kitchen materials.

Chalcopyrite [$CuFeS_2$]—the principal ore of copper

Q2.19 Name one industrial use for copper, other than that in the minting of one-cent pieces.

Quartz [SiO_2]—used in the manufacture of glass and electronic components.

Q2.20 Name a common electronic component made from quartz.

Gypsum [$CaSO_4 \cdot 2H_2O$]—used in the production of sheetrock and plaster of Paris

Q2.21 Give a common use for plaster of Paris.

Fluorite [CaF_2]—chief source of fluorine

Q2.22 Name a commercial use of fluorine that is of benefit of human health.

Sphalerite [ZnS]—the singular ore of zinc

Calcite [$CaCO_3$]—Portland cement, soil conditioner

Garnet [$Al_2(SiO_4)_3$ + other metals]—abrasives, semiprecious stones. Birthstone for one month

Olivine [$(Fe, Mg)_2SiO_4$]—silicon chips for computers

Halite [NaCl]—common salt; important source of sodium and chlorine

Q2.23 Name a use for salt other than that of a food additive.

Graphite [C]—lubricant, pencil lead

Barite [$BaSO_4$]—source of barium; the mineral is used as a weighting oil-well drilling mud

Sulfur [S]—source of sulfur

Muscovite [$KAl_2(AlSi_3O_{10})(OH)_2$]—a computer chip substrate, electrical insulation; glitter used in paint, in wallpaper, and for the ‘snow’ sprinkled on Christmas trees. Used as window panes in early houses and industrial furnaces

Kaolinite [$H_4Al_2Si_2O_9$]—an ingredient that inhibits the melting of chocolate candy. Also an ingredient in over-the-counter medication taken for the treatment of temporary stomach disorder

Talc [$Mg_3Si_4O_{10}(OH)_2$]—lubricating and drying agent.

Q2.24 Name a household product that uses talc as a principal ingredient.

3 Igneous Rocks

Topics

A. What is the definition of igneous rock? Is this a genetic definition (having to do with origin), or is it descriptive (having to do with appearance)? What is magma? What is lava? Where within Earth's interior do magmas originate (i.e., where does rock melt)?

B. What are six kinds of plutonic (intrusive) igneous rock bodies? What are the three varieties of volcanic (extrusive) igneous rock bodies?

C. What is there about the temperatures of crystallization of various minerals that facilitates evolution (change) in the composition of a magma?

D. What are the two parameters applied in the classification of igneous rocks? What two things govern texture of an igneous rock? What two things govern composition of an igneous rock?

E. Why is it that subduction results in partial melting? Why does partial melting selectively mobilize magmas rich in silica tetrahedra? How does enrichment of lava in silica tetrahedra affect the profile of volcanic cones? Is there a correlation between the degree of enrichment of lava in silica tetrahedra and the degree of explosiveness of eruptions? Why or why not?

F. What are the criteria for the identification of igneous rocks (apart from other rocks)?

G. What is one explanation for the escape of suffocating carbon dioxide from Lake Nyos?

A. *Igneous rocks—their definition and origin*

Igneous rocks defined—Igneous rocks (from Latin *ignis*, fire) are rocks that have **crystallized** from molten rock. A visual metaphor for the crystallization of minerals within molten rock is ice crystals forming within water at 0 °C (= 32 °F).

Q3.1 Is this definition genetic (i.e., having to do with a rock's origin), or is it *descriptive* (i.e., having to do with a rock's physical appearance manifest in features such as color, heft, and the particular kinds of minerals present)?

Before we examine igneous *rocks*, let's think about *molten rock*. Where molten rock spills out onto the ground it is known by the familiar name **lava**. But that same molten rock underground is called **magma**. Magma is curious in that it undergoes a name change as it emerges from a volcano (Fig. 3.1).

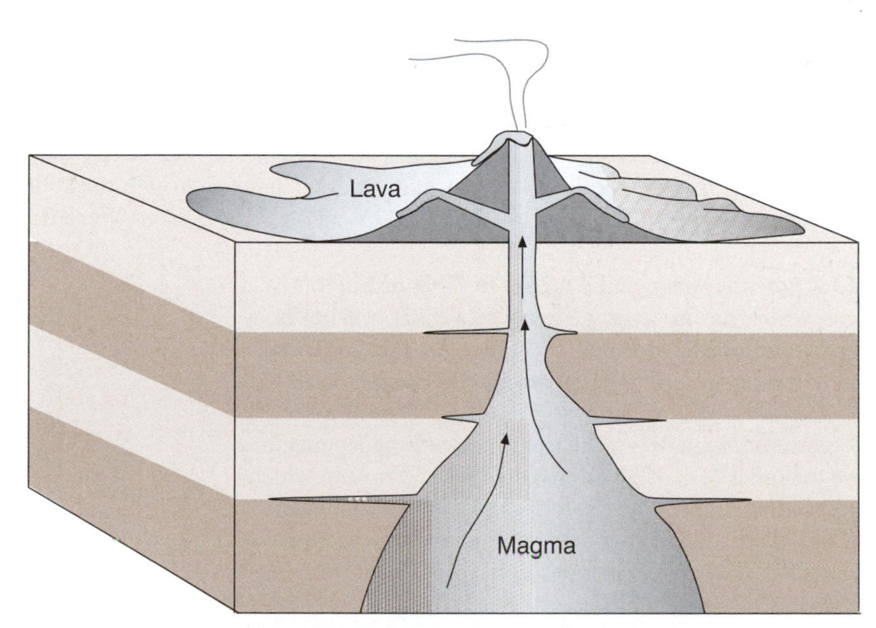

Figure 3.1 Molten rock underground is called magma, but that same molten rock above ground is called lava.

Melting of rock to form magma—Billions of years ago, when Earth was entirely liquid, heavier elements sank toward its center, and lighter elements floated on heavier elements (like oil on water). So Earth is now differentiated into layers of material that increase in density downward (Fig. 3.2). (Density is the mass of a substance relative to that of an equal volume of water, or g/cm^3.) Residual temperature from that once-molten Earth also increases downward.

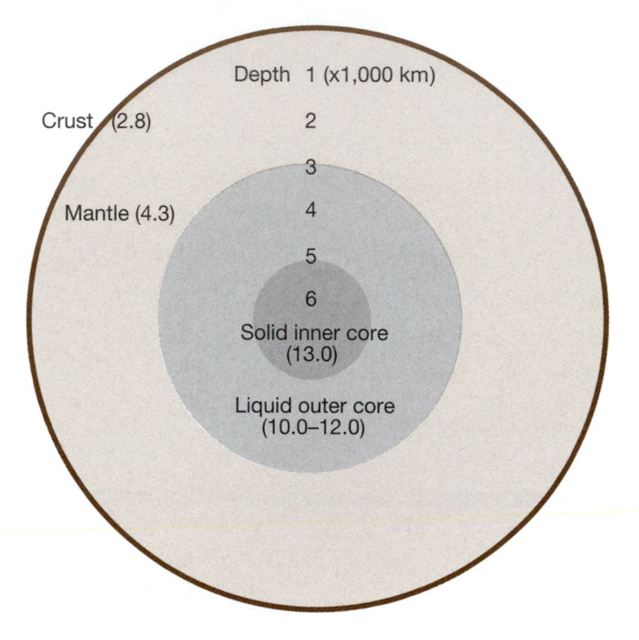

Figure 3.2 Earth consists of a series of crudely concentric layers that increase in density (shown in parentheses) downward.

Geologists have reason to believe that the outer part of Earth's core is liquid, so a beginning student might suspect that magmas and lavas are juices sapped from Earth's liquid core. But that's not the case. Actually, magmas are produced through the *melting of rock* at depths considerably less than that of Earth's core.

Q3.2 Igneous rocks make up about 80% of Earth's crust, which has an average density of 2.8. So, what is there about densities shown in Figure 3.2 that indicates that magmas are *not* juices from Earth's outer core?

The range of depths at which rocks melt to produce magmas is on the order of 50–250 kilometers (30–150 miles), which is somewhere in the upper part of the mantle or the lower part of the crust. Why such a broad range of depths? First, rocks differ in chemical composition, so different rocks melt at different temperatures—as in the case of different metals (e.g., pure lead melts at 600 °F, pure tin at 500 °F). Second, the rate of downward increase in temperature (called the **geothermal gradient**) differs from place to place (Fig. 3.3).

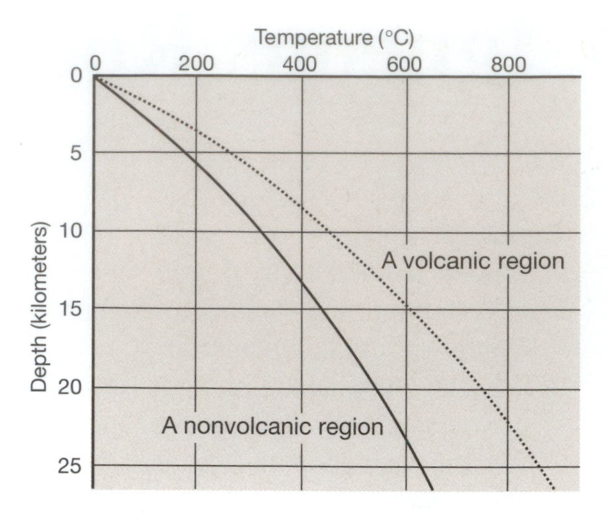

Figure 3.3 Geothermal gradients (i.e., change in temperature per depth) differ from place to place (e.g., gradients are steeper in volcanic regions). Also, geothermal gradients decrease downward.

Q3.3 Imagine a rock that melts in a laboratory oven at 800 °C. Applying the graph in Figure 3.3 for a volcanic region, at what depth (in kilometers) would you expect our imaginary rock to melt?

Your answer is probably around 22 kilometers. But what if you were told that at a depth of 22 kilometers a temperature of 1,000 °C is required to melt our rock?

Q3.4 Explain why the temperature required to melt our rock down deep is 200 °C higher than that in the laboratory. *Hints:* (a) Which is more dense, solid rock or liquid rock? (b) Qualitatively speaking, how does the weight of overlying rock affect the melting point of rock at depth (Fig. 3.4)?

Q3.5 By the way, can you name the only substance for which the solid phase is less dense than the liquid phase? *Hint:* Were this not true, Earth would be barren of life.

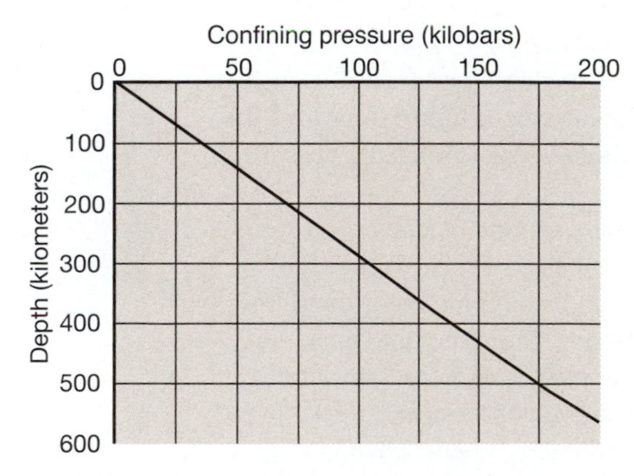

Figure 3.4 Confining pressure increases with depth. (One kilobar is approximately 1000 times atmospheric pressure at sea level.)

B. Igneous rock bodies

Igneous rocks that form underground are called **plutonic** rocks (named for *Pluto*, Greek god of the lower world). Plutonic rocks are also called **intrusive** rocks, because magmas are typically on the move, *intruding* into rocks with which they come in contact. Magmas eventually crystallize into igneous rocks because their lower densities cause them to rise upward toward Earth's cooler surface. Crystallization accompanies cooling.

Plutonic igneous rock bodies—Plutonic rocks crystallize at depths of tens of kilometers, but crustal uplift and erosion can result in the surface exposure of plutonic rocks. Because plutonic rocks are typically more resistant to erosion than are sedimentary and metamorphic rocks, plutonic rocks commonly stand above the surrounding countryside. Figure 3.5 includes a variety of plutonic rock bodies (along with a few volcanic features).

Q3.6 Is a volcanic neck intrusive, or is it extrusive? *Hint:* Examine the description of volcanic neck in Figure 3.5, and beware the adjective, volcanic.

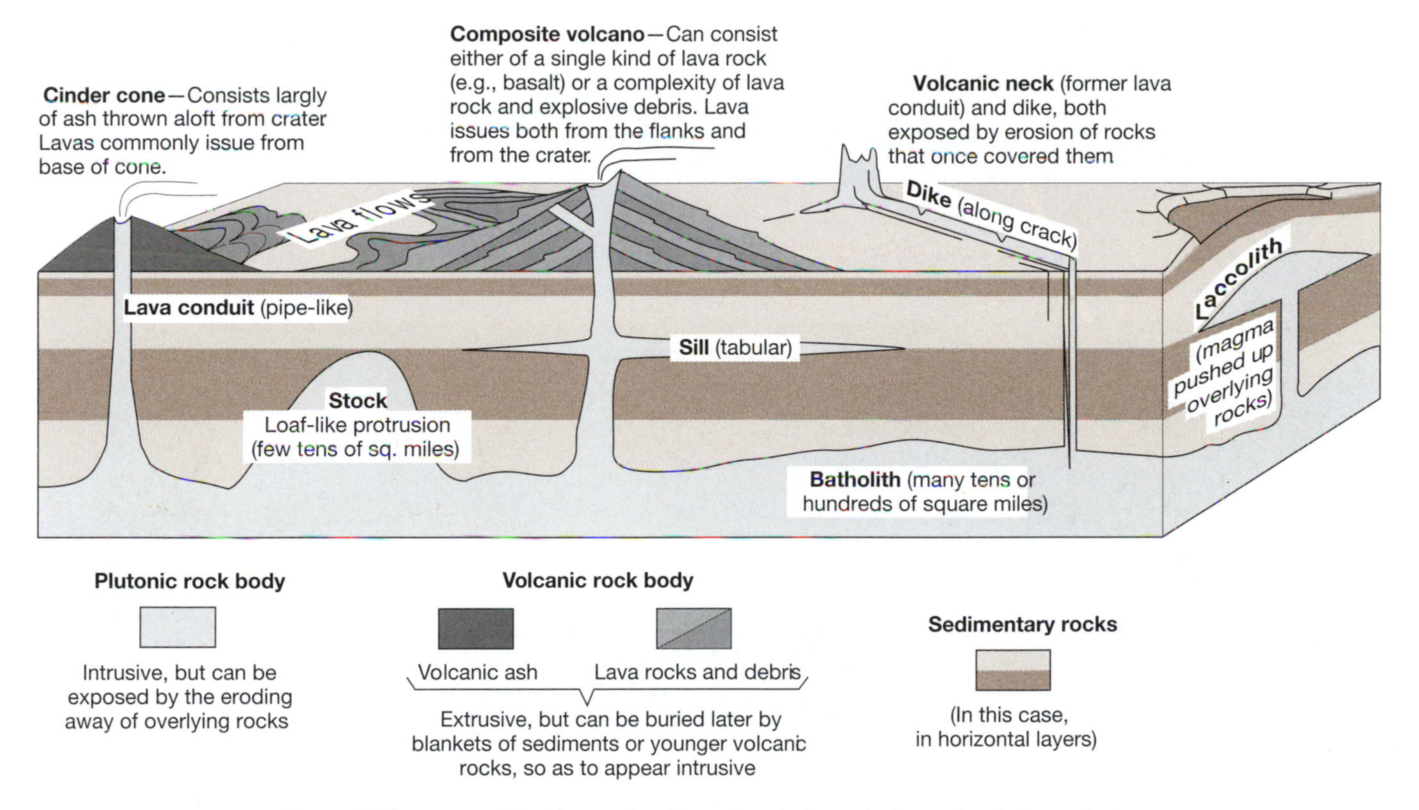

Figure 3.5 Igneous rock bodies can be either plutonic (intrusive) or volcanic (extrusive).

Viewing plutonic igneous rock bodies in satellite images

Field examples of a couple of the several plutonic rock bodies illustrated in Figure 3.5 are shown in satellite images in Figures 3.6 and 3.7.

Green Mountain, Wyoming—At Green Mountain (Fig. 3.6) a body of plutonic igneous rock is completely covered by layers of sedimentary rocks. Partial erosion has breached these layers, which are especially evident in the southeast quarter of the area.

Q3.7 What can you say about the orientation of the layers of sedimentary rocks? That is, do the layers appear to be horizontal (as in flat), or do they appear to be inclined downward and outward in all directions?

Q3.8 (A) What do you suppose was the shape (in horizontal plan view) of the magma body that pushed up the layers of sedimentary rocks that now comprise Green Mountain—circular, amoeboid (irregular), or linear (elongated)? (B) Judging from sketches in Figure 3.5, which type of plutonic igneous rock body do you suppose is buried beneath the layers of sedimentary rocks at Green Mountain?

Green Mountain, Wyoming

Sundance East, Wyoming, 7 1/2' quadrangle
N. 44° 24' 27", W. 104° 19' 30"

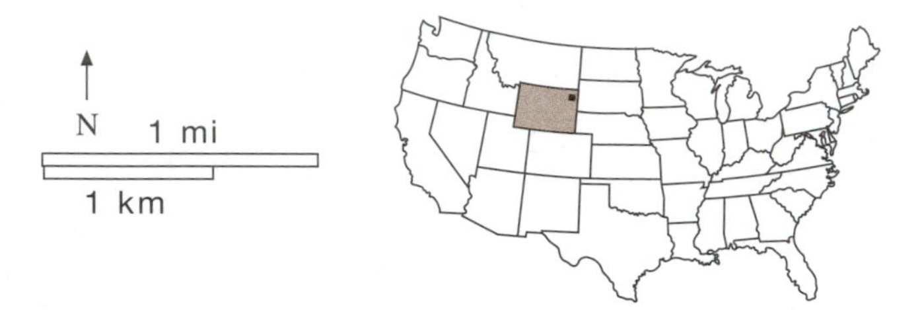

Figure 3.6 Green Mountain consists of layers of Permian and Triassic sedimentary rocks that conceal a plutonic igneous rock body below.

Stone Mountain, Georgia

Stone Mountain, Georgia, 7 1/2' quadrangle
N. 33° 48' 21", W. 84° 08' 51"

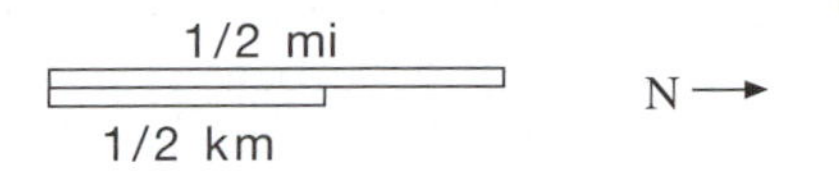

Q3.9 Granite comprising Stone Mountain crystallized from magma tens of kilometers deep within the Earth, but uplift and erosion of softer overlying sedimentary rock have since laid this pluton bare. Assuming that erosion has only slightly modified the original shape of the more durable granite body, which of the igneous rock bodies in Figure 3.5 does Stone Mountain most closely resemble?

Figure 3.7 Stone Mountain consists of a granite dome that stands 825 feet above the surrounding countryside. The largest carving in the world, that of Confederate heroes President Jefferson Davis, General Robert E. Lee, and Lt. General Thomas 'Stonewall' Jackson, is on the near-vertical north (right side in this view) face. Streaks running downslope are stains deposited by water.

Mount St. Helens, Washington
Aerial photo made *before* 1980 eruption

Mount St Helens, Washington, 7 1/2' quadrangle

Volcanic rocks are rocks that have crystallized from **lava** upon Earth's surface. Volcanic rocks are also called **extrusive** igneous rocks. Two kinds of volcanoes are included in Figure 3.5—a **cinder cone** and a **composite volcano**. Mount St. Helens is of the composite variety.

Q3.10 Features in the aerial photograph in figure 3.8 (below) are difficult to discern, but perhaps you can see that the lip of the crater has been breached by a narrow dark lava flow that has carved a furrow along its path. In what map direction did that lava flow, southeastward or northwestward? (Notice the north arrow in the legend to the upper-left of the page.)

Figure 3.8 Before the eruption of 1980, the crest of Mount St. Helens volcano was at an elevation of 9,680 feet. The low dark areas are forested, and the light-shaded high country is roughly above timberline. Patches of snow mark the highest elevations.

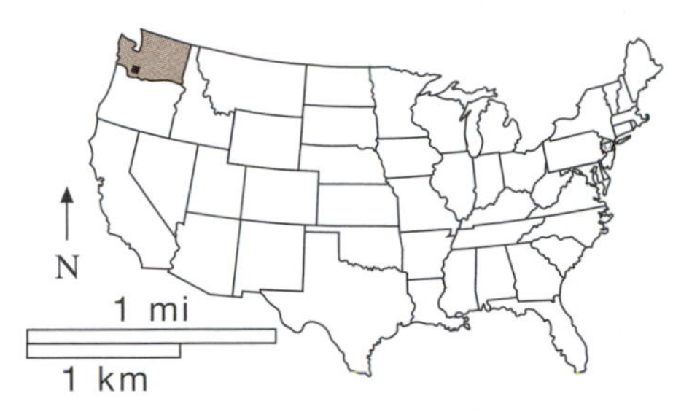

Mount St. Helens, Washington
Satellite photo made *after* 1980 eruption

Mount St Helens, Washington, 7 1/2' quadrangle
N. 46° 11' 24", W. 122° 11' 24"

Q3.11 Judging from the sizes of the before-and-after craters atop Mount St. Helens (Figs. 3.8 and 3.9), which volcanic event appears to have been the more explosive, the last event before 1980, or that which occurred in 1980? (Not a particularly difficult question.)

Figure 3.9 As a result of the 1980 eruption, the elevation of Mount St. Helens was reduced by 1,113 feet (to its present elevation of 8,567 feet). The light-colored lobate material consists of mudslides made of ash and snowmelt. Snow lies within the shaded crater.

C. Evolution of a single magma can result in a variety of igneous rock types

During the course of a magma's existence, it can change in its chemical composition—thereby producing a variety of igneous minerals and, so, a variety of igneous rocks. How does this work? Simple. The evolution of a magma is made possible by a single principle:

Different minerals crystallize at different temperatures.

Actually, that shouldn't be too surprising. Fresh water and salt water freeze (crystallize) at different temperatures, right? So it is with different kinds of minerals.

Q3.12 Incidentally, which does freeze at the lower temperature, fresh water or salt water? *Hint:* Recall that salt sprinkled on icy walkways melts the ice.

With high-temperature laboratory ovens we can show that as we lower the temperature of molten rock, some minerals crystallize earlier (i.e., at higher temperatures) and others crystallize later (i.e., at lower temperatures). But long before we had such sophisticated ovens, geologists were able to decipher the sequence of crystallization of minerals by studying the **petrography** (i.e., the fabrics) of igneous rocks. But first, a little warm-up exercise.

Q3.13 In what order were the six cards in Figure 3.10 tossed onto the table? (Careful.)

Figure 3.10 Six playing cards were tossed face-up on a table.

In the discussion of **crystal form** in Exercise 2, Minerals, it is shown that a mineral that grows free of a solid obstruction—as within a water-filled cavity or within magma or lava—it tends to develop a complete form bounded by crystal faces. But a mineral that grows against an earlier mineral will have less than a complete form, that is, its form will be *truncated* to some extent by the earlier mineral.

Earlier
Later

Q3.14 Applying the rule—*the most complete form crystallizes first, and the least complete form crystallizes last*—list the four minerals in Figure 3.11 (W, X, Y, and Z) in the order in which they crystallized. (*Caution:* The stack of cards in Figure 3.10 is 3-dimensional, with one card upon another; whereas minerals in the thin-section in Figure 3.11 have essentially no depth. View this thin-section as 2-dimensional, with minerals side-by-side.

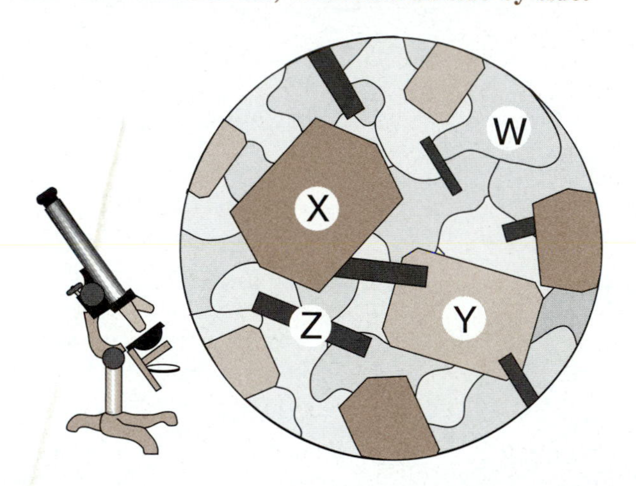

Figure 3.11 This is a sketch of a thin-section of granite as viewed with a microscope. Its thickness is approximately that of tissue paper, so it is essentially 2-dimensional, i.e, with no depth.

Q3.15 If the granite in Figure 3.11 were melted in the laboratory, what would be the order in which W, X, Y, and Z would melt? *Hint:* Imagine a pat of butter and a lump of sugar in a sauce pan. Add heat. Which melts first? Maintain the heat until both have melted, then turn off the heat. Which solidifies (crystallizes) first?

It bears pointing out that the temperature of crystallization of a mineral is essentially the same as its temperature of melting. *Analogy:* As temperature falls, pure water freezes at 0 °C (32 °F). As temperature rises, ice melts at 0 °C (32 °F).

By studying a variety of rocks, as you did with granite, geologists have been able to work out the sequence of crystallization of common silicate minerals (Fig. 3.12).

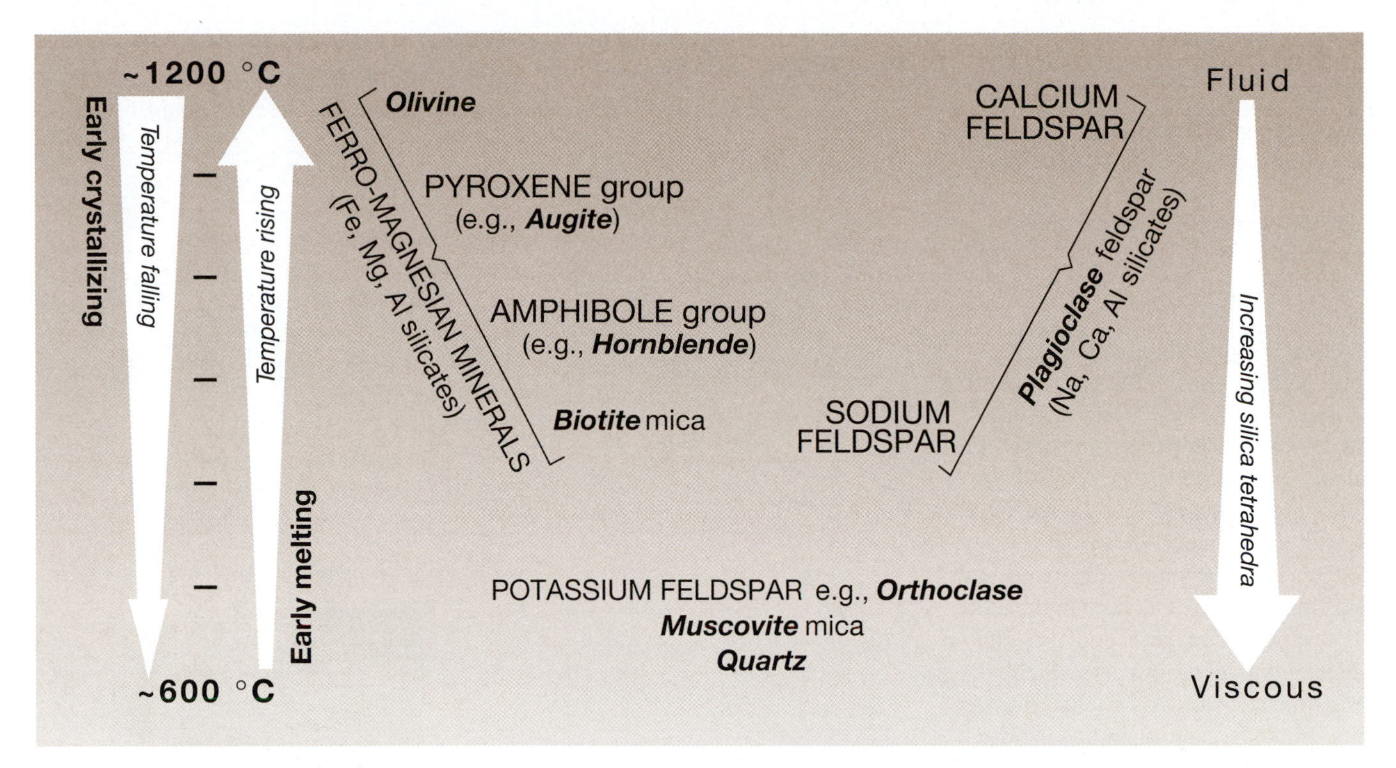

Figure 3.12 This chart shows the sequence of *crystallization* (with temperature falling) and the sequence of *melting* (with temperature rising) of the common silicate minerals. It also shows the greater amount of silica tetrahedra in low-temperature silicates, which imparts *viscosity* to magmas and lavas. *Def:* viscous, like mud; fluid, like liquid (and gas).

Partial crystallization changes composition of magma— Let's imagine a magma rising toward Earth's surface (rising because magma is *less dense* than surrounding rock). And let's imagine that partial crystallization occurs while the magma is en route to the surface (Fig. 3.13). Early formed crystals sink because minerals are *more dense* than magma.

Let's assume that the initial chemical composition of our wandering magma is that of a mixture of all the elements that comprise those silicate minerals shown in Figure 3.12.

Q3.16 Which of the ferromagnesian minerals, and which of the feldspars shown in Figure 3.12, would be the first minerals to crystallize?

Q3.17 As the high-temperature minerals continue to crystallize, how does the remaining magma change in its chemical composition—specifically with respect to the percentage of silica tetrahedra in the original magma? (Does the magma become progressively enriched in silica tetrahedra, or does it become progressively impoverished in silica tetrahedra?)

Q3.18 Judging from information in Figure 3.12, as high-temperature minerals continue to crystallize, how does the remaining magma change in its viscosity/ fluidity? Does it become more viscous, or does it become more fluid?

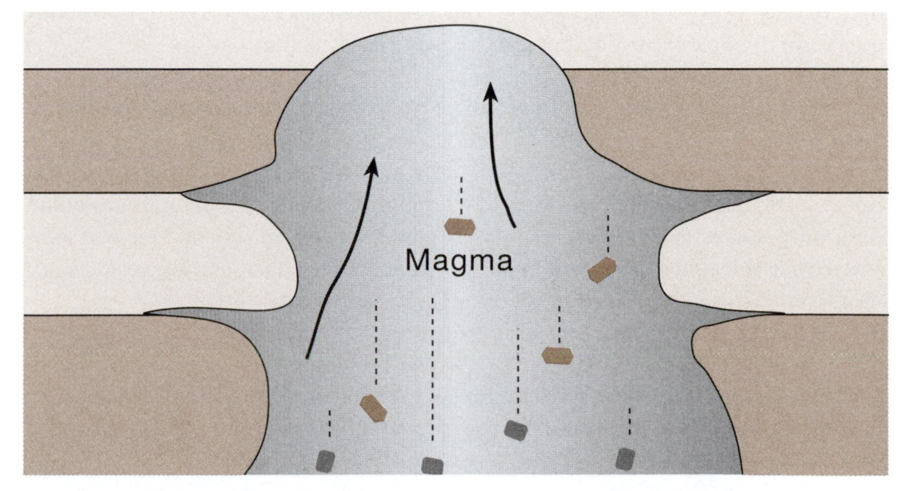

Figure 3.13 As magma rises it cools. High-temperature minerals crystallize and sink, effectively removing the elements that comprise those minerals from the magma and thereby changing the chemical composition of the remaining magma.

D. Bases for the classification of igneous rocks

Classification of igneous rocks is based on a combination of *two parameters:*

(1)
Texture, which is the visual coarseness of crystal size, and...

(2)
Mineral composition, which is the kinds and amounts of minerals present.

(1a) Texture is governed in part by *the rate at which molten rock crystallizes.* If molten rock cools slowly—which is typical of magma crystallizing to form plutonic rock—crystals grow large, resulting in coarse texture. On the other hand, if molten material cools rapidly—which is typical of lava crystallizing to form volcanic rock—crystal growth is more limited, resulting in fine texture.

Geologists, like other natural scientists, give exotic names to their subjects. Coarse-textured igneous rocks are called **phaneritic**; fine textured rocks, **aphanitic**; and igneous rocks with two populations of crystal size, **porphyritic** (Fig. 3.14).

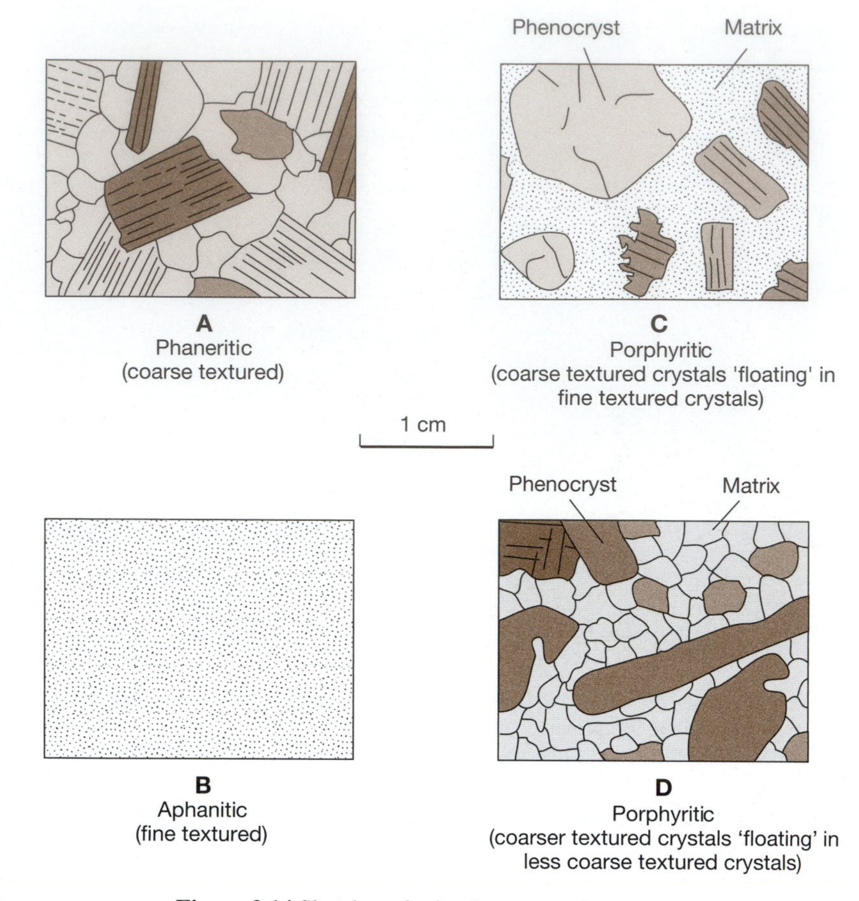

Figure 3.14 Sketches of a few igneous rock textures.

Q3.19 If coarse texture reflects a *magmatic* environment, and fine texture reflects a *volcanic* environment, what do you suppose is the history of a rock with a porphyritic texture (like that in Figure 3.14C)?

Note: A porphyritic rock need not be a mixture of fine texture and coarse texture (Fig. 3.14C). It need only be a mixture of two contrasting textures. For example, a porphyritic rock can be a mixture of coarse texture and coarser texture (Fig. 3.14D).

Special textures—Texture does not refer only to crystal size. *Special textures* in volcanic rocks serve as a basis for classification. Examples:

Some lavas cool so rapidly that elements haven't the time as it were to become organized as crystals. Instead, they solidify into *glass* ('frozen liquid') with **glassy** texture. Volcanic glass can be solid, as in the case of **obsidian**, which has a wavy conchoidal fracture like that of broken glass. Or, volcanic glass can be porous, as in the case of **pumice**. Pumice is so porous that some samples will float in water. Give it a try. Pumice has such rough surfaces that one must have keen eyesight to detect its glassy texture.

Still other volcanic rocks exhibit pea-size or larger pores called **vesicles**, which are much larger than the pores of pumice. Volcanic rock with **vesicular** texture is called **scoria**. One can see that the rock surrounding vesicles in scoria is fine textured.

Airborne volcanic ejecta (**ash** and other **pyroclastic** debris) solidifies after settling to the ground, either by welding together as hot particles or by being cemented by silica-rich groundwater. The result is fragmental **tuffaceous** texture. **Tuff** is not noticeably porous.

(1b) Texture is also governed by the viscosity/fluidity of molten rock. (Recall that silica tetrahedra impart viscosity to magma and lava.) The more **fluid** the molten rock, the greater the ease with which ions can gather into centers of crystal growth and develop into large crystals. Conversely, the more **viscous** the molten rock, the more difficult the process, so the finer the texture.

Q3.20 Given the same rate of cooling for each—which will develop coarser texture, (A) magma *rich* in silica tetrahedra, or (B) magma *poor* in silica tetrahedra? *Hint:* See again Figure 3.12 and its caption.

(2a) Mineral composition of an igneous rock is governed in part by the composition of the rock that melts to provide the magma (or lava) from which that igneous rock crystallizes (Fig. 3.15). This is the **initial composition** of a magma. *Exception:* Initial composition of a magma can differ from the composition of its parent rock if only partial melting occurs. To illustrate that point...

Q3.21. Imagine a rock consisting of all of the minerals in Figure 3.12. Heated to a temperature of 800 °C, which three minerals should melt? *Hint:* Notice the tick-marks delineating 700, 800, 900, 1,000, and 1,100 °C.

Incidentally, if it weren't for the fact that different minerals melt/crystallize at different temperatures, there would be no continents. Continents developed early in the history of the Earth through partial melting at depth and the upward migration of magma and the crystallization of lower-temperature minerals at or near Earth's surface.

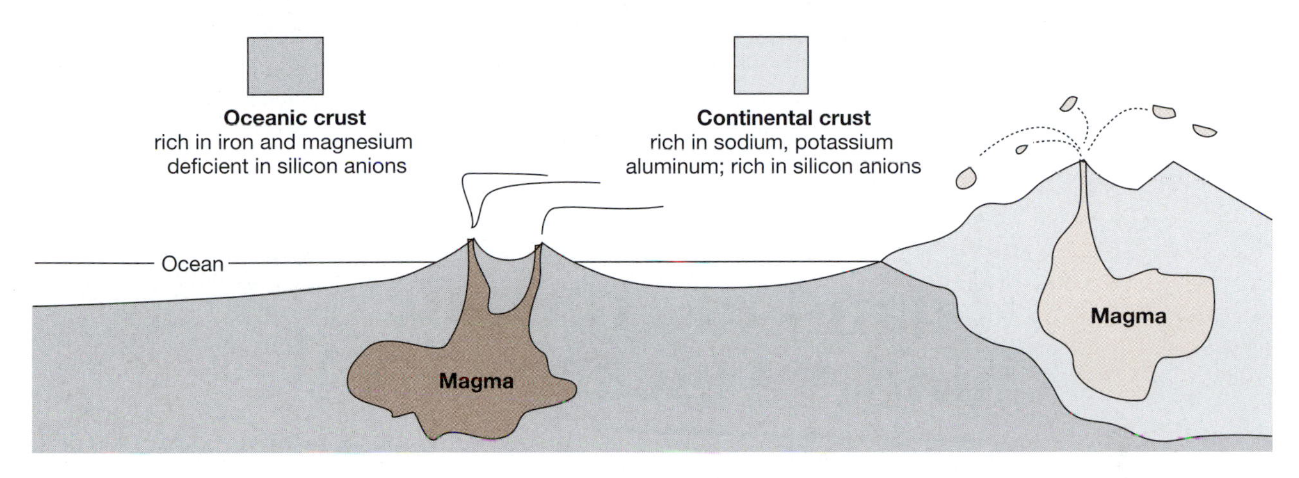

Figure 3.15 Magma that develops within oceanic crust differs in its initial chemical composition from that which develops within continental crust.

(**2b**) The second thing that affects the mineral composition of an igneous rock is the extent to which elements are removed from a magma—through **partial crystallization**—before the rock completely crystallizes. When molten rock cools, higher-temperature minerals crystallize first. Higher-temperature minerals are rich in iron and magnesium, so their crystallization causes the remaining magma to be impoverished in those elements and enriched in, say, sodium and potassium. A rock that crystallizes at a later time, and perhaps in a different place, will be enriched in sodium and potassium. This process is called magmatic **differentiation**—the *segregation* of elements that were present in the initial magma.

Q3.22 (Refer to Figure 3.16.) Reflecting on the preceding paragraph, in what way would you expect the *chemical composition* of an igneous rock crystallizing at X to differ from that crystallizing at Y?

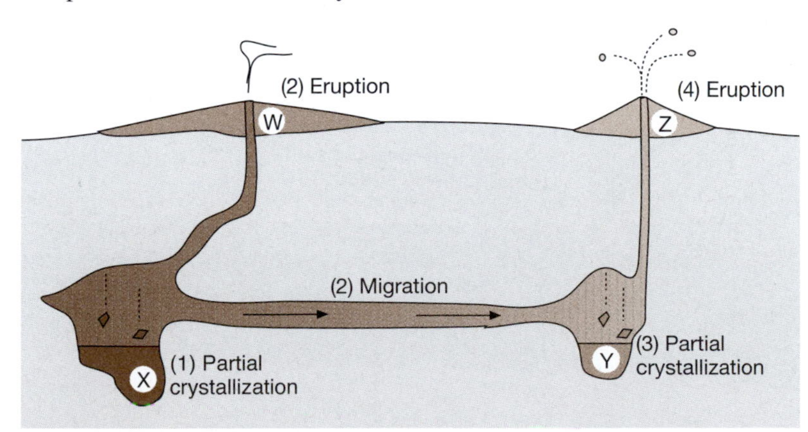

Figure 3.16 (1) Partial crystallization, (2) lateral migration, and volcanic eruption at W, (3) partial crystallization, and (4) volcanic eruption at Z result in four different kinds of igneous rocks—each with a different mineral composition and texture.

Q3.23 In what textural way would you expect igneous rock crystallizing at Y to differ from that crystallizing at Z? *Hint:* Which rock crystallizes more quickly?

Another point: Volcanic rocks *impoverished* in silica tetrahedra are less likely to form glass than are volcanic rocks *enriched* in silica tetrahedra. Recall that silica tetrahedra add viscosity to molten rock, and therefore impede crystal growth.

Q3.25 Where would you most likely find obsidian (volcanic glass), at W or at Z? *Hint:* See obsidian's compositional position in Table 3.1 on page 57.

E. Subduction and the production of viscous, explosive lavas

There are two processes that drive magmas to Earth's surface—where they erupt as lavas and ash:

One process is the accumulation of heat **within the upper mantle or deep crust** sufficient to melt rock. This can occur along sinuous belts or over circular **hot spots**. Sinuous belts, which mark the spreading apart of global plates, occur at sea as **mid-ocean** ridges. On land they occur as yawning **rift valleys**, such as that in east Africa (which explains the 2002 disastrous eruption in Goma, Congo). Hot spots are currently perking beneath Hawaii and Yellowstone National Park.

A second process is associated with **subducting global plates** (Fig. 3.17). Today's students have grown up with the concept of **plate tectonics**, so references to subduction are commonplace in the media. Figure 3.17 recently appeared in a newspaper that borrowed it from the book *Our Violent Earth*. Subduction results in partial melting, producing lavas rich in silica tetrahedra.

Q3.26 Notice the steep profile of the volcanic cone in Figure 3.17. Why so steep? *Hint:* Think composition.

Lavas rich in silica tetrahedra are also characterized by especially violent eruptions. The Cascade Range of Washington, Oregon, and California, is replete with this type of volcano—some of which are active.

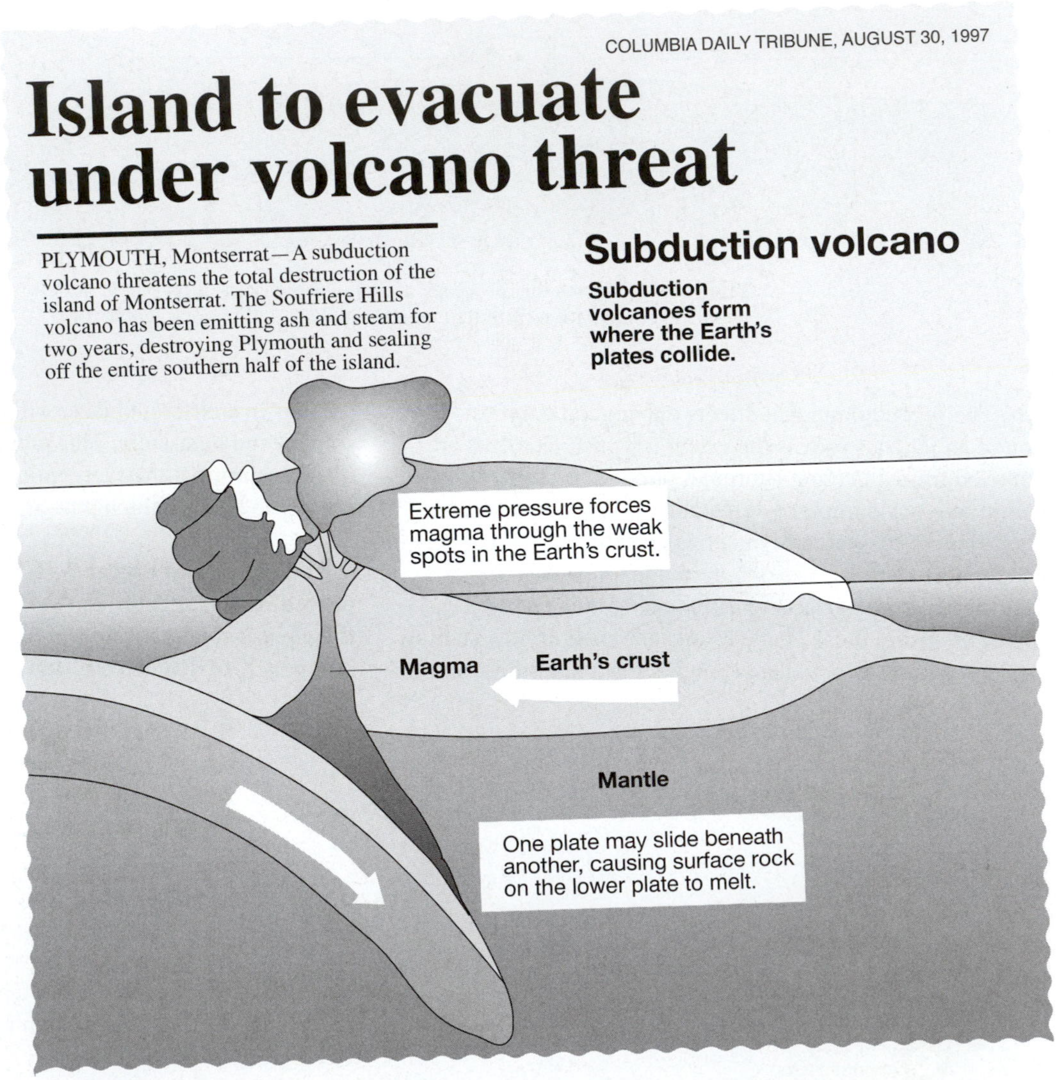

COLUMBIA DAILY TRIBUNE, AUGUST 30, 1997

Island to evacuate under volcano threat

PLYMOUTH, Montserrat—A subduction volcano threatens the total destruction of the island of Montserrat. The Soufriere Hills volcano has been emitting ash and steam for two years, destroying Plymouth and sealing off the entire southern half of the island.

Subduction volcano

Subduction volcanoes form where the Earth's plates collide.

Extreme pressure forces magma through the weak spots in the Earth's crust.

Magma

Earth's crust

Mantle

One plate may slide beneath another, causing surface rock on the lower plate to melt.

Figure 3.17 The island of Montserrat, some 250 miles southeast of Puerto Rico, is threatened by Soufriere Hills volcano, the product of subduction of a global plate.

USGS Volcano Watch Web site is at…http://hvo.wr.usgs.gov/volcanowatch/.

F. Identification of igneous rocks

Table 3.1 appears a bit daunting at first glance, but don't despair. Only a dozen or so igneous rocks are included here.

Rhyolite and andesite are grouped together under the term felsite because these rocks are too fine textured to see whether or not quartz is present—which is the basis for distinguishing between their coarse textured counterparts, granite (with quartz) and diorite (without quartz). In studying the igneous specimens provided, you will probably come to realize that the simplest guides to their identification are: (a) mineral composition, in cases of phaneritic (coarse textured) plutonic rocks; and (b) color and texture, in cases of aphanitic (fine textured) volcanic rocks. Photographs of igneous rocks follow on pages 58–59.

Table 3.1 Classification of igneous rocks

Two parameters—*mineral composition and texture*—are plotted along horizontal and vertical axes.

Mineralogic comps. / Textures (all can be porphyritic)	Felsic abundant quartz, orthoclase > plagioclase, < 15% ferromagnesian	Intermediate little or no quartz, plagioclase > orthoclase, 15–40% ferromagnesian	Mafic no quartz, no orthoclase, 10–60% plagioclase, 40–90% ferromagnesian	*Optional column* Ultramafic > 90% ferromagnesian (little if any feldspar)
PLUTONIC ROCKS (Minerals are sufficiently large for identification.)				
Phaneritic (coarse textured)	GRANITE (light colored)	DIORITE (intermediate colored)	GABBRO (dark colored)	PERIDOTITE (very dark colored, heavy)
VOLCANIC ROCKS Volcanic rocks, Group A: With the exception of possible phenocrysts, minerals are too small for identification; color may be the only guide to classification. *Where phenocrysts are present, add the modifier 'porphyritic.'*				
Aphanitic (fine textured)	RHYOLITE / FELSITE (light to intermediate colored)	ANDESITE / FELSITE (light to intermediate colored)	BASALT (dark colored)	Volcanic rocks of peridotite composition are exceedingly rare.
Volcanic rocks, Group B: Classification of volcanic rocks below is based solely on texture. (Compositional tendencies only, so the mineral-composition boundary-lines are slanted.)				
Glassy	OBSIDIAN (can be dark, even though felsic)		Volcanic glass of this composition is rare. Lava is so fluid that crystals tend to develop, even with fastest cooling.	
Vesicular (bubbly, frothy)	PUMICE (light colored; tends to be felsic)		SCORIA (dark colored; tends to be mafic)	
Pyroclastic (fragmental)	TUFF (consists of fragments less than 4 mm in size) VOLCANIC BRECCIA (consists of fragments more than 4 mm in size)			

Igneous rocks

Phaneritic (coarse textured) plutonic crystalline rocks

Granite (red)
Pink feldspars,
glassy transparent quartz,
black ferromagnesian minerals

Granite (pinkish gray)
Light gray feldspars,
glassy transparent quartz,
black ferromagnesian minerals

Diorite
Light gray and dark gray feldspars,
black ferromagnesian minerals

Gabbro
Dark gray feldspars,
black ferromagnesian minerals

Peridotite
Very dark gray feldspars,
dark greenish black ferromagnesian minerals;
noticeably heavy

Aphanitic (fine textured) volcanic crystalline rocks

Rhyolite
Pink to gray

Andesite
Light to charcoal gray

Felsites

Basalt
Black

Other volcanic textures

Pumice
White to light gray;
frothy glass; microscopic pores
(vesicles) impart a light weight

Obsidian
Red to black wispy fabrics;
dense glass; conchoidal fracture;
sharp edges

Scoria
Red to black;
pea-size and larger pores (vesicles) in
fine textured crystalline rock

Tuff
Resembles fine textured concrete;
fragments less than 4 mm in size

Volcanic breccia
Resembles coarse textured concrete;
fragments more than 4 mm in size

Mixed textures

(e.g., fine and coarse; coarse and coarser)

Porphyritic basalt
Coarse textured light gray phenocrysts
of feldspar in fine textured black basalt

G. Silent death at Lake Nyos

COLUMBIA DAILY TRIBUNE, FEBRUARY 16, 1995

Submarine combs lake that spewed deadly gas

Cameroon eager to find source of carbon dioxide

YAOUNDE, Cameroon (AP) — Convinced that another disaster lurks beneath Lake Nyos, a team of experts began a high-tech operation today to locate and eventually remove a deadly gas that emerged in 1986 to kill 1,746 villagers. The researchers hope the study of the lake bottom's geological formations will uncover the sources of the carbon dioxide that rose from the lake on Aug. 21, 1986, to suffocate victims as they slept in nearby villages.

Lake Nyos is among more than 40 crater lakes in the volcanically active mountains of northwest Cameroon. Some volcanologists believe molten rock built up steam at the bottom of the lakes and blew the toxic gas in the air.

Scientists estimated that 100 million cubic meters of toxic gas was released in the 1986 catastrophe and that about 250 million cubic meters remains.

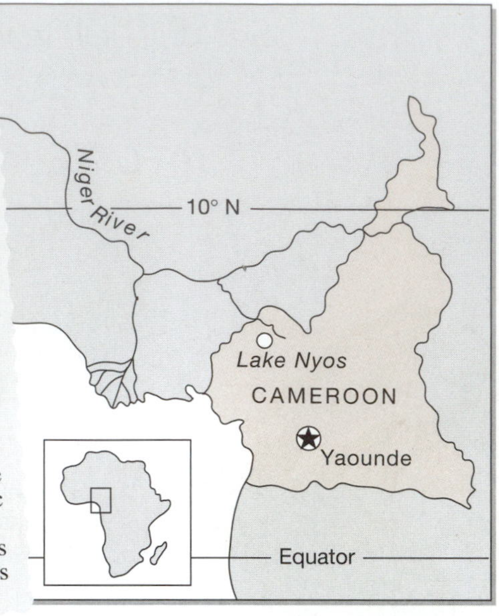

Figure 3.18 Lake Nyos, with its depth of 208 meters (680 feet), occupies the crater atop Nyos volcano at 7° N latitude.

A silent volcanic event occurred without warning in Cameroon on August 21, 1986. While residents of Nyos village lay sleeping, 100 million cubic meters of carbon dioxide escaped from Lake Nyos (Fig. 3.18) and cascaded down slope, suffocating 1,746 people as they slept. A mother awoke to find her children lying dead beside her.

Volcanoes typically emit carbon dioxide (CO_2), carbon monoxide (CO), hydrogen sulfide (H_2S), and sulfur dioxide (SO_2). These last three gases are poisonous. CO_2 simply displaces essential oxygen. Dormant volcanoes commonly continue to leak these gases via cracks and crevices from smoldering magma below. Scientists have determined that deep waters of Lake Nyos (Fig. 3.19A) absorbed CO_2 for years, and then, for reasons still debated, the CO_2-rich water abruptly rose to the surface, prompting the CO_2 to escape and cascade down the mountain (Fig. 3.19B).

A bit of limnology (i.e., the science of lakes). Lakes are *temperature layered*. More dense cold water hangs out down deep, while less dense warm water floats above. Lakes in temperate latitudes *turn over* in late autumn-early winter when surface waters grow colder—and more dense—than deep waters. But lakes in equatorial regions rarely turn over, owing to scant seasonal changes. Given the fact that Lake Nyos is at 7° N latitude, some observers doubt that it was seasonal turnover that stirred the water. But circumstantial evidence does point to turnover. Evidence: (1) The event occurred during the rainy season, when cool water is added to lakes. (2) The event occurred at night, the coolest part of the day. (3) A similar event took 37 lives near Cameroon's Lake Manoun in 1984—*at the same time of year.*

Q3.27 Three things can prompt the escape of CO_2 from carbonic acid in an opened soda pop: (1) reduction in pressure, (2) agitation, and (3) warming. Which of these three processes do you imagine played the greatest role in the abrupt escape of CO_2 from deep Lake Nyos water as it rose to the surface? Hint: Two of these processes are a bit equivocal, whereas one surely did occur.

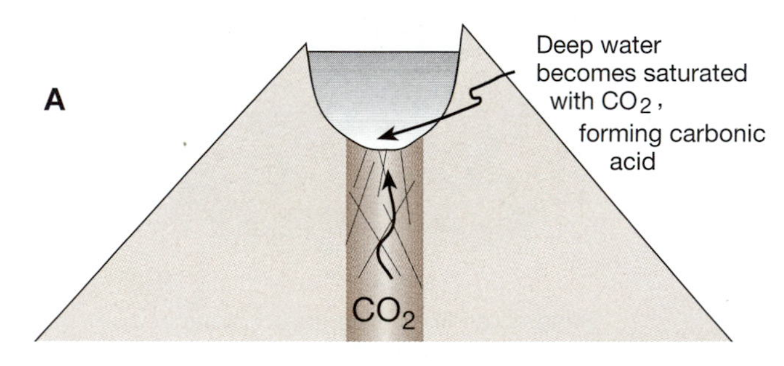

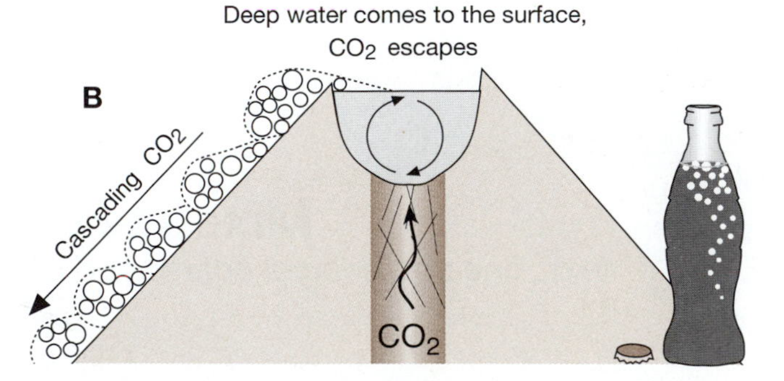

Figure 3.19 Lake Nyos. **A** CO_2 leaks from magma below and dissolves in deep water (forming carbonic acid). **B** Deep water rises to the surface, CO_2 escapes and cascades down the volcano's flank. (Note: CO_2 is approximately 1.5x heavier than air.)

4 Sedimentary Rocks

Topics

A. What is the origin of the three categories of sedimentary rocks and each of their varieties?
B. On what feature of detrital sedimentary rocks is their classification based? What are the two types of organic sedimentary rocks? What is the difference in origin between primary chemical sedimentary rock and secondary chemical sedimentary rock?
C. What is there about flat-lying sedimentary rocks that explains their typical stair-step topography?
D. What feature of ancient stream deposits is used to interpret the swiftness of the running water in which they were deposited?
E. What are some of the ten common features that enable geologists to chart the distribution of ancient lands and seas and decipher depositional environments?
F. Why do domains of England's modern political parties correlate with bedrock geology?

A. The many origins of sedimentary rocks

Sedimentary rocks consist of material derived from preexisting rocks through physical erosion and chemical weathering. Figure 4.1 charts the several paths of these two processes and the dozen or so products.

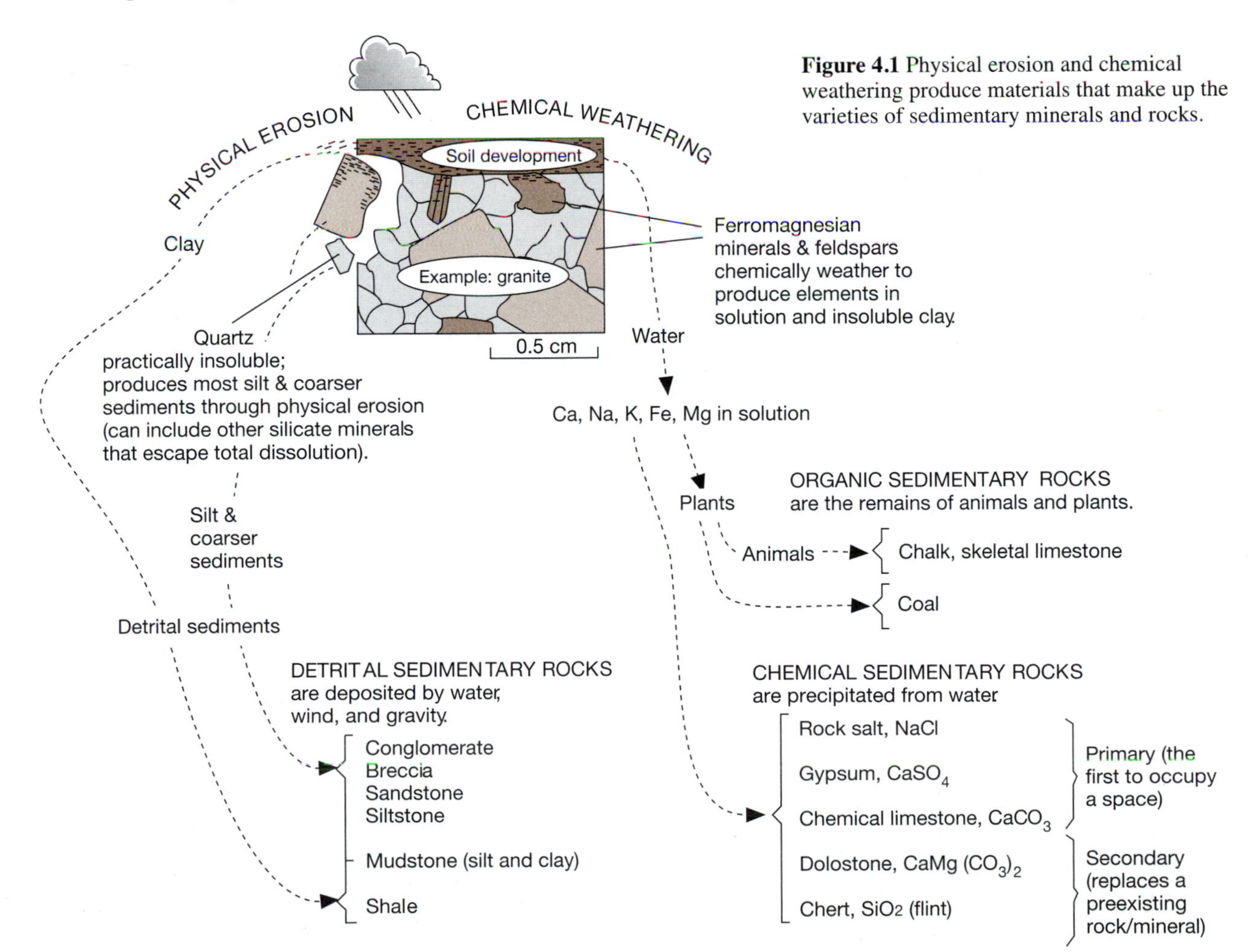

Figure 4.1 Physical erosion and chemical weathering produce materials that make up the varieties of sedimentary minerals and rocks.

(1)

Physical erosion is the erosion of rocks largely by wind and water, (Fig. 4.2A) to produce *solid* particles such as clay, silt, sand, and pebbles—which can become cemented together to form **detrital** sedimentary rocks. (Detrital: from *detritus*, fragments produced through disintegration.)

(2)

Chemical weathering is the alteration of rocks by water to produce chemical elements in solution and (commonly) residual clay (Fig. 4.2B). There can be either the complete dissolving of rock (putting all of it in solution), or the partial dissolving of certain minerals, putting some elements in solution and leaving others behind as a residue of clay minerals that comprise soil. (Clay can also be carried away by streams and be deposited elsewhere in the making of a detrital sedimentary rock called shale.) In time, elements in solution can be precipitated, either organically or inorganically, to form **organic** and **chemical** sedimentary rocks. *Organic* sedimentary rocks consist largely of the skeletal remains of plants and animals, whereas *chemical* sedimentary rocks are precipitated directly from water.

Q4.1 (A) Apart from the composition of the rock itself, what climate do you imagine is more conducive to physical weathering? (B) How about chemical weathering? *Hint:* Clues are evident in graphics of Figure 4.2 below.

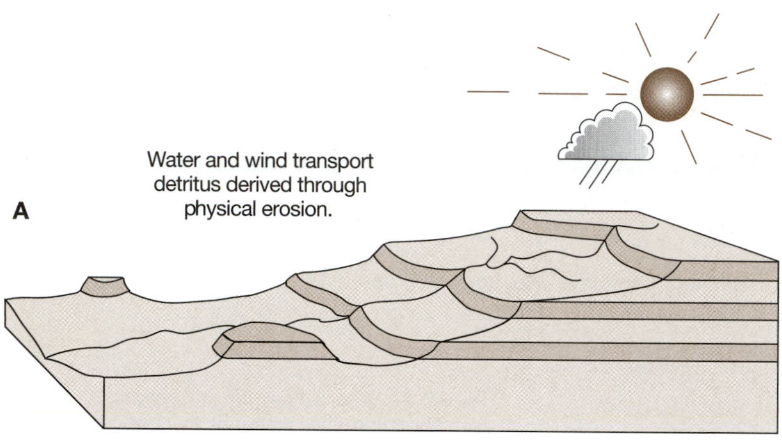

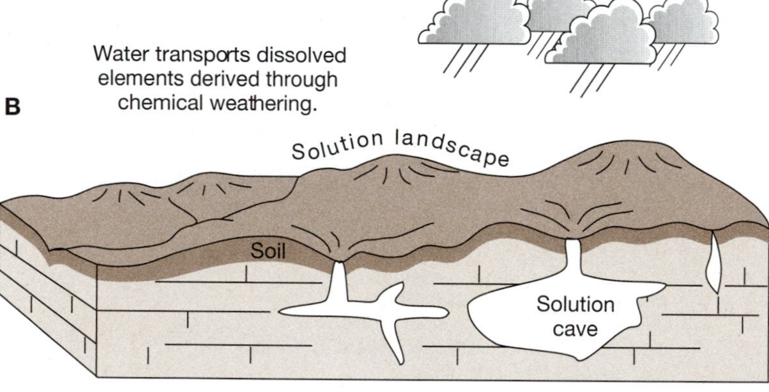

Figure 4.2 A Physical erosion produces solid debris that is eventually deposited in layers and cemented to become detrital sedimentary rocks. **B** Chemical weathering is the dissolving of elements by surface and subsurface water, leaving a residue of clay-rich soil. Those elements in solution can later be extracted from water in the development of organic and chemical sedimentary rocks.

Lithification (Gr. *lithos*, stone) includes myriad processes whereby different materials—ranging from solid particles to elements in solution—contribute to the development of sedimentary rocks. Lithification is a complex subject, but four processes pretty well cover the field (see **Processes** in the table below): *compaction, cementation, precipitation, and recrystallization*. While all four of these processes likely play some role in all cases of lithification, some are more important than others in specific processes. For example, compaction plays a huge role in converting clay to shale, whereas cementation is paramount in lithifying sands into sandstones.

Processes	change these	into these
Compaction	clays	shales
Cementation	pebbles, detrital sands and silts, skeletal sands	conglomerates, sandstones, siltstones, fossiliferous limestones
Precipitation	chemical sediments	chemical sedimentary rocks
Recrystalllization	skeletal lime sediments	chalks

B. Classification of sedimentary rocks

Table 4.1 on page 68 presents a classification of sedimentary rocks. This table, along with photographs of 22 rocks on pages 69–71, serve as a guide to identifying specimens provided by your instructor. The discussion that follows addresses some of the points outlined in Table 4.1.

Detrital Sedimentary Rocks

Called **clastic** (i.e., fragmented) and **terrigenous** (i.e., of the Earth), **detrital** sedimentary rocks are classified according to the sizes of their grains. In 1922, Professor Chester K. Wentworth, of Iowa State University, borrowed names of sedimentary particles from our lay vocabulary, proposed formal size limits for each, and suggested corresponding detrital rock names (Fig. 4.3). Don't be put off by the quantitative grain-size limits. This classification is applied to detrital sedimentary rocks in the field on the basis of *estimates* only.

Q4.2 (A) Based on your life experiences, what's your guess as to which of the four rocks listed in Figure 4.3 most resembles concrete? (B) How about pottery?

Grain name	Grain size	Detrital sedimentary rock
gravel	2 mm or larger	CONGLOMERATE
sand	1/16 to 2 mm	SANDSTONE
silt	1/256 to 1/16 mm	SILTSTONE
clay	1/256 mm or less	CLAYSTONE OR SHALE
A mixture of siltstone and claystone is called mudstone		

Figure 4.3 Wentworth's size classification of detrital sedimentary rocks.

The unaided eye can hardly detect the 1/256-millimeter boundary between silt and clay, so one must be guided by the general appearance and feel of a fine-grained detrital rock. Generalized textural features of detrital sedimentary grains are:

gravel—grains are obviously pebble-size.

sand—grains range from the size of match heads down to pinheads or thereabouts. Sandstone feels sandy to the touch—like sandpaper.

silt—grains are hardly visible to the unaided eye, but siltstone feels abrasive to the touch, like the fine side of a nail file. Gritty to the teeth.

clay—not abrasive; smooth; slick when wet. Pasty to the teeth.

A mixture of silt and clay is called **mud**.

Still another rock is called **breccia**, a rock in which pebble-size grains are *angular* in shape. Rock fragments within breccia have not been transported by water, otherwise they would be rounded. So one wonders if breccia should be included in the category of detrital sedimentary rocks—given the fact that breccia fragments most commonly reflect the breaking of rock either through faulting or through the collapse of cave roofs. Still other breccias are volcanic in origin.

Organic sedimentary rocks

Two groups of organic sedimentary rocks are separated on the basis of the *kind* of organic material that comprises each (Fig. 4.4 below):

(1)

Organic limestones consist of fragments of sea shells cemented together. There are three common varieties of organic limestones:

• *Fossiliferous limestone* (aka *bioclastic limestone*; aka *skeletal limestone*) consists of shell fragments that are clearly visible to the unaided eye. Fragments are cemented together either by transparent (glassy) calcite or by what was in the beginning lime mud.

• *Coquina* consists of shell fragments lightly cemented by thin coatings of calcite cement. Coquinas are very porous.

• *Chalk* consists of the microscopic skeletal remains of protozoans, mollusks, and calcareous algae.

(2)

Coals consist of the **organic carbon** residue of plants that were buried with sediments and later compacted and distilled by the heat and pressure of burial. Decaying organic compounds (the residue of plants and animals) combine with oxygen in air and water and escape as carbon dioxide—much like the visible burning of wood, but more slowly. If oxygen is insufficient to complete this slow burn, some carbon remains in the sediments where it might someday be mined as coal. Burning of the coal releases the heat energy that would have been released at the time the plants died, had there been sufficient oxygen. Petroleum (oil and gas) originates through much the same process.

Q4.3 Imagine that the gasoline that you are burning in your car this week was refined from petroleum that was produced from Jurassic-age rocks in southern Arkansas. Question: What is the *primary* source of the energy that has been reclaimed to power your car? Explain. *Hint:* What is the 'heat engine' that drives the origin of coal?

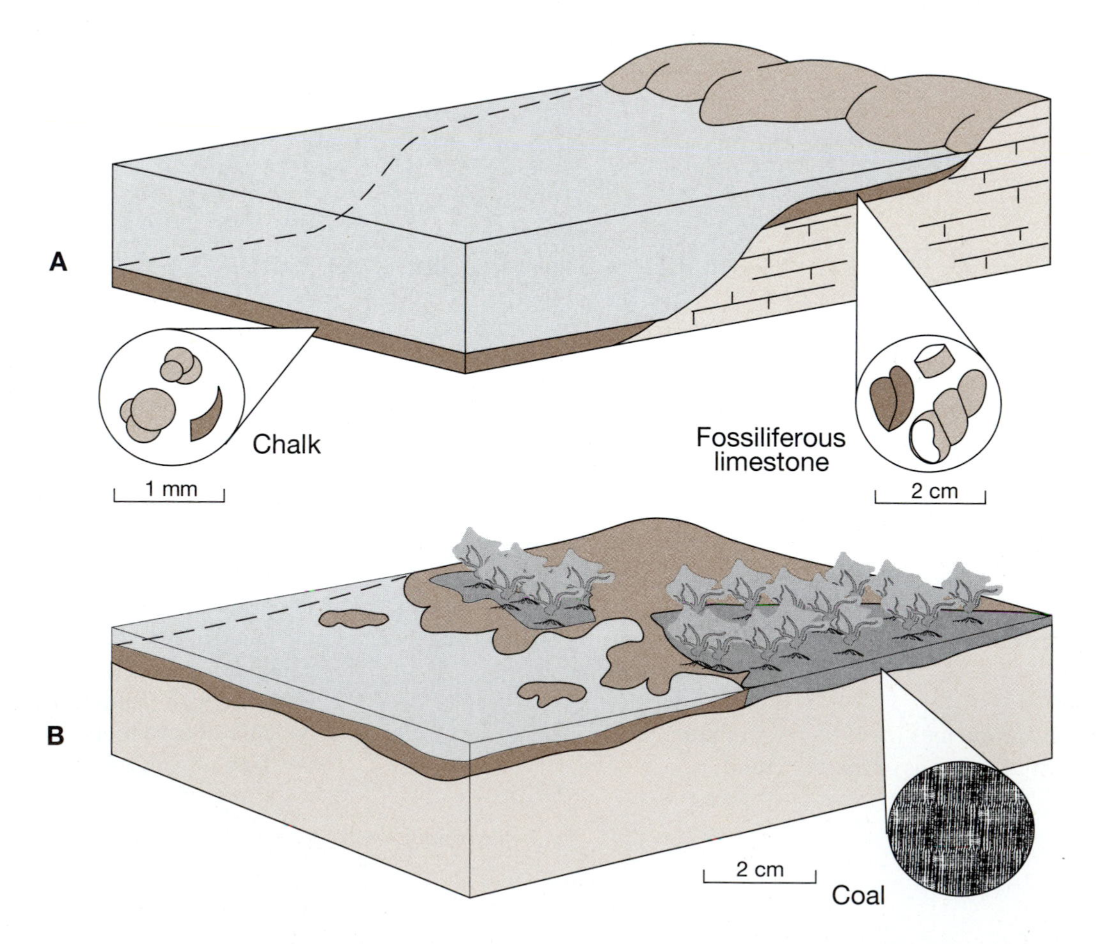

Figure 4.4 Organic sedimentary rocks. **A** Deposition of organic limestones. Fossiliferous limestones accumulate in shallow waters that occur on submarine shelves marginal to land. Chalks are more reflective of deeper, calmer waters farther offshore. **B** Coals record swamp conditions, like those of the Florida Everglades, where peat—the precursor of coal—is accumulating today.

Chemical sedimentary rocks

The five kinds of chemical sedimentary rocks covered in this exercise fall into two groups: **primary** and **secondary**.

Primary chemical sedimentary rocks are precipitated on rock surfaces and in open spaces within rocks at or near Earth's surface. Examples are rocks precipitated within streams, lakes, and oceans as a result of *evaporation* (Fig. 4.5A below)—and rocks precipitated within caves as a result of the *loss of carbon dioxide from water*—which causes water to become less acidic (Fig. 4.5B). Primary chemical sedimentary rocks include the following three varieties described here (1 through 3):

(1)

Rock salt is an 'evaporite' mineral that is precipitated within seas and lakes where water has evaporated to a salinity approximately *ten times* more salty than that of ordinary seawater.

(2)

Rock gypsum is precipitated within seas and lakes where water has evaporated to a salinity approximately *four times* that of ordinary seawater.

Q4.4 Progressive growth of the barrier sand bars in Figure 4.5A should produce an increase in the salinity of lagoon water, which should be recorded in the rock record by: (a) a layer of salt on top of a layer of gypsum, or (b) a layer of gypsum on top of a layer of salt?

Hint: **Drawing from the two points, (1) and (2), made in the preceding column, which is more soluble in seawater, salt or gypsum? With increasing evaporation and salinity, which would be the first to *precipitate*, salt or gypsum? With decreasing evaporation and decreasing salinity (i.e., a freshening of the water), which would be the first to *dissolve*?**

(3)

Chemical limestone consists of calcium carbonate that is precipitated (a) from cave waters as **travertine**; and (b) in shallow seas as lime mud, which forms a rock called **micrite**. Sand-size carbonate grains resembling white fish eggs are called **ooids** (Gr. öo, egg). These grains, which develop in shallow seas and saline lakes, can become cemented to form a rock called **oolite**.

Secondary chemical sedimentary rock is the second (or later) rock to occupy a particular space. What does that mean? That means that secondary chemical sedimentary rock develops through the chemical change ('replacement') of earlier rock by elements transported by water (Fig. 4.6). Two examples (4 and 5):

(4)

Dolostone is a rock that consists of dolomite produced by the replacement of limestone by the activity of magnesium in water (Fig. 4.6A).

(5)

Chert (flint) is rock that consists of micro-crystalline quartz. Chert nodules are produced by the spotty replacement of limestone by silicon dioxide (quartz) (Fig. 4.6B). Chert can also develop from an accumulation of siliceous parts of microscopic marine animals and plants (e.g., radiolarians and diatoms).

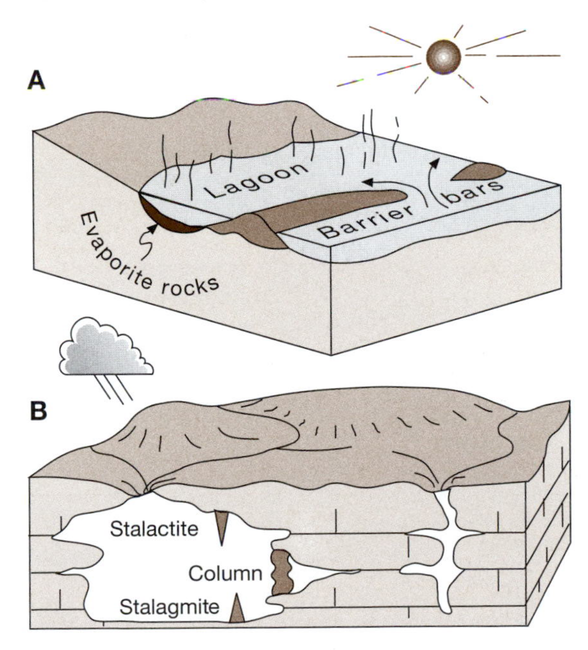

Figure 4.5 A Evaporation of water and the precipitation of minerals (e.g., salt and gypsum) in an arid lagoon. Sand bars restrict exchange of water between the lagoon and open sea.

B Precipitation of cave deposits caused by the loss of carbon dioxide from cave water, which reduces salinity of the water.

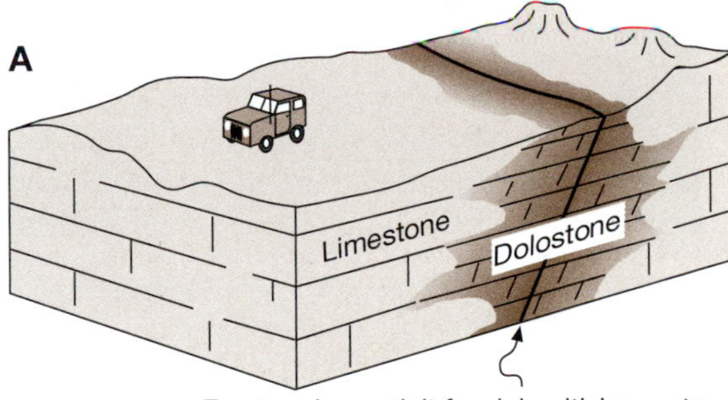

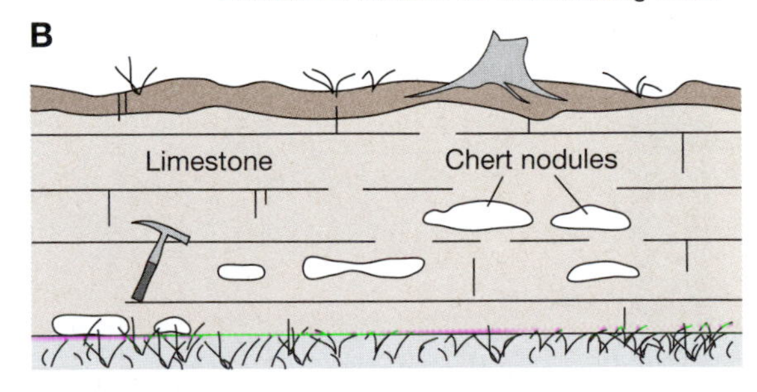

Figure 4.6 A Massive replacement of limestone by dolostone, in this case by magnesium-bearing water moving along a fracture in the rocks.

B Spotty replacement of limestone by chert (aka flint).

Table 4.1 Classification of sedimentary rocks

DETRITAL	particle size		physical properties	rock name	
	coarse (>2 mm) pebbles and cobbles		abraded (rounded) pebbles in sandstone; resembles concrete	CONGLOMERATE	
			angular fragments show no signs of abrasion	BRECCIA	
	medium (1/16–2 mm) like coarse sandpaper	SANDSTONES	dark sand-size rock fragments impart a salt-and-pepper appearance; usually rich in clay	GRAYWACKE SS.	
			abundant grains of feldspar are typically weathered to dull white or dull pink; commonly with some amount of clay	ARKOSE SS.	
			relatively pure content of quartz grains; some examples are porous and permeable	QUARTZ SS.	
	fine (1/256–1/16 mm)		gritty to the touch; grains barely detectable with the unaided eye, like fine sandpaper	MUDSTONE (silt + clay)	SILTSTONE
	very fine (<1/256 mm)		smooth to very slightly gritty; grains not detectable with the unaided eye; slick when wet		CLAYSTONE
			smooth to slightly gritty; grains hardly detectable with the unaided eye; fissile (i.e., breaks into sheets); slick when wet	SHALE	

ORGANIC	constituents		physical properties	rock name
	calcite shell fragments	LIMESTONES	shells packed together and cemented with transparent calcite, or 'floating' in cemented opaque lime mud	FOSSILIFEROUS (= bioclastic, = skeletal) LIMESTONE
			shells packed together and slightly cemented with calcite; very porous; appearance of corn flakes	COQUINA
			shells too small to be seen with the unaided eye; chalky or powdery; typically white to light gray to light brown	CHALK
	plants		brown; fibrous, soft, porous; plant fragments visible	PEAT
			brownish to brown-black; crumbly, but harder than peat	LIGNITE 'SOFT COAL'
			brown to black; nonporous; sooty; commonly spotted with yellowish sulfur compounds	BITUMINOUS COAL
	siliceous animals and plants		micro-crystalline quartz (can't scratch with steel); conchoidal fracture, sharp edges; possibly 'ghosts' of microscopic fossils	CHERT (aka 'flint')

CHEMICAL	minerals		physical properties	rock name
	halite	EVAPORITES	cubic crystals of halite with cubic cleavage usually apparent; hardness 2.5; salty taste	ROCK SALT
	gypsum		soft (hardness 2); very fine crystals impart sugary appearance; usually white or pink	ROCK GYPSUM
	calcite	LIMESTONES	coarsely crystalline calcite; typically with yellow red, and brown bands resembling tree rings	TRAVERTINE
			fine-grained character resembles that of mudstone; slightly conchoidal fracture; slightly sharp edges; scratched with steel	MICRITE LIMESTONE
			sand-size grains of calcite called ooids; white to gray; a broken ooid exhibits concentric banding about a nucleus	OOLITE LIMESTONE
	dolomite		texture ranges from that of mudstone to that of travertine; can appear sugary	DOLOSTONE
	quartz		micro-crystalline quartz (can't scratch with steel); conchoidal fracture, sharp edges; possibly 'ghosts' of macro-fossils	CHERT (aka 'flint')

Sedimentary rocks

Detrital

Conglomerate
Varicolored;
like concrete with rounded pebbles

Limestone Breccia
Varicolored;
like concrete with angular fragments

Graywacke Sandstone
Gray to dark gray or greenish;
salt-and-pepper appearance;
commonly with considerable clay

Arkose Sandstone
Gray to red;
abundant feldspar grains; some clay

Quartz Sandstone
White to gray to red or yellow;
mostly quartz grains; little clay

Siltstone
Varicolored;
microscopic grains like fine sandpaper

Claystone
Varicolored;
slick when wet

Mudstone
(mixed siltsone and claystone)
Varicolored;
gritty between the teeth

Shale
Varicolored; parallel flat clay grains cause the rock to be fissile (i.e., breaking into sheets); soft (dull 'thud' when struck)

Organic

Coquina
Tints of brown, yellow, orange;
shells weakly cemented by clear calcite, therefore very porous (as in corn flakes)

Fossiliferous Limestone
Gray to brown;
shells cemented with clear calcite, or 'floating' in cemented lime mud

Chalk
White to gray to brown;
soft (i.e., powdery or chalky);
microscopic fossils perhaps visible

Lignite
Brown to black;
'soft coal'; crumbly

Bituminous Coal
Brown to black;
'hard coal'; sooty to shiny, common spots of yellow to orange sulfur

Chert
White to gray;
very hard, conchoidal fracture with sharp edges;
some examples are translucent

Chemical

Rock Salt
Clear, white, gray;
cubic crystals and cubic cleavage;
scratched with nail; salty taste

Rock Gypsum
White to pink;
fine to coarse textured; scratched
with fingernail; fine textures
appear sugary

Travertine
Yellow, red, brown;
color-banded appearance

Micrite variety of Limestone
Gray to brown;
fine textured; conchoidal fracture;
possible scattered fossils

Oolite variety of Limestone
White, gray, beige;
millimeter-size spherules

Chert
White to gray;
very hard, conchoidal fracture
with sharp edges; possible visible 'ghosts' of fossils

Dolostone
Gray to brown;
fine to coarse textured; finer textures appear sugary, coarser textures appear sandy

C. Sedimentary rocks

Sedimentary layers—Sedimentary rocks typically consist of layers that differ in rock type, and therefore differ in their resistance to weathering and erosion. So sedimentary rocks commonly display a stair-step topography (see again Figure 4.2A on page 64). Gentle slopes and flatlands develop on easily eroded shales, whereas steep slopes and cliffs usually mark more resistant sandstones and limestones.

Figure 4.7 is a Google Earth image of Digital Globe satellite photography of an area in Kansas where Permian-age shales and limestones exhibit stair-step topography.

Q4.5 Draw a topographic profile (a side-view of the landscape) along the line A–B indicating the relative heights of the three hills in your profile. Identify the most obvious layer within this shale-limestone succession and show its position underground along the line of topographic profile.

A

B

Image © 2008 DigitalGlobe

Flat-lying sedimentary layers

Pottawatomie Co., Kansas

Tuttle Creek Dam, Kansas, 7 1/2' quadrangle
N. 39° 18' 07", W. 96° 35' 40"

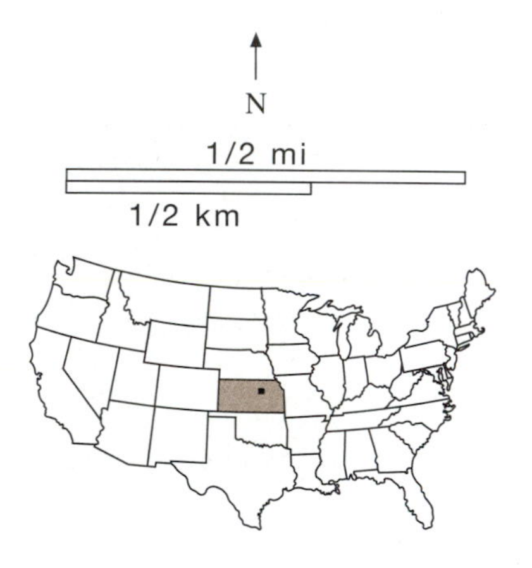

Figure 4.7 Shale and limestone layers of Permian age in Pottawatomie County, Kansas. The shale weathers to a light color; the limestone is marked by dark vegetation.

False sedimentary layers—A variety of stresses within the earth produce fractures in rocks called **joints**, which typically occur in 'sets' within which many joints exhibit a similar orientation (like the grain within wood). Joints are important because they are commonly pathways for groundwater and oil.

Figure 4.8 is a Google Earth image of Digital Globe satellite photograph of Canyon Lands NP. Mesozoic-age sandstones and shales lie horizontally—like layers in a cake. Hilltops consist of sandstone that exhibits steep ridges bounded by joints along which water has eroded the sandstone (see the topographic profile sketched along line A–B). At first glance these ridges appear to be sedimentary layers turned on edge, but they're actually fin-like ridges bounded by joints.

Q4.6 Drawing on the foregoing discussion in this exercise, what kind of sedimentary rock do you suspect probably underlies sandstones that cap the hills in Figure 4.8?

Q4.7 Judging from the orientation of the fin-like ridges in Figure 4.8, what is the approximate orientation of the joints that bound them—i.e., roughly north-south or east-west?

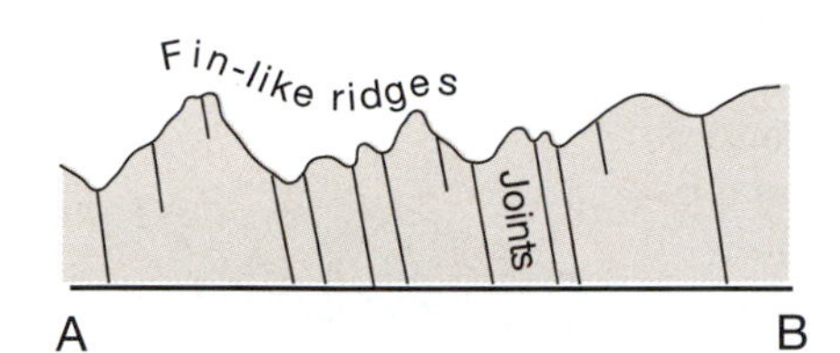

Vertical joints in horizontal sedimentary layers Canyon Co., Utah

Springdale East, Utah, 7 1/2' quadrangle
N. 37° 13' 16'', W. 112° 52' 34''

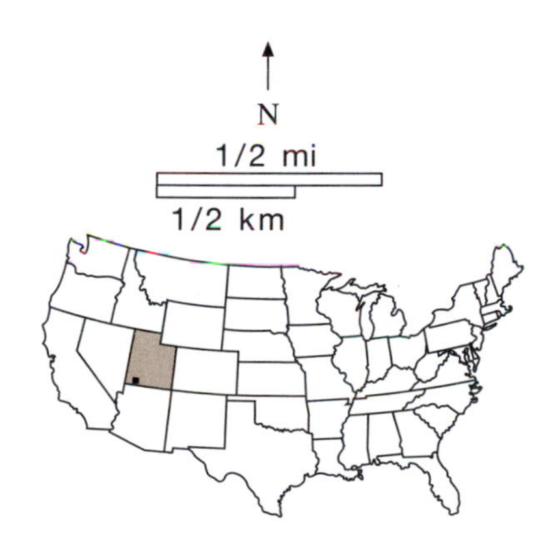

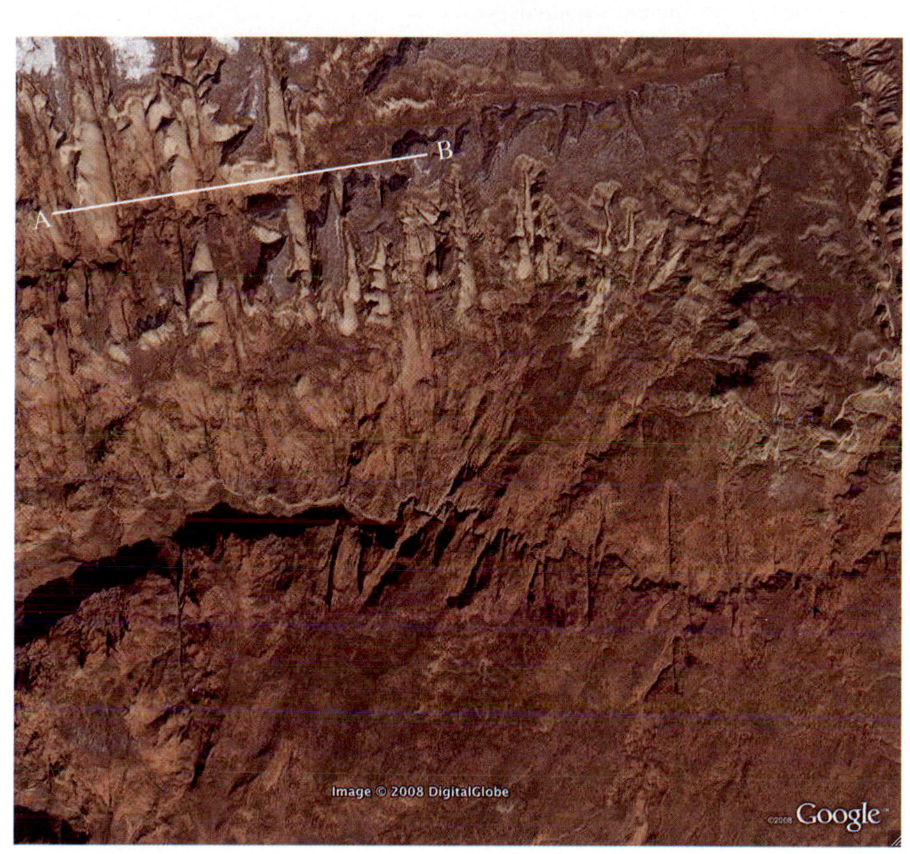

Figure 4.8 Canyon County, Utah. Layers of Mesozoic-age sandstone and shale have been broken by parallel joints (i.e., fractures).

D. Deciphering depositional history of a stream deposit

Streams erode (e.g., carving valleys), and streams deposit (e.g., building vast bodies of sand and gravel, ancient examples of which yield oil and gas). Deciphering the history of development of an ancient stream deposit is a common activity in geology.

One question about the origin of an ancient stream deposit has to do with the swiftness of the stream or river that deposited it. Put another way, does the stream deposit reflect a babbling brook, or does it record a roaring torrent? In general, the coarser the grain size of a detrital sediment, the stronger the current of water that deposited it. For example, an accumulation of clay-size sediment records deposition in still water, whereas an accumulation of pebbles records deposition by swift-flowing water. Silt, sand, and mud record intermediate flow velocities.

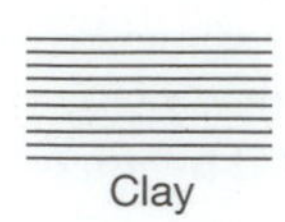

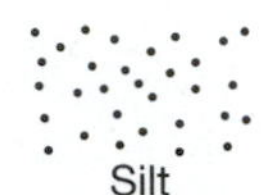

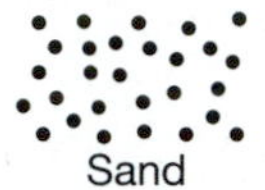

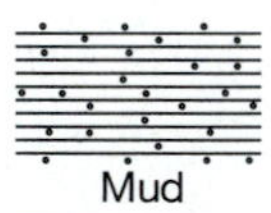

On page 79 there is a drawing (Fig. 4.13) of a hypothetical 50-meter vertical section of an ancient layered stream deposit representing 100 years of deposition. A legend of patterns in Figure 4.13 identifies the different grain sizes, and their vertical succession records how the velocity of stream flow differed over time.

Q4.8 In the column to the right of the vertical section in Figure 4.13, plot a vertical *graph* that charts the changes in stream velocity during the time required for the accumulation of the entire 50 feet of vertical record—coarsest grained (recording the strongest currents) to the right, and finest grained (recording the weakest currents) to the left. The graph for the bottom two layers has already been plotted.

E. Charting ancient lands and seas, and deciphering depositional environments

Sedimentary rocks provide features that enable geologists to chart the distribution of ancient lands and seas and decipher depositional environments. The number of such clues is limitless, but ten of the more common ones appear in Figure 4.9. A brief description of its environmental setting accompanies each feature.

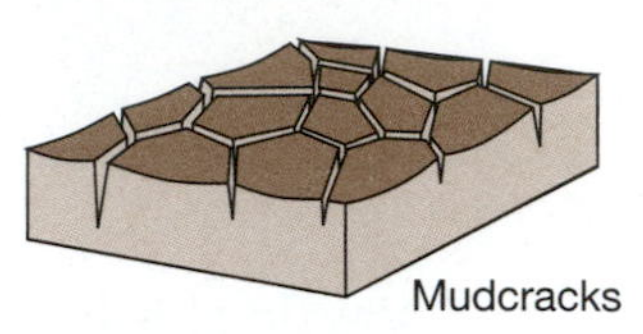

Figure 4.10 on page 76 is a map of 14 locations where environmental features have been identified and recorded within a fictitious marine limestone of Cretaceous age near Austin, Texas.

Q4.9 Your task is to draw a paleogeographic map (directly on the Figure 4.10 map) that is consistent with all 14 depositional features labeled on the map. Your map should include:

- **a shoreline (label land, label sea)**
- **a sandy beach**
- **a tidal flat**
- **an area of sand bar development**
- **an area of deep water**
- **a lagoon**
- **the direction (indicated with an arrow) of at least one current**

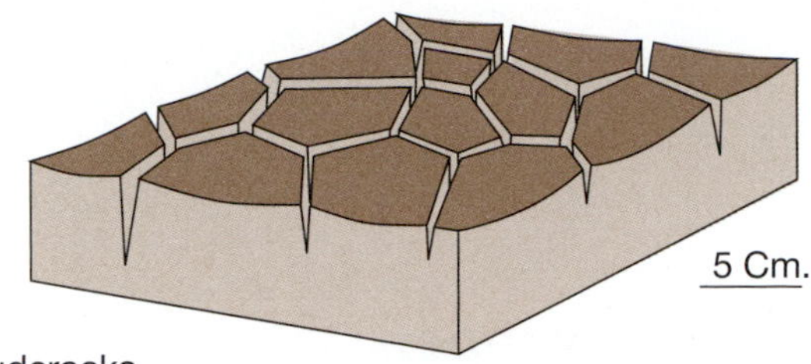

Mudcracks.
Periodic wetting and drying—as on a tidal flat.

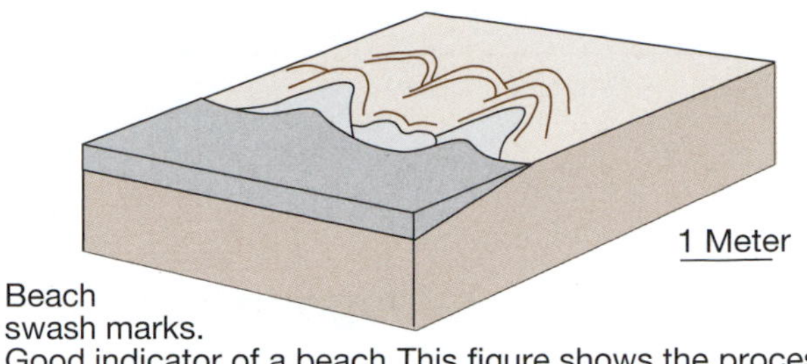

Beach
swash marks.
Good indicator of a beach. This figure shows the process.

Raindrop
impressions.
Periodic exposure—as on a beach (if sand)
or a tidal flat (if mud).

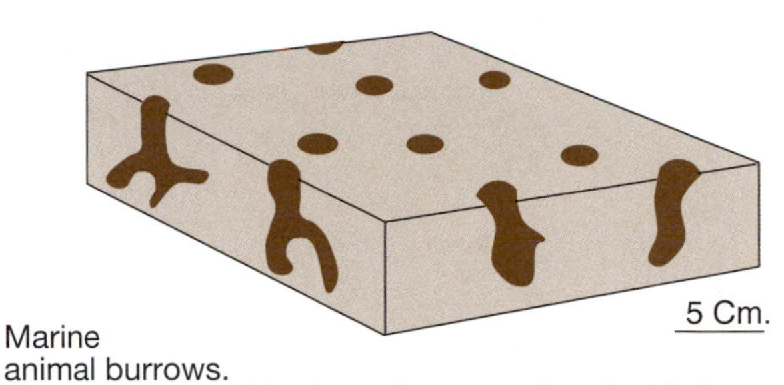

Marine
animal burrows.
To some extent ubiquitous, but very abundant in lagoons.

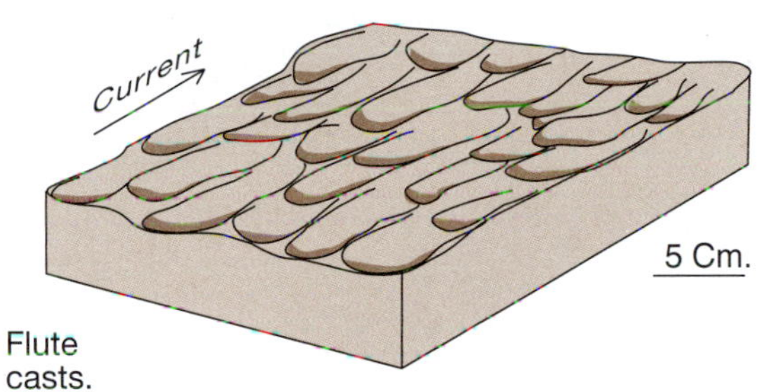

Flute
casts.
Marine flutes are carved in deep-sea sediments by water
driven by submarine landslides.

Animal tracks.
Ubiquitous on land.

1 Meter

Planar cross-bedding.
Indicative of beaches.

Current

5 Cm.

Ripple mark.
Typical of silt and sand. Rare in mud because mud is
largely a suspension sediment. Good indicator of
current direction.

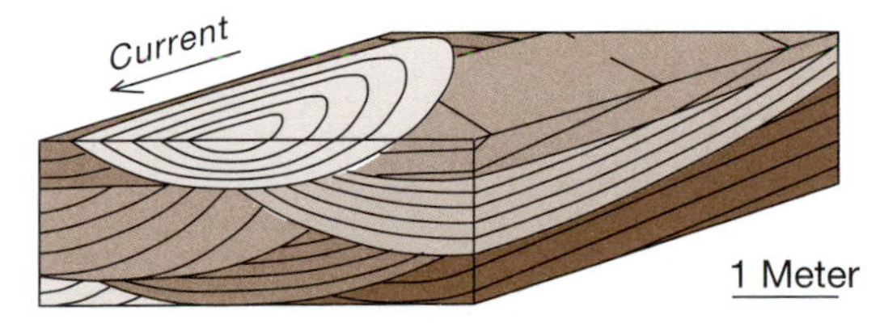

Trough cross-bedding.
Commonly associated with sand bar development.

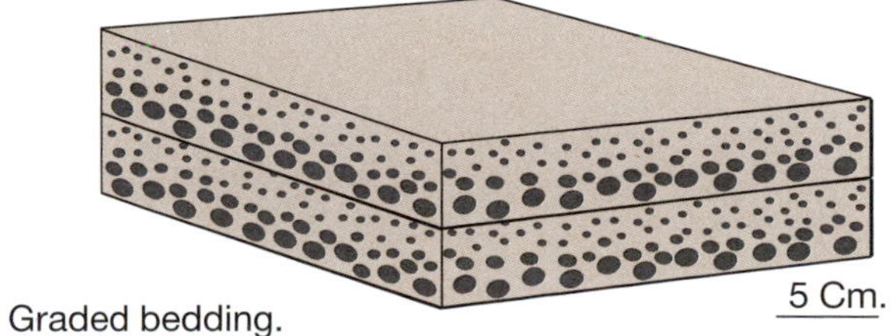

Graded bedding.
One origin is deposition of sediment suspended by
submarine landslides. As current slows, finer-grained
sediment is deposited.

Figure 4.9 These are clues to depositional environments. Although a number of these features occur in both continental and marine settings, for the purpose of completing this exercise, each is labeled with a setting applicable to an area with a seacoast, an adjacent shallow sea, and deep ocean water beyond.

(Student's name) (Day) (Hour)

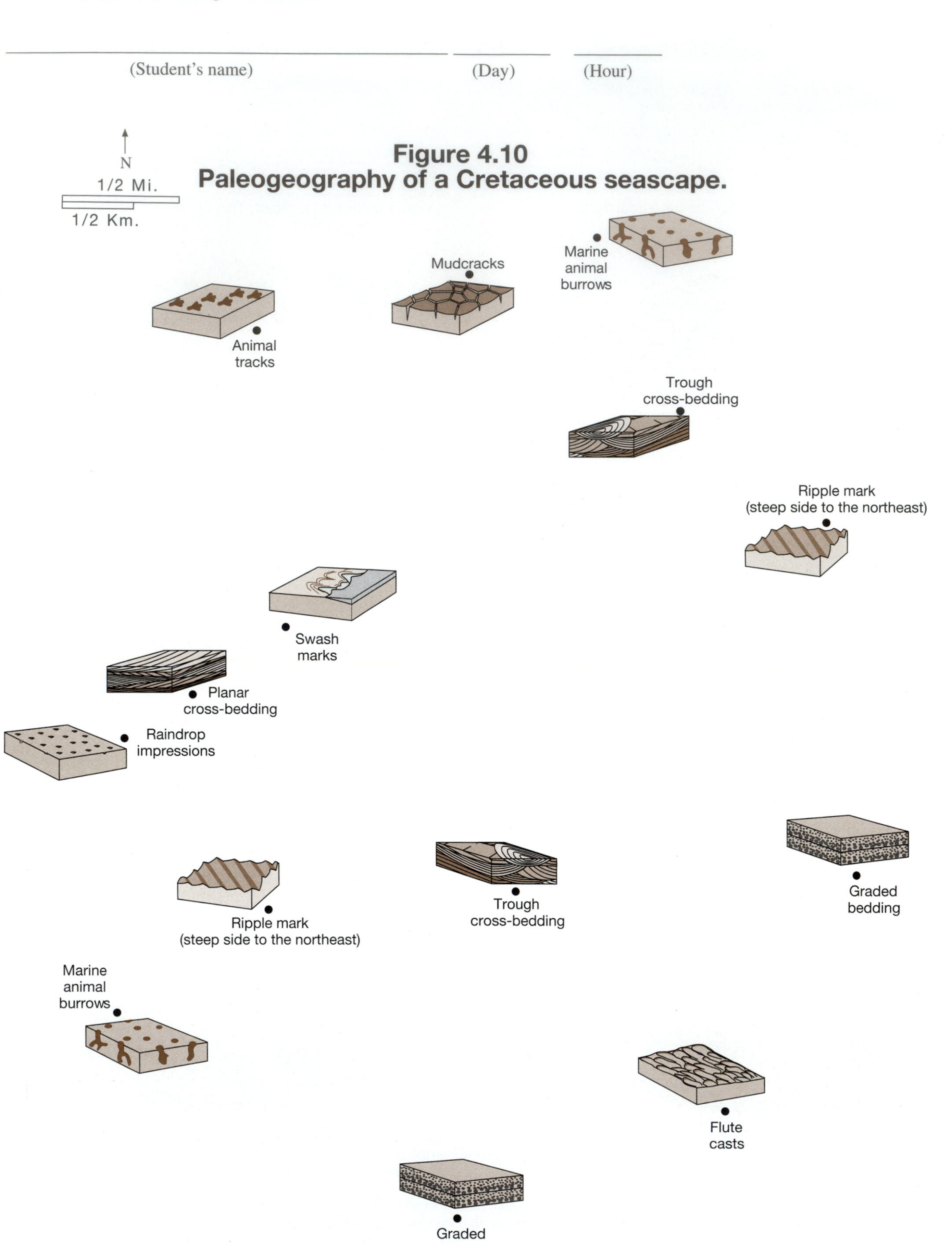

Figure 4.10 Paleogeography of a Cretaceous seascape.

F. Sedimentary rocks in society

Although igneous and metamorphic rocks make up the bulk of continental crust, sedimentary rocks cover 80% of the *surface* of continents. This veneer of sandstone, limestone, and shale is important for many reasons, including being the source of *petroleum, aluminum, bricks, concrete, construction board, fertilizers, salt, sulfur, and uranium*—to name a few. Sedimentary rocks also provide their share of stories about how geology has influenced the course of human history. For example, fertile soils—which to a large extent reflect the mineralogy of underlying sedimentary rocks—have influenced patterns of migration in America.

British geographers have called attention to the correlation between England's domains of modern political parties and bedrock geology. Figure 4.11 shows the distribution of Cretaceous and Carboniferous Systems in England. The Cretaceous System is characterized by low, flat-lying layers of chalk typified by the White Cliffs of Dover, the same geologic situation as that across the channel at Normandy, France, site of the Allied invasion that led to the end of World War II. Much like in our own Gulf Coastal Plain, farming and ranching in England's Cretaceous have been the rule for centuries, with large entrepreneurial landowners. In contrast, the Carboniferous System is characterized by sandstones and shales of the Midlands that are rich in the vast coal deposits that fueled the birth and development of the 19th-century Industrial Revolution.

Q4.10 Figure 4.12 shows the distribution of district councils controlled by two unlabeled parties. One party is the Labour Party—the party of labor unions and big government. The other party is the Conservative Party—the party of private ownership and minimal government intervention. (A) Which party's domain do you suppose is labeled Party X in Figure 4.12? (B) Which party's domain do you suppose is labeled Party Y? *Hint:* Key terms are labor unions, big government, private ownership, minimal government intervention.

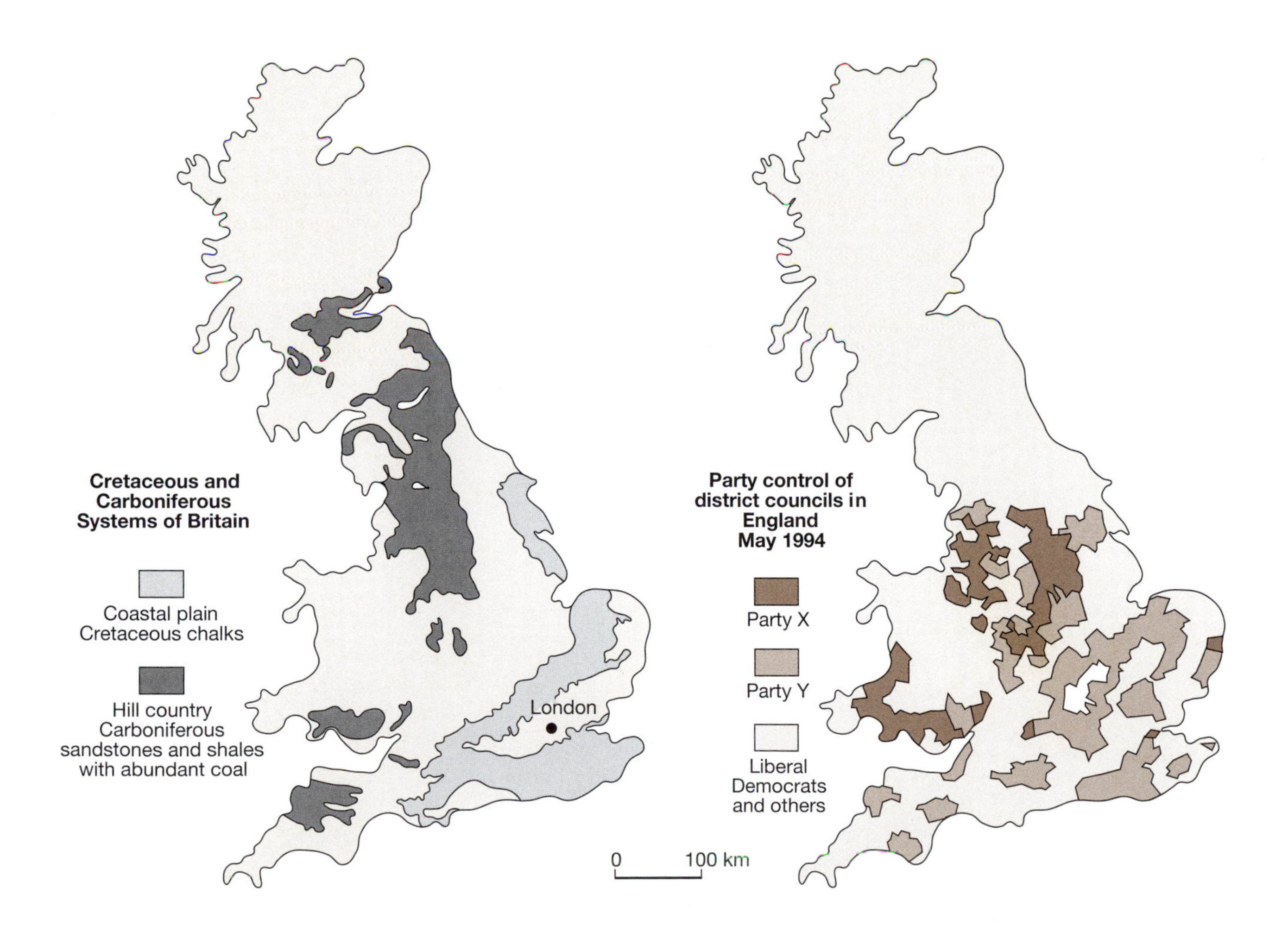

Figure 4.11 This is a partial geological map showing the distribution of Cretaceous and Carboniferous Systems in England.

Figure 4.12 This map shows the distribution of unlabeled domains of Labour and Conservative political parties.

Intentionally Blank

5 Metamorphic Rocks

Topics

A. What are two metamorphic changes that affect minerals?
   What are two metamorphic changes that affect rock fabric?

B. What are the three variables on which the classification of metamorphic rocks is based?

C. What are the three changes that occur in the evolution of anthracite, a metamorphic coal?
   What is the origin of sulfur in coal? How does the burning of coal generate acid rain?

D. What are the two metamorphic settings, and what factor is emphasized in each?

E. What is the meaning of the term metamorphic grade? What are index minerals? What are isograds? What are isobars? How does a map of atmospheric pressure indicate the direction and velocity of wind?

A. *Metamorphic processes*

Metamorphism is the mineralogic and/or physical changing of rock caused by high levels of heat and/or pressure—but without the melting of the rock. (If the rock were to melt, the eventual product would be called igneous.)

Mineral changes—two types

Changes in mineral composition during metamorphism are of two types: (1) original minerals change into new minerals; and (2) original minerals recrystallize, resulting in changes in the shapes and sizes of those minerals.

(1)

Development of new minerals—Unusually high temperatures can cause a recombination of elements, producing new minerals. Figure 5.1 illustrates the process of recombination affecting the common sedimentary rock—calcite-cemented sandstone—in which calcite and quartz recombine to produce a new mineral, wollastonite.

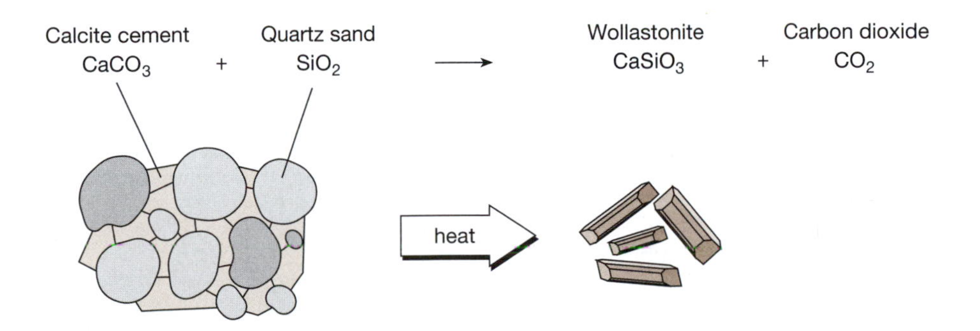

Figure 5.1 Under elevated temperature, calcite and quartz in a calcite-cemented sandstone can recombine into the mineral wollastonite plus carbon dioxide.

(2)

Recrystallization of pre-existing minerals—Unusually high temperatures can also cause the recrystallization of minerals. For example, a limestone (or dolostone) can recrystallize to produce calcitic (or dolomitic) **marble** (Fig. 5.2). Sandstone can also recrystallize, producing **quartzite** (Fig. 5.3).

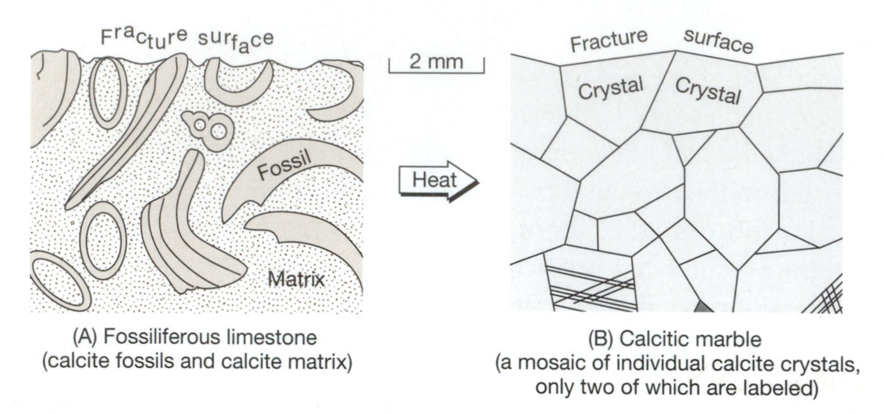

Figure 5.2 A fossiliferous limestone (A) can recrystallize to form the metamorphic rock calcitic marble (B). Notice the several changes that are apparent in B. (These are magnified views of sawed surfaces that are in the plane of the page.)

Q5.1 (Ref. Figure 5.2) (A) Judging from the changes that are evident in these two before-and-after illustrations, do you suppose that the geologic age of calcitic marble could be determined from its fossil content (as can be done with fossiliferous limestone)? (B) Why or why not?

Q5.2 (Ref. Figure 5.2B) Account for the straight boundaries (that mark flat surfaces) of individual crystals at the fracture surface of the calcitic marble. *Hint:* Check out the properties of calcite in Table 2.3 on page 37.

Q5.3 (Ref. Figure 5.3) Explain the difference between the fracture surface of quartz sandstone and that of quartzite. *Hint:* Check out the properties of quartz in the lower one-half of Table 2.3 on page 36.

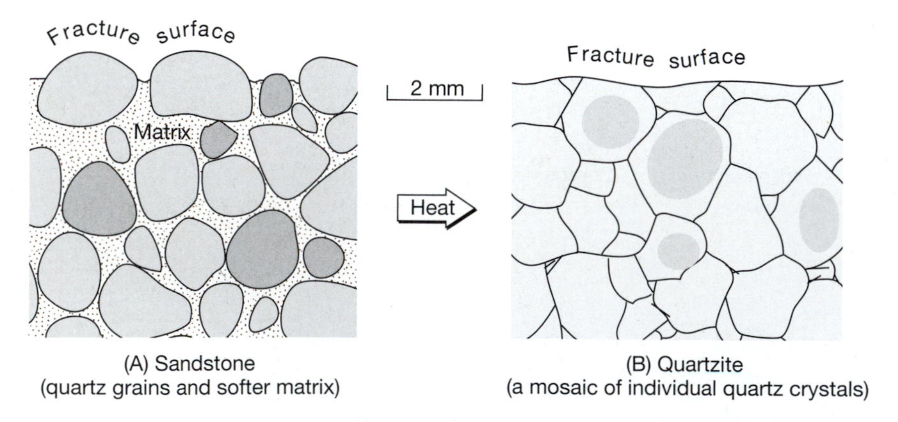

Figure 5.3 Sandstone (A) can recrystallize to form the metamorphic rock quartzite (B). 'Ghosts' of original quartz sand grains (see the cloudy centers of crystals) might be apparent in quartzite. (These are magnified views of sawed surfaces that are in the plane of the page.)

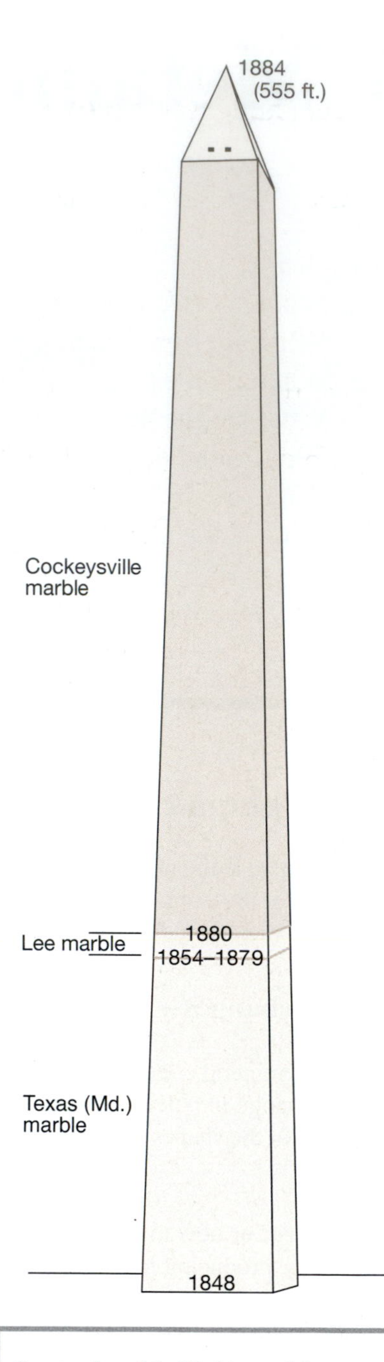

Construction of the Washington Monument in the 19th century marked the change from sandstone to marble as the stone of choice in buildings and monuments of Washington, D.C. This landmark was begun in 1848 using marble from the Texas, Maryland quarry, but a lack of funds halted construction in 1854. Interest was rekindled in 1879, and construction was resumed with Lee marble. But Lee marble proved to be too costly, so Maryland's Cockeysville marble was used from 1880.

Building Stones of our Nation's Capital is a pamphlet produced by U.S. Geological Survey and distributed by U.S. Government Printing Office: 1998—673-048 / 20115.

Physical changes—two types

In addition to changes in mineralogy, metamorphic rocks can also develop new fabrics in response to pressure (Fig. 5.4). There are two types of pressure-induced fabrics—**lineation** and **foliation**.

Type 1

Lineation (French *lignage*, line), is a linear fabric. Example: stretched pebbles in metaconglomerate so as to be elongated (Fig. 5.4A). The long axes of stretched pebbles collectively comprise a lineation.

Type 2

Foliation (French *foillage*, leaf), is a tabular fabric. There are two kinds of foliation: One (Figs. 5.4B, C) is manifest in the parallel orientation of plate-like minerals (e.g., clays and micas), which imparts rock cleavage (not to be confused with mineral cleavage). This is called *slaty cleavage* in **slate** and *schistosity* in **schist**. Both slates and schists develop from shale, but schists result from greater pressure. The hardness and smooth cleavage of slate make it ideal for roofing shingles, floor tile, chalkboards, and pool tables.

The other type of foliation (Fig. 5.4D) is manifest in the segregation of light minerals and dark minerals into layers, imparting a banded appearance. This rock is called **gneiss**. Gneiss does not exhibit rock cleavage as do slate and schist.

Q5.4 Imagine pebbles in a metaconglomerate that have been stretched (elongated) and flattened by pressure, so as to exhibit *three* axes of different lengths. Do such strained pebbles exhibit lineation, foliation, or both?

Q5.5 Examine the before-and-after fabrics (and captions) in Figure 5.4B and explain how slaty cleavage can be distinguished from the fissile sheet-like breaking of shale. *Hint:* Notice the sedimented sand grains and the two captions.

Before | **After**

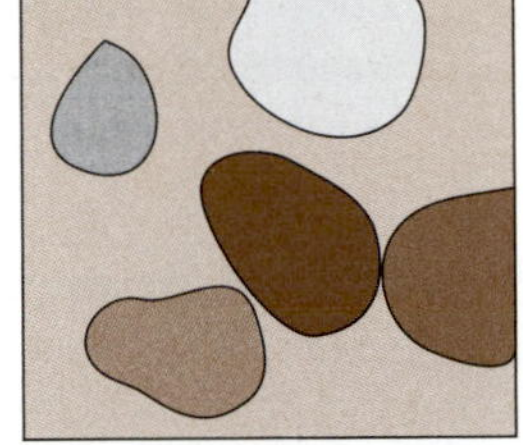

Conglomerate… (irregularly shaped pebbles)

A
Stretched pebbles
Parallel long axes

…changes to…

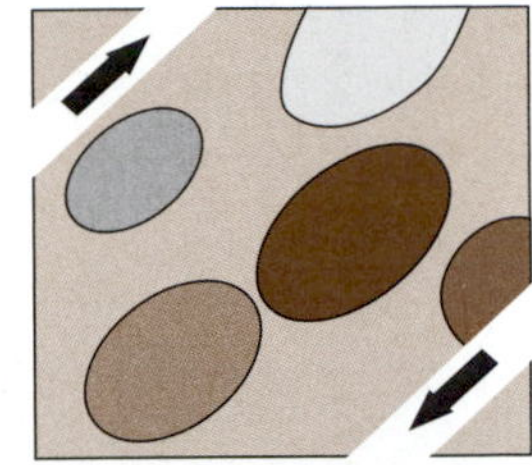

…metaconglomerate (stretched pebbles)

Shale (sheet-like breakage parallel to sedimented layers)

B
Slaty cleavage
Parallel clay minerals impart smooth cleavage

…changes to…

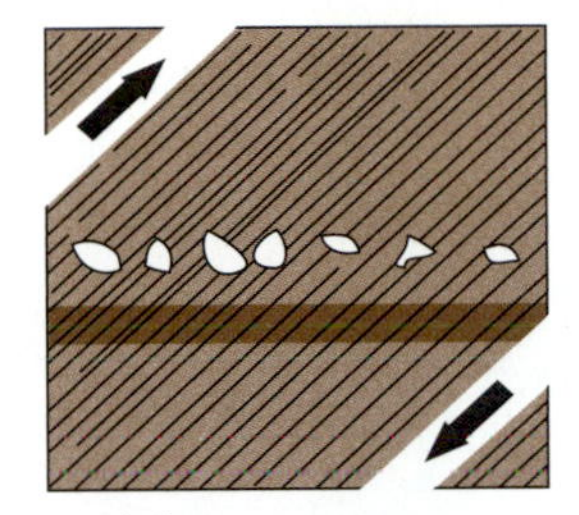

…slate (slaty cleavage at an angle to sedimented layers)

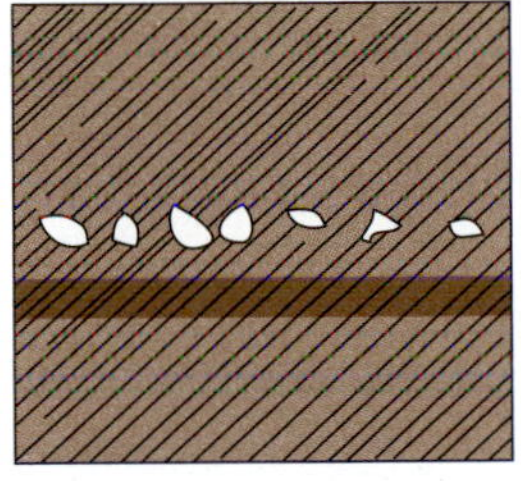

Slate…

C
Schistosity
Parallel micas impart crude rock-cleavage

…changes to…

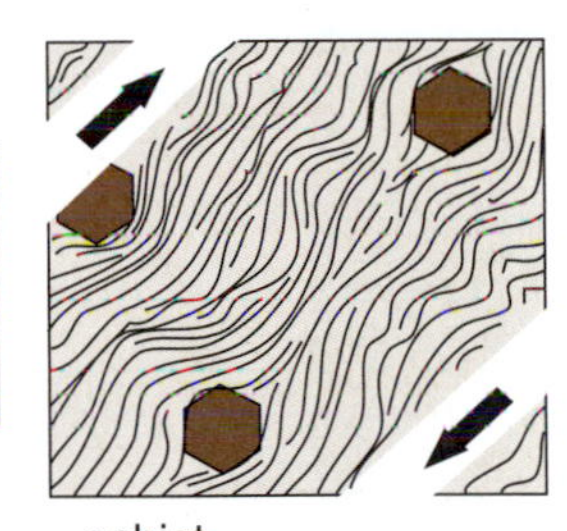

…schist (with knotty crystals)

Granite…

D
Gneissic banding
Parallel bands of segregated minerals

…changes to…

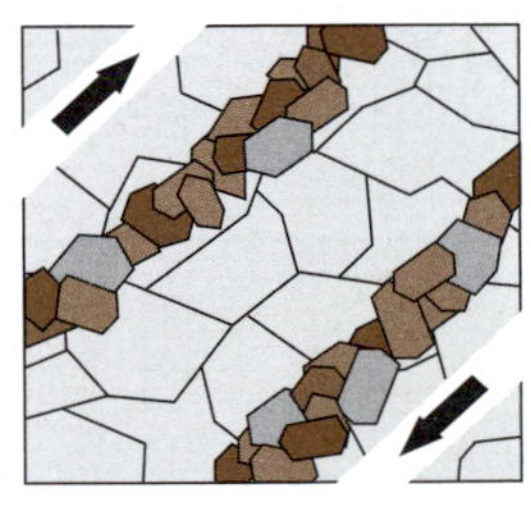

…gneiss

Figure 5.4 Metamorphic fabrics result from shear pressures (bold arrows). The segregation of light and dark minerals into bands in gneiss is not well understood.

B. Classification of metamorphic rocks

The classification of metamorphic rocks is based on three variables: *Fabric*, *texture*, and *mineralogy*. The organization of Table 5.1 (below)—along with photographs that follow on page 85—illustrates these variables.

Accessory minerals—A variety of new minerals can occur in any metamorphic rock as crystals called **porphyroblasts** (see knotty crystals in the schist of Figure 5.4C on page 83). Common porphyroblasts are hornblende and garnet.

Table 5.1 Classification of metamorphic rocks

	fabric and appearance	texture	mineralogy	rock name
FOLIATED	rock cleavage owing to parallel orientation of clays (in slate), clays and micas (in phyllite), and micas (in schist)	very finely crystalline; may exhibit slight sheen; more compact than shale (goes *'clank'* when struck with metal object, whereas shale goes *'thud'*)	appearance of clay, but very finely micaceous	SLATE
		finely crystalline; micas hardly discernible but impart a sheen or luster	silicates, especially micas	PHYLLITE
		coarsely crystalline; crinkled micas discernible	light to dark silicates, e.g., chlorite, mica, and hornblende; garnets are common	SCHIST
	minerals segregated in layers (banded), but no rock cleavage	coarsely crystalline; appearance of granite	light-colored quartz and feldspar; dark ferromagnesian minerals	GNEISS
WEAKLY FOLIATED	weakly foliated at best; subtle rock cleavage reflects parallel orientation of plate-like minerals	coarsely crystalline	dark silicates, abundant amphiboles	AMPHIBOLITE
	stretched (lineated) and possibly flattened (foliated) pebbles so pebbles can have three axes of different lengths, reflective of three-dimensional stress; breaks through pebbles, rather than around them as in conglomerates	pebble-grained	composition of pebbles variable, as in conglomerate	META-CONGLOMERATE
NONFOLIATED	appearance of basalt	finely crystalline	silicates	HORNFELS
	looks like sandstone, but grains are fused together, so it breaks through original sand grains, rather than around them	finely to coarsely crystalline	quartz	QUARTZITE
	sugary sparkle imparted by cleavage of calcite or dolomite crystals	finely to coarsely crystalline	calcite or dolomite	MARBLE
	shiny black; conchoidal fracture	super-fine textured, like tar; glossy	metamorphosed plant material	ANTHRACITE

Metamorphic rocks

Foliated (layered)

Phyllite
Various tints of gray;
fine textured parallel micas
impart a sheen

Slate
Black, gray, green, maroon;
hard (a sharp 'clank' when struck);
rock cleaves to produce flat surfaces;
sedimented layers form angles with cleavage

Schist
Silver to black sparkle;
parallel micas are crinkled;
possible chlorite, hornblende, garnets

Gneiss
Colors and texture of granite;
light and dark minerals are
segregated into layers

Weakly foliated

Amphibolite
Black and dark tints of gray;
coarse textured;
some parallelism among platy minerals;
abundant amphibole (e.g., hornblende)

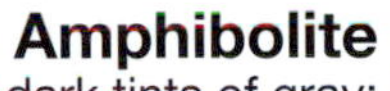

Nonfoliated

Quartzite
Varicolored;
suggestive of sandstone, but less gritty

Marble
White, gray, pink;
fine to coarse textured;
cleavage of crystals imparts sparkle
to fracture surfaces

Hornfels
Dark gray to black;
fine textured appearance of basalt
but commonly spotted

Anthracite coal
Black to brown; high gloss; conchoidal fracture

C. Anthracite and other coals

Anthracite coal is included in most classifications of metamorphic rocks (see the last item in Table 5.1 on page 84). Anthracite—the 'hardest' of three basic **grades** of coal—is the end member in a continuum of coal development that begins with coal's precursor, peat. Figure 5.5 and its caption illustrate the progression of changes in coal as it evolves from peat to anthracite, namely (a) compaction, (b) loss of water, and (c) increase in the percent of carbon.

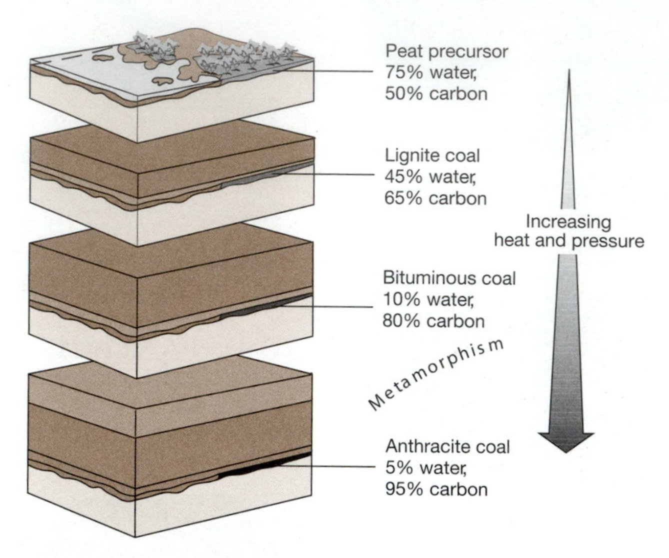

Figure 5.5 The evolution of coal is in response to increased heat and pressure that results from increased depth of burial. Because of limitations of scale, the degree of compaction is not shown here, but to illustrate—an interval of peat 100 ft. thick compacts to a mere 10 ft. as it is converted to bituminous coal.

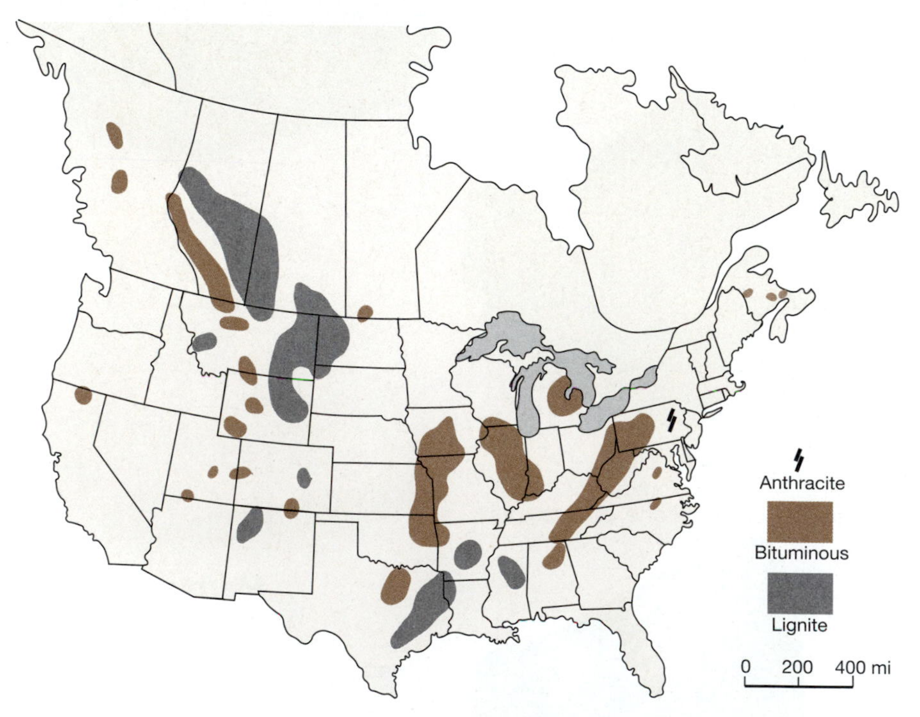

Figure 5.6 Coal deposits of North America. Notice the scarce occurrence of anthracite (only in the folded Appalachian Mountains of Pennsylvania). *(From U.S. Geological Survey)*

Coal grade signifies a coal's heat producing capacity. The higher the carbon content, the higher the heat value. In British thermal units (BTUs) per pound, lignite produces 7,000; bituminous, 12,000; and anthracite, 14,000. (One BTU = the heat required to raise the temperature of 1 lb of water 1°F.)

Q5.6 Approximately how many tons of lignite would be required to produce the amount of heat generated by one ton of anthracite coal?

Some 60% of our nation's electricity is generated in coal-fired power plants, thanks to an abundance of domestic coal (Fig. 5.6). But—as with oil and gas—the burning of coal pollutes our atmosphere with dreaded greenhouse gases CO and CO_2. Also, the burning of coal releases sulfur into the atmosphere, which combines with moisture to produce **acid rain**.

Source of sulfur in coal: Plants contain only about 1% sulfur, but bacteria decomposing woody material can obtain oxygen from *sulfate* ions. This reduction of sulfate (i.e., its loss of oxygen) liberates sulfur, which combines with iron to form FeS. Elemental players:

SO_4 (reduction) → S / O

$S + Fe \rightarrow FeS_2$

Q5.7 Find the name of this iron sulfide mineral at the very top of Table 2.3 on page 36.

This iron sulfide goes into the power plant with coal, and the sulfur goes to make the sulfuric acid in acid rain.

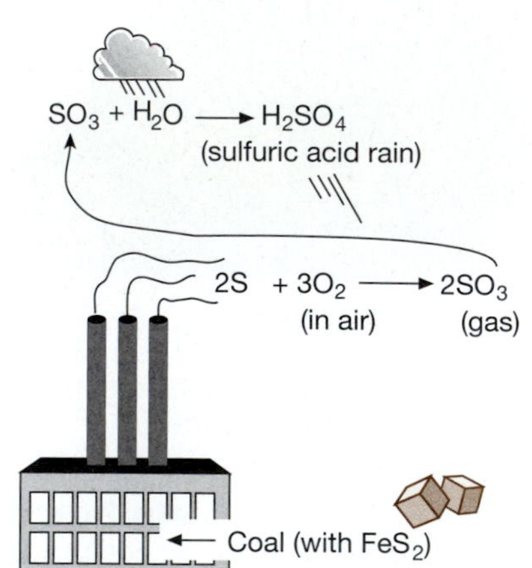

D. Plate tectonic settings of metamorphic rocks

Contact Metamorphism

The term **contact metamorphism** refers to metamorphism at or near the boundary with a magma. Here, the emphasis is on *heat*, which can cause *recrystallization* of minerals, producing **marble** (from limestone or dolostone), **quartzite** (from sandstone), and **hornfels** (from claystone, shale, or mudstone). A geologic setting for contact metamorphism is shown in Figure 5.7A in the context of **plate tectonics**.

Regional Metamorphism

The term **regional metamorphism** refers to metamorphism produced by pressure associated with development of folded (wrinkled) mountain ranges. Here, the emphasis is on *pressure*. Folded mountains develop where global plates collide, for example, along the west coast of South America, which accounts for the lofty Andes Mountains (Fig. 5.7B). Associated pressure—which can be either shearing or vice-like compression—can produce (1) stretched (lineated) and flattened (foliated) pebbles (in **metaconglomerate**); (2) slaty cleavage (in **slate** and **phyllite**); (3) schistosity (in **schist**); and (4) mineral layers (in **gneiss**). Weakly foliated varieties of marble, quartzite, and hornfels—marked only by wispy stringers of color differences—can also develop in folded mountain ranges.

Q5.8 Name another region, in addition to that of South America shown in Figure 5.7B, where one should expect to find lineated and foliated metamorphic rocks. *Hint:* Topography should be your guide here, so think of another linear (chain-like) mountain system.

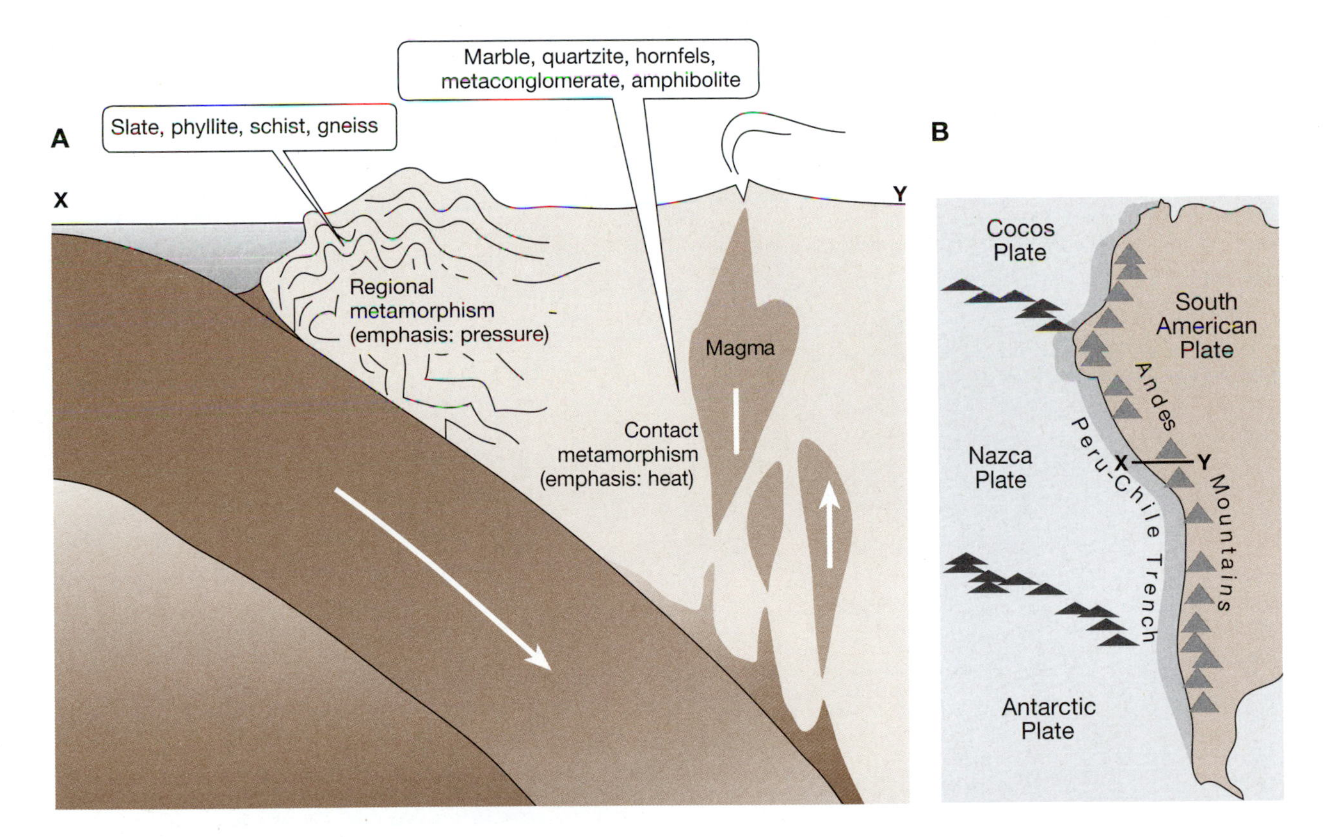

Figure 5.7 A. This is a geologic cross-section drawn along line X-Y on the map in **B**. Here, metamorphism is a consequence of the collision between the Nazca and South American plates. *Contact metamorphism* has been produced by the heat of magma arising from the partial melting of the descending Nazca plate. *Regional metamorphism* is a result of the folding of the Andes Mtns.

A metamorphic landscape as viewed in Google Earth

It is difficult to recognize metamorphic rocks in air-photos because the diagnostic fabrics of slate, schist, marble, et al. are too small in scale. But there is an area in North America where metamorphic rocks are so well-exposed (for lack of vegetation) that the setting of their origin is evident in satellite photographs (Fig. 5.8).

Q5.9 (Refer to Figure 5.8.) (A) What metamorphic setting appears to be reflected here—contact or regional? *Hint:* Does this landscape appear to have been merely 'baked,' or does it appear to have been folded as well? (B) Which types of metamorphic rocks would you expect to find here?

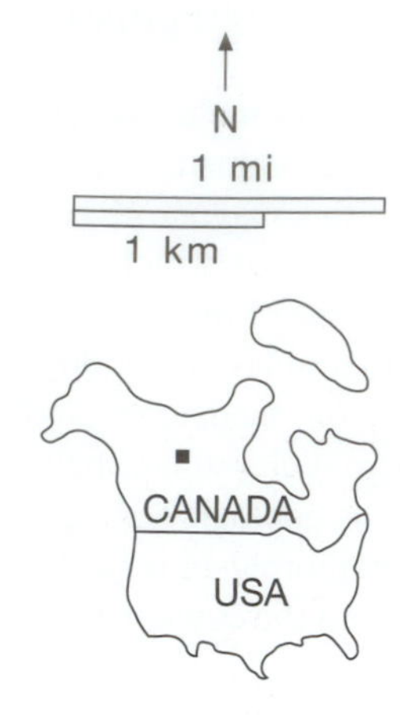

Metamorphic rocks Northwest Territories, Canada

N. 62° 29' 58", W. 114° 04' 28"

Figure 5.8 A metamorphic landscape of Precambrian rocks northeast of Yellowknife, Northwest Territories, Canada is apparent in this Google Earth image of Digital Globe photographs. Vegetation is sparse in this landscape because of sulfuric acid pollution produced by a metal-refining smelters upwind to the west.

E. Metamorphic grade and isograd maps

During metamorphism, the higher the temperature and pressure, the higher the **grade** of metamorphic rock (Fig. 5.9).

Consider the metamorphism of shale. The chemical composition of the metamorphic product (e.g., slate, phyllite, or schist) has a chemical composition essentially like that of the ancestral shale, but chemicals have become *redistributed* so as to produce new minerals at the expense of old ones. The particular minerals that develop reflect specific levels of temperature and pressure. Although the range of differences in temperature and pressure is a continuum, a half-dozen or so grades of metamorphic rocks have been categorized and given the names of diagnostic **index minerals**. These index minerals are shown in the context of temperature, pressure, and depth in Figure 5.10.

Q5.10 According to information in Figure 5.10, (A) which index mineral develops at 500°C temperature and 6 kilobars pressure? (B) At approximately what depth of burial (in kilometers) might these two values occur?

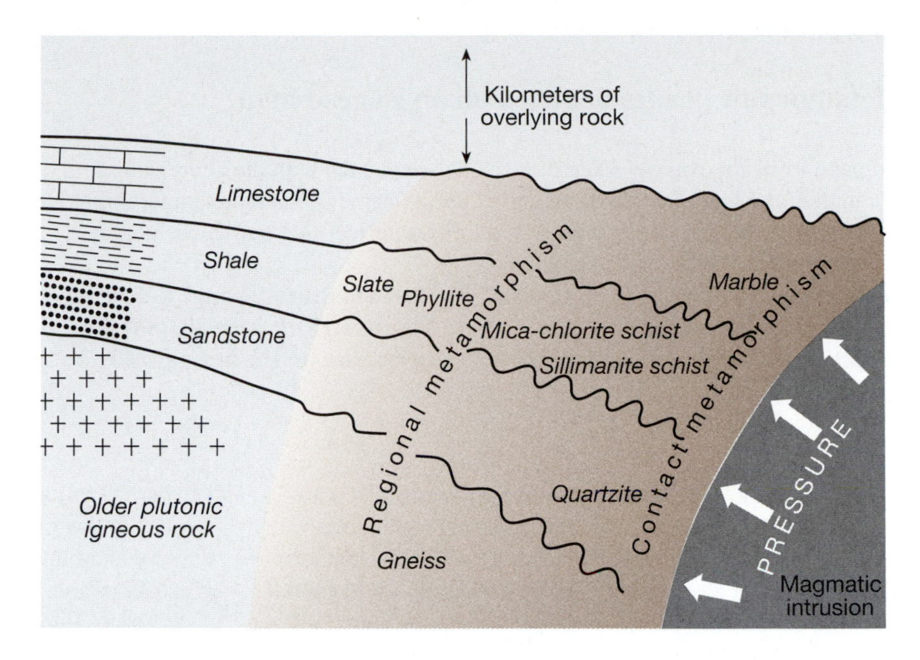

Figure 5.9 The closer to a source of heat, and the deeper within a stress-producing setting, the higher the grade of metamorphic rock.

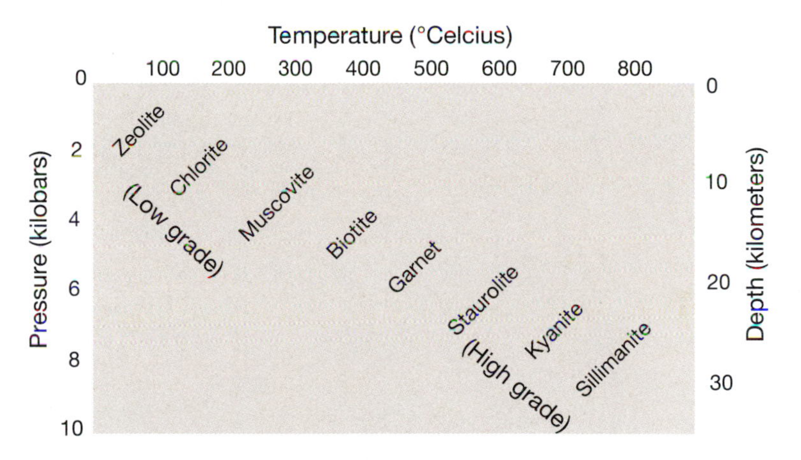

Figure 5.10 Index minerals—guides to metamorphic grades— are plotted here in their relative positions of *temperature*, *pressure*, and *depth*. This chart is generalized in that there is a range of temperature/pressure/depth values under which any particular mineral develops. (*Note:* One kilobar equals approximately 1,000 times the atmospheric pressure at sea level.)

Metamorphic grades depicted on an isograd map

On page 91 of the Answer Page there is a map of the Lake Superior region of Wisconsin and Michigan where Precambrian-age rocks have been metamorphosed (Fig. 5.12). A number of field localities where rocks were collected (for later analysis) have been labeled, each with the abbreviation of the name of the index mineral that characterizes that particular locality.

Q5.11 Do the following: Referring to Figure 5.10 on page 89, plot the temperature (to the nearest 100 degrees Celsius) at each of these localities in Figure 5.12, and then draw lines connecting minerals of equal metamorphic grade. (These lines of equal grade are called isograds.) Finally, draw a circle around the area of maximum grade.

Useful analogy?—A map of metamorphic isograds is a bit like a kind of weather map that you might see in the media. More common weather maps show the distribution of temperatures, along with weather icons, but a **barometric map** shows the distribution of air pressure at ground level (Fig. 5.11). Such maps can alert you to wind conditions, in case you're interested in going sailing or flying a kite. Air moves from higher pressures to lower pressures. And, the steeper the gradient (i.e., the closer the **isobaric lines** are to one another), the greater the wind velocity. So, from a cursory examination of a weather map, one can see where it is breezy and where it is calm.

Q5.12 According to Figure 5.11, where was wind stronger on October 14, 1997—in north-central Canada or in the American mid-continent?

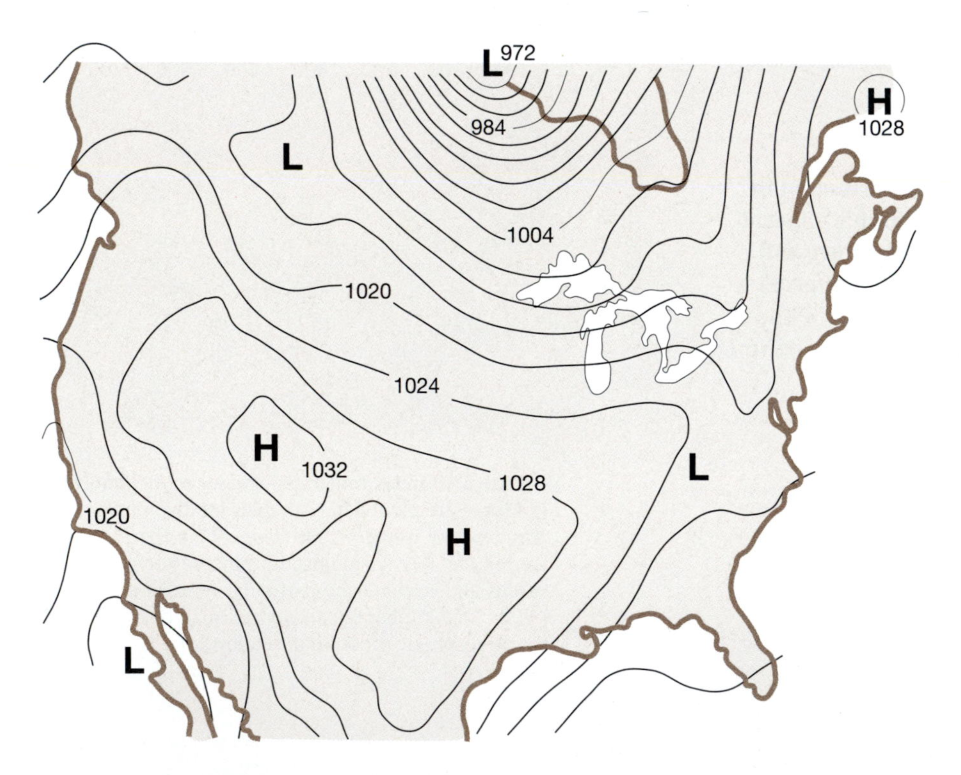

Figure 5.11 This is a simplified barometric map of North America plotted at 7:00 a.m. *EST* on October 14, 1997. The isobar interval is 4 millibars. (One millibar = 1/1,000 of atmospheric pressure at sea level.)

6 Weathering

Topics

A. What is the definition of weathering? What is regolith?

B. What is physical weathering? How does physical weathering facilitate chemical weathering? What are four examples of physical weathering, and what is the definition of each?

C. What are six examples of chemical weathering, and what is the definition of each? What are a couple of health problems associated with clay eaters of the rural South? What ore of an industrial metal is produced by leaching? How can oxidation destroy structures and threaten lives? How does pyrite form within lake or marine sediments? What two problems can be caused by the oxidation of pyrite? How do caves and landscapes in limestone country form? What are the two reasons why caves form at the top of the saturated zone? How is it that a cave appears to migrate from below the water table to above the water table?

D. What is the definition of soil? What are three general factors that determine the character of any soil? How did the distribution of soils, which reflect differences in bedrock on which they develop, influence ethnic emigration of farmers in Fayette County, Texas? What is the geologic setting of champagne in France? How have the dynamics of soil development been assessed in the Glacier Bay National Park, Alaska, area?

A. Weathering defined

Definition—Everyone knows about weathering. It's the fading and peeling of paint. It's the deterioration of masonry buildings and monuments. It's the wear and tear on *all* things exposed to 'the elements.' Geologically speaking, weathering is the physical and chemical alteration of rock and sediments through the effects of air, moisture, heat, cold, and organic matter.

Weathering consists of all those processes that convert rock into **regolith**—the loose, unconsolidated surface material that rests on the bedrock from which it developed (Fig. 6.1). In engineering parlance, regolith is earth material that can be moved with a bulldozer or front-end loader. (If explosives are required, the material is likely to be called *bedrock*.) While this is a practical definition of regolith, it includes unconsolidated *sediments* that are not products of *in situ* weathering, and, so, are not part of the regolith.

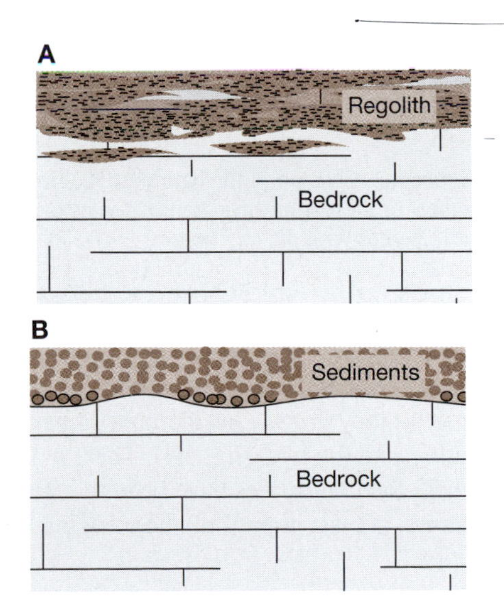

Figure 6.1 A Bedrock overlain by regolith. **B** Bedrock overlain by sediments.

Genetically, regolith is *dynamic* in that it is constantly in a state of observable *change*, whereas bedrock and sediments are, by comparison, relatively stable.

Q6.1 The nature of the boundary with bedrock serves to distinguish regolith from sediments. Describe the difference in the boundaries between (A) bedrock and regolith, and (B) bedrock and sediments, in Figure 6.1.

The bad and the good—A downside of weathering is repainting your home and refurbishing masonry. But there's an upside to weathering, most importantly the production of **soil**, without which Planet Earth would be barren.

There are two broad types of weathering, (a) **physical weathering** and (b) **chemical weathering**.

B. Physical weathering

Definition—*Physical weathering* is simply the breaking of rocks—making little rocks out of big rocks. Because chemical weathering (which we will get to shortly) is a *surface* process, physical weathering facilitates chemical weathering by increasing a rock's surface area-to-volume ratio (Fig. 6.2).

Q6.2 In Figure 6.2, if A is one unit area, how many unit areas are in B? In C? In D?

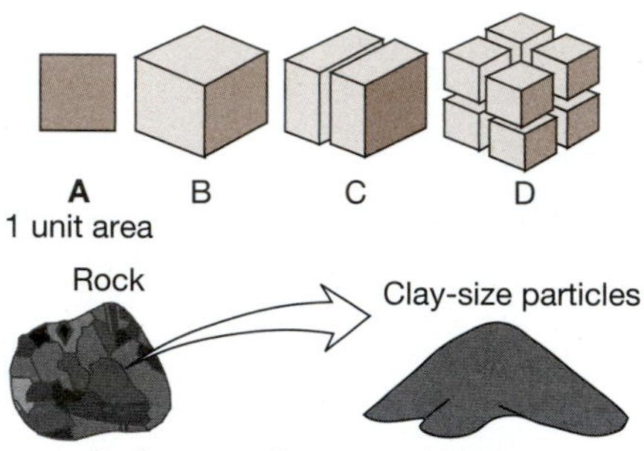

Figure 6.2 Breakdown of rock increases its surface area-to-volume ratio. A rock broken down into clay-size particles increases in surface area approximately 10,000 times.

Salt and sugar are granulated so that they dissolve more easily in foods. (Can you imagine waiting for a cube of halite to do its thing sitting atop a fried egg?) There are several mechanisms in nature that work toward granulating rocks. Four geologically significant mechanisms are:

- exfoliation
- root wedging
- frost wedging
- landslides

You can probably envision the last three of these four mechanisms of physical weathering, but the first item, *exfoliation*, might be new to you.

Exfoliation

Igneous rocks that crystallize from magmas at depths of tens of kilometers do so under immense pressures. In cases where such rocks are later uplifted and stripped (through erosion) of their overlying burden, the release of pressure results in expansion and 'blistering' of the rock into scaly sheets. These sheets are eventually stripped away by erosion as well, thereby sculpting such igneous bodies into domes. This kind of physical weathering is called **exfoliation** (Latin *folio*, leaf). At least one skin-care product is advertised as causing your face to exfoliate, i.e., flake off, revealing the real you. Granite is especially susceptible to exfoliation. Well-known examples of granite domes include Stone Mountain, Georgia and Enchanted Rock, Texas. (Stone Mountain appears in the satellite photograph on page 49.) Swarms of multiple domes occur in Yosemite National Park, California; at Rio de Janeiro, Brazil (Fig. 6.3); and at Pusan, Korea.

Addendum—Chemical weathering that changes silicate minerals into clay minerals is accompanied by swelling, and so perhaps contributes to the exfoliation process.

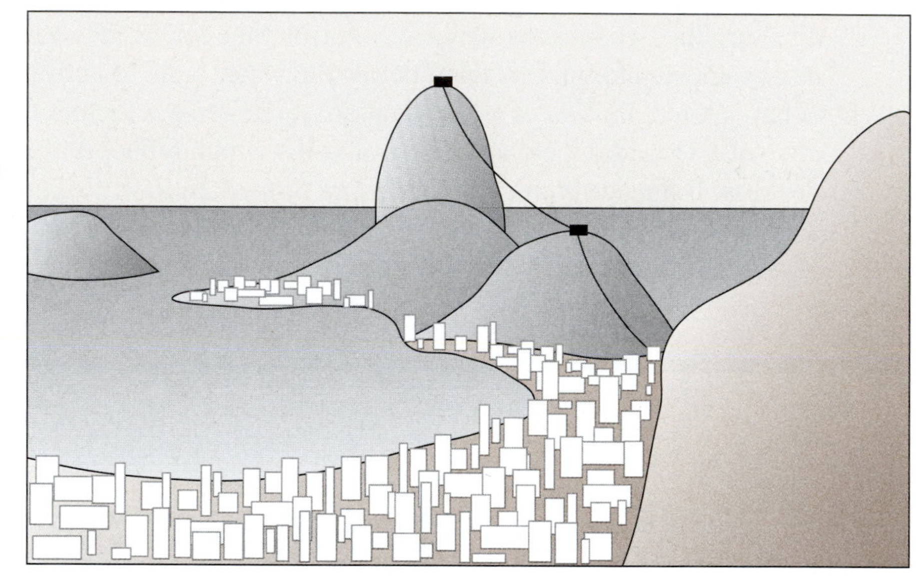

Figure 6.3 A sketch of the harbor at Rio de Janeiro, Brazil from an airplane window. The highest of these imposing granite domes is called Sugarloaf. A cable car connects two of these domes with the city below.

Root wedging

Tree roots can find their way into cracks and crevices within rocks and pry them apart as they grow. Considerable damage can be done to sidewalks, driveways, patios, and the like (Fig. 6.4). There have been cases where fatal landslides have been thought to have been caused, at least in part, by the wedging action of tree roots.

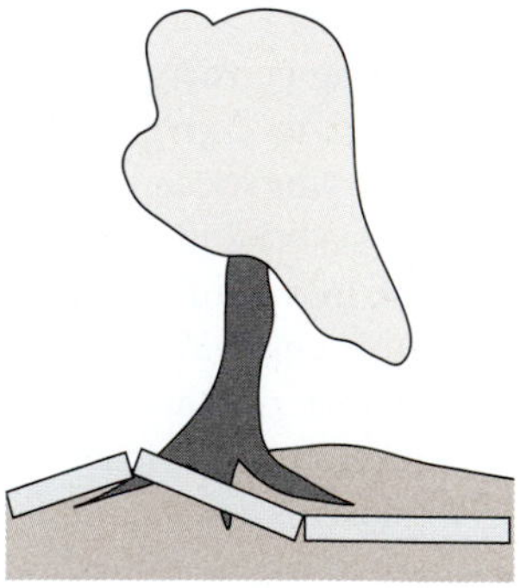

Figure 6.4 Tree roots have the capacity to disrupt sidewalks, driveways, streets, and parking lots as well as breaking rocks.

Frost wedging

H_2O is peculiar in that it is less dense in its solid state (ice) than it is in its liquid state (water). So water expands upon freezing—approximately 9%. If this were not true, our world would be uninhabitable because ocean basins would freeze solid and chill Earth to intolerable levels. That's the good news about water/ice. The bad news is that the expansion of water upon freezing bursts unheated water pipes and automobile radiators low on antifreeze.

The freezing of water is an effective mechanism of physical weathering. Like tree roots, water finds its way into cracks in rocks, then, upon freezing, wedges the rock apart (Fig. 6.5).

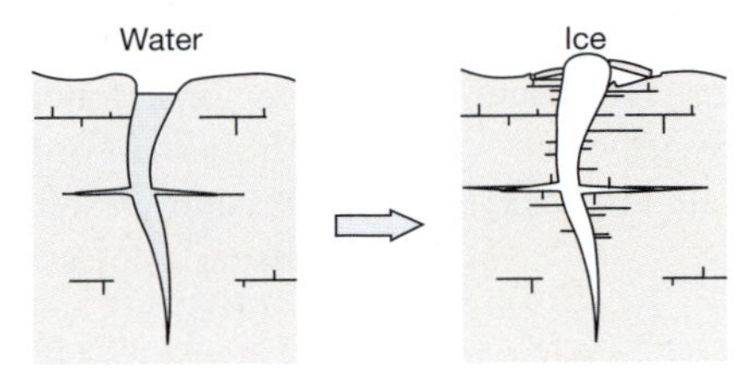

Figure 6.5 Water expands upon freezing with sufficient force to shatter rocks.

The *frequency* of freezing and thawing is important, so frost wedging is most effective where the number of yearly freeze-thaw cycles is high, say, in a climate where water freezes at night and melts during the day.

Q6.3 (A) Where would you expect frost wedging to be most effective during winter months—in coastal Ecuador, in Antarctica, or in Ohio?

The effect of frequency in the freeze-thaw cycle can be apparent from a car window. North of the Tropic of Cancer, shadows invariably occur on the north sides of objects (Fig. 6.6)—which brings to mind a familiar orienteering axiom, 'Moss grows on the north sides of trees.' The reason: The north sides are the shady sides, and so retain more moisture for thirsty mosses. In our Western Hemisphere, the Tropic of Cancer runs through central Mexico, so all of our conterminous 48 states are north of that latitude.

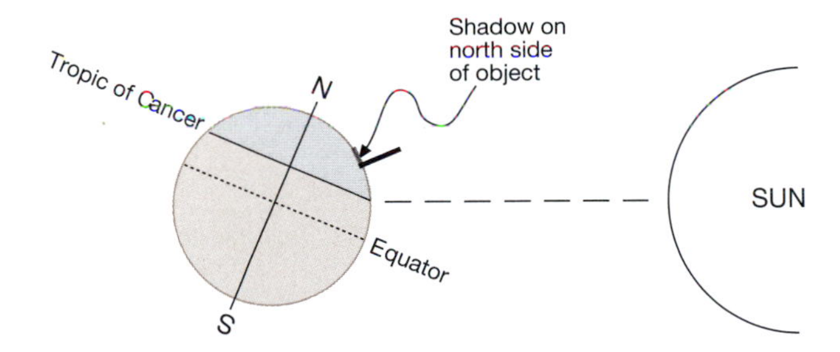

Figure 6.6 The Tropic of Cancer is the latitude where on June 22 the sun is directly overhead (i.e., at its *zenith*). North of that latitude shadows are invariably on the north sides of objects.

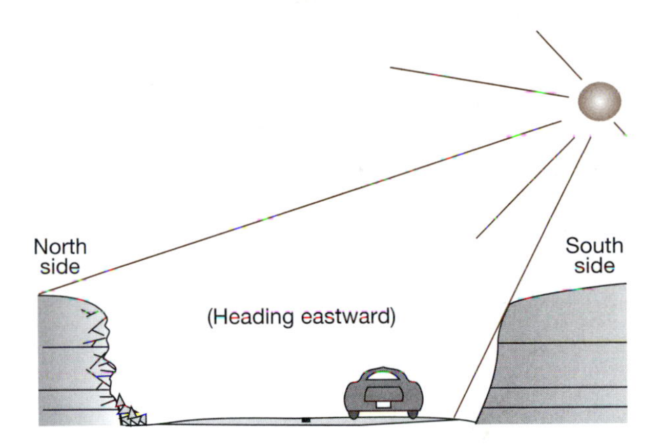

Figure 6.7 Rocks exposed in roadcuts decades old commonly exhibit a greater degree of frost wedging on the *north* sides of *east-west* highways than on the south sides.

Q6.4 Rocks exposed in roadcuts on the north sides of east–west highways are commonly more weathered (rubbly) than those on the south sides (Fig. 6.7). Why is that? *Hint:* What would happen in the situation illustrated in Figure 6.7 when rocks are wet and the daytime high temperature is, say, 30 °F on a cloudless day?

Q6.5 Less obvious is the greater degree of weathering that might be apparent on the east sides of north-south highways (Fig. 6.8). What is the reasonable explanation for that? *Hint:* A clue is imbedded in Figure 6.8.

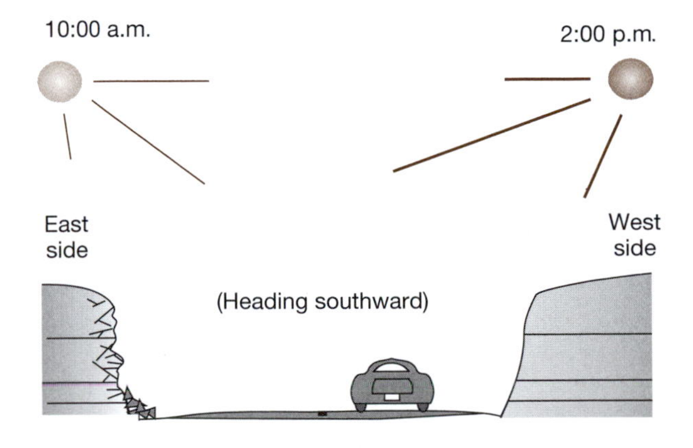

Figure 6.8 Rocks exposed in roadcuts a few decades old commonly exhibit a greater degree of frost wedging on the *east* sides of *north-south* highways.

C. Chemical weathering

Its importance—As stated on the first page of this exercise (page 93), the most important product of weathering in general, and of chemical weathering in particular, is *soil.* Two ounces of average rock are required to provide, through chemical weathering, the necessary phosphorus, iron, and calcium to produce sufficient soil–to produce sufficient grass–to produce the cow–to produce one pound of beef (Fig. 6.9).

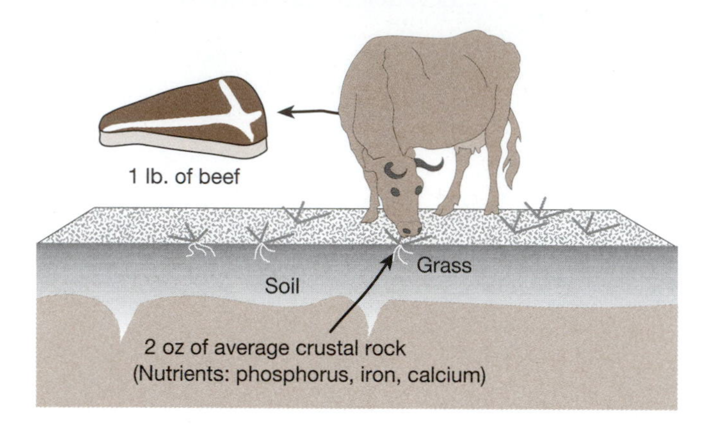

Figure 6.9 Chemical weathering derives the necessary phosphorus, iron, and calcium in one pound of beef steak from two ounces of average rock of Earth's crust.

Definition—Chemical weathering is the chemical alteration of minerals, a broad subject that can be subdivided into the following five processes:

• hydrolysis • leaching • oxidation • reduction • dissolution

Hydrolysis

Hydrolysis (the *splitting of water* into H^+ and OH^-) occurs when water reacts with a silicate mineral to form both soluble products and relatively insoluble products. The soluble products include the following family of metal ions:

potassium (K^+)
sodium (Na^+)
calcium (Ca^{++})
magnesium (Mg^{++})
iron ($Fe^{++, +++}$)

...plus the relatively insoluble family of clays ($AlSiO_n$), which are negatively charged.

The hydrolysis reaction can be written in generalized form as follows, with **Me** indicating unspecified metal ions:

Me = unspecified metal ion (K, Na, Ca, Mg, Fe)

$$\mathbf{Me}AlSiO + H_2O \longrightarrow \mathbf{Me} + OH + HAlSiO_n\overset{?}{\mathbf{Me}}$$

(Silicate) (Water) (Dissolved ions) (Clay, with or without **Me**)

About the hydrolysis of orthoclase feldspar—Notice in the reaction at the bottom of the left column that unspecified metal might or might not remain in residual clay. This depends on the thoroughness of hydrolysis. In regions of moderate rainfall (e.g., the U.S. Midwest), hydrolysis of orthoclase feldspar fails to put all of the potassium in solution, so residual clay contains some of the feldspar's potassium (Fig. 6.10). Call this 'common clay.'

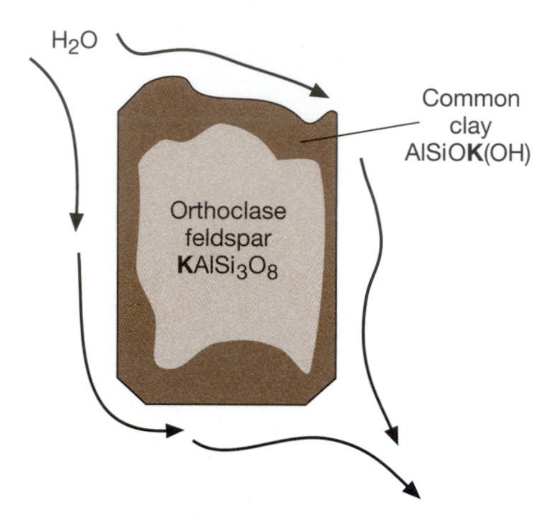

Figure 6.10 Generalized hydrolysis of an orthoclase feldspar crystal in a region of moderate rainfall. Residual common clay retains some of the potassium in the orthoclase.

In contrast, in regions of high rainfall (e.g., southeastern United States) hydrolysis of orthoclase is more thorough, and a variety of naturally refined clay (**kaolinite**) is produced (Fig. 6.11).

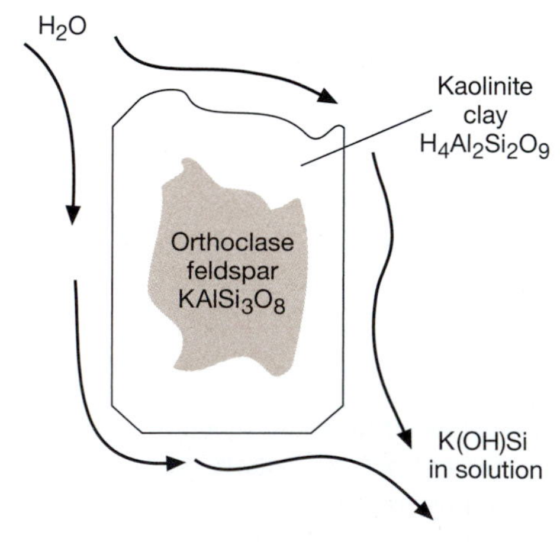

Figure 6.11 Hydrolysis of an orthoclase feldspar crystal in a region of high rainfall converts the feldspar to kaolinite clay.

Q6.6 What is the singular difference (apparent in Figures 6.10 and 6.11) in the chemical composition of common clay and kaolinite?

Clay eaters of the rural South
Georgia kaolinite has been produced for generations for uses in innumerable industries—including those of adhesives, ceramics, glass, rubber, paint, paper, ink, pharmaceuticals, plastics, rubber, and whitewash. The most recent story about kaolinite is from the pesticide industry. Mixed with liquid pesticide and sprayed on crops, kaolinite serves as a matrix for more even distribution and for more holding power.

But clearly the most curious use of kaolinite has been that by 'clay eaters of the rural South.' The practice known as **geophagia** (earth eating) goes back at least as far as Plato, who wrote of Greek women who ate clay. But don't snicker (no pun intended). Kaolinite is used in chocolate candy to minimize melting and in a variety of over-the-counter remedies for stomach disorders, including patented Kaopectate®, Di-Gel®, Rolaids®, Mylanta®, and Maalox®. It's not surprising that one of the common complaints of clay eaters is occasional constipation.

The most common clinical problem with clay eaters is anemia. Kaolinite is so impoverished in metal ions that it plucks what it can from one's blood. Also, it can coat the intestines, thereby interfering with the absorption of essential nutrients.

Q6.7 Judging from the list under Hydrolysis on facing page 96, what important metal ion is likely to be plucked from the bloodstream by kaolinite? *Hint:* This is a common 'vitamin supplement' claimed by manufacturers to restore the health of __?__-poor blood.

ARKANSAS GAZETTE-DEMOCRAT, NOVEMBER 1, 2000

Mineral coating could cut chemical use

Clay-based substance protects crops from insects, diseases, heat.

KEARNEYSVILLE, W. VA.—Orchards, vineyards, and even row crops someday could take on a ghostly appearance that would help protect from insects, diseases, sunburn, and heat stress.

Agriculture Research Service scientists have covered fruit trees, vines and vegetable crops with a white, reflective coating of a specially processed type of clay called kaolin that protects the plants at the Appalachian Fruit Research Station in Kearneysville, according to the U.S. Department of Agriculture.

About the hydrolysis of plagioclase feldspar—Plagioclase includes sodium and calcium in place of potassium (Fig. 6.12).

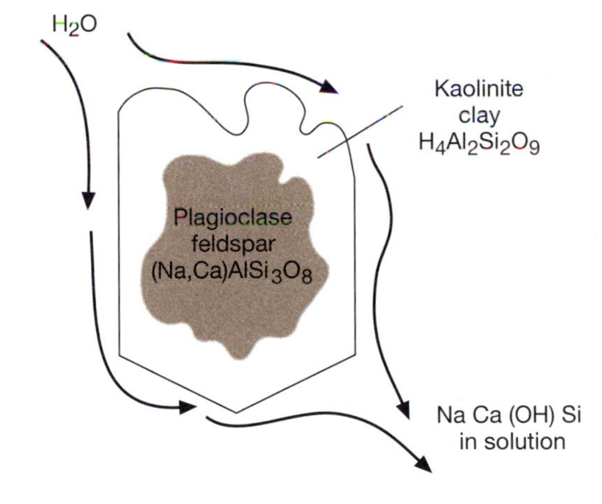

Figure 6.12 Sodium and calcium are dissolved from a plagioclase feldspar crystal to form kaolinite.

Q6.8 Given a granite (with its mix of plagioclase and orthoclase) that is subjected to moderate-to-high rainfall, the plagioclase is more likely to form kaolinite than is the orthoclase. So which metal ion(s) appear to be the more soluble—sodium and calcium, or potassium?

An interesting site for observations on clay eaters is at http://www.newhouse.com/archive/story1c012502.html

Leaching

Leaching—another common variety of chemical weathering—is the removal, by water, of soluble elements and compounds from bedrock and regolith. These dissolved elements, which occur in all natural waters, make the water 'hard,' and can impart an unpleasant taste.

In areas of high rainfall and warm climate, leaching removes the more soluble elements and concentrates the less soluble ones as thick red soils called **laterites**. Some laterites contain valuable minerals. For example, in some tropical areas, *goethite* (FeO·OH) is mined for iron.

Clearly the most important economic mineral in lateritic soils is gibbsite [$Al(OH)_3$], the definitive mineral in the rock **bauxite** (pictured on page 41)—the ore of aluminum. Today we obtain most of our bauxite from tropical regions of South America, but during World War II, when shipping lanes in the Gulf of Mexico were threatened by German submarines, Arkansas bauxite, which formed as a soil some 50 million years ago, was mined and smelted.

Q6.9 How does bauxite differ in composition from kaolinite? *Hint:* What element present in kaolinite has been leached in the development of bauxite?

Mother Nature's problem in making bauxite is getting silicon out of the silicate mineral. Other metal ions are soluble under *acid* conditions, but silicon is soluble under *alkaline* conditions. So it requires unusually wet and warm conditions to leach silicon. The task in central Arkansas was made easier by the fact that the bedrock on which the bauxite developed consists of **syenite**—a silica-poor relative of granite. Not only is there little or no quartz in this hybrid igneous rock, but the feldspars are low in silicon as well.

Oxidation

The good and the bad—Oxidation and hydration of iron contribute to colorful landscapes, but these same processes can destroy structures and threaten lives (Fig. 6.13).

Oxidation of iron to form **hematite** (aka rust) is…

$$4Fe + 3O_2 \longrightarrow 2Fe_2O_3 \text{ (red hematite)}$$

Hematite occurs in intense hues and tints of red. It has been the paint pigment of choice for Native Americans and others around the world.

When water combines with hematite, **limonite** forms, adding hues of yellow to Nature's palette. That reaction is…

$$Fe_2O_3 + H_2O \longrightarrow Fe_2O_3 \cdot H_2O \text{ (yellow limonite)}$$

Q6.10 Rocks in fresh roadcuts are commonly gray to brown in color, but after a few decades reds and yellows begin to appear. Why? *Hint:* What elements in the atmosphere and surface water are at work here?

Q6.11 These reds and yellows are especially apparent along fractures within rocks. Any idea why? *Hint:* Consider fractures as pathways for an important substance.

USA TODAY, MAY 22, 2000

Investigators detect corrosion in collapsed bridge

Walkway packed before collapse

CHARLOTTE, N.C.—State officials said Sunday that they found signs of corrosion in the steel bands in a pedestrian walkway that snapped as hundreds of people were leaving a stock-car race Saturday night. There were no fatalities, but 170 people remained in critical condition Sunday.

Figure 6.13 Corroded (oxidized) cables were blamed by some for contributing to the collapse of a pedestrian walkway at a North Carolina race car track.

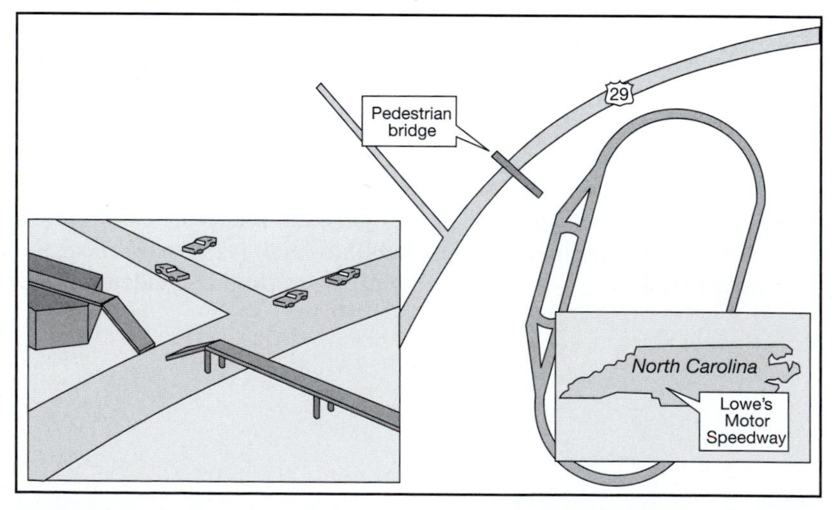

Similar claims of metal corrosion were made in the case of the I-35 bridge collapse in Minneapolis, Minnesota in 2007. (For numerous videos, Google: *bridge collapse minnesota*)

Reduction

Reduction of iron—Decaying organic matter requires oxygen for its slow 'burn.' If there is insufficient free oxygen (within water and/or air), decaying organic matter can pluck oxygen from *hematite* (Fig. 6.14), thereby **reducing** the iron oxide and turning it gray.

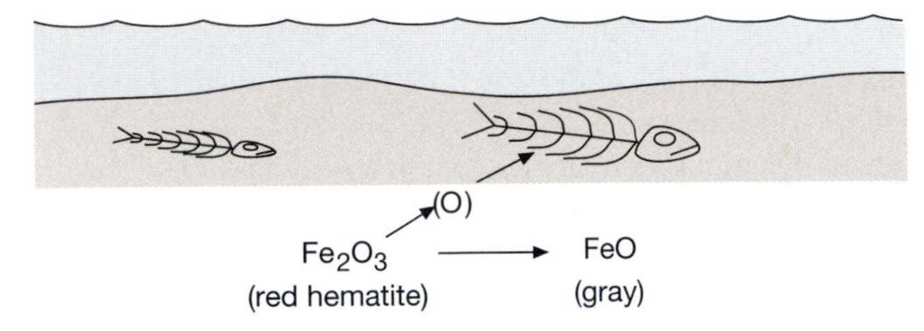

Figure 6.14 Within lake and marine sediments, decaying organic matter can acquire essential oxygen through the reduction of iron oxide.

Q6.12 Why do you suppose sedimentary rocks crowded with fossils are seldom red in color? *Hint:* What is the postmortem (L. *post mortem*, after death) history of a fossil?

Reduction of sulfate—Decaying organic matter can also obtain essential oxygen by reducing *sulfate* ions, thereby liberating iron, which then combines with sulfur to form pyrite (Fig. 6.15).

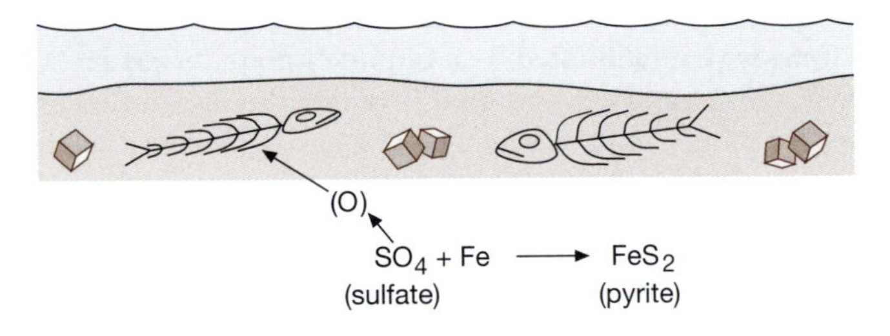

Figure 6.15 Within environments like that in Figure 6.14, decaying organic matter can also reduce sulfate in its search for essential oxygen.

Q6.13 Coal deposits are commonly rich in pyrite. Why is that? *Hint:* How does coal form? (This topic is covered in the exercise *Metamorphic Rocks* on page 86.)

Oxidation of pyrite—sulfuric acid problems

Building discoloration—When pyrite within sedimentary rocks is exposed to the atmosphere, it oxidizes to sulfuric acid and hematite (Fig. 6.16).

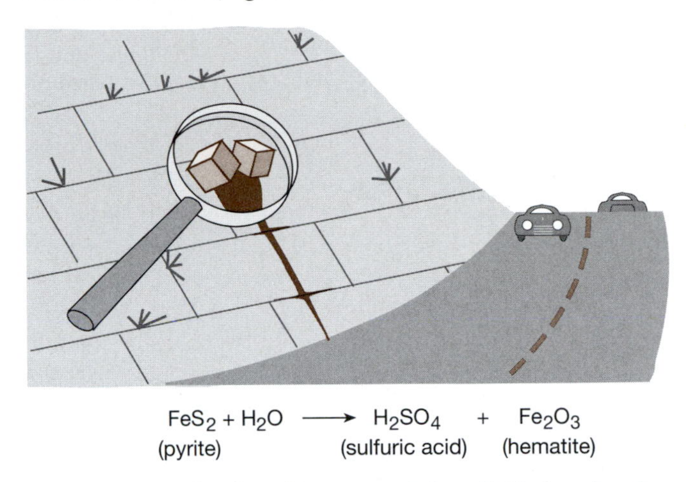

Figure 6.16 Geochemistry from a car window. Oxidation of pyrite within this limestone has left a tell-tale streak of red hematite. The sulfuric acid has either evaporated or gone into the nearest creek.

If pyrite is overlooked in the choosing of building stones (and of gravel for concrete), unsightly hematitic stain can result (Fig. 6.17). Epoxy sealer applied to the pyrite, along with acid cleaning of the rock, can mitigate the problem.

Figure 6.17 This decorative limestone has become stained with hematite that resulted from oxidation of a pod of pyrite.

Acid mine waters—Some ponds within abandoned coal strip-mines exhibit exceedingly high amounts of sulfuric acid, owing to the oxidation of pyrite (Fig. 6.18). The pH of one sample of water oozing from such a pond was 2.3.

Q6.14 Guess which item has a pH of precisely 2.3: tomatoes, lemon juice, baking soda, or beer? The answer is on the following page. Don't peek until after you have tried.

Figure 6.18 The pH of water oozing from coal strip-mine ponds can kill fish when it accidentally finds its way into streams.

Did you guess correctly the answer to Q6.14 at the end of preceding page 99?

pH value

0 hydrochloric acid (HCl)
battery acid

1

2 stomach acid (1.0–3.0)
lemon juice (2.3)

3 vinegar, wine, soft drinks, beer
orange juice, some acid rain

4 tomatoes, grapes

banana (4.6)
5 black coffee, shaving lotions
bread
normal rainwater
6 urine (5–7)
milk (6–6)
saliva (6.2–7.4)
7
pure water
blood (7.3–7.5)
8 egg white (8.0)
seawater (7.8–8.3)

9 baking soda
phosphate detergents
Clorox, Tums
10
soap solutions
milk of magnesia
11 household ammonia (10.5–11.9)
nonphosphate detergent

12 washing soda (Na_2CO_3)

hair remover
13
oven cleaner

14 sodium hydroxide (NaOH)

Dissolution (and cave development)

The biggest natural history story about dissolution is the dissolving of limestones to form caves. The very landscapes associated with caves have been carved in large part through the process of dissolution. But first, a qualifier: Gypsum (as in Almería Province, Spain) and salt (as along Dead Sea shores) are even more soluble than limestones, but gypsum and salt comprise a small fraction of land areas. Usually, when one thinks solution landscape, 'limestone country' comes to mind.

Our subject, dissolution, begins with carbon dioxide (CO_2), which occurs in two vast reservoirs—atmosphere and soils.
Supplying carbon dioxide to the atmosphere are (as you probably know):

- forest fires
- volcanoes
- decay of organic matter
- burning of organic fuels (coal and petroleum)
- respiration by animals

Supplying carbon dioxide to a soil reservoir is (as you might know):

- respiration by plant roots and soil fauna

Carbon dioxide combines with atmospheric moisture to form carbonic acid (H_2CO_3)—the same compound that imparts the tangy character of soda pop.

About respiration by plants—Surely you know about *photosynthesis*—that remarkable process whereby plants synthesize inorganic compounds into organic compounds in the presence of sunlight. But do you know about *respiration* by plants? Yes, respiration. It's true that plants are known for (a) taking in carbon dioxide, (b) making carbohydrates, and (c) expelling oxygen; whereas animals are known for (a) taking in oxygen, (b) burning carbohydrates, and (c) exhaling carbon dioxide. This is the grand *symbiosis* that keeps the biosphere perking. But plants also respire. In fact, plant roots only respire, thereby contributing to the soil CO_2 reservoir. Root-produced carbon dioxide combines with soil water to add to the carbonic acid in rainwater (Fig. 6.19).

Q6.15 Dissolution of limestone is facilitated by a humid climate. What are the two things associated with humid climates that promote the dissolution of limestone? *Hint:* These two things, both of which can be easily observed from a car window, are illustrated in Figure 6.19.

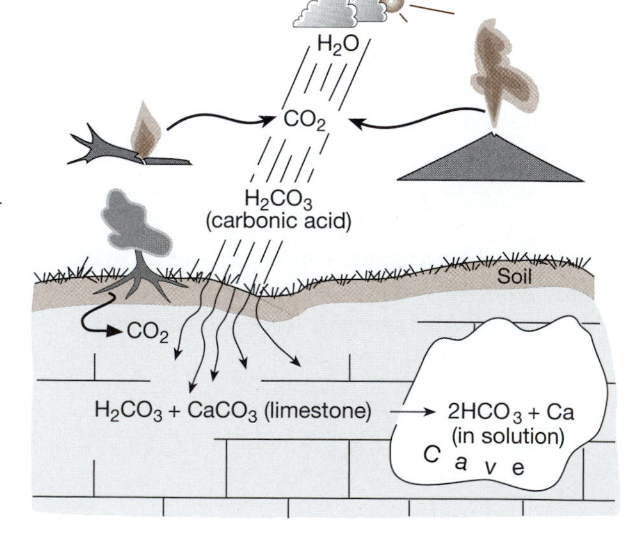

Figure 6.19 Carbonic acid reacts with limestone (which consists of the mineral calcite) to produce bicarbonate ions and calcium in solution—forming caves and shaping landscapes. (The soil on a dissolution landscape consists of a residue of insoluble material within the limestone; e.g., clay and other insolubles.)

Cave evolution

Where caves form—Caves commonly develop at the top of the saturated zone, just beneath the water table (Fig. 6.20). There are two reasons for this. The first reason has to do with the distribution of carbonic acid.

Q6.16 (A) Where within the saturated zone do you suppose the concentration of carbonic acid is greater—shallow or deep? (B) Why do you say that? (C) Which of the two 'histograms' in Figure 6.20 appears to represent the distribution of carbonic acid—the gray or the brown?

Before describing the second reason for why caves develop just beneath the water table, imagine adding a tablespoon of table salt (NaCl) to a glass of water, stirring the water vigorously to thoroughly dissolve the salt, and then letting this salt water stand overnight.

Q6.17 (A) On the following morning, what would you imagine to be the distribution of the sodium and chlorine ions within the water? *Hint:* Dissolved ions have mass, just like solids. (B) Where within the water in the glass should it be easier to dissolve additional salt, shallow or deep? (C) Which of the two histograms in Figure 6.20 do you suppose represents the distribution of dissolved ions—the gray or the brown?

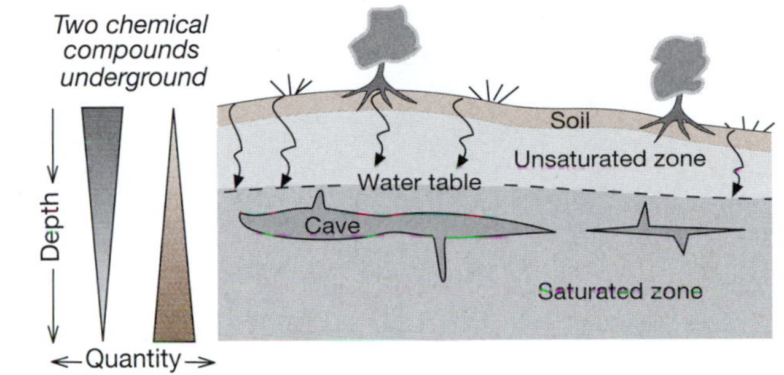

Figure 6.20 Two variables, plotted as histograms, account for the fact that caves most commonly develop at, or near, the top of the saturated zone, i.e., just beneath the water table.

Caves empty, caves fill—Dissolution that *produces* caves occurs within the saturated zone, whereas the *filling* of caves—through the precipitation of cave deposits (aka *speleothems*)—occurs within the unsaturated zone (Fig. 6.21).

Q6.18 How is it that a cave is born within the saturated zone, but then somehow finds its way into the unsaturated zone? *Hint:* The answer is evident in Figure 6.21.

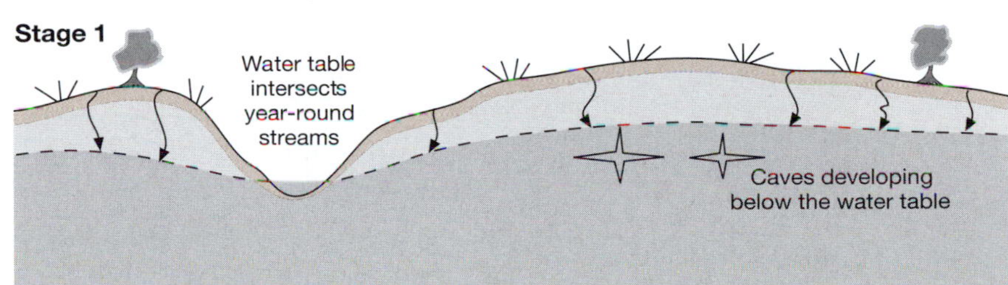

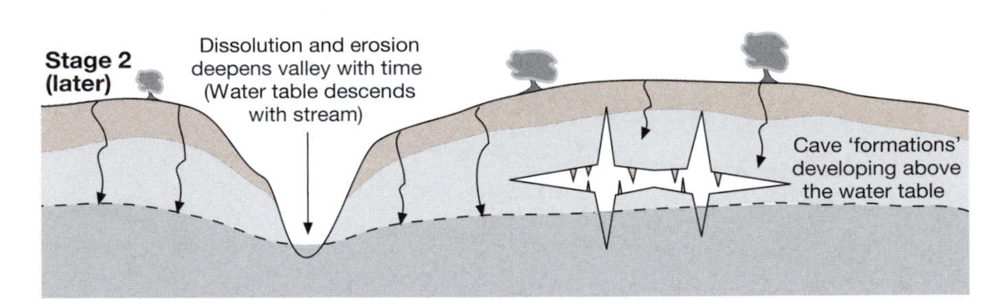

Figure 6.21 Two stages showing the evolution of a cave, from its incipient development to its being occluded with cave deposits.

Halite crystallizes from evaporating salt water. So is this the way in which calcite cave deposits develop (i.e., through the evaporation of cave water)? Probably not. Humidity within most caves approaches 100%, so evaporation is insignificant. Instead, a chemical reaction is believed to be involved.

On the facing page we saw that the *dissolution* of limestone (calcite) begins with the *adding* of CO_2 to water to produce carbonic acid. So what might happen if CO_2 were *removed* from calcite-rich carbonic acid—like bubbles escaping from soda pop? The escape of CO_2 drives the following reaction to the right, resulting in the **precipitation** of calcium carbonate (calcite):

$$2HCO_3 + Ca \quad CO_2 + H_2O + CaCO_3$$

Q6.19 Agitation (as in shaking) liberates CO_2 from soda pop. So what one or two actions of cave water do you think might be accompanied by agitation of the water? *Hint:* We're looking for a simple cause of turbulence…as in the case of rain water.

D. Soils and human geography

Soil defined—There is no topic that universally impacts humans more than the subject of soil. Not only do soils act as storehouses for water and nutrients essential for plants—which provide food—but soils also play a role in maintaining air quality and filtering contaminants from groundwater.

Q6.20 Name two uses of plants in addition to food.
***Hint:* You can probably see both from your desk.**

Soil and regolith—As pointed out on page 93, the term *regolith* conveys the connotation of *process*, specifically the progressive breakdown of bedrock through various weathering processes. Soil is the most dynamic part of the regolith, commonly exhibiting several *horizons* (Fig. 6.22) in various stages of development. Soil scientists define soil as *layers of weathered, unconsolidated material that contains organic matter sufficient for supporting plant life*, whereas others simply define soil as *that part of the regolith that supports rooted plants*. The limits and character of a regolith are important to an engineer preparing a construction site, whereas the limits and character of a soil are important to a farmer.

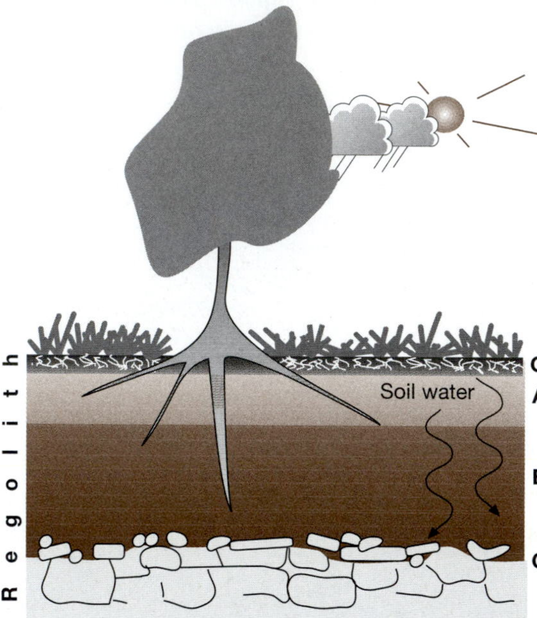

Figure 6.22 A humid region typically supports soil with four horizons (O, A, B, and C). Boundaries separating horizons are transitional in nature.

Soil horizons

O Organic material (topsoil)

A Leaching by downward percolating water

B Accumulation of iron oxides, aluminum oxides, calcium carbonates

C Fragments in part physically—and in part chemically—weathered from bedrock

COLUMBIA DAILY TRIBUNE, APRIL 20, 2002

Soil mapping project complete

5,000 types in state, 100-year study finds

JEFFERSON CITY, MO (AP)—After more than a century of digging into the matter, researchers have identified and mapped all 44.6 million acres of Missouri's soil.

The key finding: Missouri has about 5,000 different types of soil. Soil maps are used by contractors because some soils are more likely to crack concrete foundations. Farmers also use the maps to plan crops. Under state law, soil types also are used in the assessment of agricultural property.

Soil surveys—Soil scientists recently reported that there are 5,000 different types of soil in the state of Missouri. (This gives you some idea of how seriously soil scientists view their work.) Even reducing that number down to the *12 major soil orders* involves more detail than an elementary geology course warrants. So we will look at only a few of those orders.

USDA, Natural Resources Conservation Service/Soil Survey Division is at www.nssc.nrcs.usda.gov.

Four factors in soil development

Judging from the fact that there are 5,000 varieties of soils in Missouri, there must be innumerable factors that determine the character of any one soil. Dominant factors include:

(1) Bedrock—What you begin with influences what you end up with.

Q6.21 Recall the importance of a certain bedrock (described in the second column under Leaching on page 98) that was important in the development of the only bauxite ore in the United States. Name that rock.

(2) Climate—Differs with elevation, latitude, and the proximity of seacoasts. Climate explains differences among the three soil profiles in Figure 6.23.

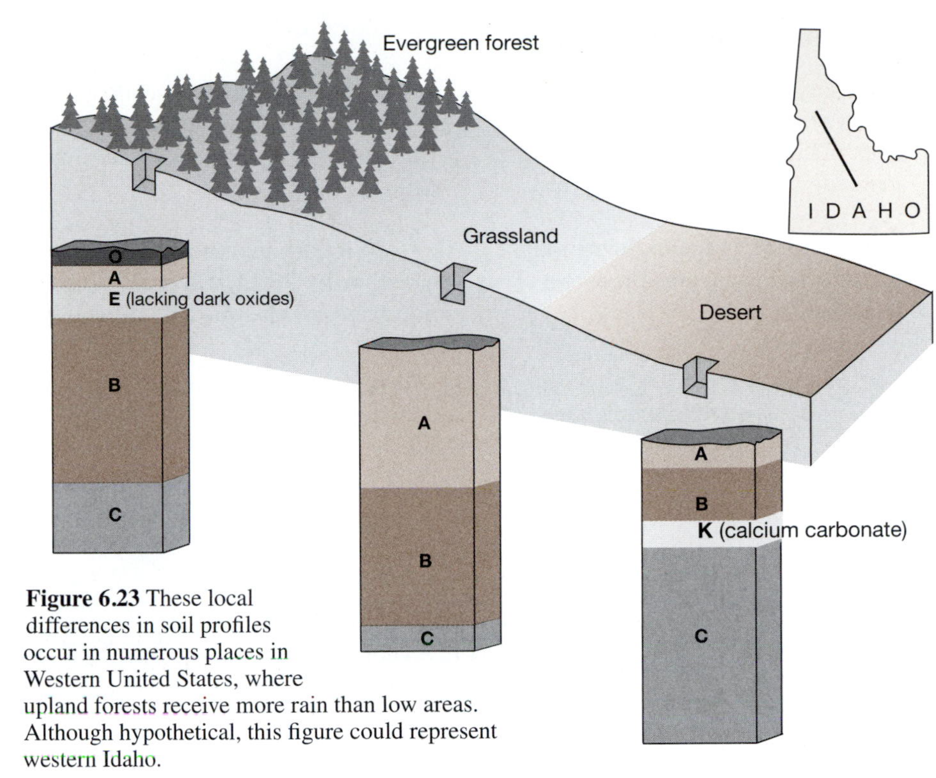

Figure 6.23 These local differences in soil profiles occur in numerous places in Western United States, where upland forests receive more rain than low areas. Although hypothetical, this figure could represent western Idaho.

On a larger scale, climate explains the difference in soils of Eastern United States as compared with those of western states (Fig. 6.24). Because of high rainfall, eastern states are characterized by **high-aluminum** soils. (Again, recall the leached laterite soils described on page 98.) In contrast, western states are characterized by **high-calcite** (i.e., calcium carbonate) soils.

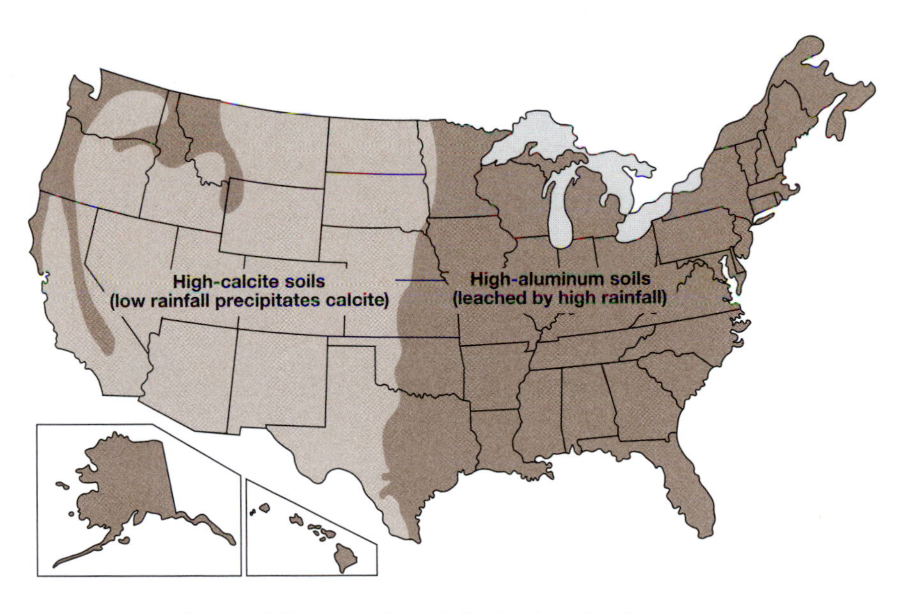

Figure 6.24 This continental distribution of soil types reflects a difference in rainfall. Eastern states are relatively humid, whereas western states are relatively dry.

Q6.22 Ref. Figure 6.24. Why the patchy occurrences of high rainfall in the northwest? *Hint:* Look again at the caption to Figure 6.23.

(3) Slope—Steep slopes lose soil through erosion.

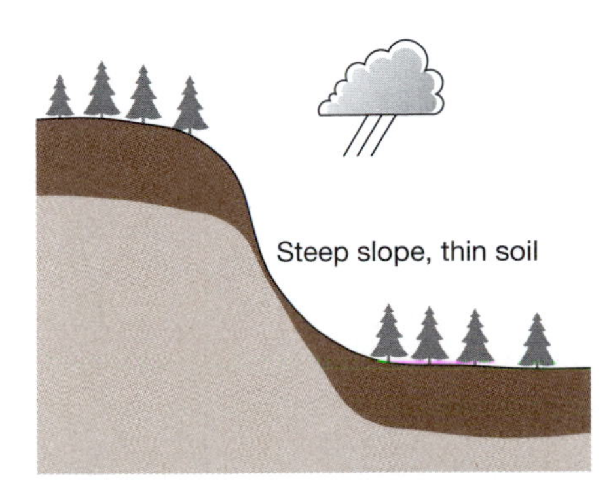

(4) Time—Soil development is a process. Given a particular set of conditions, the longer the time, the thicker the soil.

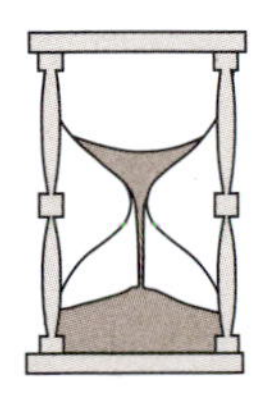

Soils and human geography

A cursory examination of the general highway map of Fayette County, Texas, (Fig. 6.25) reveals two areas that differ markedly in the abundance of roads.

Q6.23 Identify the highway map feature that separates these two areas from each other.

FYI—As you might surmise, the greater abundance of roads correlates with a greater degree of agricultural development.

Q6.24 Relative to the map feature asked for in Q6.23, where, in broad map terms, is the more fertile soil in Fayette County (NW, NE, SW, SE of that map feature)?

In the area of more fertile soil, examine the names of towns and the name of the church near Dubina.

Q6.25 (A) What ethnic group of immigrant farmers appears to have arrived first in Fayette County? (B) What ethnic group arrived second?

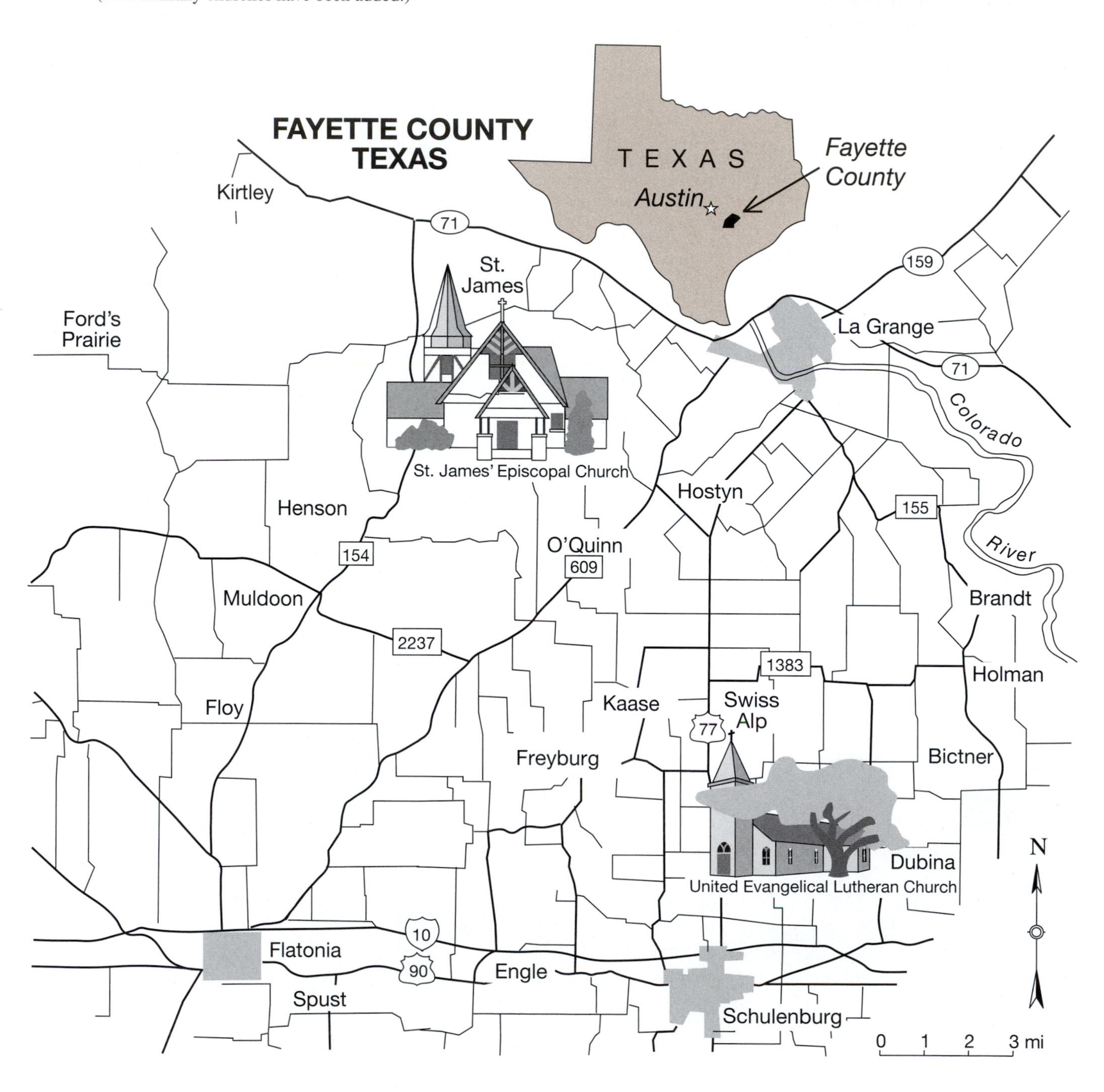

Figure 6.25 The general highway map of Fayette County, Texas, reveals a story of ethnic immigration. (Two of many churches have been added.)

Bedrock of Fayette County

(Ref: Figure 6.26.) The fertile Oakville Formation is characterized by thick soil that ranges from neutral to moderately alkaline. Water-holding capacity is high. In contrast, the Catahoula Formation exhibits moderately thick to thin soil developed on weakly to strongly cemented volcanic ash. It ranges from moderately acid to neutral. Water-holding capacity is low to moderate.

This history of immigrations in the Fayette County, Texas region might be covered in any number of courses in addition to geology (e.g., soil science, human geography, rural sociology, anthropology).

Geology humor? A University of Texas geology student once quipped, "You can map the Oakville Formation on the distribution of Lutheran churches."

The '**Painted Churches of Texas'** do in fact provide a must-see attraction.

Google: *Painted Churches of Texas*

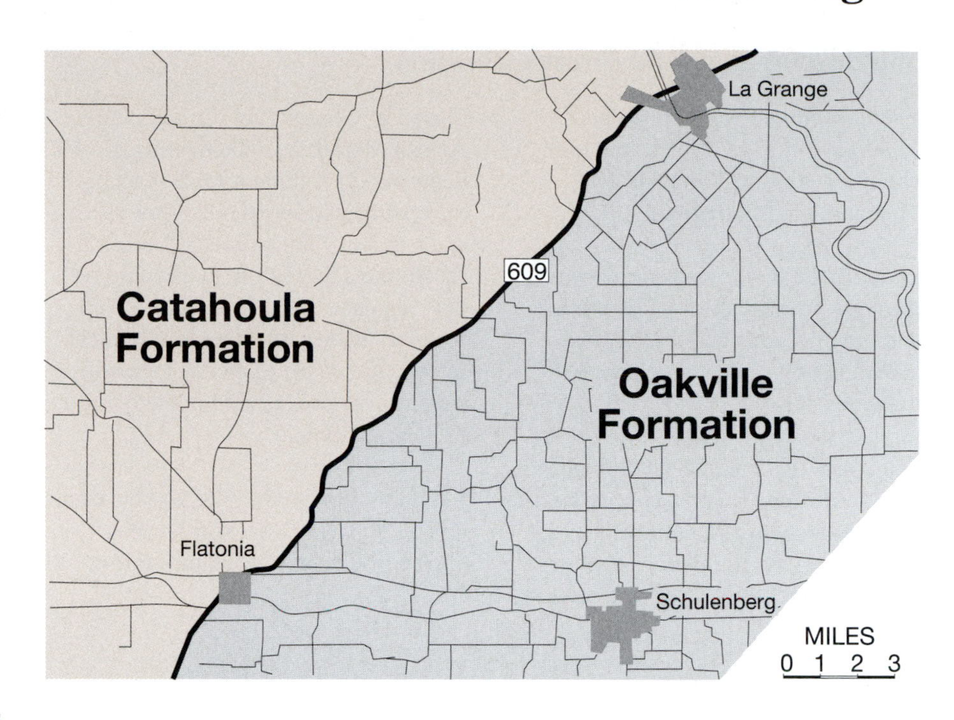

Figure 6.26 The bedrock geology of Fayette County, Texas, consists of the Catahoula Formation blanketed in part by the Oakville Formation. Both tilt gently into the Gulf of Mexico Basin, so on this geology map we view their up-turned edges.

The geology of champagne

In the famous Champagne region of France (Fig. 6.27) a particular bedrock nutrient is a welcomed supplement.

A part of French Law known as *Appellation contrôlée* requires wine distributors to be faithful to the region of origin when labeling products. So one can be sure that French champagne comes from the northeast margin of the Paris Basin, where Cretaceous chalks are rimmed by bluffs of Tertiary clay and lignite coal. Grapevines prefer the superior drainage of the Cretaceous chalk, while at the same time benefiting from a particular nutrient washed down from Tertiary lignite above.

Q6.26 Judging from the description of lignite in Figure 5-5 on page 86, what could this elemental nutrient possibly be? *Hint:* It is a basic building block in organic compounds.

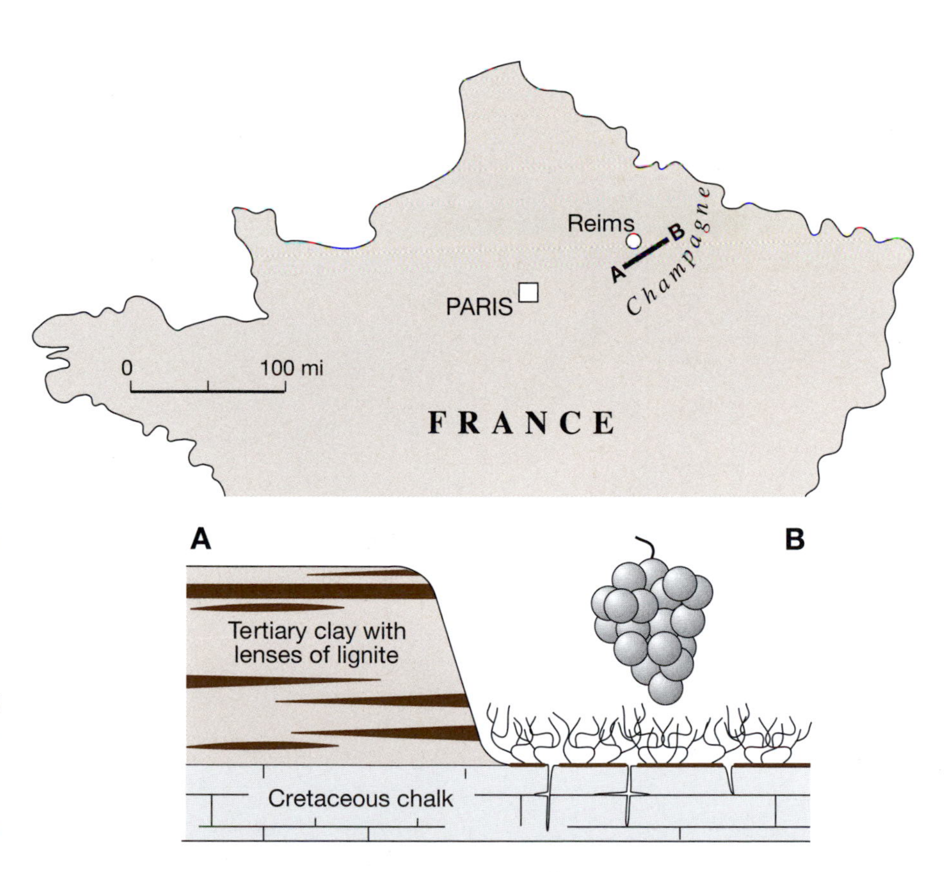

Figure 6.27 It is said that champagne grows with its feet in the Cretaceous and its head in the Tertiary. Vineyards benefit from good drainage in the Cretaceous chalk, plus an essential nutrient washed down from Tertiary lignite.

Soil development and forest succession

There is another variable in soil development in addition to the four variables examined on page 103. That variable is **substrate**.

Plants make their own bed as it were through the production of carbonic and organic acids and through the accumulation of decaying organic matter.

Dynamics of substrate development can be illustrated with a succession of cores collected from within soils developing on the heels of a retreating glacier in Glacier Bay National Park, Alaska (Fig. 6.28). The melting glacier deposited its sediment load, forming a carpet of glacial *till* (i.e., clay, silt, sand, gravel). The porous till has acted much like a regolith in facilitating soil development. So, despite the cold climate of Glacier Bay, the development of soil, and of the plants that the soil supports, has been rapid in the context of geologic time:

- Within a few years—'A' soil horizon.
- After 50 years—'B' soil horizon.
- After 100 years—A mature forest.

Q6.27 After 30 years of development, (A) which horizon was the thickest? (B) Which horizon was yet to develop?

Q6.28 Judging from their relative thicknesses, which horizon appears to be the most essential for the development of the mature spruce-hemlock forest—O, A, or B?

Figure 6.28 This is a cross-section based on a series of four cores collected from soil profiles near Glacier Bay NP, Alaska. This is a snapshot of the development of soil during the past 250 years. (Look back at Figure 6.22 on page 126 for descriptions of soil horizons if needed.)

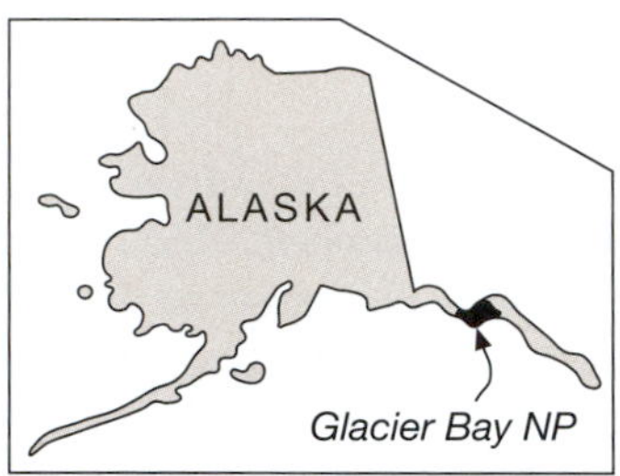

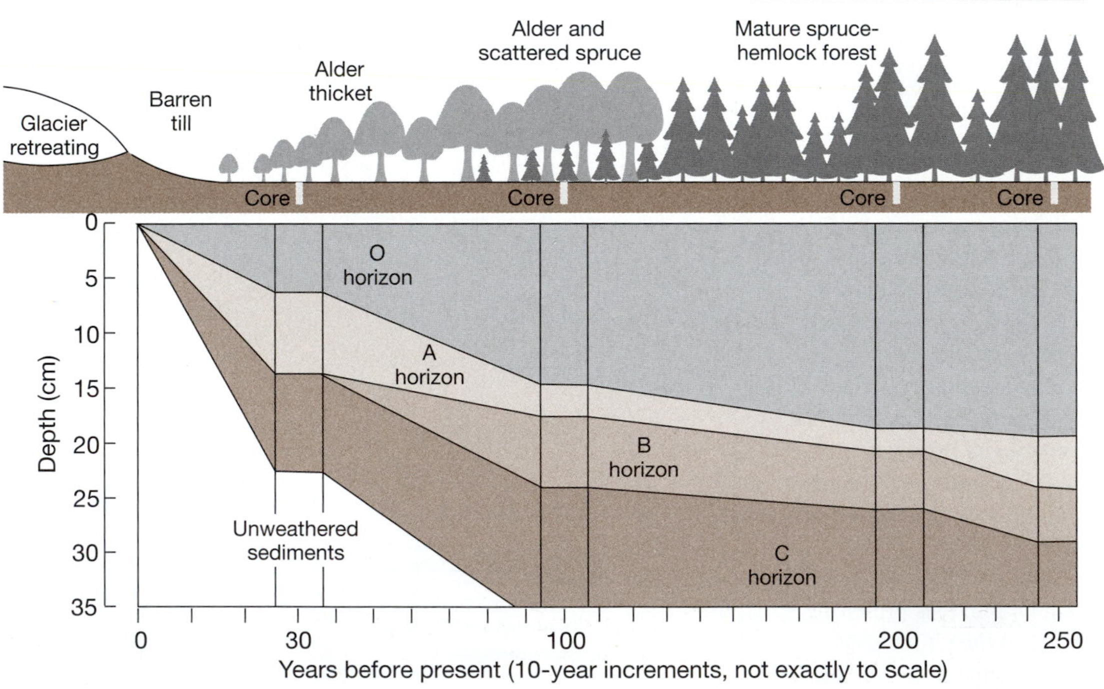

(Student's name) (Day) (Hour)

(Lab instructor's name)

ANSWER PAGE

6.1 (A)

(B)

6.2

6.3

6.4

6.5

6.6

6.7

6.8

6.9

6.10

6.11

6.12

6.13

6.14

6.15

6.16 (A)

(B)

(C)

6.17 (A)

(B)

(C)

6.18

6.19

6.20

6.21

6.22

6.23

6.24

6.25 (A) (B)

6.26

6.27 (A) (B)

6.28

7 Plate Tectonics

Topics

A. What are the two lines of early evidence suggesting continental drift. What was Pangaea, and who was the person who developed the theory of its existence?
B. What is the brief history of thought concerning magnetism and Earth's magnetic field? How does an electromagnet work? What is the evidence suggesting that Earth's inner core rotates faster than that part of Earth that surrounds it?
C. How does the field of paleomagnetics corroborate the 'drifting of continents?'
D. What is the reason why sea-floor topography includes (a) mid-ocean ridges, and (b) guyots extending away from mid-ocean ridges into progressively deeper water?
E. What is paleomagnetic chronology, and how does it document sea-floor spreading?
F. What is subduction? What evidence led Hugo Benioff to envision subduction under chains of islands parallel to deep-sea trenches?
G. What is the sense of motion of the sea-floor across a transform fault? What is the geographic setting of a particularly troublesome transform fault in North America?
H. What is a hot spot? What chain of volcanic islands is the product of hot spot dynamics?

A. Early evidence for 'continental drift'

An apparent fit—Dutch map maker Abraham Ortelius postulated in his *Thesaurus Geographicus* (1596) that the Americas were once joined with Europe and Africa and were "...torn away...by earthquakes and floods."

The earliest before-and-after portrayal of this theory was by geographer Antonio Snider-Pellegrini in 1858 (Fig. 7.1).

Q7.1 What is there about the graphic portrayal of the drifting of continents in Figure 7.1 that reflects a nationalistic ego on the part of French geographer Antonio Snider-Pellegrini? *Hint:* Check latitudes.

By the early 1900s a number of European geologists accepted the idea of continental drift, but American geologists rejected the idea for lack of a *plausible mechanism*. A particular historical geology textbook of the 1950s devoted but a single paragraph to the theory of 'continental drift.' But now **plate tectonics** (Gr. *tekton*, to work) is integral to the science of geology.

Why this change in American attitude? *Answer*: Because discoveries of the second half of the 20th century provided an explanation for how continental drift works. How did these discoveries occur—discoveries that ushered in a new paradigm about how we view the Earth?

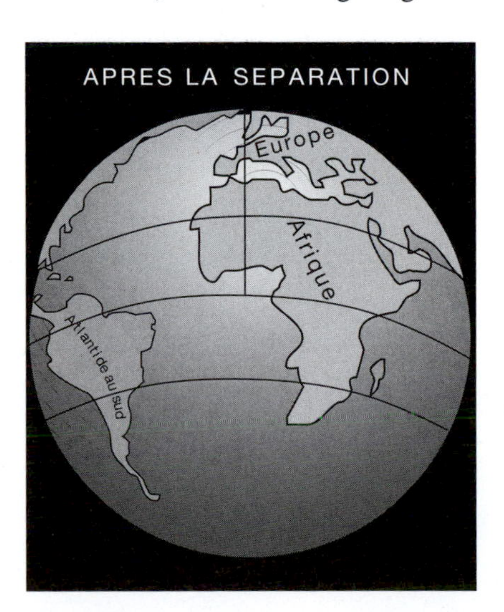

Figure 7.1 Antonio Snider-Pellegrini, a French geographer, illustrated his vision of the separation of the Americas from Europe and Africa in these avant and apres (before-and-after) maps of the world.

Alfred Wegener's Pangaea—While recuperating from wounds suffered as a German soldier in World War I, Alfred Wegener, a planetary astronomer by training, synthesized numerous lines of geologic and paleontologic evidence indicating former togetherness of present-day continents. For example, Wegener documented the fact that there is a remarkable sameness among fossil plants and animals in South America and Africa (Fig. 7.2). This, along with other discoveries, prompted Wegener to envision a supercontinent that somehow broke into pieces—pieces that have since drifted aimlessly about the globe as present-day continents.

Q7.2 Identify the five land masses (A–E) in Figure 7.2 with their present names.

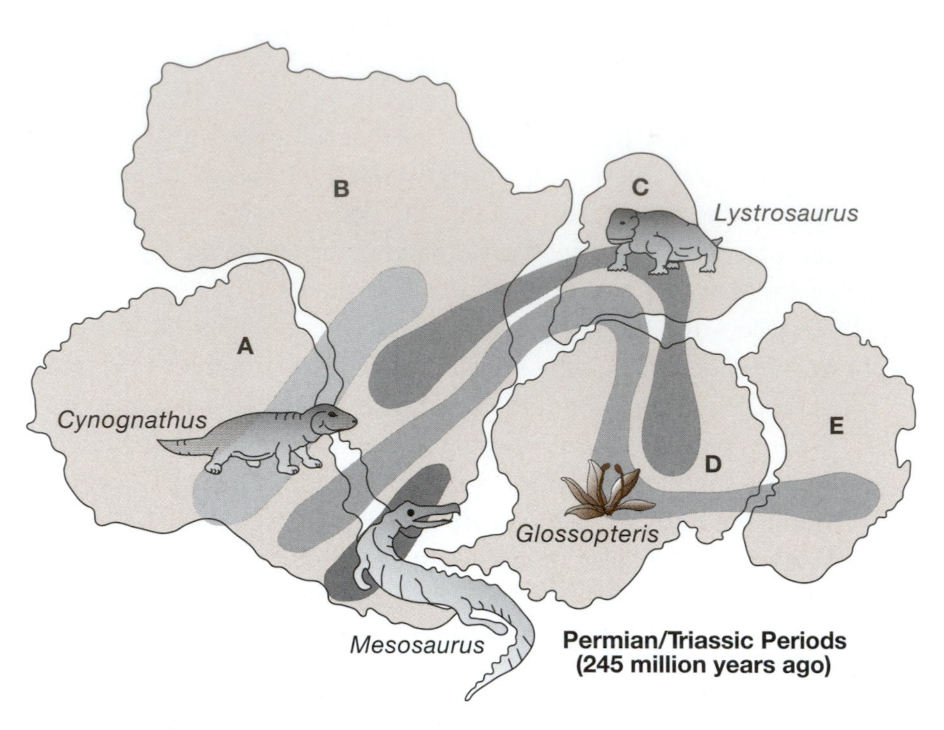

Figure 7.2 This map, which illustrates zoogeography recorded in the fossil record, shows a former togetherness of four of today's continents plus one subcontinent.

Wegener argued that such things as fossil ferns in Antarctica and ancient glacial deposits in Africa must surely indicate that these continents have not always been at their present latitudes. He eventually concluded that *all* present-day continents were united during the Late Paleozoic and Early Mesozoic Eras. He called this ancient supercontinent **Pangaea**—meaning *all lands* in Greek (Fig. 7.3). He published his grand synthesis in 1915 in a paper titled *Die Entstehung der Kontinente und Ozeane* (The Origin of Continents and Oceans).

Q7.3 (A) What present-day land areas can you identify in Alfred Wegener's Pangaea (Fig. 7.3)? (B) Wegener's map shows a couple of mere slivers of water within his map of Pangaea that have since grown to be two of our three largest oceans. Name those two oceans. (C) The vast ocean surrounding Pangaea has since shrunk in size, but persists as the largest of present-day oceans. Name that largest of today's oceans.

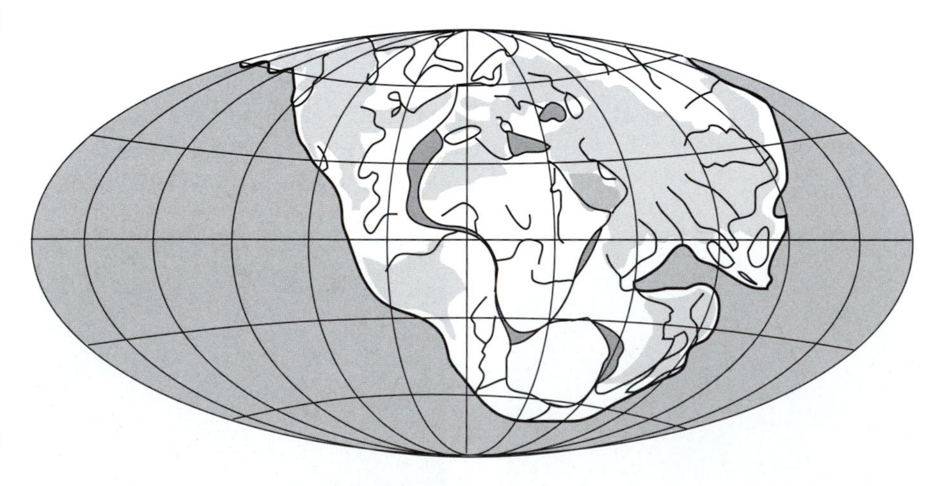

Figure 7.3 Alfred Wegener's 1915 map shows the supercontinent Pangaea as he believed it appeared during late Carboniferous time.

B. Earth's magnetic field

The dipolarity of a magnetic field—As early as the 7th century B.C. Greeks had discovered that lodestone (rock that consists of iron-rich minerals within veins, or *lodes*) attracts bits of iron-rich metals. They referred to such lodestones as *magnets*—a name taken from the ancient city of Magnesia, where there is an abundance of lodestone.

The mineral that accounts for the magnetic character of lodestones is the oxide of iron called magnetite (Fe_3O_4). A weaker magnetic quality is exhibited by another oxide of iron, hematite (Fe_2O_3).

Q7.4 According to the two formulas above, which has the greater *proportion* of iron, magnetite or hematite?

By the 11th century A.D. the property of lodestones had led to an awareness of Earth's **dipolar** magnetic field, prompting Arab and Persian sailors to construct mariner's compasses by simply floating pieces of lodestone on wood or cork (Fig. 7.4).

Figure 7.4 This medieval mariner's compass consists of a piece of lodestone on a board floating in a vessel of water. Stars mark the lodestone's magnetic dipoles.

It was then discovered that an iron needle becomes magnetized when in contact with lodestone and made to behave like lodestone in Earth's magnetic field. Magnetized needles proved to be much more suited to compass construction than lodestone, so compasses soon pervaded the world of exploration. A mariner's compass enabled Vasco da Gama to sail around Africa to India in 1497–98; and, it was a simple magnetic compass that enabled Meriwether Lewis and William Clark to chart observations made by their Corps of Discovery in 1804–06 (Fig. 7.5).

Figure 7.5 Lewis and Clark used a compass such as this to guide them in their trek from the Mississippi River to the Pacific Ocean.

Shape of Earth's magnetic field—In the 13th century Petrus Peregrinus, while experimenting with lodestone fashioned into a sphere, discovered that a magnetized needle tends to orient itself perpendicular to the sphere's surface at either of two opposite poles—thereby indicating a **magnetic** axis. Also, mariners were beginning to notice that compass needles behave in a similar manner in response to Earth's magnetic field. In 1600 Sir William Gilbert elaborated this phenomenon in his treatise, *De Magnete* (Fig. 7.6).

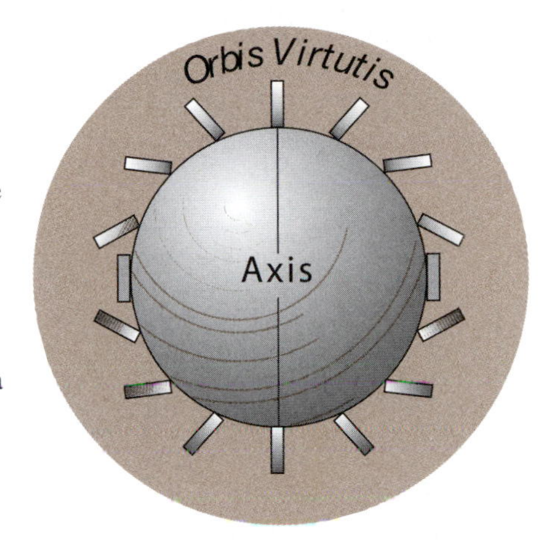

Figure 7.6 Gilbert delineated the shape of a lodestone's magnetic field by observing the orientation of tiny bar magnets placed over the surface of a lodestone sphere.

Gilbert postulated that the shape of the magnetic field of a simple bar magnet (Fig. 7.7) is like that of Earth—prompting him to theorize that Earth's magnetic field is due to a permanently magnetized metallic interior. But it was eventually discovered that temperatures within Earth's interior are far in excess of temperatures that allow for magnetic properties of metals. So Gilbert's explanation of Earth's magnetic field had to be revised.

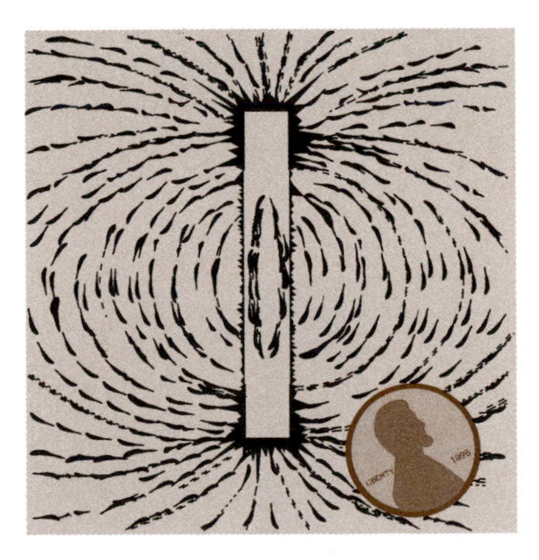

Figure 7.7 Iron filings sprinkled on a sheet of paper resting on a bar magnet align themselves along lines of magnetic force. In the early 19th century Karl Gauss defined one unit of magnetic intensity as one line of magnetic force per square centimeter.

Q7.5. The shape of Earth's magnetic field is most like which of the following fruits?

Pear

Orange

Apple

Tangerine

Electromagnetism—Scottish physicist James Maxwell discovered in the 1860s that electricity and magnetism are not separate phenomena, but different manifestations of a single phenomenon—**electromagnetism**. A flow of electricity produces a magnetic field (Fig. 7.8A); and, a loop of wire within a magnetic field produces a flow of electricity (Fig. 7.8B). Each of these two phenomena can be viewed as the reverse of the other. This is the basis for the **dynamo** principle (Gr. *dynamis*, power).

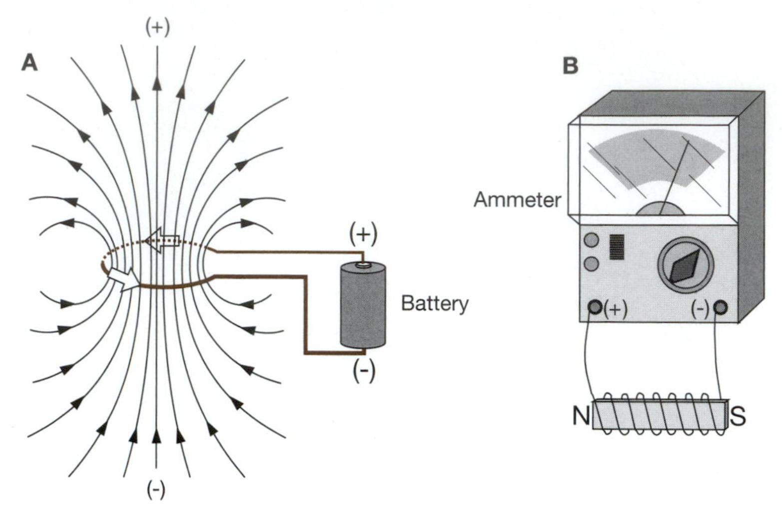

Figure 7.8 A The flow of electricity produces a magnetic field—the force that drives an electric motor. **B**. A magnetic field produces an electric current within a wire coil, boosted in strength by rotation of that coil—the force that drives an electric generator.

Q7.6 (A) Name a couple of household devices that convert electricity to mechanical power. (B) Name a couple of energy sources that convert mechanical power (motion) to electricity.

The orientation of the polarity of an electromagnet—an essential consideration in manufacturing—is perpendicular to the axis of a coiled electrical wire. The *right-hand rule* (Fig. 7.9) enables a manufacturer to predict the positive (+) and negative (-) poles of an electromagnetic axis. Incidentally, as concerns the behavior of a compass needle, north-seeking equates with *positive*, and south-seeking equates with *negative* (Fig. 7.8B).

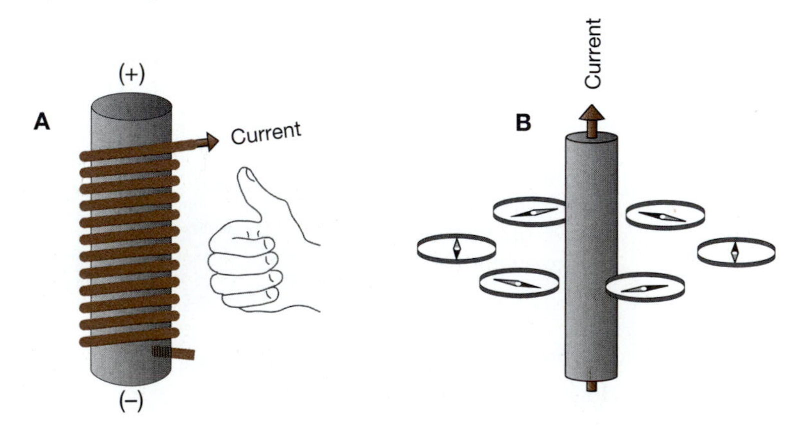

Figure 7.9 A The *right-hand rule*. If you hold your hand in a way such that the direction down the length of your fingers simulates the direction of current through a coiled wire, your extended thumb indicates the positive (+) pole of the magnetic field. B. Given a current flowing along a straight wire, north-seeking (black) ends of compass needles align themselves as illustrated.

Q7.7 Some household electric drills have a reverse switch. When that switch is thrown, the drill rotates in the opposite direction. How do you suppose that switch affects the drill's circuitry?

Q7.8 Judging from the specifics of the right-hand rule, in what direction is the current flowing in the power lines shown in Figure 7.10—toward your lower-right on the page, or toward your upper-left on the page?

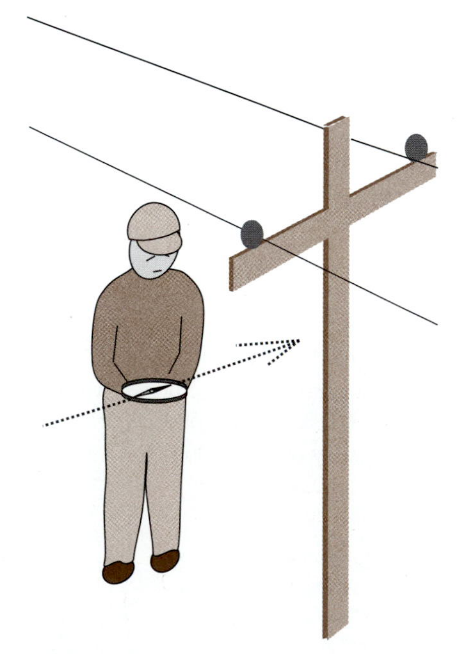

Figure 7.10 Beware of transmission lines when using a compass. The magnetic needle of a compass will orient itself perpendicular to the direction of current flow and give a false impression of magnetic north.

Earth's magnetic field—Twentieth century studies of **seismic waves** produced by earthquakes provided an understanding of Earth's anatomy—a bit like ultrasound providing clinical views of human anatomy. From seismic studies we now know that Earth consists of an inner solid core, an outer liquid core, a mantle, and a crust.

Given (a) the liquid state of the outer core, and (b) changes in the orientation and strength of Earth's magnetic field over time, we believe that **convection** (Fig. 7.11) of electrically charged particles within the liquid core accounts for Earth's magnetic field. But details have remained a mystery until recently.

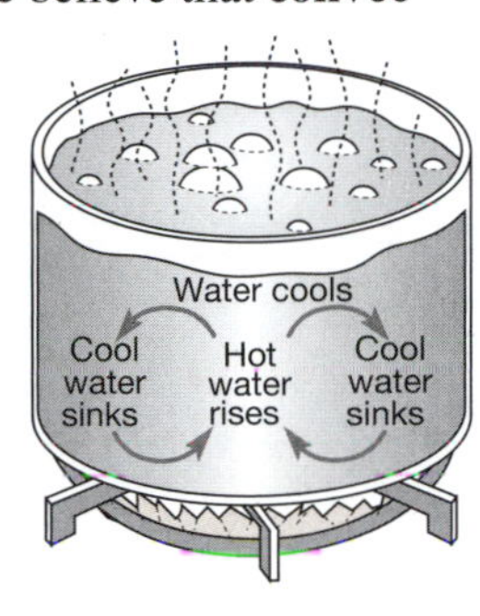

Figure 7.11 As within a boiling kettle, convection within Earth's liquid core results in hotter, less dense material rising and cooler, more dense material sinking.

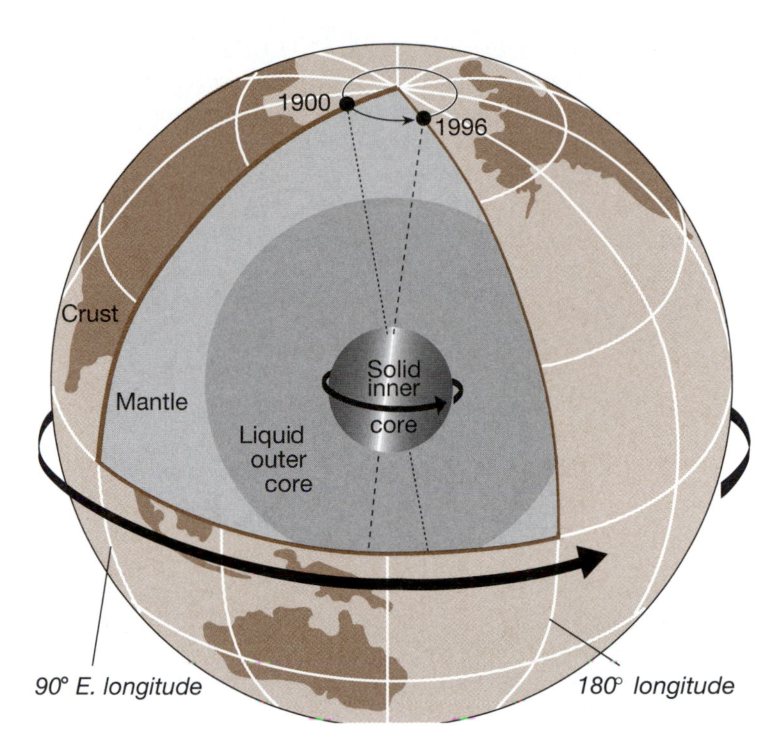

Figure 7.12 A projection of the inner core's high-velocity pathway migrates in the direction of Earth's rotation (e.g., shown here from a point in the year 1900 eastward to a point in the year 1996). This indicates that the inner core is rotating faster than surrounding material.

Q7.9 Speaking of convection—scientists divide Earth into four 'geospheres:' the *biosphere* (all plants and animals), the *lithosphere* (all rock materials), plus two other spheres in which convection is a defining quality. Name those two other spheres. *Hint:* These are tangible Earth surface spheres.

In 1995 G.A. Glatzmaier and P.H. Roberts designed a computer simulation of a **geodynamo** that explains the dynamics of Earth's magnetic field over time. Their model presumes that the inner core rotates faster than does surrounding Earth material (Fig. 7.12). This is analogous to a dynamo in that the more slowly rotating liquid outer core simulates a stationary magnetic field, while Earth's faster-rotating inner core simulates a rotor. This was a radical prediction, but their idea motivated two geophysicists, X. Song and P. Richards, to look for evidence in support of the idea of a super-rotating inner core.

Because of the heterogeneous composition of Earth, the velocity of a seismic wave depends to some extent on the route along which it travels. Seismic routes are like highways; some are slower, some are faster (Fig. 7.13).

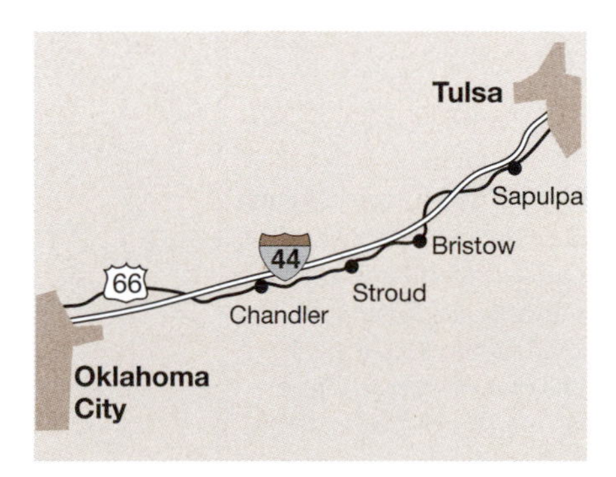

Figure 7.13 A graphic metaphor: Legendary Route 66 is slow-going, whereas I-44 is fast.

In their study of past seismic records, Song and Richards discovered a seismic route with exceptional high velocity that is crudely parallel to the inner core's rotational axis. Seismic waves traveling along this high-velocity pathway from the Southern Hemisphere intersect the Northern Hemisphere at 80°N latitude and advance over the surface of Earth through time, indicating that the inner core is indeed rotating faster than surrounding Earth material.

Q7.10 Judging from the longitudinal gain of the projection of the solid core's high-velocity seismic pathway between 1990 and 1996 (Fig. 7.12), what is the approximate annual advance over Earth's surface in degree(s) of longitude per year(s)? *Hint:* Do the simple math involving the difference in longitude markers and the difference in year markers.

For additional information about Glatzmaier and Roberts, see *Core Convection and the Geodynamo* at… http://ees-www.lanl.gov/IGPP/Geodynamo.html

Magnetic declination—So, Earth's magnetic field arises from a complexity of Earth's rotation and convection within its liquid outer core. So, in a perfect world magnetic poles would align with geographic poles, but such is not the case. North magnetic pole deviates approximately 11° from the geographic pole (Fig. 7.14).

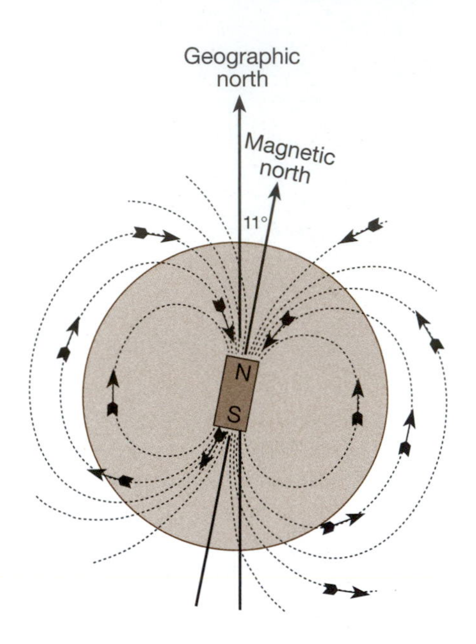

Figure 7.14 Earth's magnetic axis is inclined approximately 11° from its geographic axis.

As compasses began to be widely used in the 15th century in conjunction with celestial navigation, it became apparent that a compass does not always point exactly toward geographic north (as indicated by Earth's north star, Polaris). And, the discrepancy between the two directions, called **declination**, differs with location of the observer (Fig. 7.15). Legend has it that Columbus's men threatened mutiny during their historic voyage of 1492 because the disagreement between magnetic north and celestial north appeared to change as they sailed westward. Some of his sailors took this to be a bad omen and a mandate to turn back.

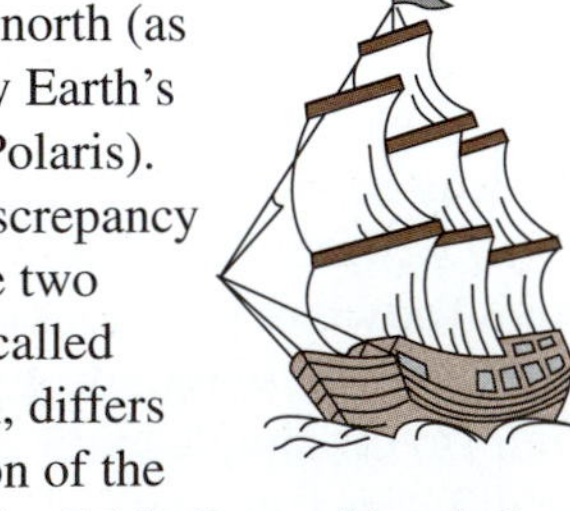

Q7.11 Consistent with details in Figure 7.15, as Columbus's fleet sailed westward the discrepancy between magnetic north and celestial north changed. Did the discrepancy increase or did it decrease?

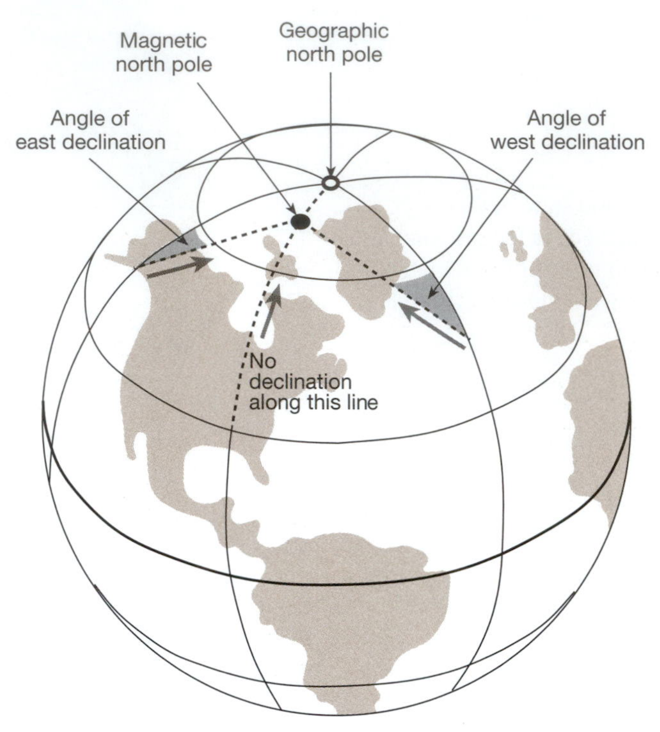

Figure 7.15 Earth's magnetic pole lies within the vicinity of Canadian Arctic islands some 1,290 km from north geographic pole. As illustrated here, the magnitude of magnetic declination differs as a function of longitude.

Modern vehicles with compasses come with instructions for adjusting the compass for magnetic declination (Fig. 7.16). For example, if one were to purchase a vehicle in New York, with the compass set for that state, and then move to California, an adjustment of some 30° would be necessary to restore accuracy to that compass. The manufacturer simplifies symbolism by converting degrees of declination to zone numbers.

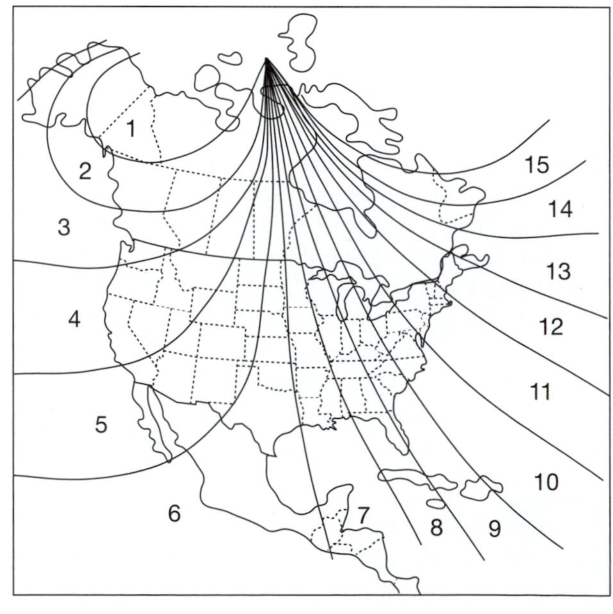

Figure 7.16 This is a figure from the owner's manual of a an automobile in which magnetic declination is divided into zones 1–15. The owner simply enters the number of his/her zone, and the compass becomes electronically adjusted for local declination.

Magnetic inclination—(Revisiting the shape of Earth's magnetic field.) With the advent of global navigation it was soon discovered that in the Northern Hemisphere the north-seeking end of a compass needle points not only northward, but to some degree *downward* as well; whereas in the Southern Hemisphere the north-seeking end of a compass needle points to some degree *upward* (Fig. 7.17). This is known as **magnetic inclination**.

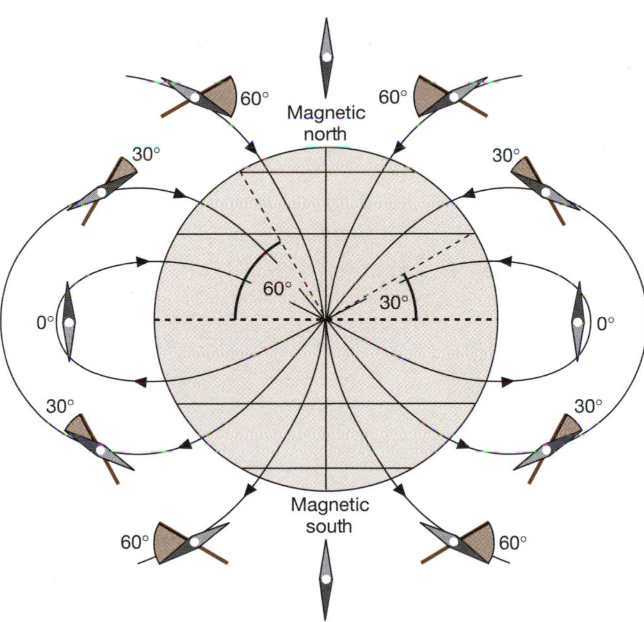

Figure 7.17 The north-seeking ends of compass needles (shown black in this figure) dip downward in the Northern Hemisphere and upward in the Southern Hemisphere.

In Figure 7.18 a Brunton compass—the 'Swiss Army knife' of the field geologist—is held with its face in a vertical plane. As can be seen from this figure, the needle dips downward in the northern hemisphere and upward in the southern hemisphere.

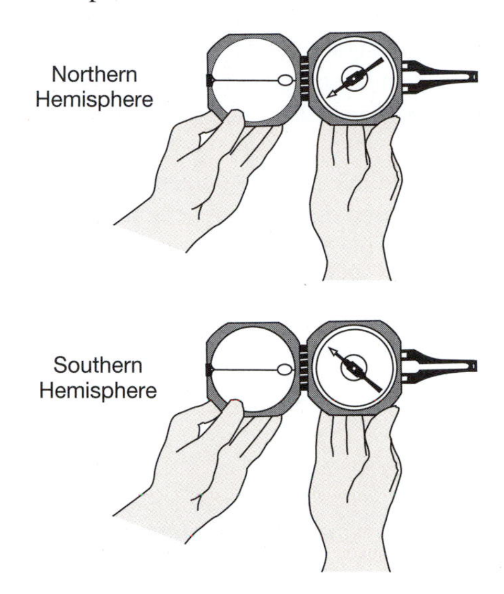

Figure 7.18 This figure shows two views of a Brunton compass with its face held in a vertical plane. Notice how the inclination of the north-seeking needle differs with the hemisphere.

Q7.12 (A) Explain the difference in the orientations of the inclined compass needles in Figure 7.17. (B) Toward which cardinal direction is the observer facing in both views: north, south, east, or west?

A compass like that illustrated in Figure 7.19 is designed for measuring magnetic inclination. The angular value of inclination does not equate numerically with the latitude of the observer (as does the altitude of our north star, Polaris), but, with the aid of tables, the latitude of the observer can be determined from the degree of magnetic inclination.

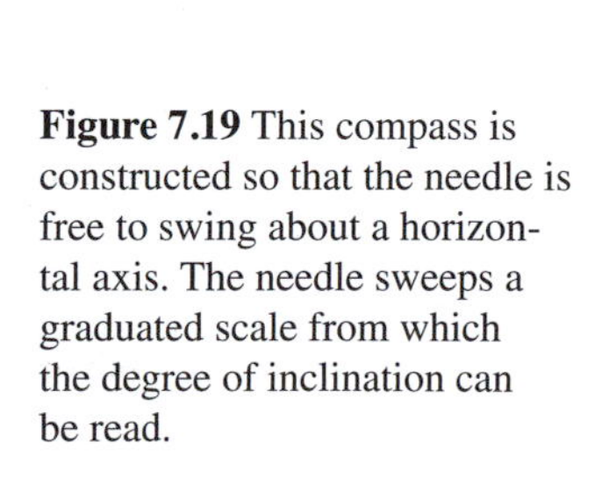

Figure 7.19 This compass is constructed so that the needle is free to swing about a horizontal axis. The needle sweeps a graduated scale from which the degree of inclination can be read.

Q7.13 In Figure 7.19, (A) Where is the observer located? (B) Again, toward which cardinal direction is the observer facing—north, south, east, or west?

C. Paleomagnetics corroborates continental drift

Paleomagnetism—As early as the late 19th century physicists discovered that certain iron-bearing minerals within igneous rocks appear to have acquired magnetic fabrics imprinted by Earth's magnetic field. In modern parlance, an igneous mineral's magnetic field is a clone of Earth's magnetic field—yes, with the same orientation. Thus was born the study of **paleomagnetics** (Gr. *paleo*, ancient).

One of the most notable of early paleomagnetists was Pierre Curie, who discovered that the magnetic field acquired by magnetite, for example, does not develop immediately upon crystallization from molten material, but at a temperature hundreds of degrees Celsius below its temperature of crystallization. This temperature of magnetic imprinting is called the **Curie point** (Fig. 7.20).

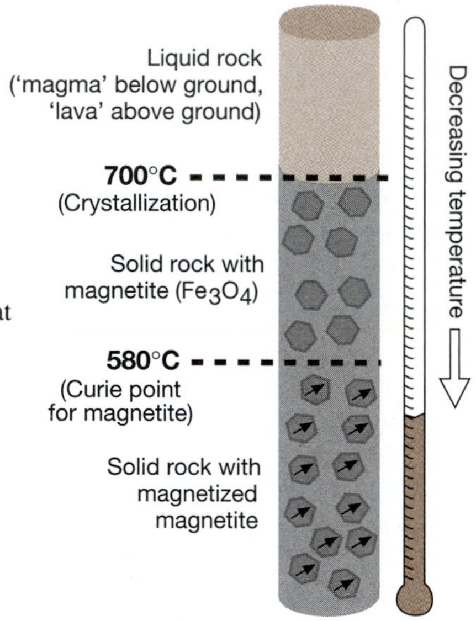

Liquid rock ('magma' below ground, 'lava' above ground)

700°C (Crystallization)

Solid rock with magnetite (Fe_3O_4)

580°C (Curie point for magnetite)

Solid rock with magnetized magnetite

Decreasing temperature

Figure 7.20 Crystallization occurs at one temperature, but magnetization occurs at a lower temperature—the Curie point.

Q7.14 (A) What would be the first change to occur to a bar magnet if you were to place it in a laboratory furnace capable of melting any mineral and turned the thermostat to the maximum? (B) How about the second change?

Plotting the location of Earth's north magnetic pole—Inasmuch as a magnetic needle not only points northward, but also inclines as a function of latitude, one can compute the location of Earth's north magnetic pole from measurements made at a single location (Fig. 7.21).

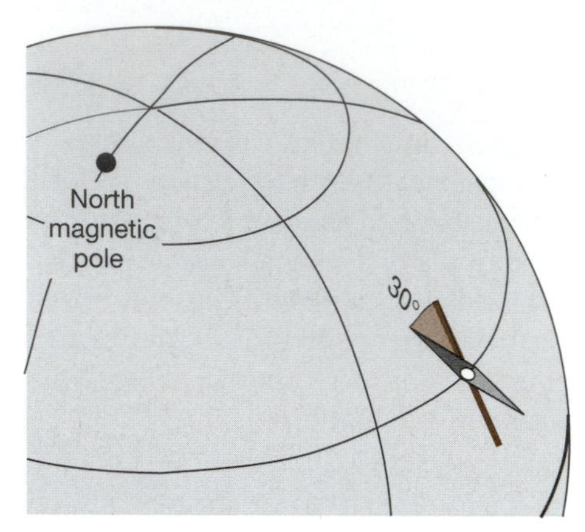

Figure 7.21 From the behavior of the compass needle at one locality, the location of Earth's north magnetic pole can be computed and plotted on a globe.

Past locations of Earth's north magnetic pole appear to have changed through time—In 1956, Cambridge University physicists discovered that paleomagnetic measurements made on rocks of various ages in Europe appear to record different positions of north magnetic pole over the course of millions of years (Fig. 7.22).

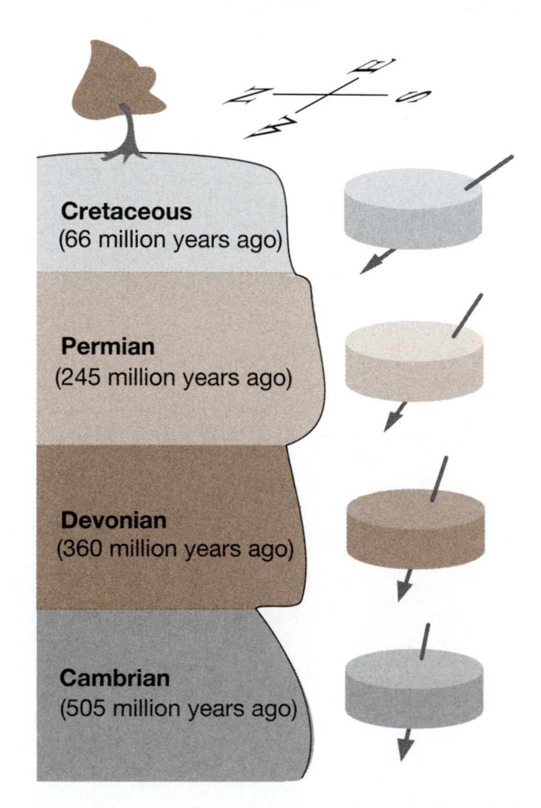

Figure 7.22 Directions and inclinations of magnetic polarities recorded by rocks of different ages suggest changes in the orientation of Earth's magnetic field through time. (*Schematic*)

'Polar wandering' curves—The plotting on a map of past positions of Earth's north magnetic pole from field measurements such as those schematically illustrated in Figure 7.22, describes a path on Earth's surface that is referred to in geologic literature as a **polar wandering curve**. A more accurate term would be apparent polar wandering curve—for reasons that will become clear.

Two theories—Two end-member theories presented themselves as explanations for the apparent polar wandering curves plotted from data gathered in Europe by Cambridge physicists (e.g., Fig. 7.23):

(1) Theory #1: Europe has remained stationary, while Earth's north magnetic pole has migrated across the northwestern Pacific Ocean.

(2) Theory #2: Earth's north magnetic pole has remained stationary, while Europe has moved about.

The scientific method, in its simplest form, consists of three components: *observation, theory, and testing (of the theory)*. Here we have an observation and two possible theories that might explain that observation. So let's test the two theories, one against the other, and choose the winner.

Q7.15 Imagine that you are an Earth scientist writing a grant proposal to the National Science Foundation, requesting a budget that would enable you to test the validity of one of the foregoing two theories over the other. What would a logical test be? *Hint:* A clue lies in the paragraph, 'Two theories,' in the column to your left. (Try to answer this question before turning the page.)

Figure 7.23 This apparent polar wandering curve was derived from the plotting of data gathered from field measurements made in Europe by Cambridge physicists.

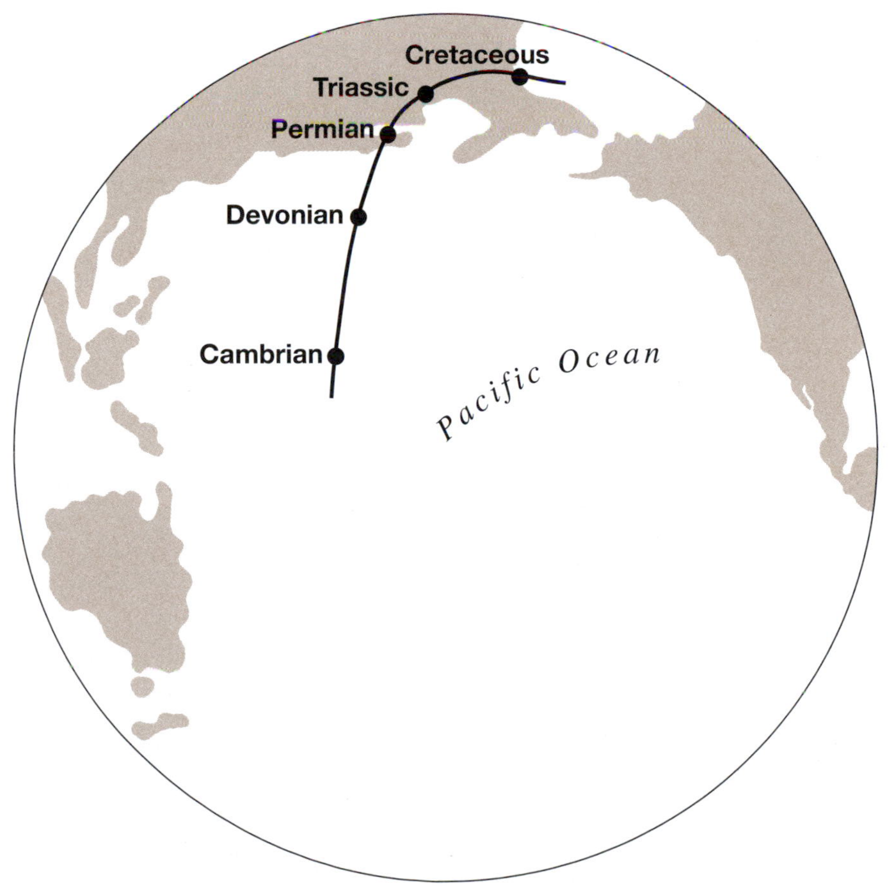

A test of the validity of the apparent polar wandering curve (continuing from page 117). If your answer to Q7.15 was, "Plot an apparent polar wandering curve from paleomagnetic measurements made on a *second* continent," good for you! That's exactly what geologists did (Fig. 7.24).

Figure 7.24 This is the apparent polar wandering curve derived from plotting paleomagnetic measurements made in North America.

To make a precise visual comparison of the curve for Europe with that of North America, remove this page from your manual and view Figure 7.23 on the previous page with back-lighting. (Figure 7.23 should appear to be superposed on top of Figure 7.25.)

Q7.16 (A) Do the two apparent polar wandering curves match when plotted from different continents? (B) During which of these five geologic periods were Europe and North America most different in their positions within Earth's magnetic field?

So Theory #2 on page 117 proved to be correct. That is, Earth's north magnetic pole has remained stationary, while Europe has drifted about the globe. North America, too.

So, paleomagnetists could chart the movement of continents about the globe, but they continued to have little or no idea as to the mechanism whereby such 'wandering' could occur.

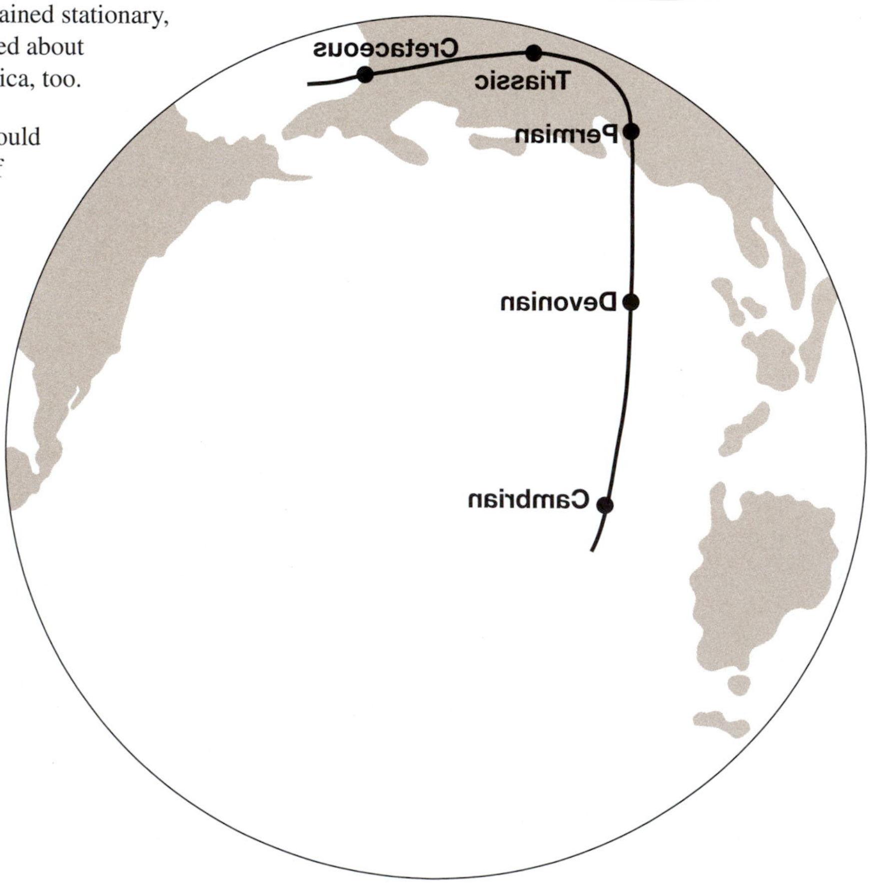

Figure 7.25 This is a reversed image of Figure 7.24, for viewing behind back-lighted Figure 7.23.

D. Mid-ocean ridges and sea-floor spreading

Early ideas—Arthur Holmes, of Durham University, suggested in 1929 that **convection within Earth's mantle** is the driving force behind continental drift.

Harry Hess, of Princeton University, charted physical evidence for Holmes's idea while serving as Captain of an assault transport in the Pacific during World War II. Hess conducted surveys using echo-soundings as time allowed, and it was these soundings that he interpreted as reflecting a convective process he called **sea-floor spreading**.

But first, a bit of background about the early understanding of the physiography of the Pacific Ocean seafloor.

A century before surveys of the Pacific sea-floor were undertaken, Charles Darwin studied the evolution of Pacific volcanic islands (Fig. 7.26), which he subdivided into atolls, fringing reefs, and barrier reefs. Based on limited field observations, Darwin correctly interpreted the chronological order in which these stages of island-reefs develop.

Q7.17 Figure 7.26 illustrates Charles Darwin's classification of reefs. Your task: Place these three kinds of reefs in their correct evolutionary succession—oldest first, youngest last? (Incidentally, there is as much revealed in Figure 7.26 about the anatomy of Pacific reefs as young Darwin saw.)

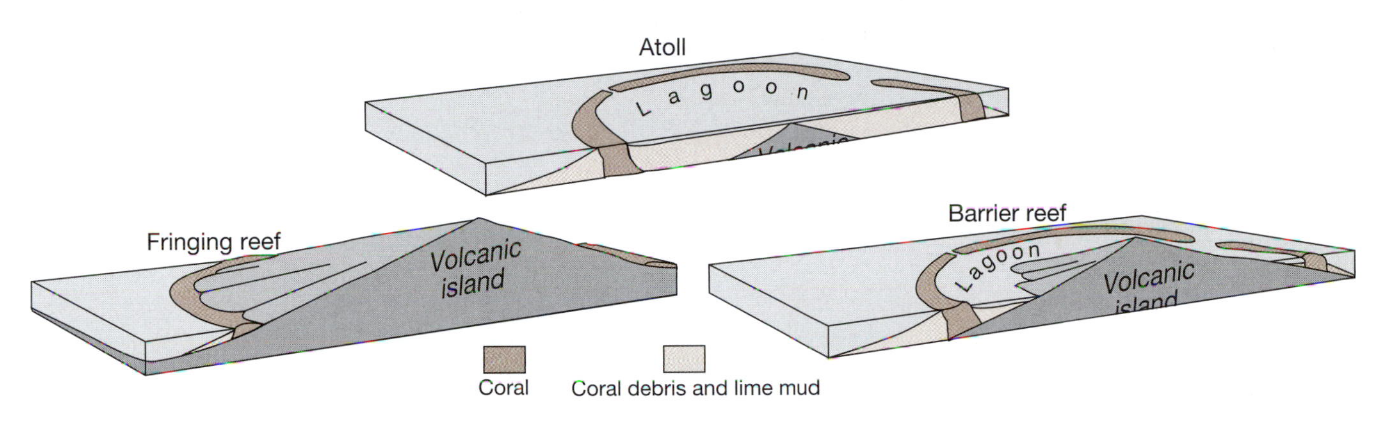

Figure 7.26 Charles Darwin believed that three forms of reef-cloaked Pacific volcanic islands do not reflect three different origins, but, instead, three different stages in the evolution from a single origin.

Guyots suggest sea-floor spreading—Harry Hess's echo-sounding surveys revealed flat-topped seamounts that he named **guyots**. He correctly interpreted guyots as being drowned atolls—prompting him to envision the moving away, and downward, of seamounts from earlier sites atop mid-ocean ridges (Fig. 7.27). But he stopped short of including continents in his convective recycling of ocean crust. Incidentally, since Hess's time, rates of sea-floor spreading have been shown to be on the order of a few *centimeters* per year, whereas rates of the descent of guyots into deeper water are on the order of a few *millimeters* per year. (You will use these concepts later in this exercise).

Q7.18 What do you suppose explains the shrinking of ocean crust—and, so, the deepening of water—as the crust moves away from volcanoes? *Hint:* Think temperature and density.

Figure 7.27 Harry Hess's interpretation of guyots as drowned atolls led him to envision sea-floor spreading—ocean crust being generated through volcanic action along mid-ocean ridges and then moving away (and downward) like giant conveyor belts (arrows).

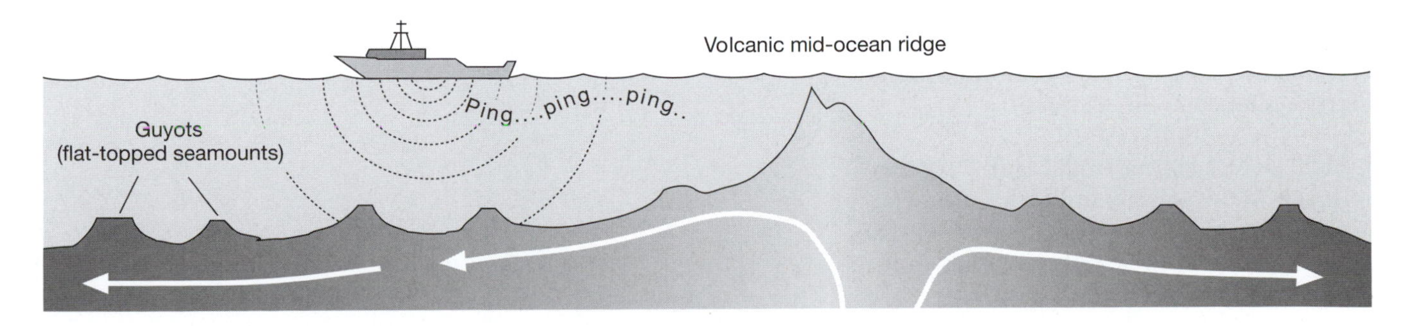

E. Paleomagnetics corroborates sea-floor spreading

Magnetic reversals—Early in the 20th century Bernard Brunhes, of France, and Motonari Matuyama, of Japan, discovered that magnetic polarities of some ancient volcanic rocks are the reverse of Earth's present magnetic polarity (Fig. 7.28). They correctly inferred that Earth's magnetic poles were reversed at the time magnetic minerals within such rocks crystallized from lavas and cooled below their Curie points.

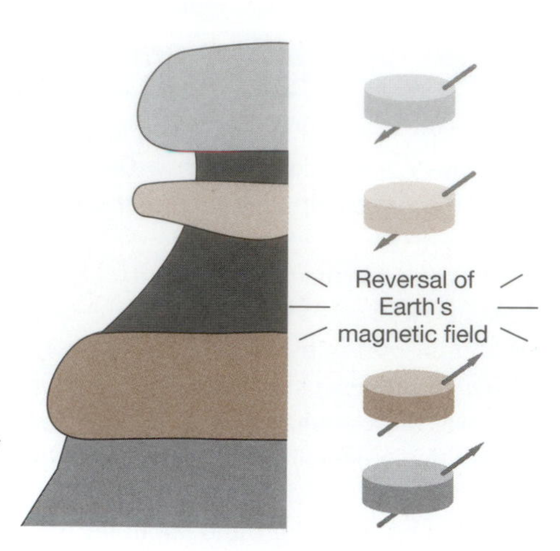

Figure 7.28 A schematic rock record of a reversal in the polarity of Earth's magnetic field.

Land-based paleomagnetic chronology—In the early 1960s Alan Cox, Richard Doell, and Brent Dalrymple used reversals in magnetic polarity recorded within layered volcanic rocks, along with radiometric ages of those rocks, to establish a paleomagnetic chronology (Fig. 7.29). It became apparent that the vertical accumulation of volcanic rocks serves as a '**tape recorder**' of changes in Earth's magnetic polarity.

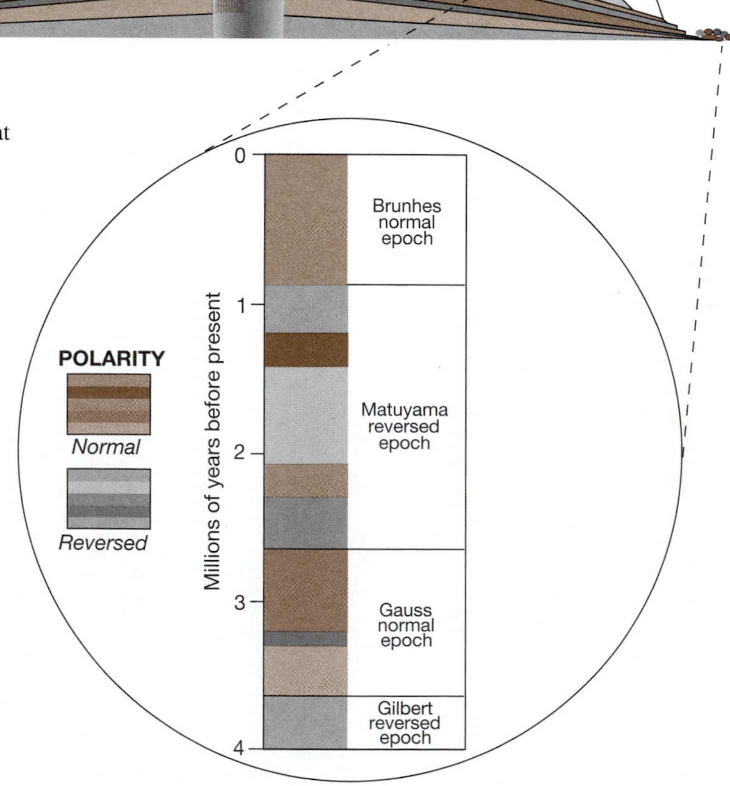

Figure 7.29 The first attempt at constructing a paleomagnetic time scale extended back to only 4 million years before present. Workers have since extended this time scale back hundreds of millions of years.

Q7.19 Labels for the four paleomagnetic epochs extending back to 4 million years BP (Fig. 7.29) bear the names of notable scientists. What did these four particular scientists have in common? *Hint:* All four of these men have been mentioned in this exercise.

Q7.20 Does it appear from Figure 7.29 that reversals in the polarity of Earth's magnetic field occur separated by equal periods of time, or do they appear to be rather random in their temporal occurrence?

Paleomagnetics goes to sea—By the early 1960s exploration of seafloors included airborne magnetic surveys across mid-ocean ridges. In such a survey, each flight line produces a paleomagnetic profile that is a graph of magnetic polarity and intensity manifest within sea-floor sediments and underlying ocean crust (Fig. 7.30A). From a series of profiles, a paleomagnetic map can be constructed (Fig. 7.30B).

British geologists Frederick Vine and Drummond Matthews noticed that magnetic maps that include mid-ocean ridges exhibit a curious striped appearance, which is even more apparent in colored enhancements (Fig. 7.31).

Q7.21 Vine and Matthews noticed three simple things about these stripes—things that proved to be critical to their interpretation. What are those three things? *Three hints:*

(1) What is the *directional trend* of stripes relative to that of the ridge?

(2) What is the order of color-coded magnetic stripes on one side of the ridge relative to that on the other side?

(3) How do the *widths* of stripes compare with the *durations* (in years) of color-coded equivalent epochs in the chronology of Figure 7.29 (page 120)?

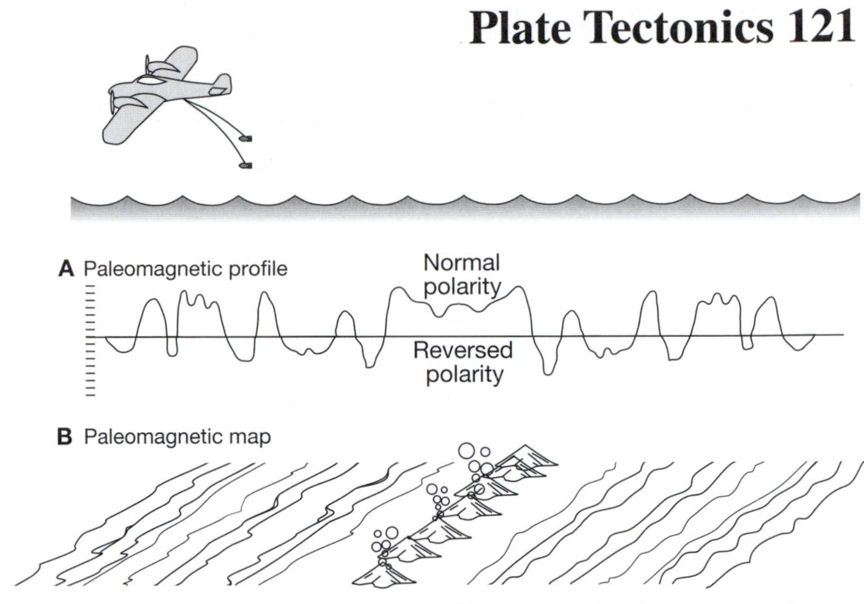

Figure 7.30 **A** This is an aerial paleomagnetic profile across a volcanic mid-ocean ridge. B This is a paleomagnetic map synthesized from multiple paleomagnetic profiles, which exhibits a puzzling striped appearance. Each stripe-like map unit represents an area of sea-floor characterized by a combination of magnetic *polarity* and magnetic *intensity*.

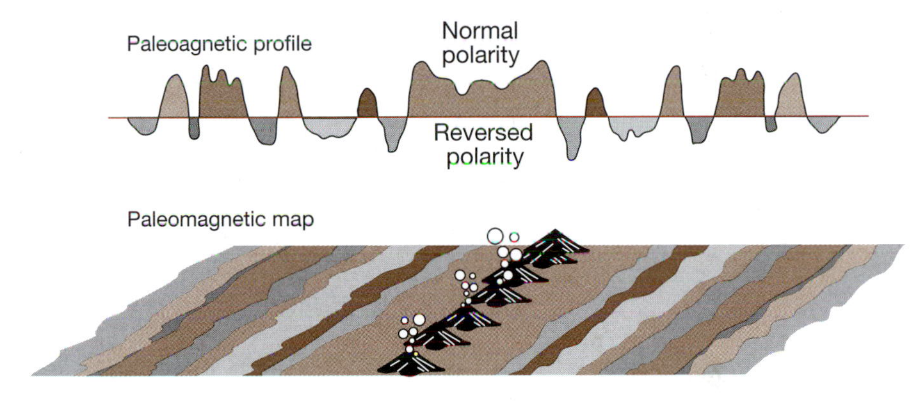

Figure 7.31 This is Figure 7.30 enhanced with color.

Integration of sea-floor paleomagnetics and land-based chronology—The three observations made by Vine and Matthews led them to turn the paleomagnetic time scale invented by Cox et al. on its side, and, voilá, they discovered that its pattern matches that of the puzzling stripes on paleomagnetic maps associated with mid-ocean ridges. This leap in imagination produced a snapshot of the convection envisioned by Holmes and the sea-floor spreading envisioned by Hess.

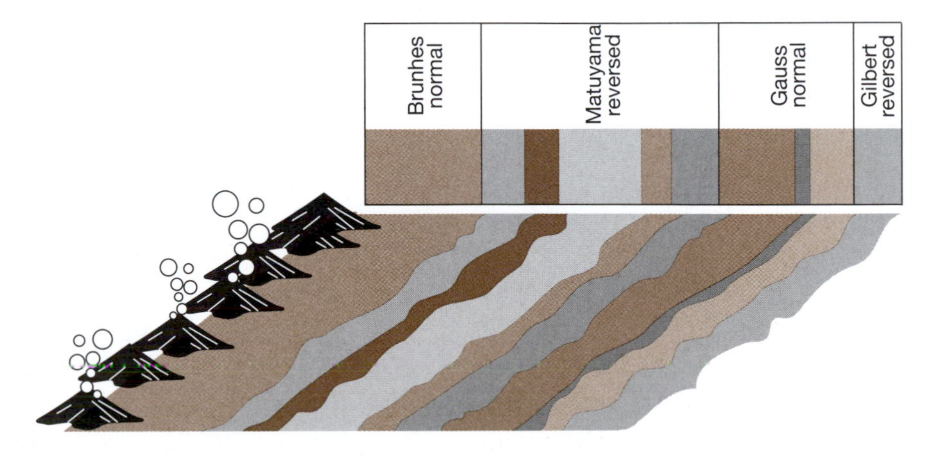

Figure 7.32 Vine and Matthews correlated the paleomagnetic record of the sea-floor with the paleomagnetic tape recorder on land and concluded that the seafloor, too, is a 'tape recorder' of paleomagnetic events.

F. Ocean trenches and subduction of crust

If ocean crust is growing (by way of volcanic activity along mid-ocean ridges), is Earth expanding?

Anatomy of Earth's interior—By 1900 seismologists had subdivided Earth's interior into crust, mantle, and core—with boundaries believed to be marked by some combination of (a) chemical/mineral *composition* and (b) *state* (e.g., solid crust, solid mantle, and liquid core). That anatomy was soon refined to include a liquid outer core and a solid inner core (recall Figure 7.12 on page 113).

Seismologists further observed that the brittle fracture of rocks (events signaled by earthquakes) occurs no deeper than approximately 720 km (450 miles), indicating that below that depth stress is relieved not through brittle fracture, but through plastic flow. This indicated that there is another boundary, well within the mantle, that separates a solid interval above, the **lithosphere** (Gr. *litho*, stone), from a plastic interval below. In 1914, Joseph Barrell proposed that this plastic zone be called **asthenosphere** (Gr. *astheno*, weak). So there are *two* schemes of Earth anatomy—one based on state and one based on composition (Fig. 7.33).

Note: This definition of lithosphere differs from the earlier definition of lithosphere in Q7.9 on page 137 as one of four members of the geosphere family.

Emphasis: state	Depth	Emphasis: composition
Lithosphere (solid)		Crust (low-density minerals)
	10–30 km.	
	100 km.	
Asthenosphere (plastic)		Mantle (high-density minerals)
	350 km.	
Mesosphere (solid)		
	2883 km.	
Outer core (liquid)		Core (very high-density minerals)
	5410 km.	
Inner core		
	6371 km.	

Figure 7.33 Dual schemes of Earth's internal anatomy—one emphasizing state, the other emphasizing composition.

Q7.22 With reference to subdivisions within Earth's anatomy shown in Figure 7.33—what is the one boundary held in common by the two schemes?

Subduction at ocean trenches—While Vine and Matthews were documenting sea-floor spreading at mid-ocean ridges, seismologists were already drawing conclusions having to do with the distribution of earthquake foci (Fig. 7.34) beneath deep-sea trenches of the Pacific Ocean.

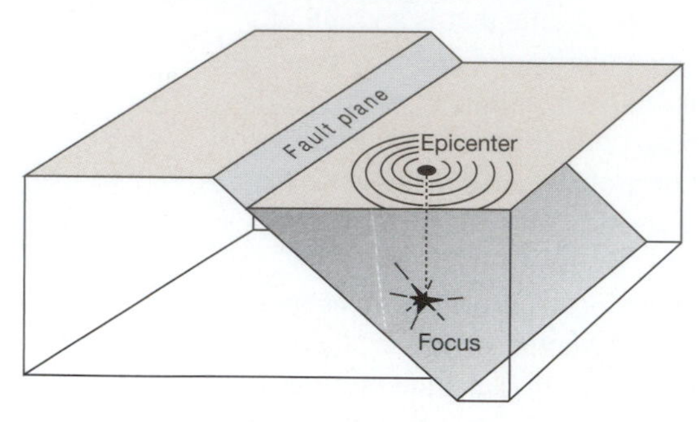

Figure 7.34 An earthquake reflects abrupt movement at some point on a fault plane. This point is called the focus of the earthquake. Of more interest to human beings is the point on the ground nearest the focus, which is called the epicenter.

Hugo Benioff, of Caltech, noticed *two things* about the distribution of earthquake foci beneath deep-sea trenches of the western Pacific (e.g., the Tonga trench)—in addition to their simply being deep (Fig. 7.35).

Q7.23 What do you suppose those *two things* are? (A) One has to do with the shape of space delineated by earthquake foci. (B) The other has to do with distribution of that space relative to the trench and volcanic islands. *Hint:* See details in the caption to Figure 7.35.

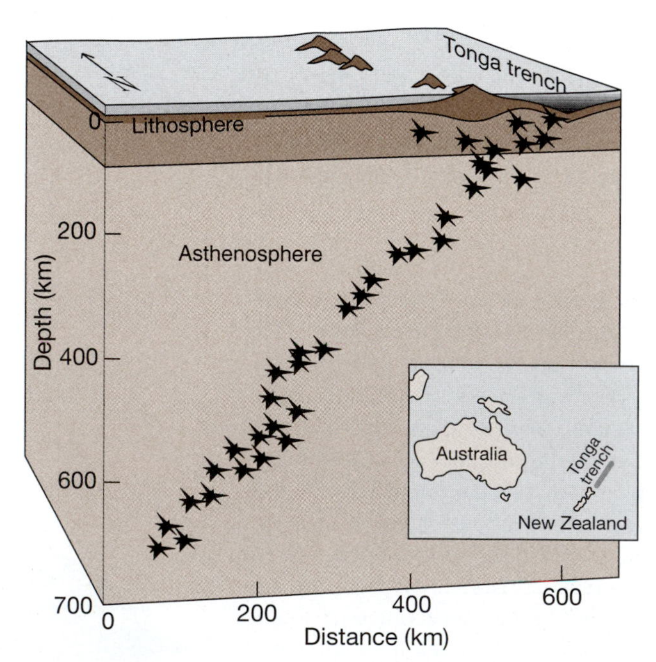

Figure 7.35 This is a section drawn perpendicular to the Tonga Trench. This same distribution of earthquake foci would be repeated throughout an infinite number of sections drawn parallel to this one—indicating its three-dimensional shape.

The two observations called for in Q7.23 prompted Benioff to envision descending lithosphere (accompanied by brittle fracturing signaled by earthquakes)—a process he called **subduction** (Fig. 7.36). Heating of descending lithosphere at depths of around 700 km converts lithosphere into asthenosphere. Both types of global plate boundary—spreading ridge and subduction trench—occur off the west coast of South America (Fig. 7.37).

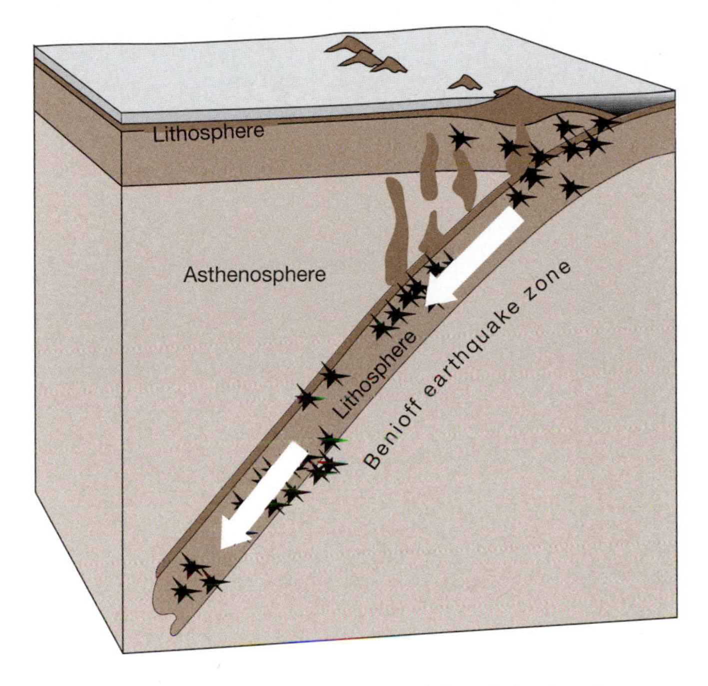

Figure 7.36 Hugo Benioff envisioned the subduction of lithosphere beneath Pacific trenches. Partial melting of the descending lithosphere creates magmas that ascend and emerge as lavas—building volcanic islands that occur on the subduction-side of a trench.

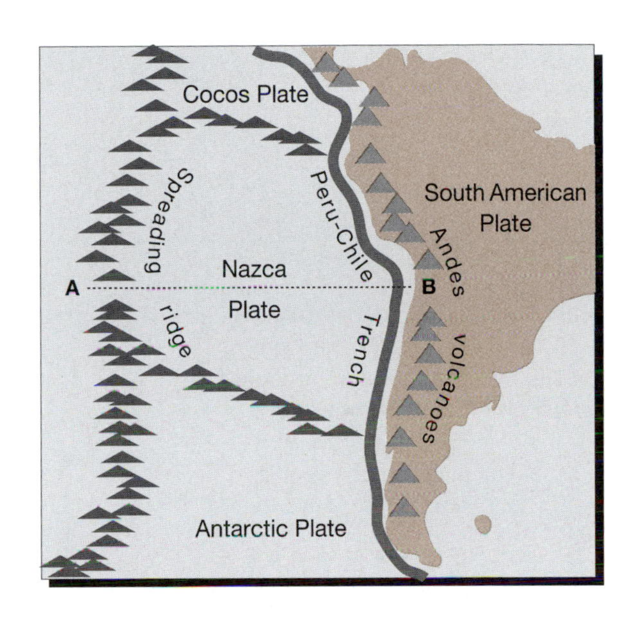

Figure 7.37 The Nazca plate is bounded on the west by the East Pacific Rise spreading ridge and on the east by the Peru-Chile subduction trench.

Now you know the answer to the question at the beginning of section F (page 122): Is Earth expanding? The answer is *no*.

The process of subduction completes the circle of Holmes's convection model. And, subduction facilitates Hess's sea-floor spreading. In summary, solid lithospheric plate is generated by volcanic activity along a spreading ridge and then moves away to where it descends (i.e., is subducted) by convection within the asthenosphere. The zone of subduction is marked by a deep-sea trench. This cycling and recycling is the essence of plate tectonics. In some instances a plate carries chunks of continental crust along piggyback. And so there you have it—*the mechanism for continental drift!*

Q7.24 Draw a geologic cross-section through the Earth along line A-B of Figure 7.37, including the kinds of information shown in Figure 7.36.

Q7.25 At the Peru-Chile trench (Fig. 7.37) the Nazca plate is subducting beneath the South America plate. Why not the reverse? Why doesn't the South American plate descend beneath the Nazca plate? *Hint:* Think in terms of compositons of continental crust and oceanic crust.

G. Transform faults complete the triad

If Earth were flat our story having to do with the mechanics of plate tectonics might end here. But Earth is not flat. It's spherical. And so in 1965 Canadian J. Tuzo Wilson theorized that in order for spreading ridges and subduction trenches to interact on our spherical Earth, a third type of plate boundary must exist—that of the **transform fault** (Fig. 7.38).

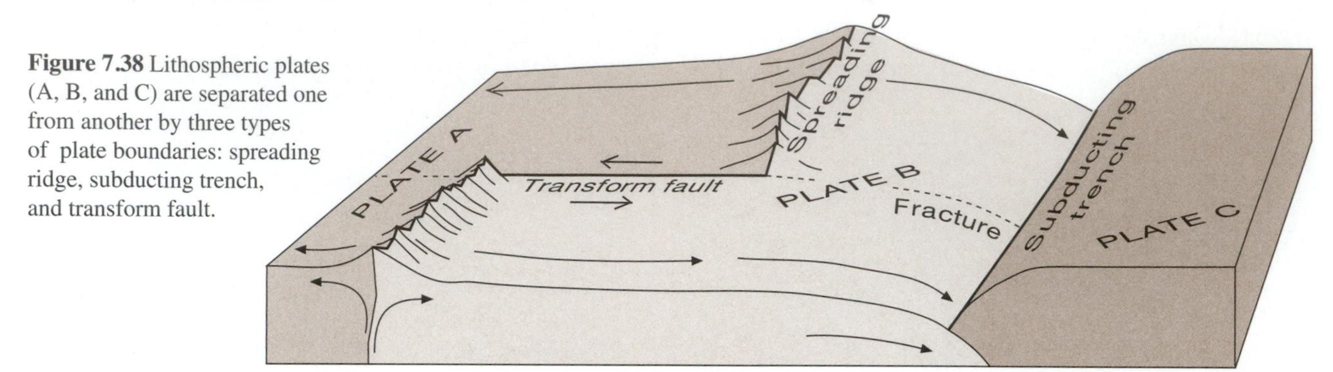

Figure 7.38 Lithospheric plates (A, B, and C) are separated one from another by three types of plate boundaries: spreading ridge, subducting trench, and transform fault.

Movement along the San Andreas Fault—Transform faults are far from being incidental. A particularly troublesome transform fault occurs as a segment of the boundary between the North American plate and the Pacific plate—the San Andreas fault of California (Fig. 7.39).

Movement along the San Andreas Fault is such that coastal California (including Los Angeles) is moving northwestward relative to northern and eastern California (including San Francisco). Movement along the San Andreas occurs, as in all faults, as random and local slippages signaled by earthquakes. Judging from the magnitude of offset along the San Andreas of rocks of known geologic age, it is believed that the rate of northwestward movement of the Los Angeles block is approximately 5 centimeters per year.

Figure 7.39 Motion along the San Andreas Fault in Southern California (arrows), is bringing Los Angeles and San Francisco toward each other at a rate of approximately 5 centimeters per year.

Q7.26 San Francisco is 625 kilometers northwest of Los Angeles. At the approximate rate of convergence of 5 centimeters per year, how many years will it be until the two cities are side-by-side (arguing about which is a suburb of which)?

A world map—Other geoscientists soon corroborated Wilson's theory, and, in time, a world map of major plates emerged that includes all three types of global plate boundaries (Fig. 7.40).

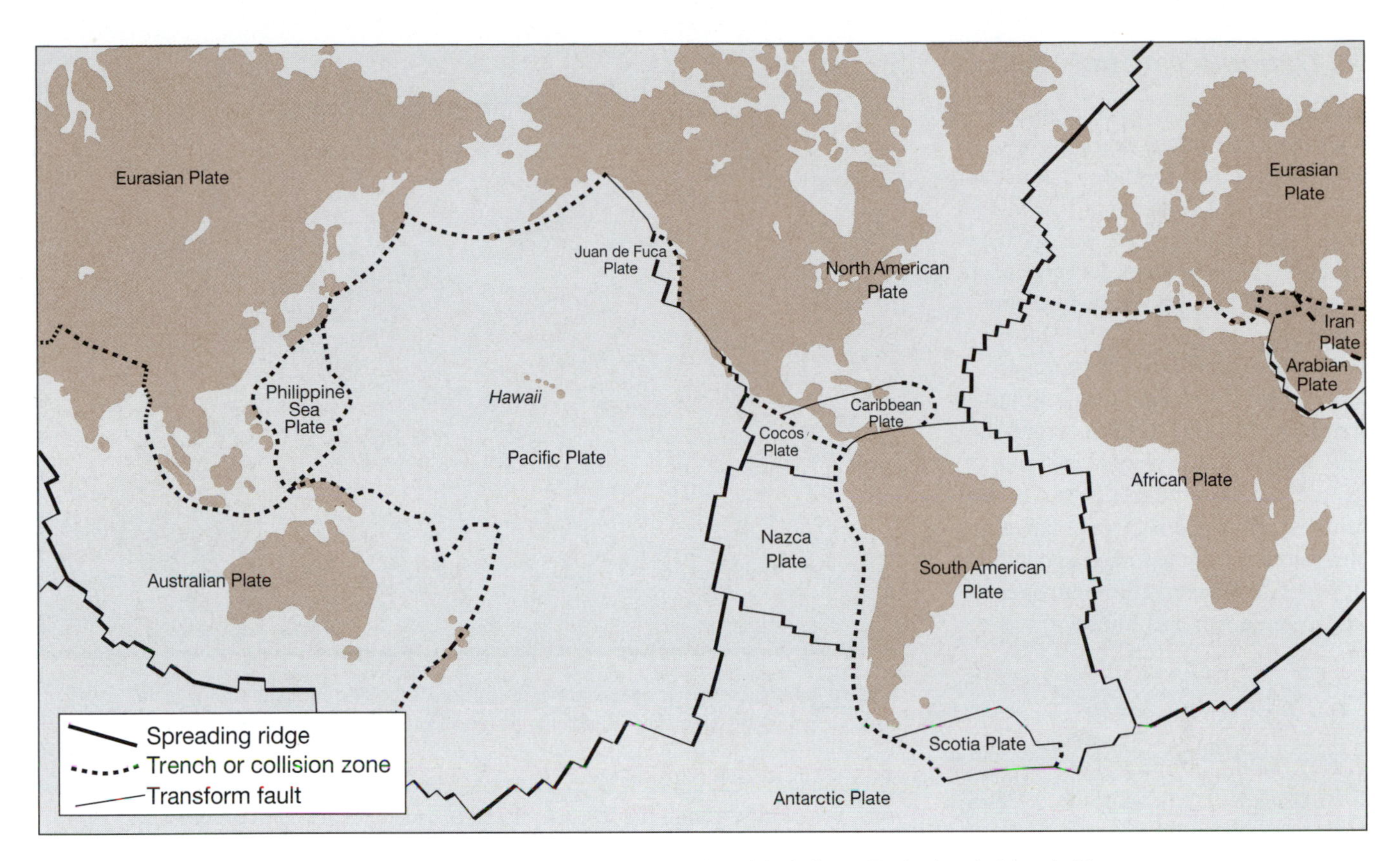

Figure 7.40 Schematic representation of Earth's major global plates. Each plate is bounded by some combination of spreading ridges, subducting trenches, and transform faults.

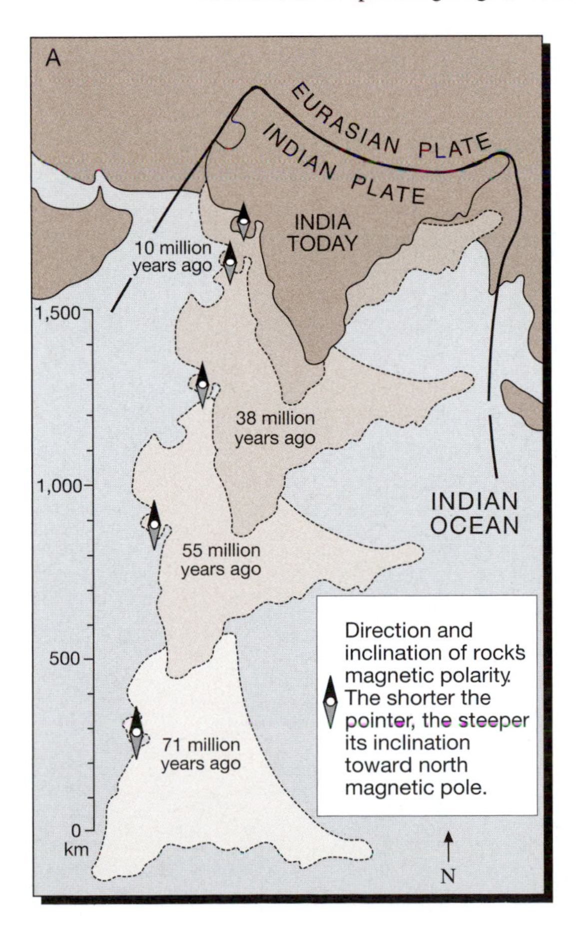

Continental drift: India, a case history—One of the most cited case histories in plate tectonics literature is that of the subcontinent of India's having departed the Southern Hemisphere and traveled northward to eventually collide with Eurasia (Fig. 7.41), thereby pushing up the lofty Himalaya Mountains and Tibetan Plateau—'the rooftop of the world.' A visual metaphor: Kicking wrinkles in the edge of a carpet.

Figure 7.41A Paleomagnetic measurements reveal former latitudes and orientations of India. (Shorter 'needles' indicate steeper inclinations of polarities at higher latitudes.) Ages of the volcanic paleomagnetic samples were determined radiometrically. **B** This is a geologic cross-section from south-to-north through the Indian-Eurasian plate boundary.

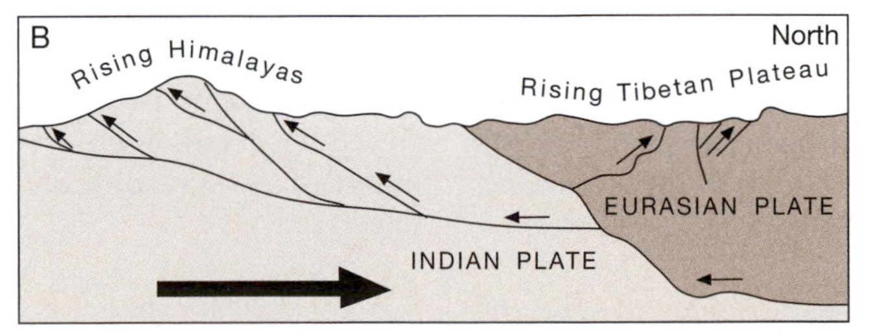

Q7.27 Figure 7.41A includes both a map scale and the number of years-before-present of earlier positions of the subcontinent of India. From these data, compute the average rate (in centimeters per year) of northward movement of India during the past 71 million years.

H. Dynamics of hot spots

Volcanoes typically occur along spreading ridges and subducting trenches, but how about those volcanoes that occur *within* plates, e.g., those of Hawaii (Fig. 7.40, page 125)? Again, we have J. Tuzo Wilson to thank for our understanding of this curious phenomenon.

Wilson envisioned the Pacific plate moving northwestward over a relatively stationary magma chamber within the asthenosphere. Occasional eruption of the magma has produced a trail of volcanic islands—the Hawaiian Islands—downstream from the magma chamber (Fig. 7.42). Wilson referred to such a magma chamber as a **hot spot**.

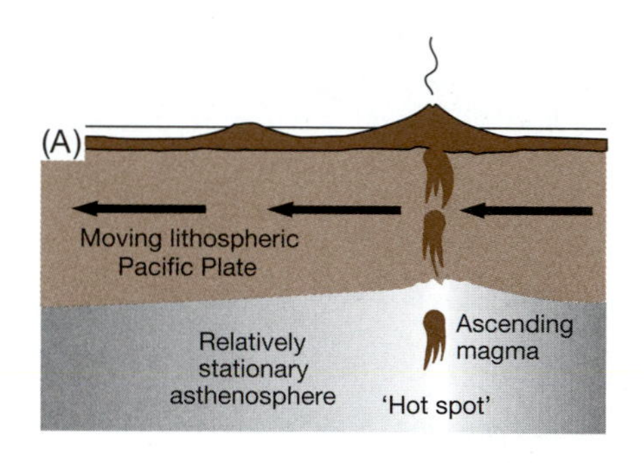

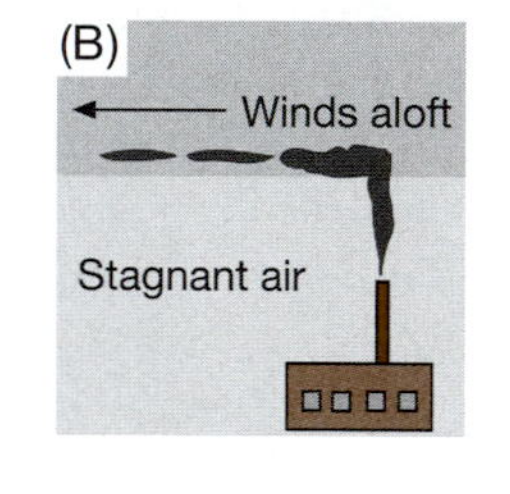

Figure 7.42 (**A**) The extinct volcano to the left of the active volcano was active while it was over the hot spot. It has since been rafted to your left on a moving lithospheric plate above a relatively stationary asthenosphere. **(B)** A graphic metaphor: Winds aloft moving over a stagnant air mass below.

Radiometric dating of the volcanic rocks that comprise individual islands within the Hawaiian chain confirmed Wilson's view of the islands' origin (Fig. 7.43).

Q7.28 Using the kilometer scale and the age provided for Oahu in Figure 7.43, calculate the rate (in centimeters per year) at which the Pacific plate is moving northwestward relative to the hot spot (which is marked by present-day volcanoes).

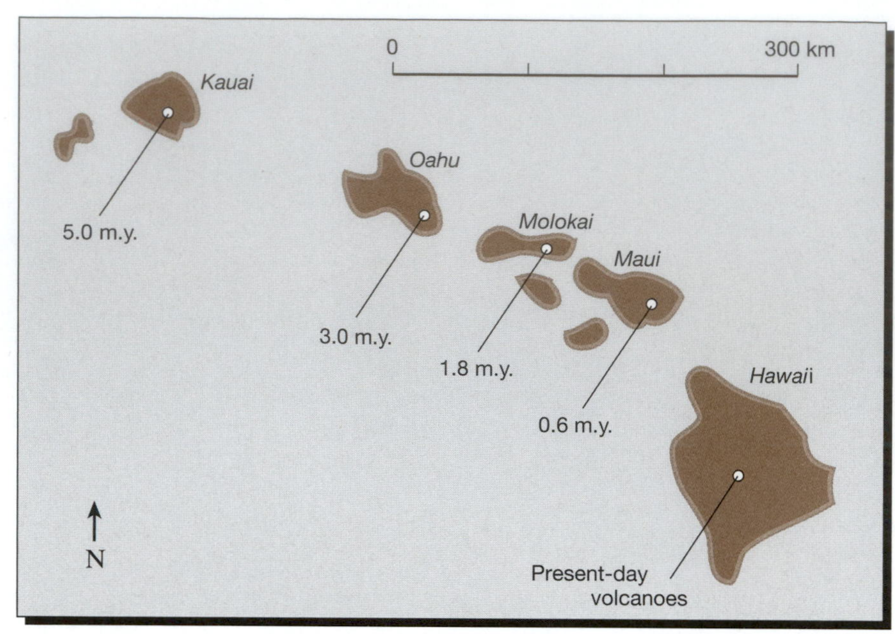

Figure 7.43 Radiometric ages (shown in millions of years) indicate that volcanic rocks that comprise the Hawaiian Islands are progressively older toward the northwest.

More evidence for the Hawaiian hot spot—The Hawaiian Island story extends northwestward through the Midway Islands to Emperor Seamounts some 5,000 feet below sea level (Fig. 7.44). Rocks dredged from atop Suiko Seamount reveal fossil corals associated with the 58-million-year-old volcanic rocks. Today, corals in this part of the Pacific are restricted to south of a line of latitude where the mean annual temperature of surface water is 22°C. And, modern corals are restricted to shallow sunlit waters for reasons of living symbiotically with microscopic algae.

Figure 7.44 The Hawaii-Midway-Émperor Seamount chain consists both of present islands and of drowned ancient islands (aka seamounts or guyots).

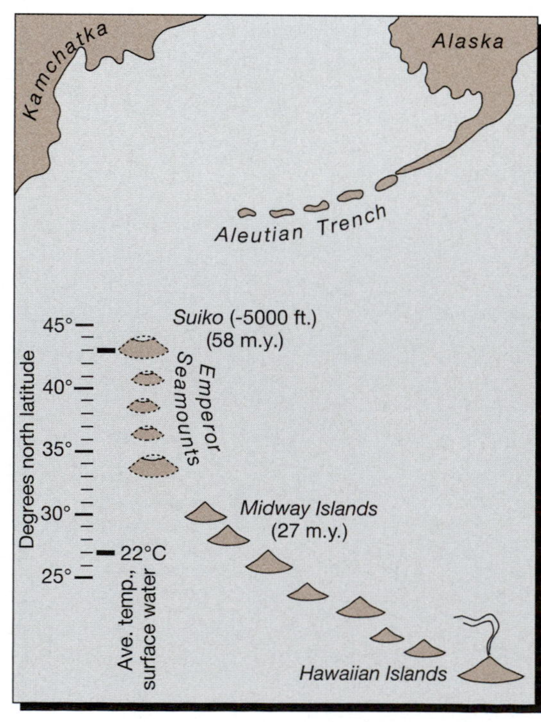

Q7.29 Assuming that Pacific water temperatures today are like those of 58 million years ago, explain both the anomalous latitude and the anomalous water depth of fossil corals atop Suiko Seamount. *Hint:* Recall the details of guyots and sea-floor spreading on page 119.

Q7.30 Describe the direction of movement of the Pacific plate during the development of the volcanic islands that comprise the Emperor Seamounts. How has that direction changed?

____________________ (Student's name) ________ (Day) ________ (Hour)

____________________ (Lab instructor's name)

ANSWER PAGE

7.1 ____________________

7.2 (A) ____________________

7.3 (A) ____________________

7.4 ____________________

7.5 ____________________

7.6 (A) ____________________

(B) ____________________

7.7 ____________________

7.8 ____________________

7.9 ____________________

7.10 ____________________

7.11 ____________________

7.12 (A) ____________________

(B) ____________________

7.13 (A) ____________________

(B) ____________________

7.14 (A) ____________________

(B) ____________________

7.15 ____________________

7.16 (A) (B) ____________________

7.17 ____________________

7.18 ____________________

7.19 ____________________

____

7.20 ____________________

7.21 (1) ____________________

(2) ____________________

____

____

(3) ____________________

____

7.22 ____________________

7.23 (A) ____________________

(B) ____________________

____

____

____

7.24

7.25 ____________________

____

____

7.26 ____________________

7.27 ____________________

7.28 ____________________

7.29 (1) ____________________

____

____

(2) ____________________

____

____

____

7.30 ____________________

____

____

____

____

____

____

____

8 Earthquakes

Topics

A. What is the physical explanation for why earthquakes occur?
B. What are the three types of earthquake waves, and how do they differ one from another? How are earthquakes recorded, and how is the distance between a seismograph station and an earthquake determined?
C. How is the location of an earthquake determined?
D. What are the two common scales of earthquake severity, and how do they differ one from another? Why aren't fields of Mercalli intensity values bounded by perfect circles?
E. How is the Richter magnitude of a distant earthquake determined? How is the moment magnitude of an earthquake measured? Why are there no earthquakes of magnitude 11.0? How does the geology of the Northridge are explain such extreme ground acceleration?
F. Why the greater number of fatalities from earthquakes in underdeveloped countries? What is the ShakeMap project? What is the "Did you feel it?" project?
G. What is the methodology in determining the location of a quake in southeast Missouri?
H. What geologic feature caused the Sumatra tsunami? What geologic feature saved countless lives in Bangladesh?

A. What causes earthquakes?

Earthquakes are produced by abrupt motion along a fault when friction that resists such motion is overcome by stress (Fig. 8.1). This is called **elastic rebound**. It's quite analogous to the bending and breaking of a green twig. Some 21 feet of abrupt adjustment along the **San Andreas Fault** generated the fateful San Francisco earthquake of 1906.

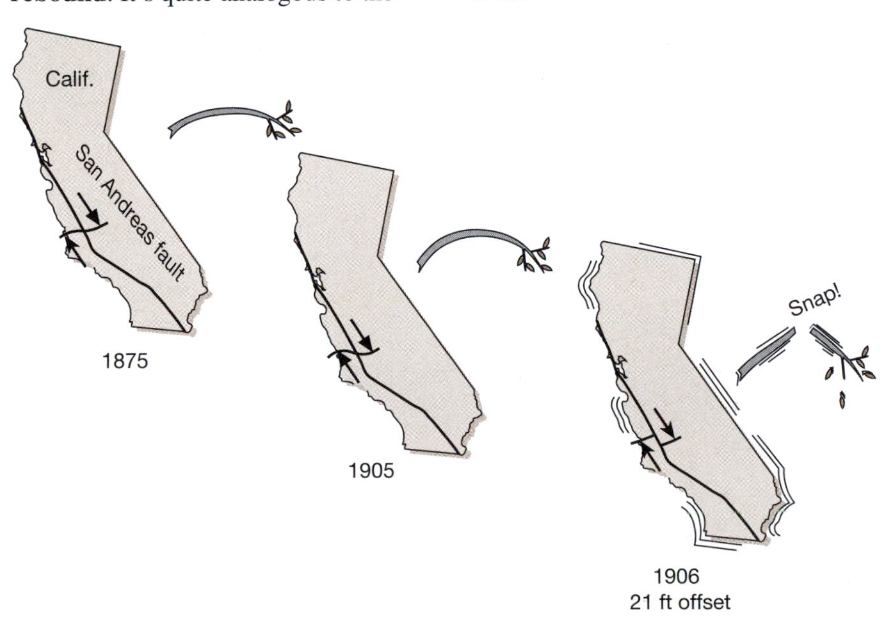

Figure 8.1 Some 21 feet of plastic strain across the San Andreas fault—exhibited here by distorted utility lines and such—accumulated before the correction of 1906.

Most earthquakes are produced by movement along a fault. But movement is not along the entire fault during any one event. Instead, movement involves a region of the fault measured over a few kilometers. The place within that region where movement first occurs is called the **focus**. But of more interest to people is the spot on the ground directly above the focus, which is called the **epicenter**. The epicenter equates with the spot where there is maximum ground motion and maximum damage (other factors being equal).

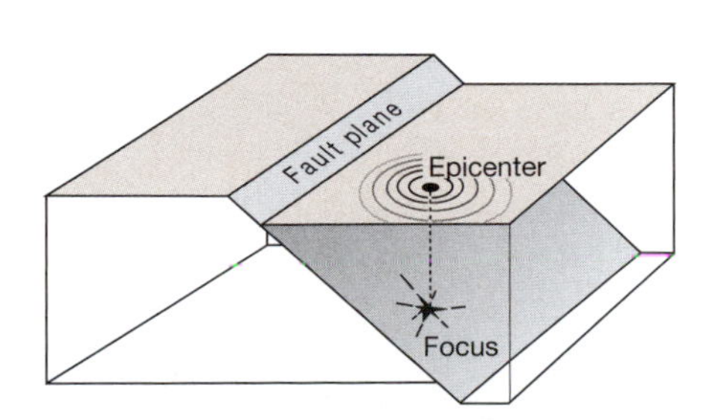

B. Earthquake waves

Good news, bad news

Good news—The P and S waves enable geophysicists to chart and decipher Earth's deep interior.

Bad news—The surface wave is the wave that destroys countless lives and untold property.

Earthquake waves—aka **seismic waves**—come in three varieties (Fig. 8.2):
1. the surface wave
2. two penetrating body waves
   (2a) the P wave
   (2b) the S wave

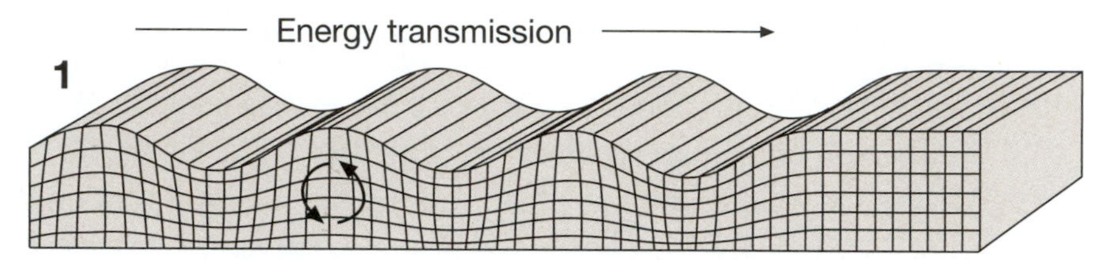

Surface wave. Rolling of rock particles (arrows) in response to passage of a moving wave. The wave form is much like that of water particles within a wave in that particle motion diminishes downward.

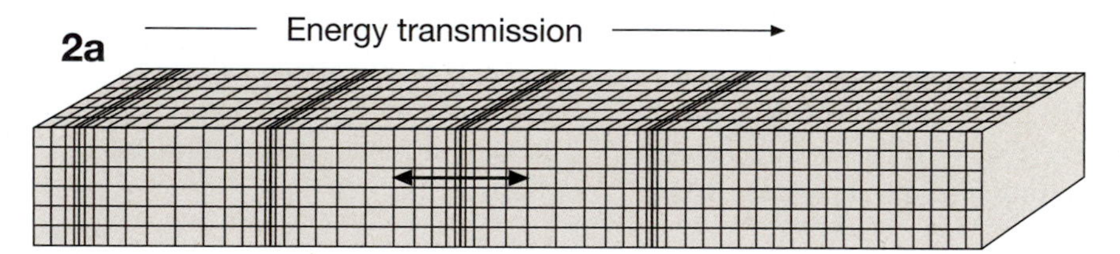

Figure 8.2 Three seismic waves consist of one surface wave and two penetrative body waves.

Primary or P wave. Motion of rock particles (arrow) is parallel to the direction of energy transmission. The wave form is that of compression and relaxation of rock, a bit like a Slinky™ toy.

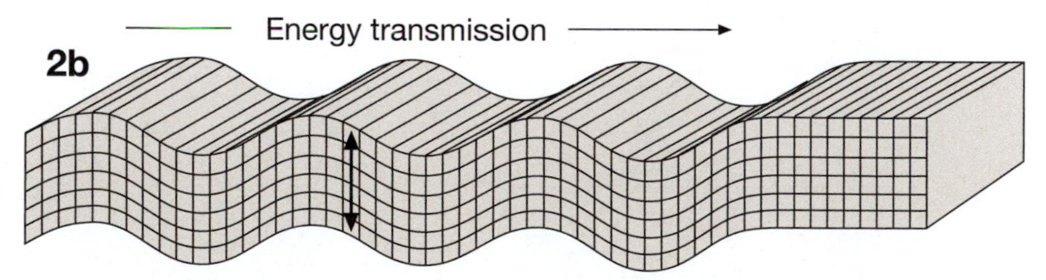

Secondary or S wave. Motion of rock particles (arrow) is perpendicular to direction of energy transmission. Standing waves result, like the distortion of a trampoline.

Recording earthquakes

Earthquakes are recorded with instruments called **seismographs**. A simple seismograph consists of a weakly anchored ink pen scribing a rotating drum that is firmly anchored to the ground (Fig. 8.3). An earthquake that moves the drum has relatively little effect on the pen, so ground motion produces a zigzag line called a **seismogram** (Fig. 8.4).

Q8.1 Judging from the seismogram in Figure 8.4, which wave appears to be the most damaging?

Determining distance to an earthquake

Speeds at which earthquake waves travel depend on the density and rigidity of rocks along their paths. But, regardless of rock character, the P wave is the swiftest, the S wave is intermediate, and the surface wave is the slowest (Fig. 8.4). Because of this difference in wave velocities, the difference in arrival times of P and S waves at a seismograph station differs as a function of the distance between the earthquake and the station (Fig. 8.5).

This relationship calls to mind the kind of questions that you encountered while taking college-entrance exams. For example, two seismology students, in separate vehicles, depart the site of a recent earthquake in southern California—at the same time—and drive to the Cal Tech campus. One drives 50 m.p.h., the other 40 m.p.h. One arrives at 4:00 P.M., the other arrives at 4:30 P.M.

Q8.2 How many miles was the earthquake from Cal Tech? *Hint:* see the bottom of Answer Page 145 for partial solution.

Figure 8.5 is a graph showing the progressive increase in separation between arrival times of P and S waves. An example described in the caption to Figure 8.5 has been plotted.

Q8.3 Determine the distance to an earthquake at a station that receives P and S waves 5.0 minutes apart. *Hint:* (a) Place tick marks on a scrap of paper equal to 5.0 on the minutes axis. (b) Fit that to the horizontal separation between P and S curves. (c) Read distance directly across on the distance axis.

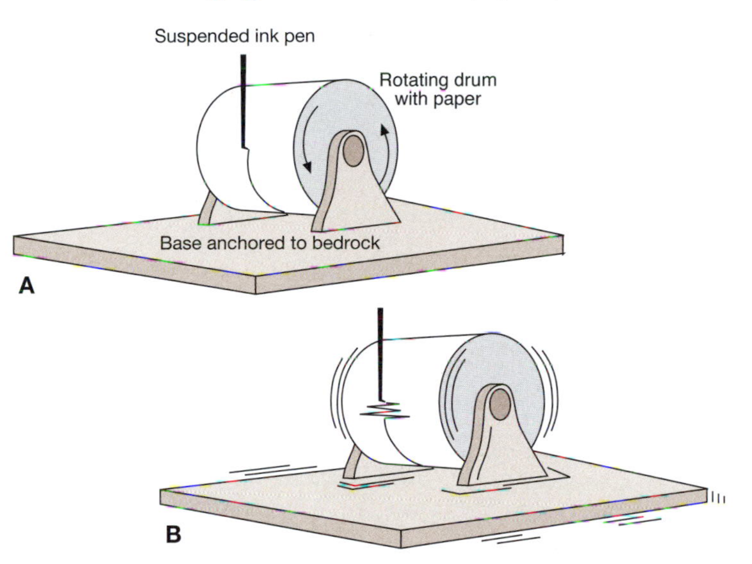

Figure 8.3 A seismograph is designed so that during an earthquake there is motion between rotating paper and an ink pen. (A) A seismograph before an earthquake. (B) A seismograph during an earthquake.

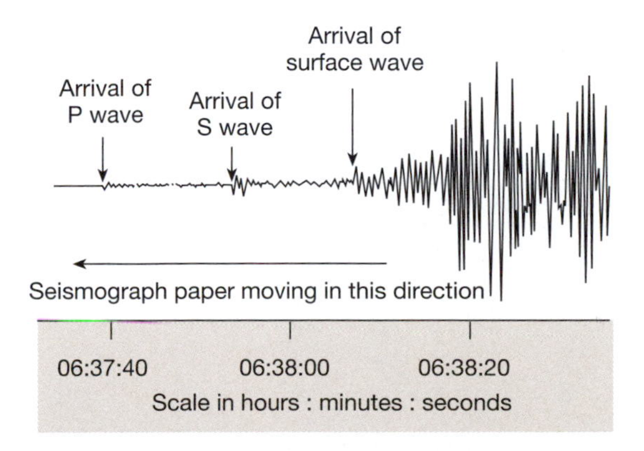

Figure 8.4 Schematic drawing of a seismogram showing arrivals of the three kinds of seismic waves. The paper was moving from your right to your left, so the earliest part of the record is to the left. The record is graduated in 20-second intervals.

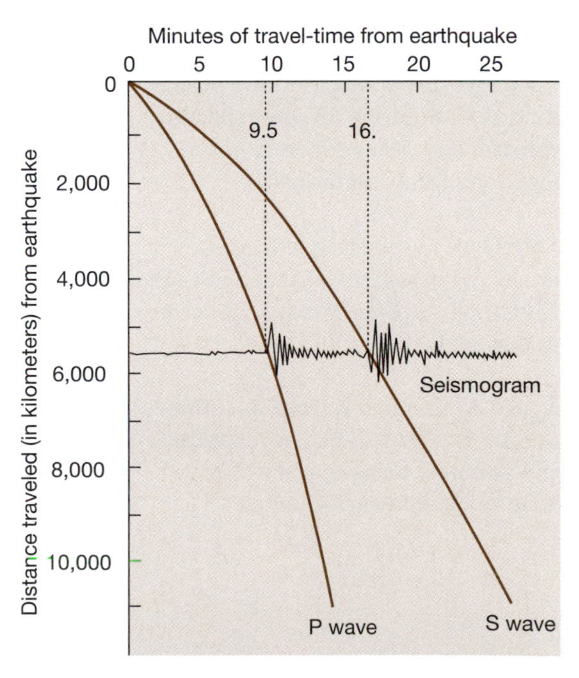

Figure 8.5 A difference in arrival times of P and S waves of 7.2 minutes (i.e., 16.7 minus 8.5) indicates the distance to the earthquake of 5,600 kilometers.

C. Locating earthquakes

Subject: At midnight on August 17, 1959 an earthquake dislodged a wall of rock in the Madison River Canyon of Montana, burying 28 campers in the valley below.

Methodology for locating that quake

Step A: A seismologist at a seismograph station in Seattle, Washington determined the distance to the earthquake, using the difference in arrival times of the P and S waves. Then the seismologist constructed around his station a circle, the radius of which is the distance between the station and the earthquake.

Q8.4 At this point, from the information in Figure 8.6A, how specific can you be as concerns the location of that earthquake?

Step B: A seismologist at a second seismograph station—that at Berkeley, California—applied the same procedure as that of the seismologist at Seattle.

Q8.5 At this point, from the information in Figure 8.6B, how specific can you now be as concerns the location of that earthquake?

Step C: A seismologist at a third seismograph station—that at Salt Lake City, Utah—applied the same procedure as that applied at Seattle and Berkeley.

Q8.6 At this point, from the information in Figure 8.6C, how specific can you now be as concerns the location of that earthquake?

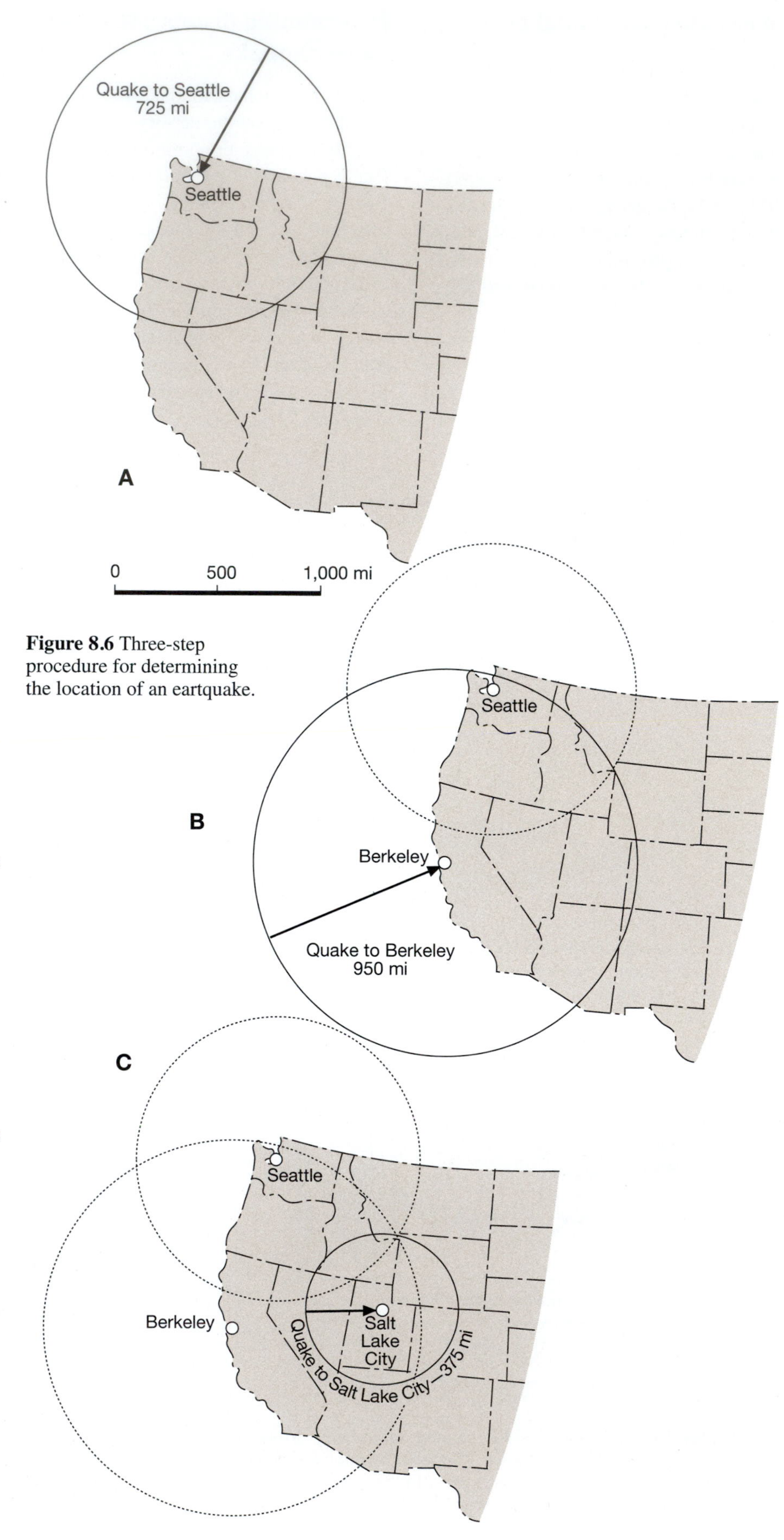

Figure 8.6 Three-step procedure for determining the location of an eartquake.

D. Earthquake scales of severity

Mercalli and Richter scales

The two most popular scales for expressing earthquake severity are those of **Mercalli intensity** and **Richter magnitude.**

In 1902 the Italian scientist Giuseppe Mercalli devised his *Mercalli intensity scale*, which assigns a Roman numeral from I through XII to an earthquake based on eyewitness accounts of the effects of that earthquake (Table 8.1). So a Mercalli assessment of energy released by an earthquake is subjective, rather than quantitative. Also, bear in mind that Mercalli intensity diminishes with distance away from an epicenter, so it is only locally applicable.

Table 8.1 is an abbreviated and condensed version of the Mercalli intensity scale. For a complete and modified Mercalli intensity scale, see the web site at the bottom of this page.

Table 8.1
Abbreviated Mercalli intensity values and effects

X–XII	Total and major damage
IX–X	Major damage
VII–VIII	Everybody runs outdoors; moderate to major damage
VI–VII	Felt by all; many frightened and run outdoors; minor to moderate damage
IV–V	Felt by most people; slight damage
III	Felt indoors
I–II	Usually detected only by instruments

Q8.7 The effect of value III in Table 8.1 is 'Felt indoors.' Why specify indoors? Why should eye witness accounts indoors differ from eye witness accounts outdoors? *Hint:* Consider the surroundings.

The study of seismograms experienced a memorable moment in 1935 when Charles Richter devised his *Richter scale of earthquake magnitude*, which he related to the energy released at an earthquake's focus.

The measure of ground motion on a seismogram that is used to access Richter magnitude is the *amplitude of the S wave*.

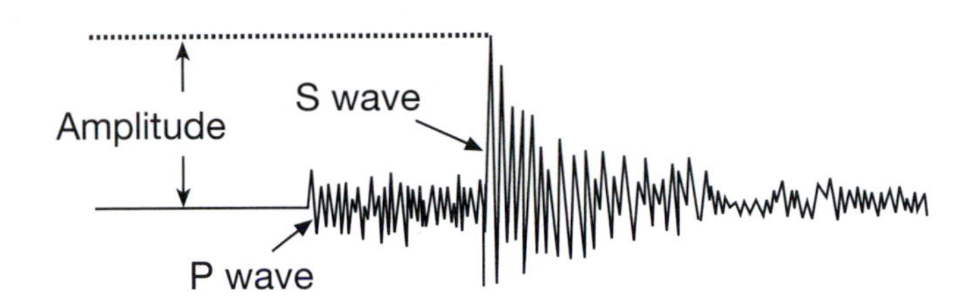

The Richter scale consists of numbers ranging from less than 0 (negative numbers) to 8.0 (Table 8.2). The scale is logarithmic in that the amplitude of the recorded S wave increases tenfold for each unit increase in Richter magnitude. For example, ground motion produced by a magnitude 5.0 quake is ten times larger than that produced by a magnitude 4.0 quake.

Q8.8 How much greater is ground motion generated by a magnitude 4 quake than a magnitude 2 quake?

Table 8.2

Richter magnitude	TNT equivalent	Example (approx.)
-1.5	6 oz	Breaking a rock on a lab table
1.0	30 lbs	Large blast at a construction site
1.5	320 lbs	
2.0	1 ton	Large quarry or mine blast
2.5	4.6 tons	
3.0	29 tons	
3.5	73 tons	
4.0	1,000 tons	Small nuclear weapon
4.5	5,100 tons	Average tornado
5.0	32,000 tons	
5.5	80,000 tons	Little Skull Mtn., Nev. quake, 1992
6.0	1 million tons	Double Spring Flat, Nev. quake, 1994
6.5	5 million tons	Northridge, Calif. quake, 1994
7.0	32 million tons	Largest thermonuclear weapon
7.5	160 million tons	Landers, Calif. quake, 1992
8.0	1 billion tons	San Francisco, Calif. quake, 1906
8.5	5 billion tons	Anchorage, Alaska quake, 1964

Modified Mercalli intensity scale is at USGS site http://www.aeic.alaska.edu/Input/lahr/magnitude/mm.html

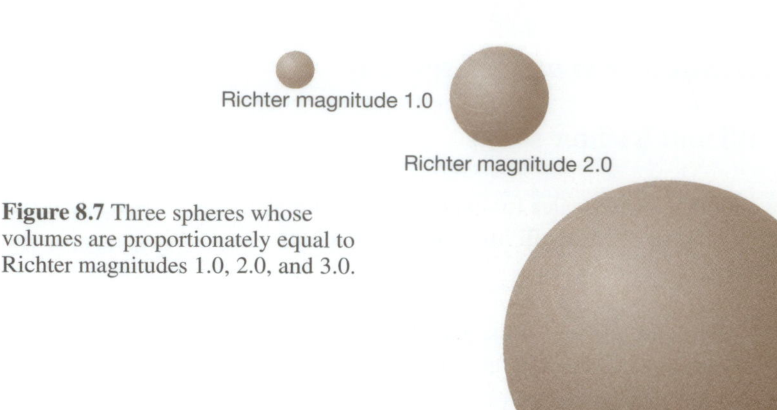

Figure 8.7 Three spheres whose volumes are proportionately equal to Richter magnitudes 1.0, 2.0, and 3.0.

An earthquake's total energy

As pointed out on the previous page, the Richter scale is logarithmic in that the amplitude of the S wave increases tenfold for each successive unit. You might view this as energy transmitted along a line from the focus of the earthquake to the seismograph. But how about the *total energy* transmitted in all directions? It turns out that the total energy increases by a factor of 30 for each successive unit on the Richter scale.

To illustrate, Figure 8.7 shows three spheres whose volumes are proportionate to total energy released by three earthquakes with Richter magnitudes of 1.0, 2.0, and 3.0. In comparison, the total energy released by the 1906 San Francisco earthquake (Richter magnitude 7.8) would be represented by a sphere with a diameter of over 100 feet!

It is difficult to equate Mercalli intensities with Richter magnitudes (Table 8.3) because Mercalli intensity depends in part on the nature of earth materials along the energy path. Weakly consolidated ('soft') rocks are more easily distorted by seismic energy than are well-consolidated ('hard') rocks. View it as an energy budget. An earthquake can spend its energy either by greater distortion over a lesser distance, or by lesser distortion over a greater distance. It's like kicking the edge of a carpet with someone standing on it a couple of feet away versus kicking the end of a 30-ft board with some standing on it at its end. This is illustrated in a real-world example in Figure 8.8.

Q8.8. Why aren't the fields of Mercalli intensity values in Figure 8.8 bounded by perfect concentric circles (as in a target pattern), rather than in these spatter shapes?

Table 8.3

Abbreviated Mercalli intensity values and effects		Richter magnitude values
X–XII	Total and major damage	8
IX–X	Major damage	7
VII–VIII	Everybody runs outdoors; moderate to major damage	6
VI–VII	Felt by all; many frightened and run outdoors; minor to moderate damage	5
IV–V	Felt by most people; slight damage	4
III	Felt indoors	3
I–II	Usually detected only by instruments	2

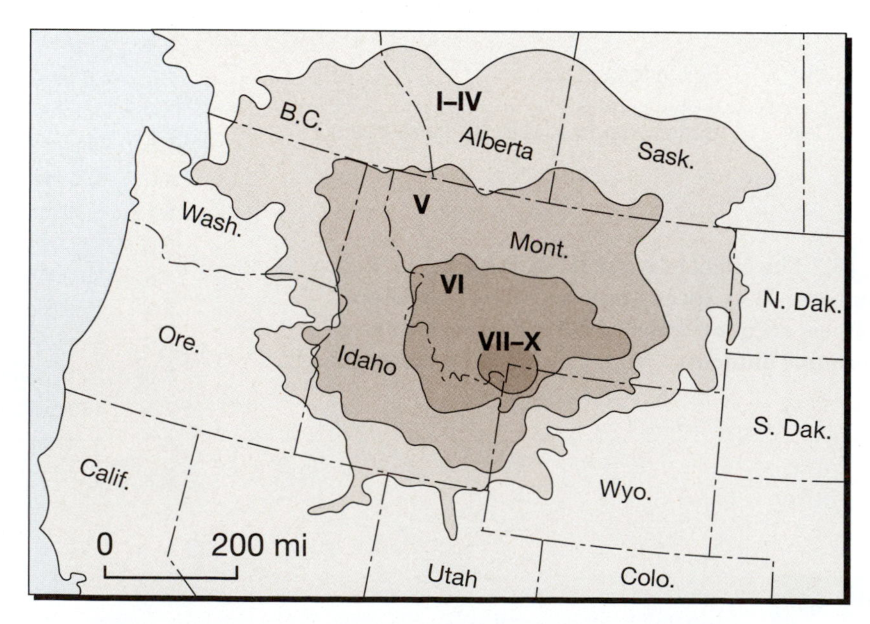

Figure 8.8 Distribution of fields of Mercalli intensity values associated with the 1959 Madison River Canyon earthquake studied on page 132.

E. Determining the Richter magnitude of a distant earthquake

For any earthquake, the farther away from the seismograph station, the less the zigzag deflection produced on the seismogram (i.e., the less the earthquake is felt at that station). Even so, it is possible for a single station to assess the magnitude of a distant earthquake by using the nomogram in Figure 8.8. The procedure follows:

Step 1 Measure the difference in arrival times of the P and S waves (expressed in increments of seconds).

Step 2 Measure the amplitude (height from base line) of the S wave on the seismogram (expressed in millimeters).

Step 3 Lay a straightedge on the nomogram so as to intersect the values for arrival time difference and amplitude.

Step 4 Read the magnitude directly from the Richter magnitude scale.

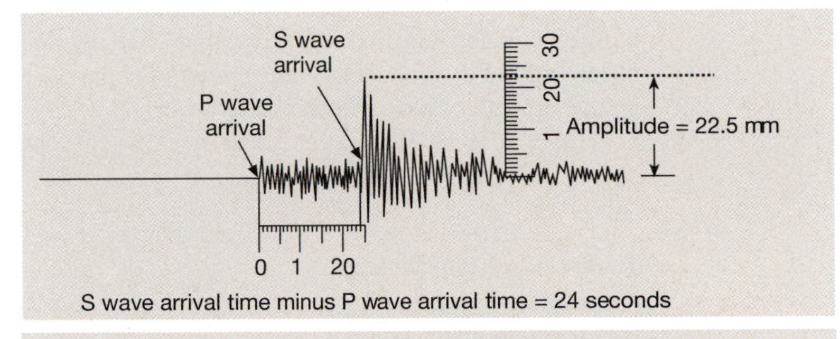

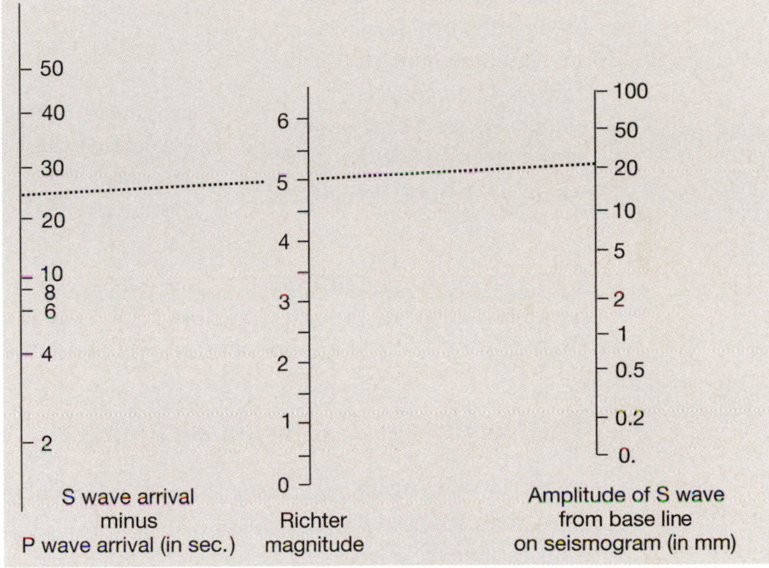

Figure 8.8 Nomogram for solving for the Richter magnitude of a distant earthquake. The example is that of an earthquake producing a difference in arrival times of P and S waves of 24 seconds and an amplitude of S wave of 22.5 millimeters. Richter magnitude is 5.0. *(Nomogram courtesy of the California Institute of Technology)*

Q8.10 Using the nomogram above, determine the Richter magnitude for the three earthquakes listed below (A, B, and C). When finished, transcribe your answers to an identical table on Answer Page 145.

S arrival minus P arrival	Amplitude of S wave	Richter magnitude
(A) 8 seconds	20 millimeters	____________
(B) 8 seconds	0.2 millimeters	____________
(C) 6 seconds	10 millimeters	____________

Moment magnitude—the seismologists' choice

Seismologists have abandoned Richter magnitude in favor of **moment magnitude** (**M_w**) in an effort to better understand earth tremors—including those produced by human activities (e.g., nuclear explosions). Moment magnitude is proportional to the amount of displacement ('slip'), *multiplied* by the area of rupture, *multiplied* by the strength of the particular earth materials.

There can be some ambiguity in communicating Richter values and moment values, because the two designations are similar in numerical value. Table 8.4 compares Richter and **M_w** values for large earthquakes that have occurred within the past two centuries.

Table 8.4

Earthquake	Richter magnitude	Moment magnitude
Chile, 1960	8.3	9.5
Alaska, 1964	8.4	9.2
New Madrid, 1812	8.7 (est.)	8.1 (est.)
Mexico City, 1985	8.1	8.1
San Francisco, 1906	7.8	7.7
Loma Prieta, 1989	7.1	7.0
San Fernando, 1971	6.4	6.7
Northridge, 1994	6.4	6.7
Kobe, Japan, 1995	7.2	6.9

Q8.11 As indicated in Table 8.4, there appears to be an upper limit to Richter and moment magnitudes at around 8.5. Why is this? Why not at around 10.5? *Hint:* See the metaphor of elastic rebound in Figure 8.1 on page 129.

The Northridge, California earthquake

At 4:31 A.M. on Monday, January 17, 1994 there occurred the most disastrous earthquake in recorded history of the Los Angeles area. The Northridge earthquake, which measured 6.7 (M_w), resulted in a displacement of 11.5 ft. —only one-half the displacement of the 1906 San Francisco earthquake, but the Northridge quake produced some of the most severe *upward ground motion* ever recorded.

Ground acceleration—*the quantitative measure of ground motion that occurs during an earthquake*—is expressed as a percent of gravity. (Think of ground acceleration as a 'heave ho' force.) Ground acceleration associated with the Northridge quake was 1.8 g (i.e., 180 percent of gravity) horizontally and 1.2 g (120 percent of gravity) vertically. At a value of 1.0 g vertical acceleration, objects not tied down are tossed into the air. In the case of the Northridge quake, ground acceleration destroyed buildings and bridges, and 61 people were killed. Another 9,000 people were injured, and the total damage was tabulated at $20 billion. A count of failed buildings produced an estimate of 3,000 deaths had the quake occurred during working hours, rather than before dawn.

Q8.12 What sorts of things might have toppled on top of children had they been in classrooms and lunch rooms at the time of the Northridge earthquake?

Geology of the Northridge seismic region

The San Andreas Fault has been the icon of California earthquakes since the San Francisco quake of 1906, but geologists are becoming increasingly aware of hazards associated with other faults *related* to the San Andreas.

The 1994 Northridge earthquake was produced by slippage along the Pico thrust—one of several thrust faults within the northern Los Angeles basin (Fig. 8.10). The 1971 San Fernando earthquake resulted from slippage along the San Fernando thrust, the epicenter of which is also shown in Figure 8.10B. (Simple line-traces of other thrust faults are shown in Figure 8.10B.)

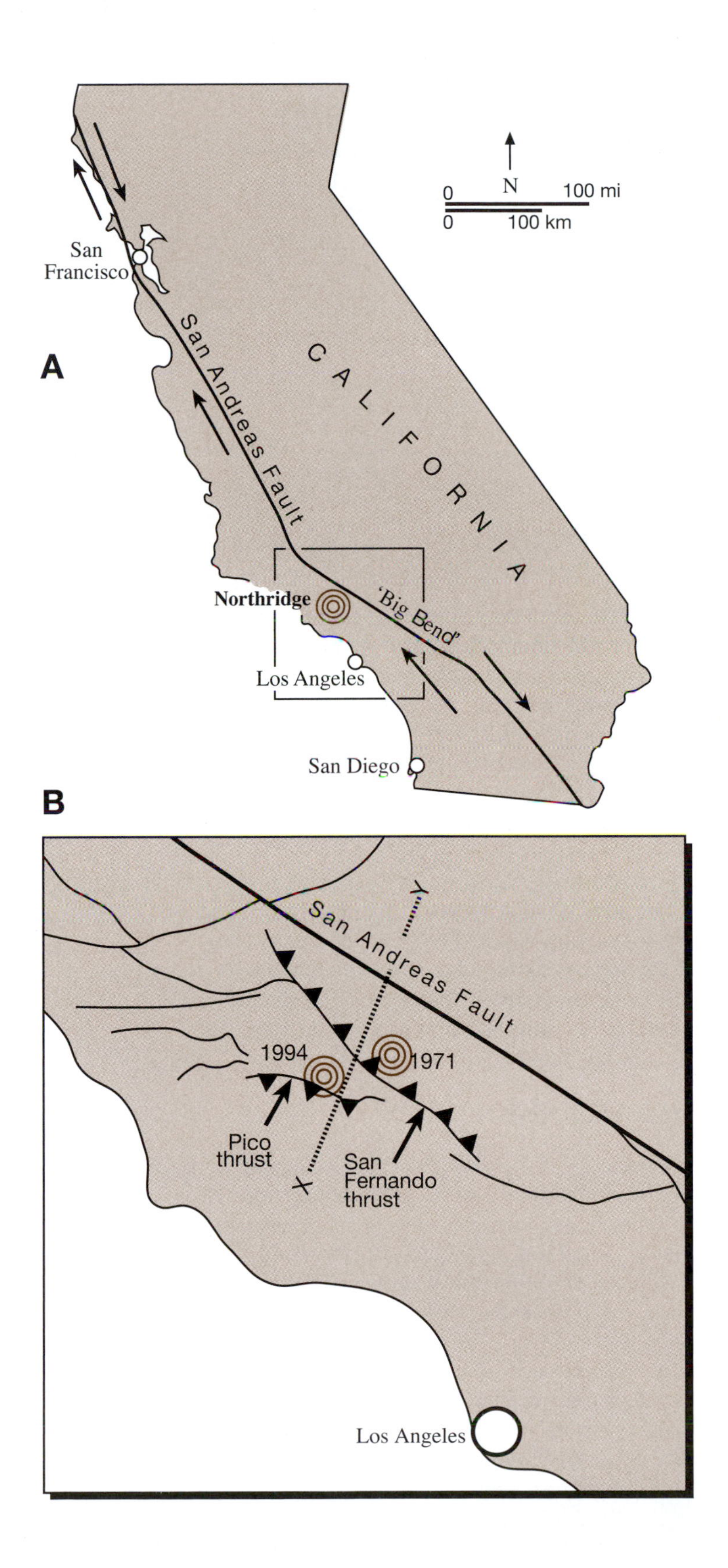

Figure 8.10 A Motion along the San Andreas Fault is lateral (arrows), but a kink in the trace of the San Andreas (aka the 'Big Bend') produces compression that results in numerous thrust faults. **B** Two thrust faults (the Pico and the San Fernando) are shown in detail, with conventional 'teeth' on the overriding block of each, indicating that the San Fernando thrust is inclined downward toward the northeast, and the Pico thrust is inclined downward toward the southwest.

Q8.13 On the Answer Page sketch a cross-section along line X-Y in Figure 8.10B. Draw each of the two thrust faults as being inclined approximately 30 degrees from an imaginary horizontal plane. (See directions of inclination of the two thrust faults in the above caption.)

F. Other earthquakes in the news

COLUMBIA DAILY TRIBUNE, MARCH 26, 2002

Afghanistan earthquake kills 1,800

10,000 homeless, aid officials say.

KABUL, Afghanistan (AP)—A powerful earthquake rocked Afghanistan and northwestern Pakistan, killing about 1,800 people and injuring 2,000, Afghan officials said today. The Afghan Defense Ministry said 600 bodies were recovered from villages still shaking from aftershocks.

"People were caught in their homes," said Nigel Fisher, a senior U.N. official in Afghanistan. U.S. Geological Survey in Golden, Colo., said it was magnitude 5.9 and centered 105 miles north of Kabul.

COLUMBIA DAILY TRIBUNE, MARCH 2, 2001

Western Washington declared disaster area

Estimates of damage break $2 billion mark.

SEATTLE (AP)—The scene left little doubt western Washington would secure an emergency disaster declaration: Cracked buildings, crushed cars and crumbled roads dominate the landscape rocked by an earthquake.

Within hours of Gov. Gary Locke's request for federal aid, President George W. Bush declared the region a federal disaster, clearing the way for low-interest loans, grants and other assistance needed to help rebuild. The declaration came yesterday as damage estimates from the magnitude 6.8 quake climbed above $2 billion.

"Earthquakes don't kill people.
Buildings kill people."

Q8.14 The 2002 Afghanistan earthquake measured 5.9 on the Richter scale and killed 1,800 people. The 2001 western Washington earthquake measured 6.8 on the Richter scale and killed only one person. Can you imagine why the huge difference in the numbers of deaths? *Hint:* It has to do with construction materials.

Some of the past century's strongest earthquakes, along with numbers of fatalities, are listed in Table 8.5.

Q8.15 The number of fatalities is not always a measure of the size an earthquake. Other forces of nature can be set in motion by earthquakes. Can you name a few? *Hint:* One consequent disaster is especially common in mountainous terrains and played a role in all of the disasters listed in Table 8.5.

Table 8.5
STRONGEST EARTHQUAKES

Some of the past century's strongest earthquakes, their locations, Richter magnitudes, and numbers of fatalities

- Dec. 16, 1920, China, 8.6, 100,000
- Aug. 16, 1920, Chile, 8.6, 20,000
- May 22, 1927, China, 8.3, 200,000
- Sept. 1, 1923, Japan, 8.3, 100,000
- July 28, 1976, China, 8.0, 242,000

The ShakeMap project

After the Northridge earthquake, the U.S. Geological Survey launched the development of project **ShakeMap**—the real-time generation of maps showing location, severity, and extent of ground shaking within seconds after the signal of an earthquake. This information goes automatically to the program's Web site and to emergency managers.

When an earthquake occurs 50 miles from downtown Los Angeles, with waves traveling at two miles per second, the network has 25 seconds to receive the data, analyze it, and broadcast it as an early warning for Los Angeles residents.

Q8.16 What would be an obvious application of an early warning system, say, during a work day? *Hint:* See the bold quotation at the top of the left column on this page. *Hint:* Think of children.

Eyewitness accounts can be more definitive than seismology in directing emergency crews to high-priority earthquake sites. USGS now has a Web site in place (see the Web address below) that allows seismologists to compile a nearly instant online map of an earthquake's intensity. This Web site received more than 17,000 reports after a quake rattled northeastern Alabama on April 29, 2003.

"Did you feel it?" A questionnaire is at http://pasadena.wr.usgs.gov/shake/ca/html/unknown_form.html

G. Determining the location of an earthquake from seismograms

The New Madrid area of the mid-Mississippi River Valley is one of the most notable seismic regions in North America. Although not measured with modern instruments, a series of three earthquakes near New Madrid, Missouri, in 1811–1812 are believed by some to have been the most intense ever to have occurred in North America in historical time. The 1811–1812 the SE Missouri earthquakes rang church bells in New England and Virginia.

Approximately 200 shocks are recorded every year in the New Madrid region. Most are detected only with seismographs, but the immediate area experiences one or two shocks every 18 months that are sufficiently strong to crack plaster in buildings.

You are asked to plot on the map in Figure 8.11 (page 140) the location of a New Madrid area earthquake using the method developed on page 132. A pencil compass is essential for accurately plotting the epicenter.

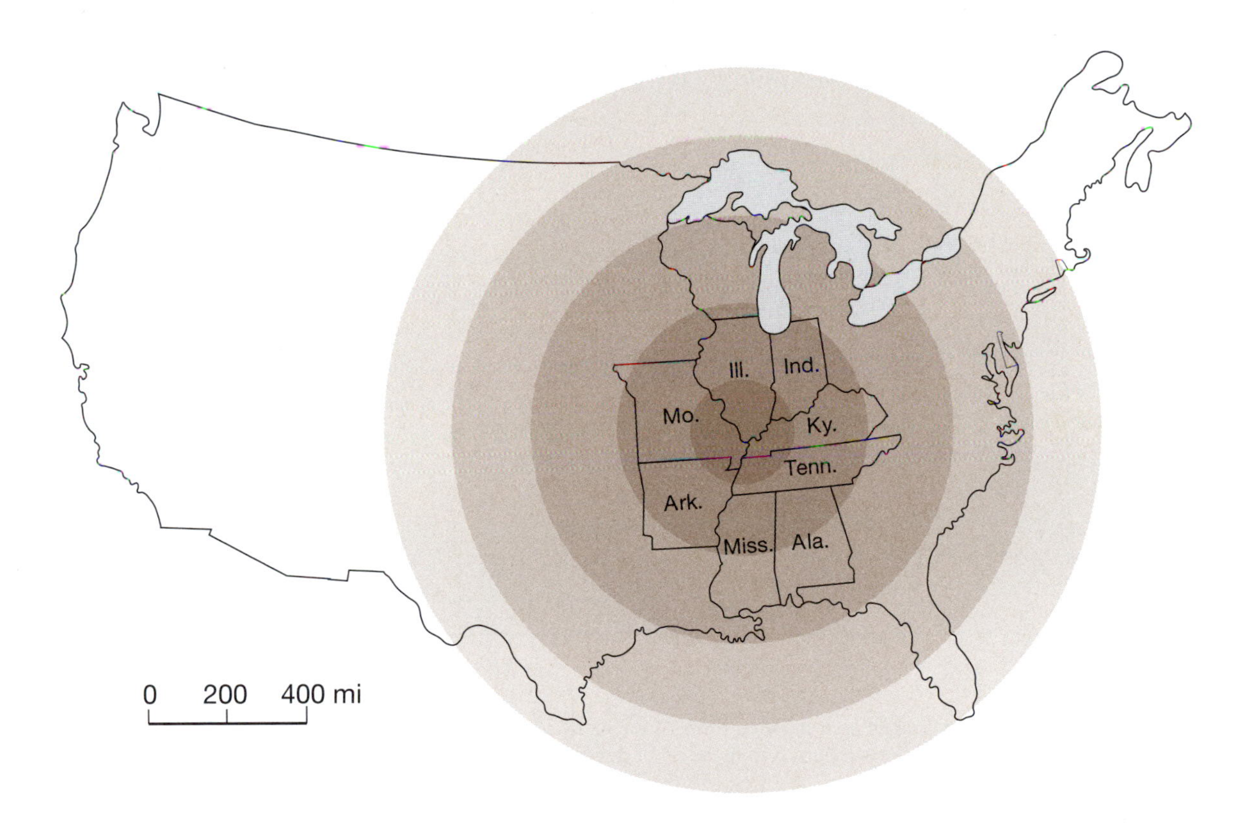

New Madrid seismic region, schematically showing the reach of its effects

Real-time earthquake information is at http://cindi.usgs.gov/hazard/earthquake.html

Locating a New Madrid earthquake

Information courtesy of Professor Brian J. Mitchell, Saint Louis University

Figure 8.11 is a map of the New Madrid seismic area showing three of the seismograph stations in the region: Powhatan, Arkansas (POW); Lennox, Tennessee (LTN); and Rosebud, Illinois (GOIL).

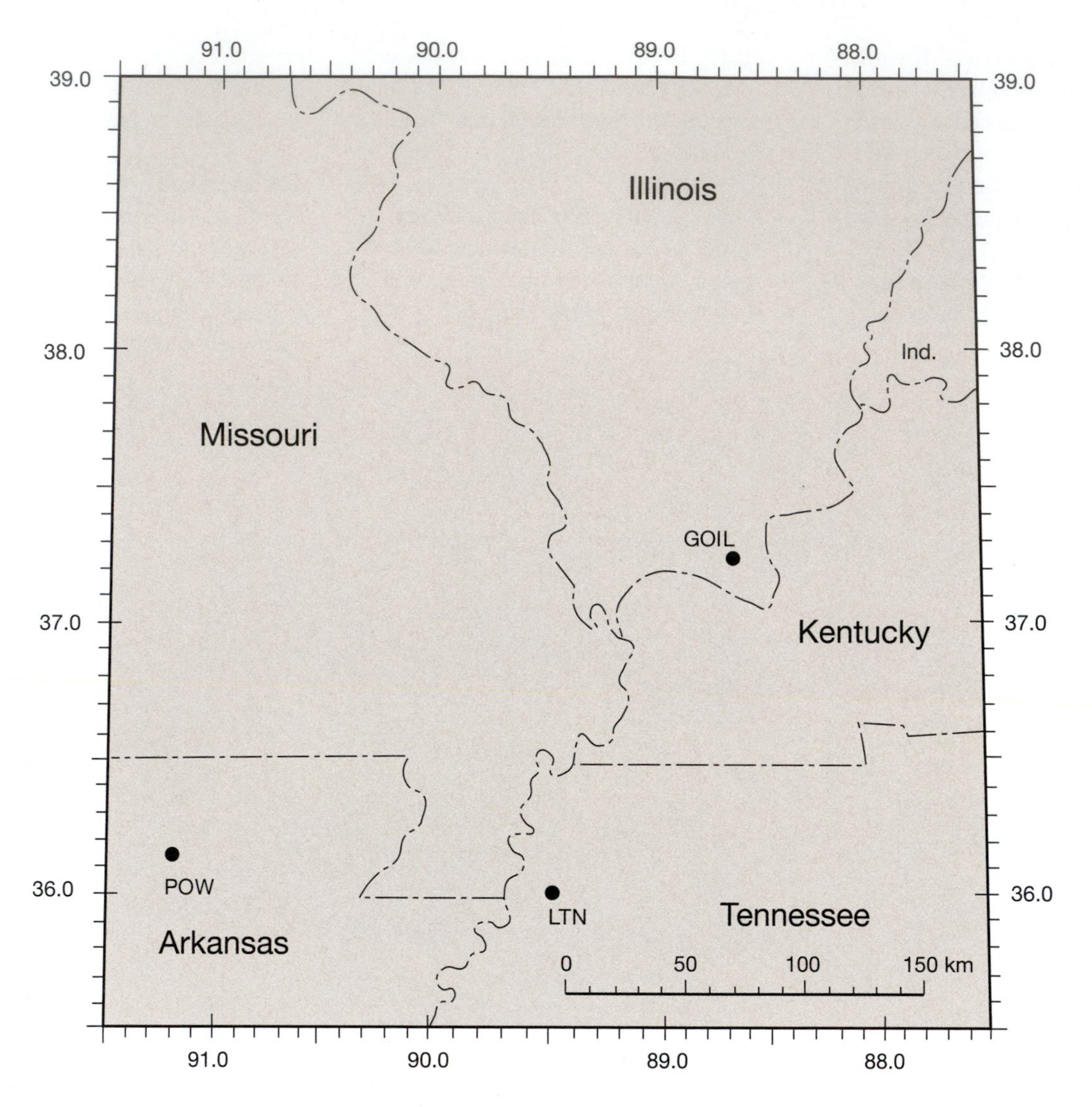

Figure 8.11 Map of the New Madrid seismic area showing three of the many seismograph stations in the region. Degrees north latitude and degrees west longitude are shown along the margins. Subdivisions of degrees are in tenths of a degree rather than in minutes of an angle.

On facing-page 141 worksheet there are three seismograms that record a Richter 3.0 earthquake that occurred in the S.E. Missouri region around 3:46 A.M. on June 19, 1987.

Q8.17 Where is the location of that June 19 quake (to the nearest tenth of a degree latitude and longitude)? To answer this question, do the following:

Step 1 Determine the arrival times of P and S waves, solve for their difference, and record on the worksheet.

Step 2 Use the graph on the worksheet on facing page 141 to solve for the distance between the earthquake and each of the three stations. Record values on the worksheet.

Step 3 Using the kilometer scale in Figure 8.11, draw a ray in any direction from each station in Figure 8.11 equal to the distance between the quake and the station.

Step 4 With a pencil compass, draw a circle around each station, using the ray as its radius. The intersection of the three circles marks the location of the earthquake.

worksheet

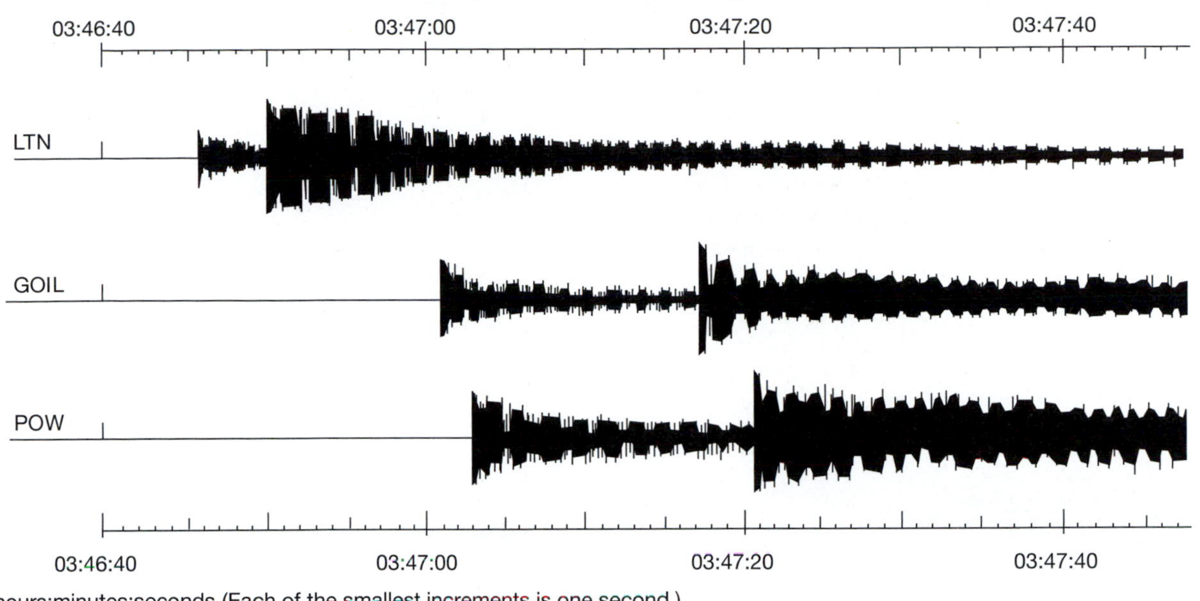

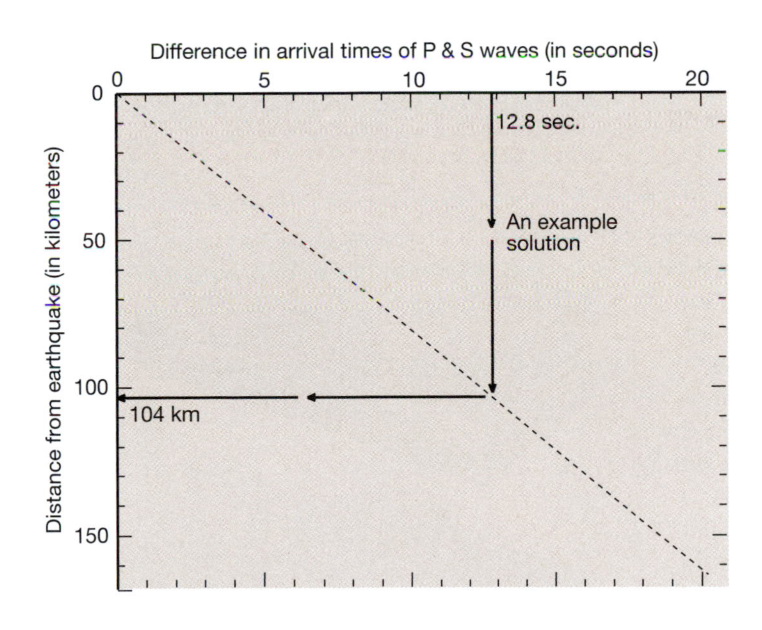

Note: This is a graph of distance from earthquake versus difference in arrival times of P and S waves. This graph differs from that in Figure 8.5 on page 131, which is a graph of distance versus separate travel times of P and S waves.

	P arrival time (hours:minutes:seconds.tenths of seconds)	S arrival time	Difference in P & S arrivals (seconds.tenths of seconds)	Distance in kilometers
LTN				
GOIL				
POW				

H. The Sumatra tsunami

Tsunami is a Japanese word meaning *harbor wave*. Perhaps the name of this devastating event derives from the facts that (a) water moving onshore tends to pile-up within harbors (i.e., the 'bottleneck effect'), and (b) there is a greater loss of life and property within harbors.

The mechanics of a tsunami are much like those of wind-driven waves (Fig. 8.12), but the two kinds of waves differ greatly in their *length, height*, and *speed* in the open sea.

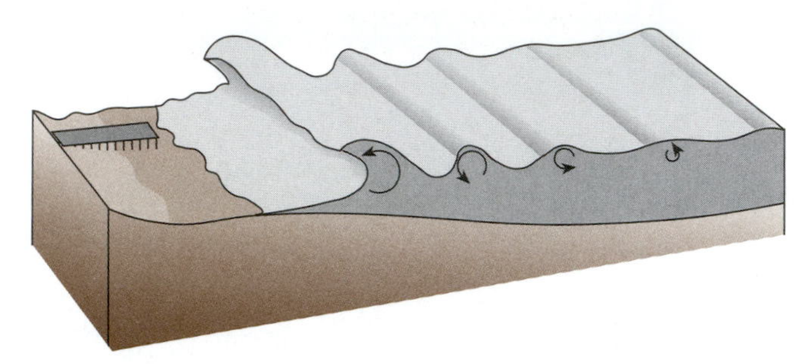

Figure 8.12 As a wave approaches shore—be it wind-driven or tsunami—drag in the shallow water causes *wave length to diminish* and *wave height to increase*. The bulging wave gathers water from both its offshore side and its shoreward side, so the initial effect of a tsunami wave is the withdrawal of water from the coast.

Q8.18 The initial withdrawal of water illustrated in Figure 8.12 lured many people to their deaths. What do you suppose lured them to the exposed sea floor?

The causes of tsunamis—Tsunamis are sea waves produced by an abrupt motion of water caused within or near an ocean by earthquakes, volcanic eruptions, landslides, meteor impacts, and nuclear detonations. The disturbance can be viewed as a shock wave that, in the open sea, can produce surface waves that travel at *speeds* of hundreds of miles per hour with *wave lengths* measured in hundreds of miles. Curiously, *wave heights* at sea are on the order of only a few feet, so little energy is spent on overcoming gravity. A passing tsunami is hardly noticed by cruise-ship passengers.

The Sumatra tsunami—On December 26, 2004, near the northwest coast of Sumatra, there occurred the largest earthquake in the global instrumental record of the past 40 years: Moment magnitude 8.3 and Richter magnitude 8.0.

The Sumatra earthquake signaled abrupt faulting along the Indian plate-Burmese plate boundary (Fig. 8.13). The Indian plate has been creeping northeastward under the Burmese plate at an average rate of a few centimeters per year, bending downward the edge of the Burmese plate (Fig. 8.14 A). On that fateful December 26th, the Burmese plate lurched upward 50–60 feet along some 1,000 miles of its length (Fig. 8.14B), sending trillions of tons of water coursing 3,000 miles westward as far as the coast of Africa where it added to the hundreds of thousands of deaths.

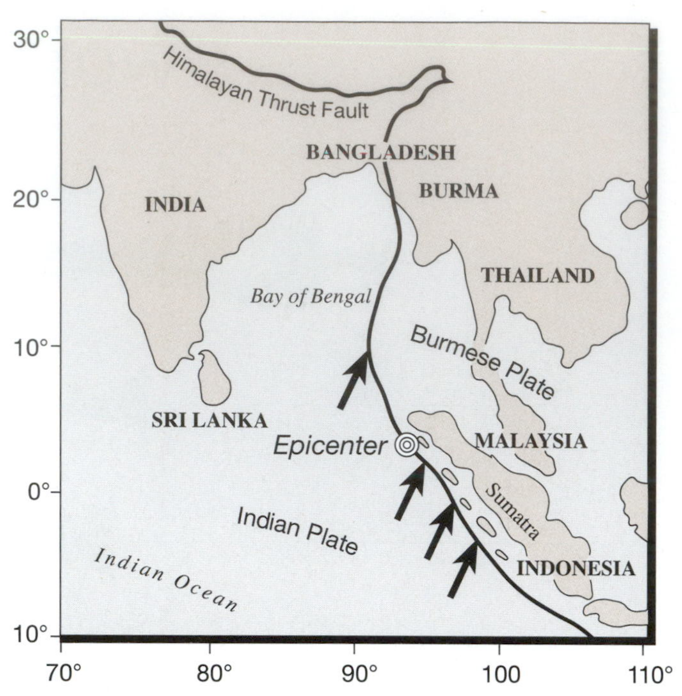

Figure 8.13 The Indian plate has been creeping northeastward under the Burmese plate at an average rate of 5 centimeters per year for millennia. *After National Earthquake Information Center (NEIC) in* Science, *v. 308, 5/20/05*

Figure 8.14 A Creeping of the Indian plate bends downward the edge of the Burmese plate. **B** Eventual rupture at the plate boundary sends the Burmese plate lurching upward, setting in motion tsunami waves across the Bay of Bengal and the Indian Ocean beyond.

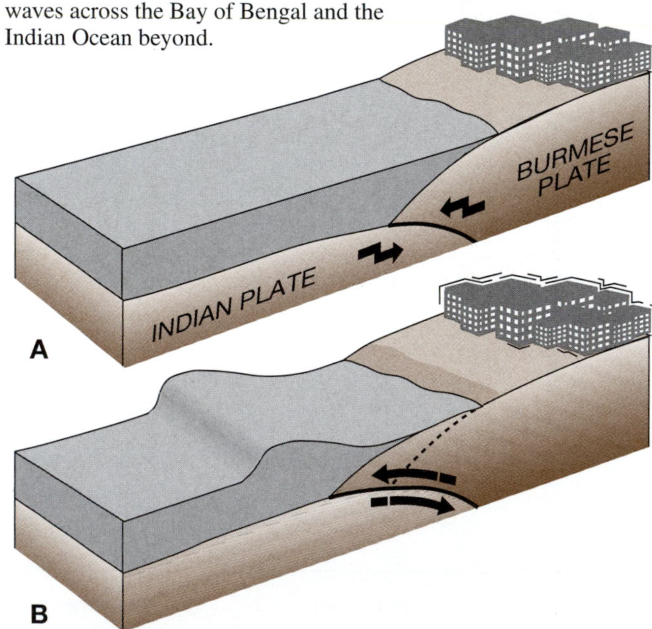

Q8.19 What is the term for the process of eventual rupture at a plate boundary? *Hint:* Recall that process as it was applied to faults on the first page of this exercise.

Scientific Background on the Sumatra Earthquake and Tsunami is at...
http://iri.columbia.edu/~lareef/tsunami/

Run-up—Tsunami **run-up** along a coast is the height of flooding, technically the vertical distance between (a) the maximum height reached by tsunami water on land and (b) average sea-level at that place.

The magnitude of run-up at a coast depends on a number of variables, including, in the case of earthquakes:

(1) The extent of crustal displacement (signaled by the earthquake)—i.e., its magnitude, depth, orientation, and length.

(2) The oceanic path taken by the shock wave—specifically the depth of water and the ruggedness of the sea floor.

(3) The configuration of the coast line, both (a) in map view (i.e., the configuration of bays and peninsulas) and (b) in vertical profile (i.e., the configuration of hills and valleys).

Variable #2 in the column at the left explains an apparent anomaly in the distribution of deaths in coastal regions shown on the map in Figure 8.15. There follows a 'rule' that applies to variable #2:

The rule: The deeper the water along the tsunami's pathway, the less energy expended by the shock wave's 'feeling' the sea floor, so the greater the energy arriving at a coast, the higher the run-up, and the greater the destruction and loss of life. And vice versa.

Q8.20 One of the curious things about the loss of life in regions surrounding the Bay of Bengal is that some 38,195 lives were lost in Sri Lanka, whereas only 2 lives were lost in Bangladesh. How could this be? It certainly couldn't be the difference in distance. *Hint:* To answer this question, you should first contour the map in Figure 8.20 on the Answer Page and then follow 'the rule.'

Figure 8.15 This map of the Bay of Bengal shows depths in meters. (Map derived from Chart I , Bottom Topography of the Oceans, in *The Oceans*, by Sverdrup, Johnson, and Fleming, 1942)

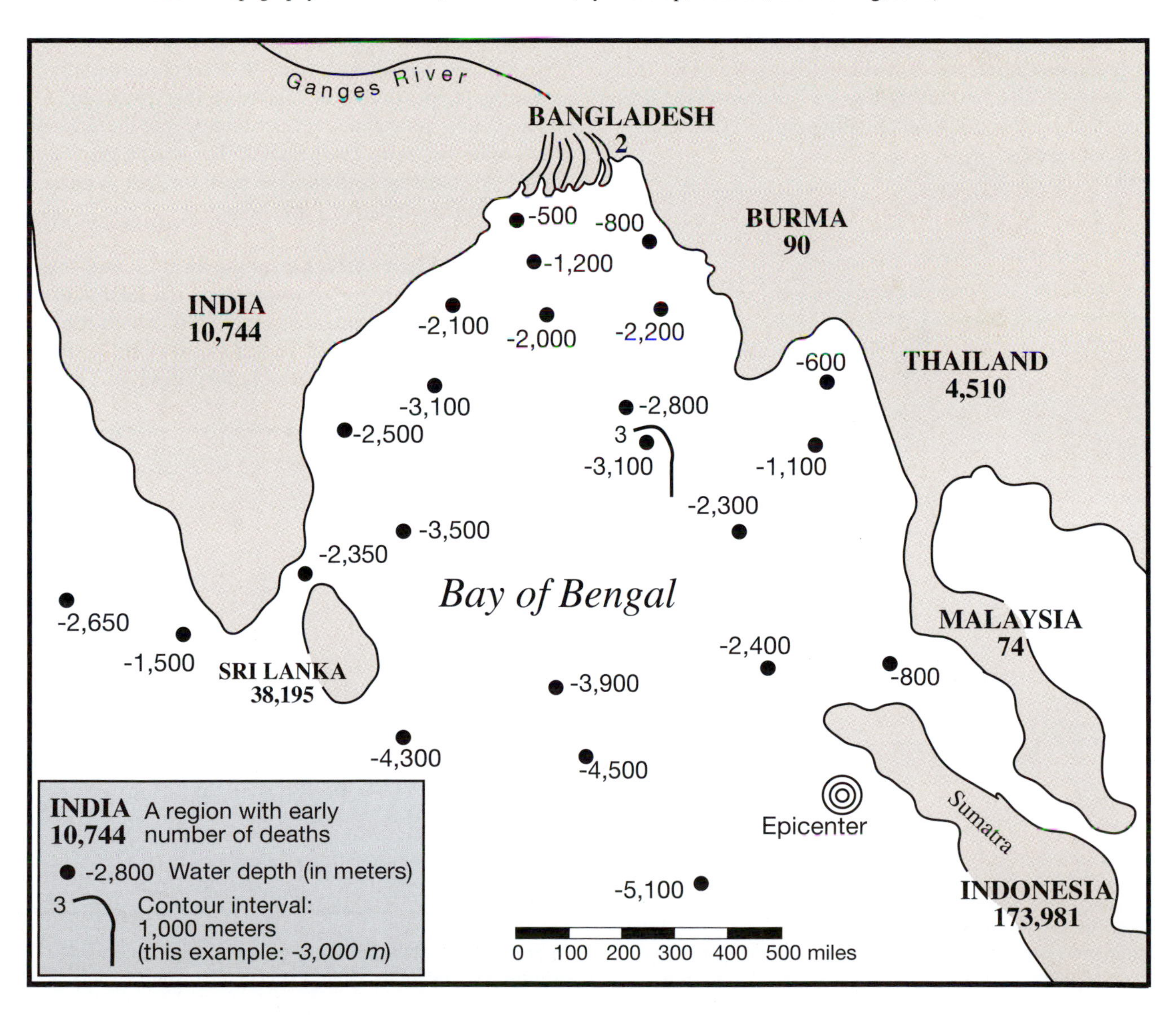

Early warning systems

Q8.21 Given the velocity of 500 miles per hour, approximately how long did it take for the Sumatra tsunami generated at the epicenter to reach the eastern shore of Sri Lanka? *Hint:* Look back at Figure 8.15 on page 143 and apply the graphic scale.

Inasmuch as most tsunamis are generated by earthquakes, there is usually time—*from 20 minutes to two hours*—for many people to climb above the likely limits of run-up. Had there been an early warning system for tsunamis in the Indian Ocean in December 2004, untold thousands of lives would have been saved.

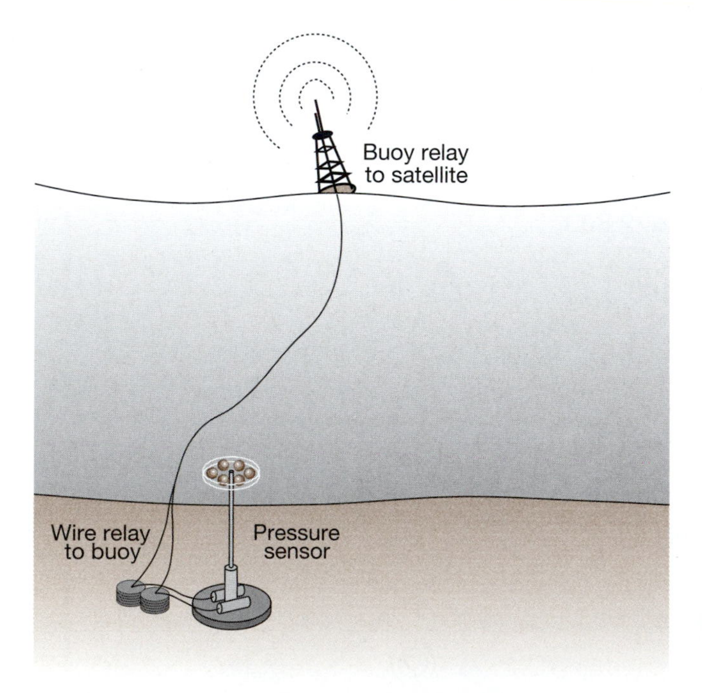

Figure 8.16 A station within a system of pressure sensors consists of (a) a sea floor sensor that receives pressure data, (b) a device that relays data to a surface buoy, which, in turn, relays data to a satellite, and (c) a management center that receives data from the satellite, processes that data, and transmits information to coastal warning facilities.

Technology—Not all oceanic earthquakes produce tsunamis, and not all tsunamis are produced by earthquakes. Inasmuch as seismic data alone can't predict tsunamis, geologists have relied on an array of gauges anchored near shore to measure wave heights around the Pacific Rim. Beginning in 1996, these water-level gauges have been supplemented by deep-ocean sensors, which measure the pressure of passing waves and transmit information to networking satellites (Fig. 8.16).

America and Japan partner in a system of seabed pressure-detectors and share data with their Pacific neighbors. This system of seven detectors, which is managed in Hawaii, cost $18 million in research and development. A similar system might now be installed for $2 million. A U.N. initiative is underway to install this technology in the Indian Ocean.

Are we Americans safe? In addition to our Northwest's subduction-plate geology, and the precarious setting of Hawaii, there are other dangers. One example: A possible landslide on the western side of the island of La Palma in the Canary Islands. It seems that a huge crack in Cumbre Vieja Volcano was opened by its 1949 eruption, thereby increasing the possibility of a landslide that could send a hundred-cubic-mile chunk of rock sliding into the Atlantic at 200 miles per hour. The height of the tsunami generated by such a submarine landslide has been forecast to be as much as 80 feet along our East Coast.

Q8.22 (Ref: Figure 8.17) Given the location and time of an earthquake, and the locations of six sea floor sensors, what would be the greatest *distance* from which the time of arrival of the tsunami at Target Coast could be forecast—500, 800, 1,000, 1,200, or 1,500 miles?

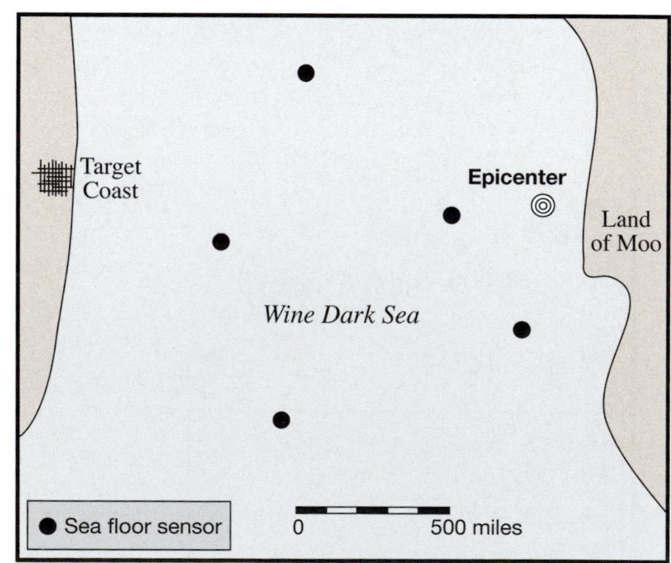

Figure 8.17 Locations of the epicenter of an earthquake and an array of five stations within a system of pressure sensors.

(Student's name) (Day) (Hour)

(Lab instructor's name)

ANSWER PAGE

8.1

8.2

8.3

8.4

8.5

8.6

8.7

8.8

8.9

8.11

8.10

S arrival minus P arrival	Amplitude	Magnitude
(A) 8 seconds	20 millimeters	
(B) 8 seconds	0.2 millimeters	
(C) 6 seconds	10 millimeters	

8.12

8.13

Note: Add arrows to the hanging walls of Pico and San Fernando thrusts to indicate the direction of ground motion during an earthquake.

8.14

8.15

8.16

Partial solution to Q8.2 (page 131) :
Note: We must use the same temporal unit—hour—in expressing both rate (i.e., speed) and time. So, the 30- minute difference in arrival times must be expressed as 1/2 (*hour*), inasmuch as rate is expressed in miles per *hour*.

Step 1: D [distance] = r [rate] t [time], so . . .
Step 2: D = 50 t [for faster vehicle], and . . .
Step 3: D = 40 (t + 1/2) [for slower vehicle]
Step 4: D = D [they drove the same distance], so . . .
Step 5: 50 t = 40 (t + 1/2)
Solve for t [time], plug it into Step 2, and solve for D.

8.17 latitude: ______ longitude: ______

8.18 ______

8.19 ______

8.20 ______

8.21 ______

8.22 ______

Question 8.20 figure

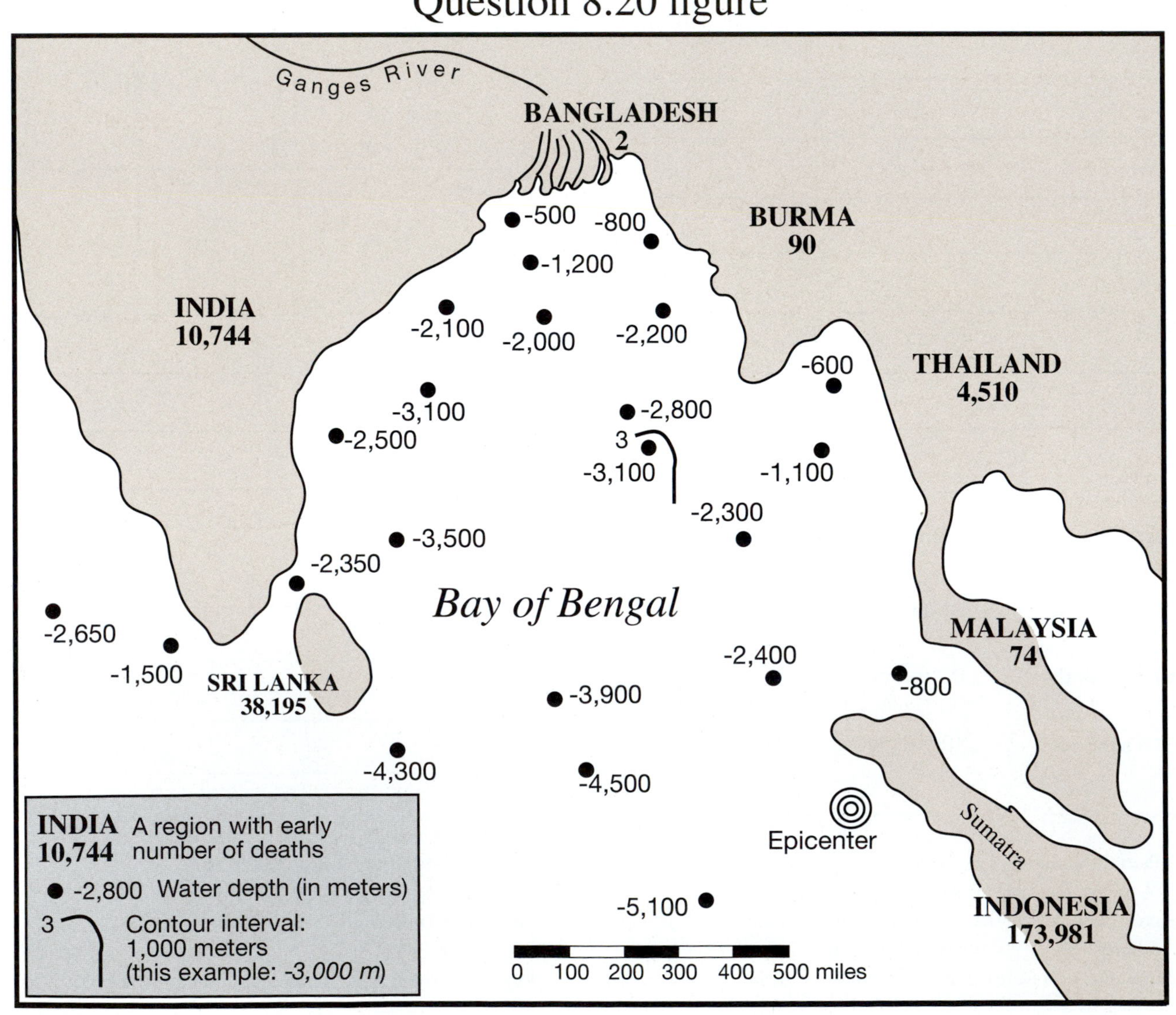

9 Streams and Rivers

Topics

A. What has been the extent of damages and loss of life from floods in the United States over the past half-century?
B. What two variables affect a stream's ability to erode? What is the origin of alluvial fans?
C. What features characterize youthful, mature, and old age landforms?
D. How do gradients differ among rivers and within river systems? What is base level?
E. What are the features that characterize floodplains? How do natural levees develop?
F. Why is it that artificial levees raise flood levels upstream? What weather pattern produced The Great Flood of 1993?
G. How do entrenched (aka incised) meanders develop?
H. What four things can occur to a drop of rain (or snow) that falls to earth in inland regions? Which two of these four events are manageable? Why does urbanization increase runoff? What is a watershed, and what is its importance in urban development?
I. How do river terraces originate? What is a notable example of a terraced river in Wyoming?
J. What is the model for development of a superimposed river that invokes alternating humid and arid climates? What is a notable example of a superimposed river in Pennsylvania?
K. What is a channelized river? Why do channelized rivers grow in their power to erode?
L. What geologic settings are revealed by dendritic and trellis drainage patterns?

A. Surface waters—some good, some bad

Streams, rivers, and lakes provide water for municipal needs, manufacturing, transportation, irrigation, electrical power, and recreation, and enhancing some of our most scenic landscapes. But there is a downside. Surface waters account for some of the greatest natural disasters known to humankind—namely floods.

Exclusive of Hurricane Katrina, **flooding** over the past half-century has accounted for an average of $3.5 billion in damages and a loss of more than 100 lives per year in the United States alone—more than any other weather-related event. Three-fourths of all presidential disaster-area designations are prompted by flooding. And yet the behavior of running water is highly predictable if a few simple principles of hydrology are kept in mind. Let's examine those principles with the aim of being better informed, should we ever find ourselves in the role of setting public policy as it applies to water management.

KANSAS CITY STAR, OCTOBER 17, 1995

LEVEE SYSTEM INCREASES LIKELIHOOD OF FLOODING

Levees cause the river to rise higher.

KANSAS CITY (AP)—Government agencies, including the U.S. Army Corps of Engineers, are rethinking the effectiveness of levees. There is a developing mood to abandon farming along rivers and letting the floodplains return to the wild.

At the same time the Corps of Engineers was feverishly rebuilding levees following the flood of 1993, several other government agencies were buying up land along the river for conversion into wetlands. The Corps was building levees to protect homes—the same homes that other agencies were buying and demolishing.

B. Dynamics of stream flow

Erosion (L. *erodere*, gnaw away) by streams removes particles from rocks, sediments, and soils and deposits those particles downstream in a variety of sedimentary features. But what are the principles that govern erosion and deposition? What are the mechanisms that change the course of a river? What causes floods?

A stream's ability to *erode* increases with (1) **velocity** and (2) **quantity of water**. Conversely, a stream is likely to *deposit* sediments when either of these variables diminishes.

(1) Velocity

Size of conduit affects velocity—It is clear to anyone who has used a garden hose to wash a driveway that the swift stream from a nozzle is more effective than the lazy stream from the mouth of that hose (Fig. 9.1). In stream parlance, conduit size is called **cross-sectional area of channel.**

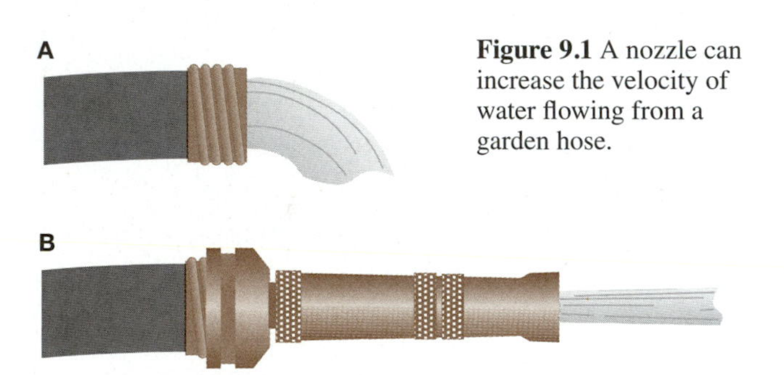

Figure 9.1 A nozzle can increase the velocity of water flowing from a garden hose.

Q9.1 The flow of water emitted from a nozzle is faster than that emitted from the open end of a garden hose. But why? *Hint:* See the lead-in to the above paragraph.

Slope also affects velocity—It is also clear to anyone who has seen a liquid spilled on a flat surface that its flow is slow and irregular, whereas flow on a steep slope is rapid and straight (Fig. 9.2). Steepness of slope is called **gradient**. The greater the gradient, the faster the flow (other things being equal).

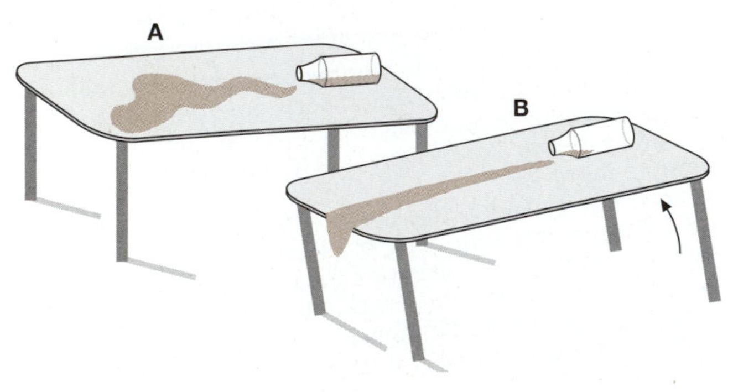

Figure 9.2 A. Low slope, lazy flow. **B**. Steep slope, speedy flow.

So, in summary, greater velocity is promoted by *small cross-sectional-area channels* and *steep gradients*.

(2) Quantity of water

The *quantity* of water in a stream or river is called its **discharge**—which amounts to the amount of water that passes a given point in a given amount of time (i.e., volume per time). In American hydrology, discharge is expressed in cubic feet per second (cfs). It is the product of velocity *(V)*, as measured in feet per second (ft/sec), multiplied by cross-sectional area *(A)*, as measured in square feet (ft^2).

$$\underset{\text{discharge (cfs)}}{Q} = \underset{\text{velocity (ft/sec)}}{V} \times \underset{\text{cross-sectional area (ft}^2\text{)}}{A}$$

Q9.2 (A) What is the velocity (in ft/sec) of a stream with a discharge of 1,000 cfs and a cross-sectional area of 100 ft^2? (B) What is the approximate velocity in mph? (C) What is the discharge in gal/sec? *Hint:* See measurements and their conversions on page *ii* in the front of this manual.

If the discharge of a stream increases (because of increased rain, snow melt, or dam failure) it figures that velocity and cross-sectional area increase. But, as an incidental factor, the two do not increase at the same rate for all streams. The reason: *Friction* associated with flow opposes velocity. And friction differs with both (1) channel shape and (2) bed roughness (Fig. 9.3).

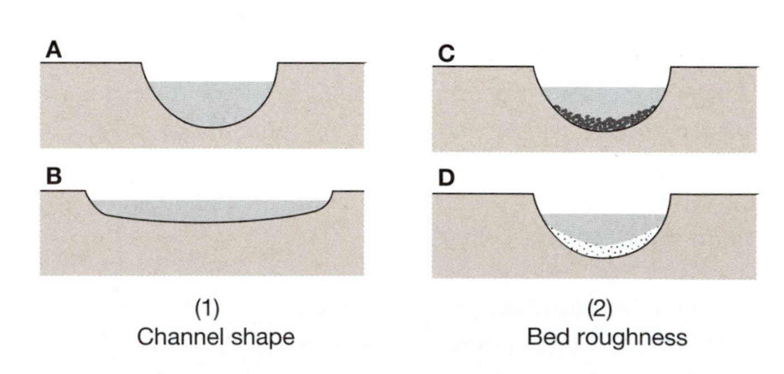

Figure 9.3 Velocity is influenced by both (1) channel shape and (2) bed roughness because each is a factor in the degree of *friction* associated with water flow.

Q9.3 In Figure 9.3, (A) which channel shape presents the greater ratio of surface area–to–water volume, thus, more friction, A or B? (B) Which bed roughness appears to exert more friction on flowing water, C or D?

Stream flow and land forms

Alluvial fans develop where streams issue from canyons onto adjacent valley floors and deposit sediments derived through the erosion of rocks upstream. Deposition of alluvial fans is triggered by a reduction in stream velocity.

Q9.4 In Figure 9.4, what *two* things account for the reduction in velocity as a stream flows from a steep, narrow valley within the mountains onto a broad nearly flat valley floor? *Hint:* Draw on the principles presented on facing page 148.

Q9.5 In an arid climate—with *dry soil* and *dry air*—what are a couple of other reasons for why this stream loses its sediment-carrying capacity?

Figure 9.4 Streams issue from narrow V-shaped canyons within the Black Mountains onto the broad floor of Death Valley. The reduction in velocity results in the deposition of sediments (called alluvium) in the form of an alluvial fan. Incidentally, the elevation of this valley floor is 264 feet below sea level, the lowest region in all of North America.

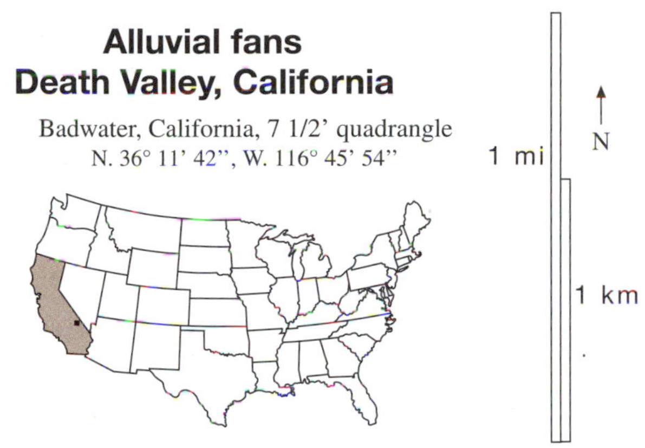

This is a satellite image of the Ouachita Mtns. of central Arkansas. Rock layers are folded ('wrinkled') such that they form east-west valleys and ridges. The Little Missouri River has carved a valley across the folds through which it flows southward. Unlike topography in Figure 9.4 above, where canyon water disgorges onto the broad valley to the west, unusual rains during the night of June 10, 2010 raised the Little Missouri River 25 feet—drowning 16 people.

Little Missouri River Flash Flood

Caddo Gap, 7 1/2' quadrangle
N. 34° 22' 20", W. 93° 42' 41"

C. The erosion cycle

Land forms, together with the rivers that shape them, are believed by some physiographers to develop in a predictable manner, passing through a succession of stages traditionally called **youthful**, **mature**, and **old age** (Fig. 9.5). This theory carries a warning label: It might not be true! Factors other than time might be at work here. For example, the amount of rainfall and vegetation in a region, along with the durability of rocks and the orientation of rock layers, might locally be more important than the passage of time.

Figure 9.6 shows a microcosm of land forms like those in Figure 9.5, plus three cross-sections showing valley profiles and a coastline with a delta. Although these land forms—youthful, mature, and old age—are developed on a grand scale in the three examples listed in Figure 9.6, any and all of the three commonly occur together within a single local area. For example, steep **tributary** streams in your neighborhood might show youthful patterns, while larger low-gradient **trunk streams** in the same area might have developed floodplains.

Q9.6 Match a stage of development in the erosion cycle—*youthful*, *mature*, or *old age*—with each of the following three quadrangle maps. Pay attention both to stream patterns and to topography as indicated by contours.

(A) La Push, Washington (page 215), the Quillayute River.

(B) Cody, Wyoming (page 161), the entire map area.

(C) Logan Pass, Montana (page 292), the entire map area.

A. Youthful stage
Stream divides are flat (i.e., much of the original flat upland remains). Valley walls are steep.

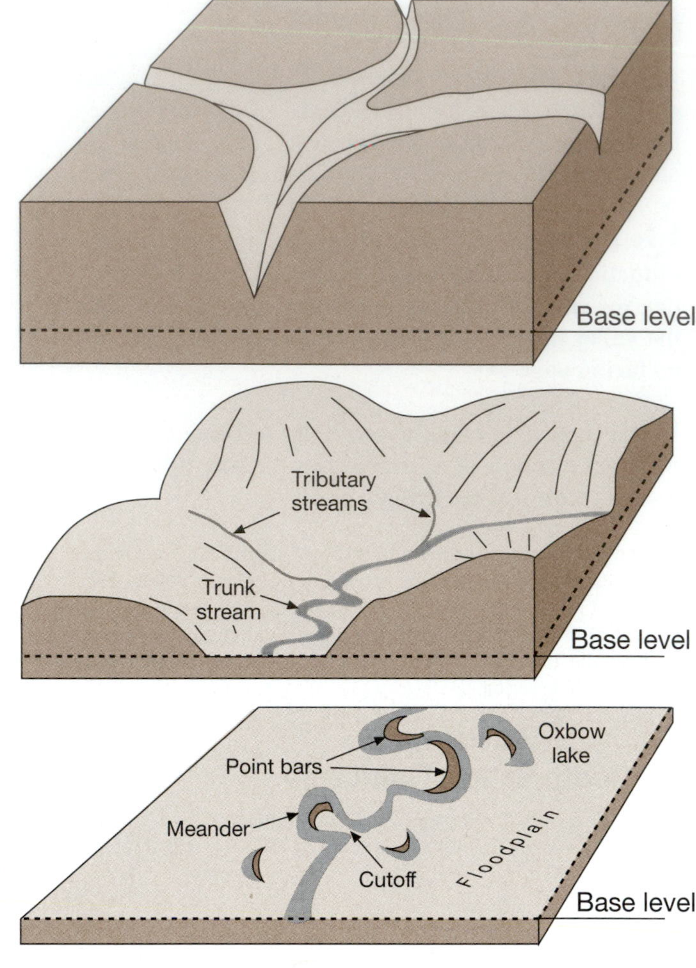

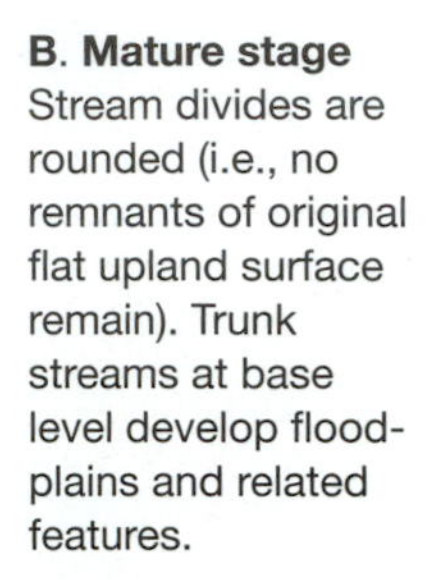
B. Mature stage
Stream divides are rounded (i.e., no remnants of original flat upland surface remain). Trunk streams at base level develop floodplains and related features.

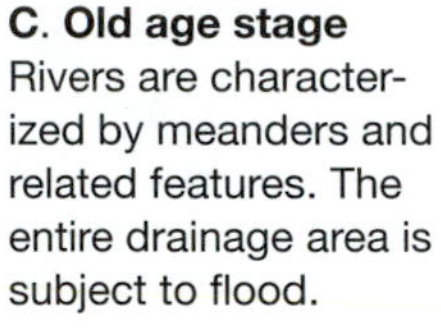
C. Old age stage
Rivers are characterized by meanders and related features. The entire drainage area is subject to flood.

Figure 9.5 The three stages of the erosion cycle—youth, mature, and old age. (Base level is defined on following page 151.)

Figure 9.6 A microcosm of all three stages of the erosion cycle—youthful, mature, and old-age—with valley profiles.

Youthful
Example: Colorado Plateau dissected by Grand Canyon

Mature
Example: Applachian Mtns.

Old age
Examples: Along major rivers and on coastal plains

Delta

Base level

A–B: Steep gradient, little in the way of floodplains

C–D: Gentle gradient, narrow floodplains

E–F: Apparently flat, expansive floodplains

D. Stream gradients

The steepness of a stream's bed is called the stream's **gradient**—which is graphically shown in a number of **long profiles** in Figure 9.7.

Q9.7 Ref: Figure 9.7. (A) How does the gradient of a river differ as a function of its length? (B) There is a characteristic curve to all long profiles. What is the term for the *shape* of such a curve? *Hint:* This term is more commonly applied to the shape of a *surface,* rather than a line.

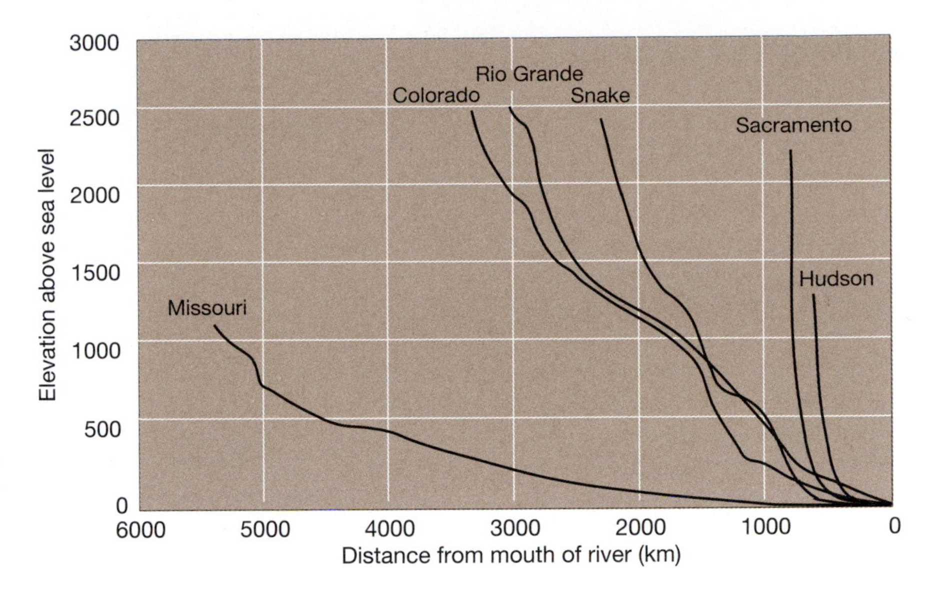

Figure 9.7 These are *long profiles* illustrating the gradients of several rivers. The average gradient of a hill-country stream (e.g., the Sacramento River of California) might reach 60m/km, whereas near the mouth of a large river such as the Missouri, the gradient might be as low as 0.1 m/km or even less.

Gradients of *tributary streams* are steeper than those of *trunk streams*.

Q9.8 Refer to the Cody, Wyoming, quadrangle map on page 161. (A) What is the gradient (in ft/mi) of Sulfur Creek in the vicinity of coordinates E-7? *Hint*: Measure the length of Sulfur Creek between contours 4900 and 5000 where they cross the creek. (B) What is the gradient of Shoshone River in the vicinity of coordinates D-4–H-3? *Hint:* Measure the length of the river between contours 4800 and 4840 where they cross the river.

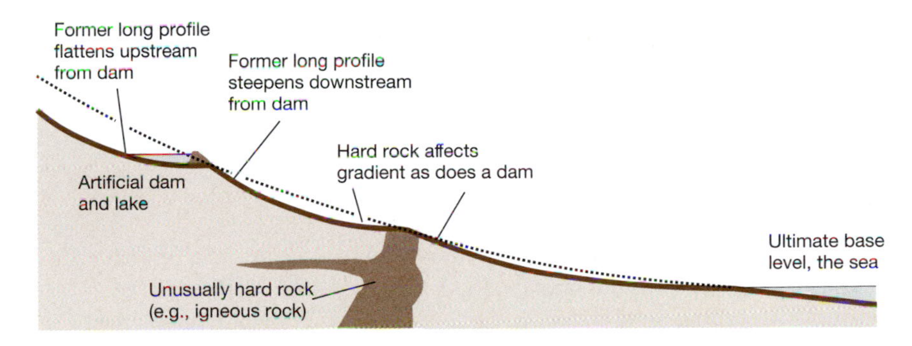

Figure 9.8 A local base level results from an artificial dam and lake. The lake checks erosion immediately upstream. Unusually hard rock within a stream course produces much the same result, with flatter gradient upstream and steeper gradient downstream.

Base levels—Gradient *decreases* upstream, and *increases* downstream, from an artificial dam (Fig. 9.8). And hard rock within a stream course can act as a natural dam, affecting stream gradient in similar ways. Such impediments to erosion are called base levels.

Natural dams as base levels—Figure 9.9 is a sketch of Johnson Shut-Ins in southeast Missouri, which owes its development to harder Precambrian igneous rocks that act a natural dam.

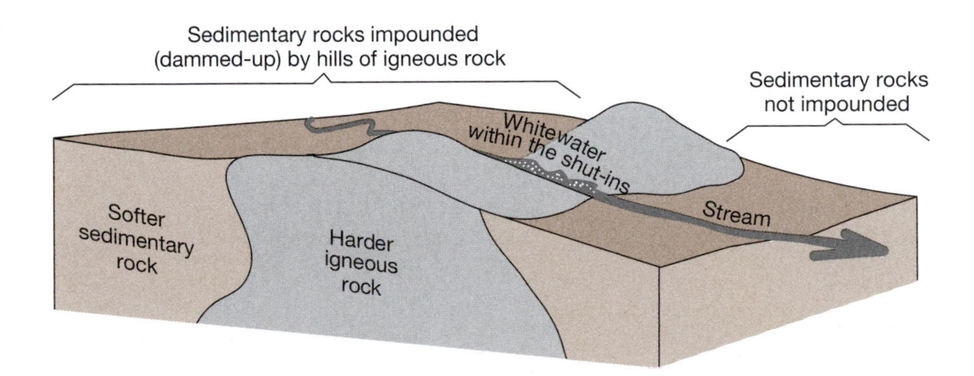

Figure 9.9 The velocity of the stream quickens as it flows through the channel of Johnson Shut-Ins, developing white water. The hard rock dam impounds softer rock upstream. The softer rock upstream can be eroded away only as fast as the valley through the harder rock dam can be deepened. Notice the difference in elevation and slope of the land surfaces above and below the shut-ins.

Q9.9 (A) What accounts for the swiftness of water within the shut-ins? *Hint:* The *two* factors at work here were presented on pages 148–149. (B) Why are there meanders above the shut-ins, but not below? *Hint:* See the last sentence in the caption to Figure 9.9.

E. Floodplain features

Meanders—A river with a sinuous course is said to **meander**, a term borrowed from the name of a looping river in Turkey. Meandering streams occur on very low slopes (e.g., those of **floodplains**), as though such streams are searching for the downhill direction.

Meanders begin as gentle curves in a river course. Then, because the water's velocity is greater on the outside of a curve than on the inside, erosion occurs on the outside, with deposition on the inside (Fig. 9.10). The result: Loops with ever-increasing sinuosity.

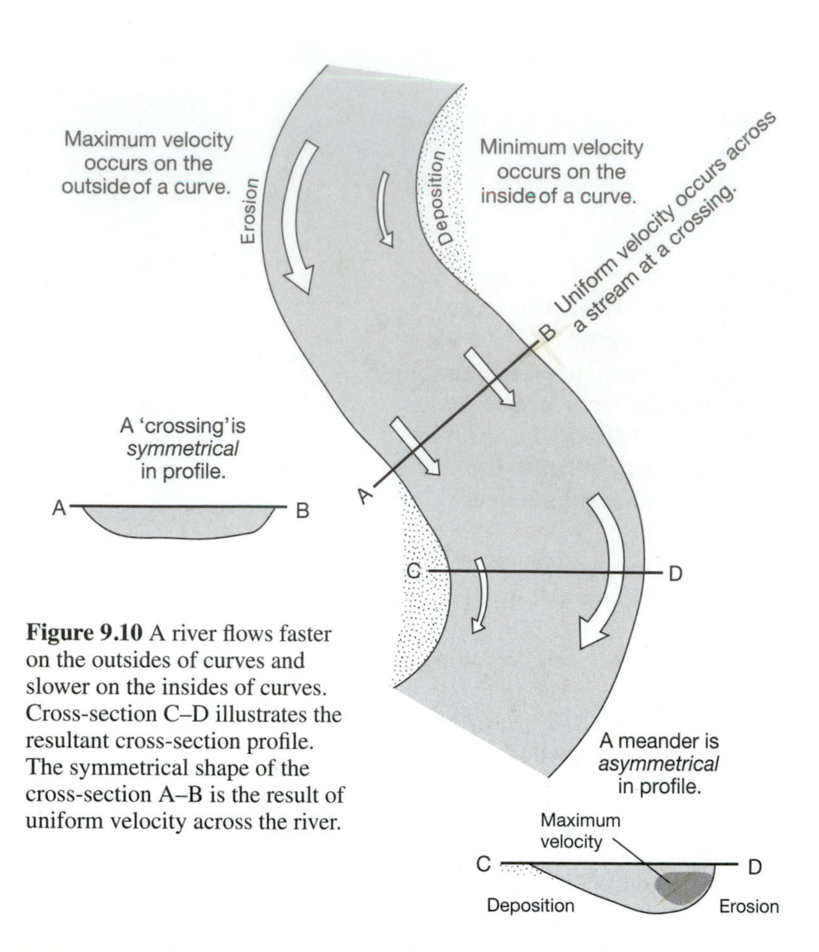

Figure 9.10 A river flows faster on the outsides of curves and slower on the insides of curves. Cross-section C–D illustrates the resultant cross-section profile. The symmetrical shape of the cross-section A–B is the result of uniform velocity across the river.

When excessive rain causes a river to swell beyond the size that can be contained within its channel, it spills out onto its floodplain. A veneer of silt and clay, deposited by receding flood waters, accumulates on top of older river-channel deposits (Fig. 9.11).

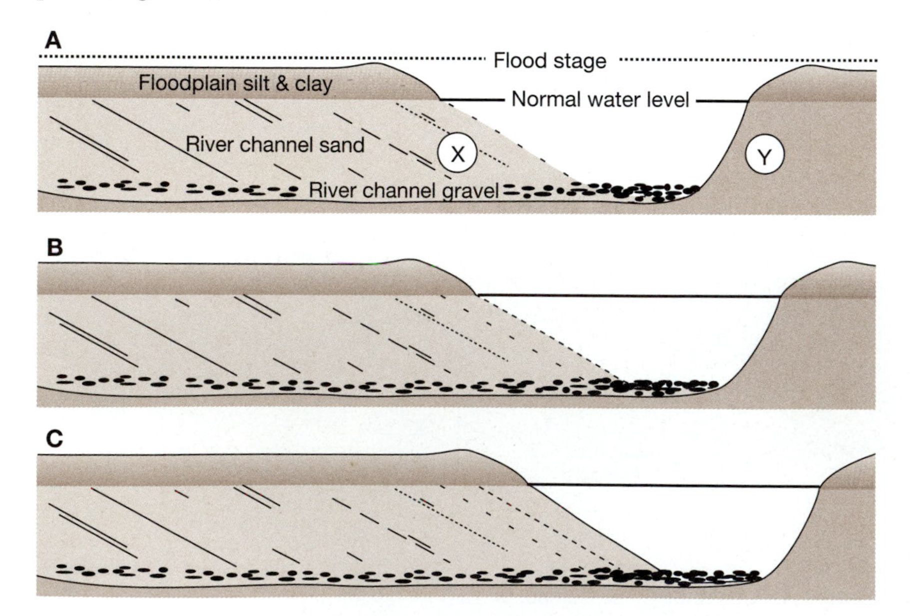

Figure 9.11 These cross-sections are time-lapse sketches of a meander migrating across a floodplain. This shows a section analogous to section C-D in Figure 9.10.

Q9.10 Ref: Figure 9.11. (A) In which direction is the meander migrating, to your right or to your left? (B) What is the age of the deposit of earth material at X relative to that at Y (i.e., which is older)?

Floodplains are good for farming and grazing, but most floodplains make for risky choices as home sites.

Q9.11 Meander scars such as those along the Ohio River on facing page 153 (Fig. 9.12) are not well understood, but one possibility is that each scar records a time of flooding. If so, approximately how many floods are recorded along the length of the black bar in this satellite photograph?

Figure 9.12 The Ohio River at the Indiana-Kentucky border. The town is Henderson, Kentucky. A meander loop is marked by meander scars on its trailing (north) floodplain side. The scars consist of light-toned ridges and dark-toned troughs. A few of the troughs were occupied by dark elongate ponds at the time of this photograph.

Meander scars
Ohio River
Indiana-Kentucky

Henderson, Kentucky, 7 1/2' quadrangle
N. 36° 52' 12", W. 87° 37' 48"

Natural levees

Everyone knows about engineers constructing levees along the banks of rivers in efforts to prevent flooding in times of high water. What you might not know is that Mother Nature accomplishes much the same thing in building **natural levees**—which are ridges of sediment that accumulate along the margins of river channels on floodplains (Fig. 9.13).

Q9.12 Explain how natural levees develop. ***Hint:*** **Why should sediments be deposited the moment that a river escapes its *narrow* channel and spills onto a *wide* floodplain? Recall the explanation of alluvial fans on page 149.**

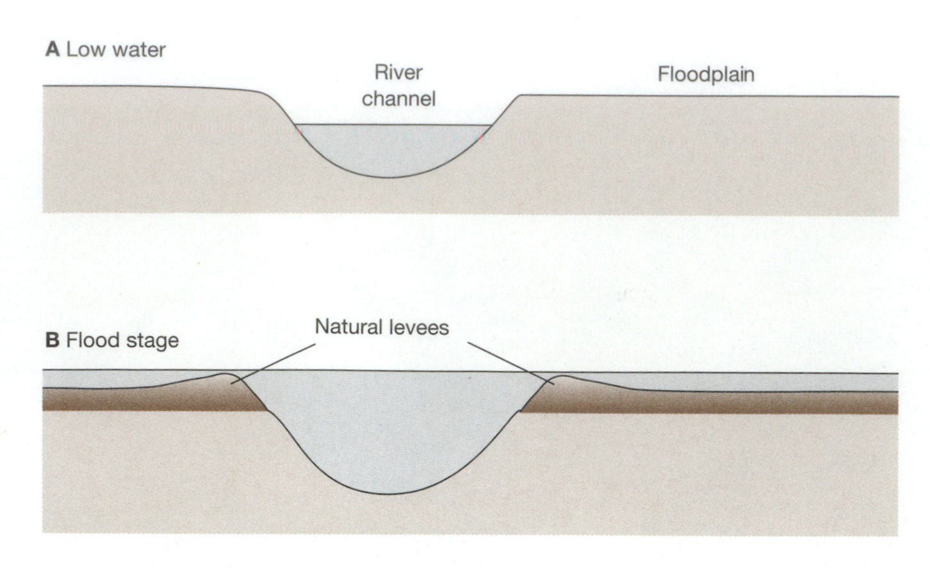

Figure 9.13 Cross-sections showing the development of natural levees during flood.

Both natural levees and artificial earthen levees are commonly covered with trees because farmers don't go there. If trees have not taken hold on their own, they are commonly planted as an additional flood-control measure.

Q9.13 With the breaching (washing out) of levees being a common problem in times of flood—especially where levees are flanked by water for weeks and months—of what particular benefit do you suppose trees might be?

In a situation where a river has topped or breached its levees, but at a water level where the levees are not fully submerged, the river appears to be *three* parallel rivers (Fig. 9.14). From an airplane window it is difficult to distinguish between an engineer's levee and a natural levee.

Levees along the Red River of the North at Grand Forks, North Dakota, failed to contain record-breaking high waters in the spring of 1997. The river inundated some 30% of the city of Grand Forks, prompting the evacuation of 50,000 residents and inflicting more than $1 billion in property damage.

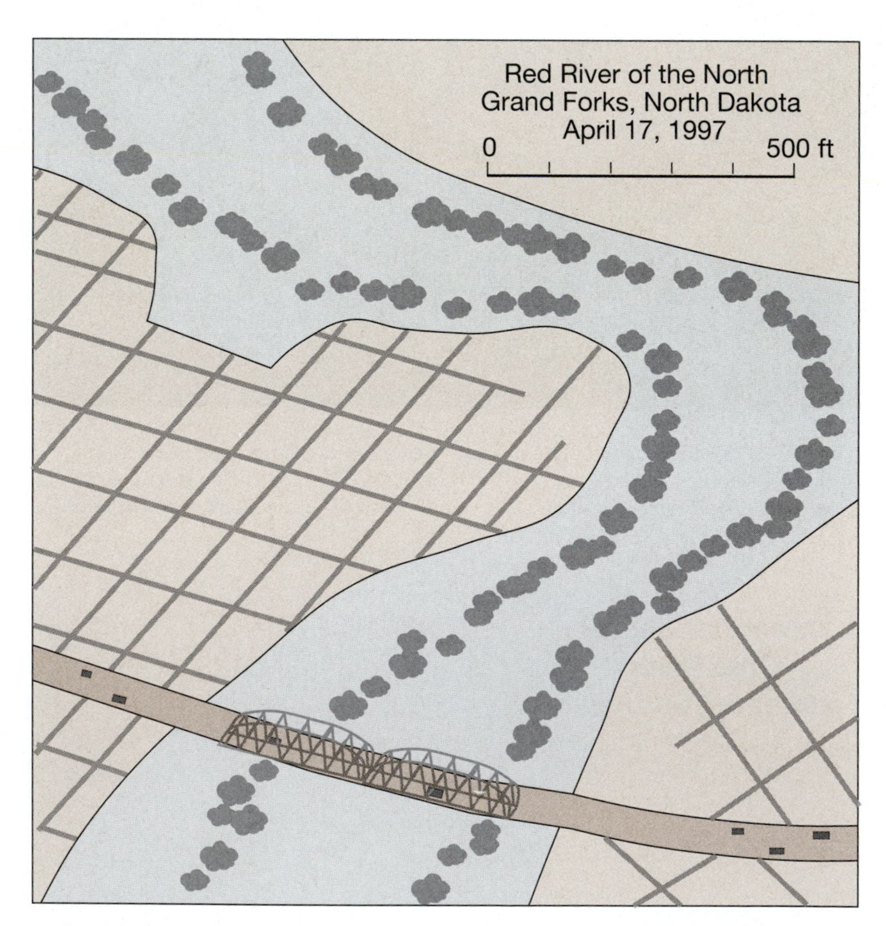

Figure 9.14 This is a schematic drawing of Grand Forks, North Dakota, and the Red River of the North in the spring of 1997. At such times of floods, tree-lined levees appear to split a river into three rivers.

F. Levees and the 'bottleneck principle'— The Great Flood of 1993

Levees are intended to contain a river within its channel during times of high water and prevent its spreading across the floodplain and inundating farmland and various developments. Most cities have ordinances prohibiting construction on floodplains, but county governments tend to be more permissive. So we continue to see warehouses, mobile home parks, ball fields, and other low-cost, high surface-area facilities on floodplains. Yes, even cemeteries. The Great Flood of 1993 unearthed 756 coffins in Ray County, Missouri, spilling the contents of some.

Two particular problems associated with levees captured public attention during the 1993 flood in the Mississippi-Missouri river basins.

First—even the most sturdy levees prove to be either too low or too weak when the 'big ones' come. And, it would appear that the false security provided by levees prompts unwise construction on floodplains.

Second—a pair of levees is to a river in flood stage as a squeeze is to a balloon. Squeeze a balloon in one place, and it swells in another place. So it is with rivers.

Q9.14 So what do you suppose happens to the height of flood waters upstream from levee constriction?

A pair of levees is simply a bottleneck. Levees are not continuous from the river's head to its mouth, so pity the poor people upstream from levees who have no levees of their own. In times of flood, they suffer higher water and broader flooding than that which would occur without the downstream levees.

The St. Louis stretch of the Mississippi River (Fig. 9.15) is an example of just such a bottleneck. Along the east side of the city, flood walls contain the Mississippi. Figure 9.16A shows how rivers upstream from St. Louis appear during normal water levels. Figure 9.16B shows how those rivers appeared during The Great Flood of 1993. Clearly, the flooding immediately upstream from St. Louis would not have been so severe had it not been for the narrow passage through the flood walls at St. Louis.

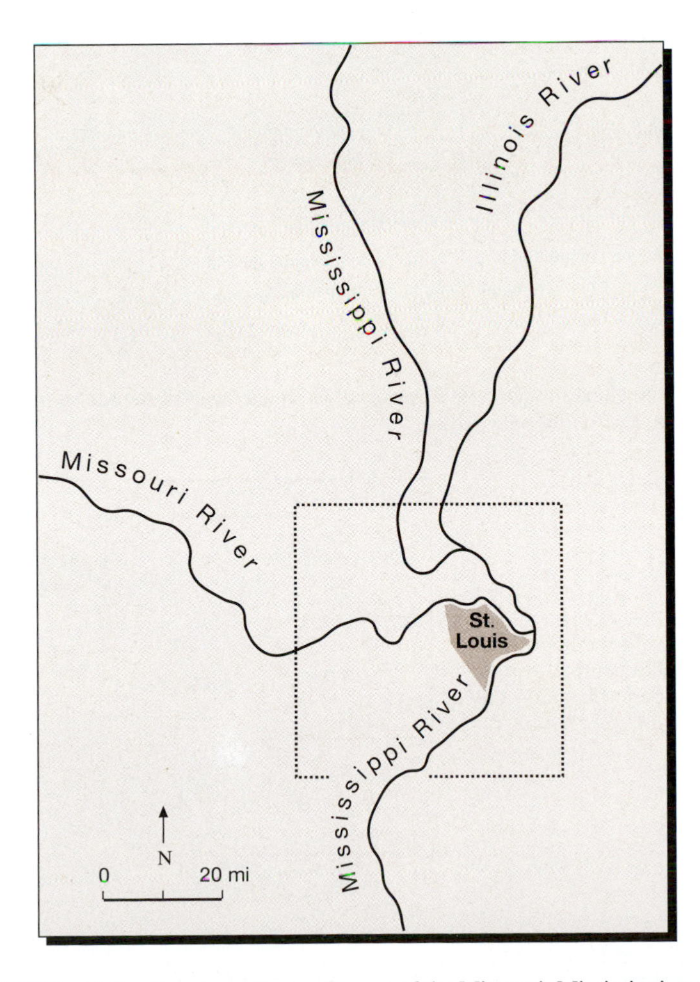

Figure 9.15 The St. Louis area, at the confluence of the Missouri, Mississippi, and Illinois Rivers—ground zero during The Great Flood of 1993. The inset is shown in Figure 9.16.

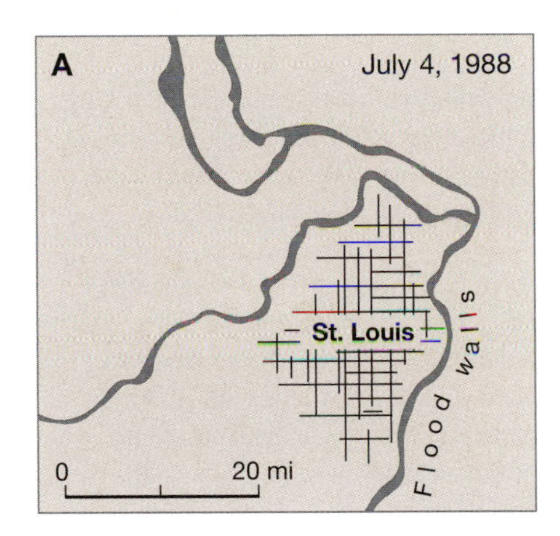

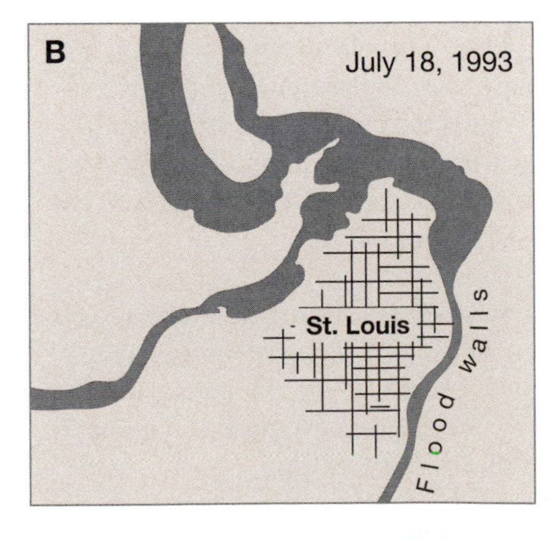

Figure 9.16 Sketches made from LandSat satellite photographs (A) *before* and (B) *during* The Great Flood of 1993.

Why The Great Flood of 1993?

First, moisture that falls as rain and snow in the Midwest is derived in large part through evaporation in the Gulf of Mexico (Fig. 9.17). (Pacific air loses its moisture long before it descends the High Plains, and the Atlantic is effectively downwind from the mid-continent.)

Warm, moist Gulf air loses its moisture-holding capacity where it encounters cool Arctic air, resulting in precipitation. Ordinarily, the boundary between the two air masses—a **weather front**—roams about the continent with the vagaries of global atmospheric circulation. But in 1993, that front was effectively stationary for *seven months!* The period of time represented by Figure 9.17 is only that of June–July 1993, but a similar weather pattern characterized the entire period from January through July of that year.

Incidentally, there's only so much Gulf moisture to go around, so while the Midwest was soaked for months, the Southeast experienced drought.

The Great Flood of 1993 occurred pretty much in the center of the Mississippi-Missouri River drainage basin (Fig. 9.18), the region in which rivers and their tributaries flow into the Mississippi in its journey to the Gulf of Mexico.

Q9.15 There's a lesson here having to do with the regional *water cycle*. What is that lesson? *Hint:* It has to do with water in the Gulf of Mexico, rainfall in the mid-continent, and the Mississippi River.

How much rain was there in the seven-month period—January through July 1993? Answer: Approximately 150% to 200% of 30-year averages at a number of places (Fig. 9.19).

National Oceanic and Atmospheric Administration (NOAA) http://www.noaa.gov/ is linked to National Weather Service.

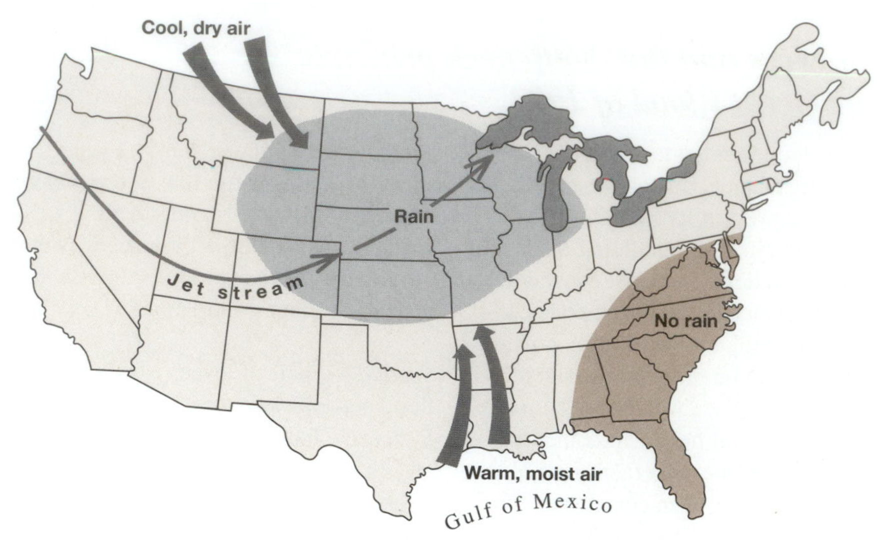

Figure 9.17 The prevailing weather pattern during June–July, 1993 was that of a stationary front, along which Gulf of Mexico moisture was condensed into rain by cool Arctic air. *National Weather Service*.

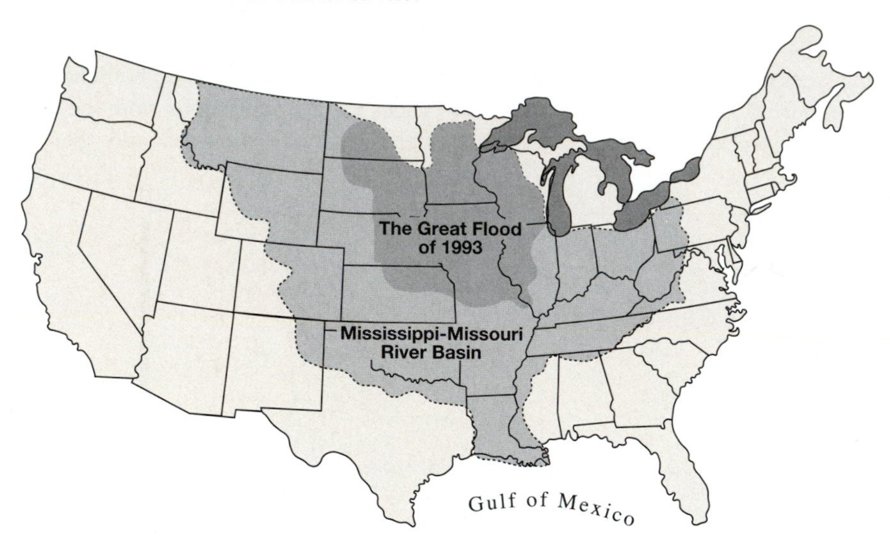

Figure 9.18 The Great Flood of 1993 occurred within the heart of the Mississippi River drainage basin. *National Weather Service*.

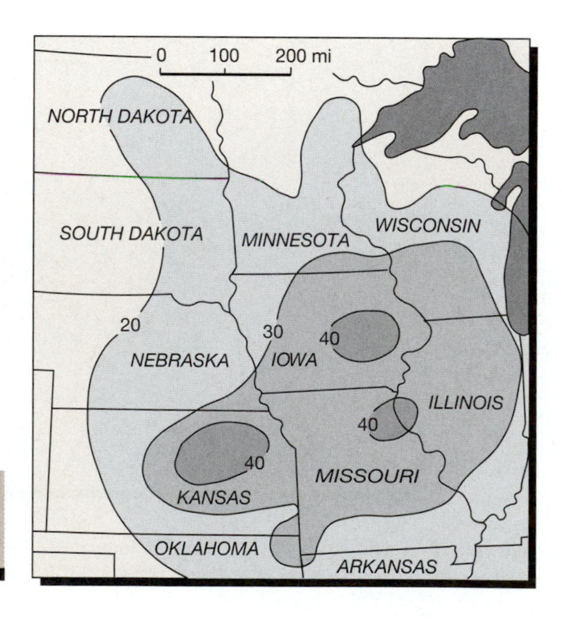

Figure 9.19 Isohyetal lines (i.e., rainfall in inches per time) for the seven-month period January–July 1993. *National Weather Service*.

G. Entrenched meanders

Not all meanders live on floodplains where they're free to roam. Some are confined as **entrenched meanders** within deep valleys (Fig. 9.20). Entrenched meanders (aka *incised meanders*) no doubt developed their sinuous patterns on floodplains, but then changed to a regime of downward erosion because either (1) the *region was uplifted*, or (2) the *base level dropped* for whatever reason. But even entrenched meanders wander as they deepen. The asymmetry of the valley apparent at the left end of Figure 9.20B indicates migration toward the outside of the meander.

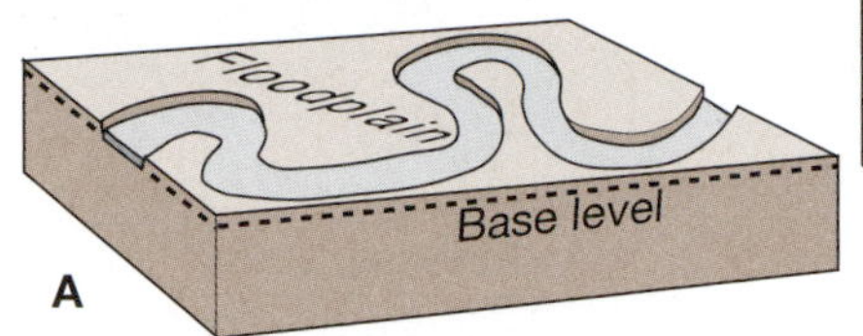

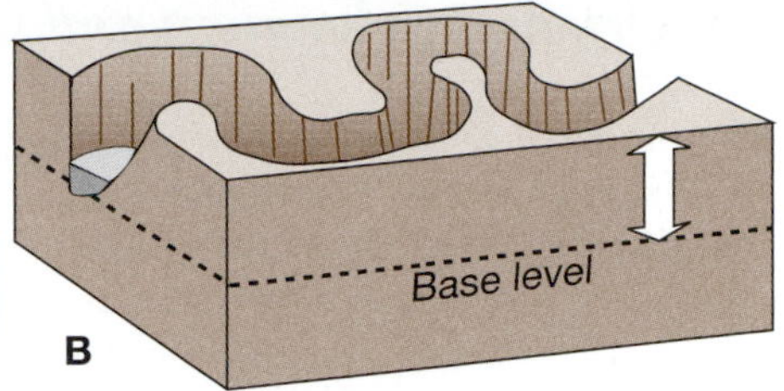

Figure 9.20 (A) Entrenched meanders develop at base level on a floodplain, but then (**B**) the floodplain and base level part company. Regional uplift is thought to explain the entrenched meanders of the geologically young Colorado Plateau, with its scenic Goosenecks of the San Juan River of southeastern Utah.

Q9.16 On the Answer Page, sketch the topographic profile along line A-B in the satellite photograph of Figure 9.21. (A) Is the valley symmetrical at the two places where the profile crosses the creek, or is it asymmetrical at those places (like that shown on one side of Figure 9.20B)? (The labels 'Cliff' and 'At water level' have been added on the photo for clarity.) (B) Does valley shape indicate that the creek has migrated laterally while becoming deeper?

Entrenched meanders
Conodoguinet Creek
Pennsylvania

Harrisburg West, Pennsylvania,
7 1/2' quadrangle
N. 40° 15' 09", W. 76° 57' 18"

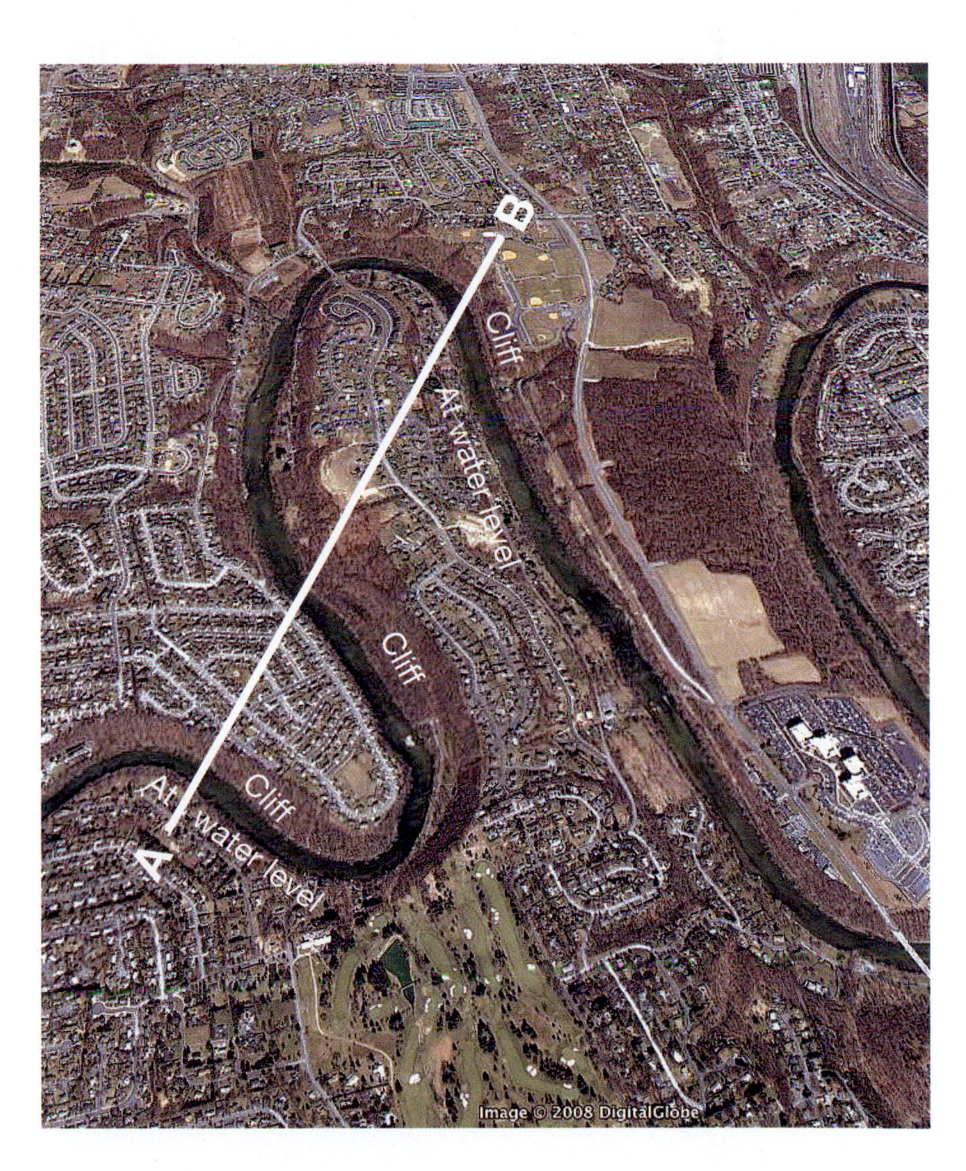

Figure 9.21 Conodoguinet Creek near Harrisburg, Pennsylvania, is an example of entrenched meanders.

H. Urban development and storm water problems

Four things can happen to a drop of rain (and snow) that falls to earth in inland regions (Fig. 9.22). Of the four, infiltration and runoff are the most manageable.

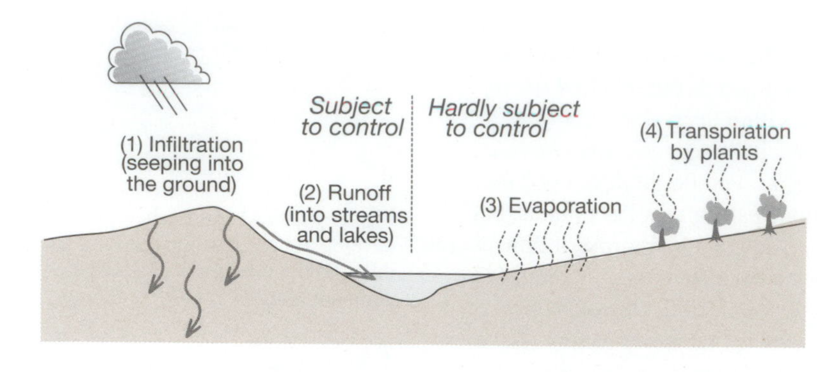

Figure 9.22 The fate of inland rain and snow.

The storm water issue

You might choose a site for your dream house in a sparsely developed setting, but you should be mindful of what's going on—or what might go on in the future—upstream before you buy. People are drawn to streams, so home sites along a babbling brook (Fig. 9.23) bring premium prices. But urban development upstream can change a babbling brook into a roaring torrent (Fig. 9.24). Enter the management of **storm water**. Incidentally, it's ironic that people who complain about damaging storm water from *upstream* are the same people who contribute to damaging storm water *downstream*.

Before

Figure 9.23 People commonly prefer a wooded slope with a creek behind.

The U.S. Geological Survey estimates that urbanization increases runoff 2–6 times that which occurs on natural terrain. Why such a broad range? Because of local effects of changes in…

- steepness of slope
- area of porous soil
- abundance of vegetation

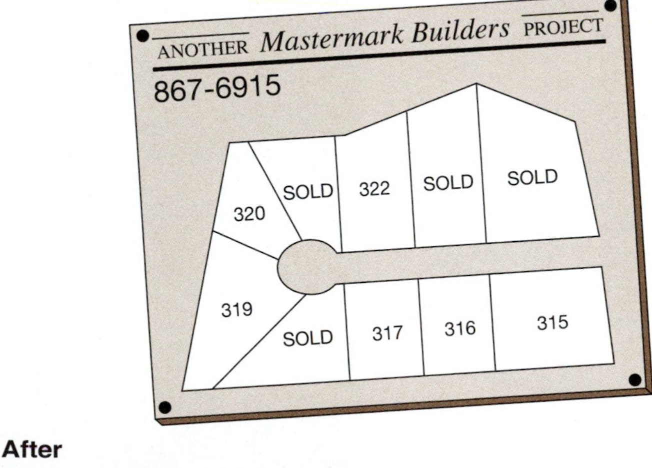

After

Figure 9.24 Runoff can increase 2–6 times following urban development upstream, changing peaceful, harmless streams into roaring destructive torrents.

Q9.17 Which of the above three factors do you suppose is the *main reason* for increased runoff and stream discharge following urbanization? *Hint:* compare Figure 9.24 with Figure 9.23.

Q9.18 Some developers attempt to mitigate storm water problems by constructing concrete or stone retention (or catchment) basins along threatening streams. But what happens during an abnormally long rainy season when retention basins become filled before 'the big rain?'

COLUMBIA DAILY TRIBUNE, MARCH 4, 2001

Watershed experts plead case for clustered homes

Runaway erosion almost washes away homes in St. Louis

COLUMBIA, MISSOURI—With watershed management becoming a hot topic in local government, the Community Storm Water Project yesterday hosted an educational workshop at which watershed experts and city officials discussed how wisely planned development can protect natural areas. It was reported that the cost of building a concrete wall to protect residents from Black Jack Creek in St. Louis exceeded the value of homes.

The watershed issue

Imagine—Your local newspaper says that a tract of land is about to be developed in your part of town. *Question*: How can you know whether your home—or access to your home—might be affected by runoff from that new development? *Answer*: It depends on whether the development and your home are within the same **watershed**, and, if so, which is downstream from which. A graphic metaphor of watershed is your average tree with its trunk, branches, and leaves and make-believe sap flowing from a spot within any leaf to the ground.

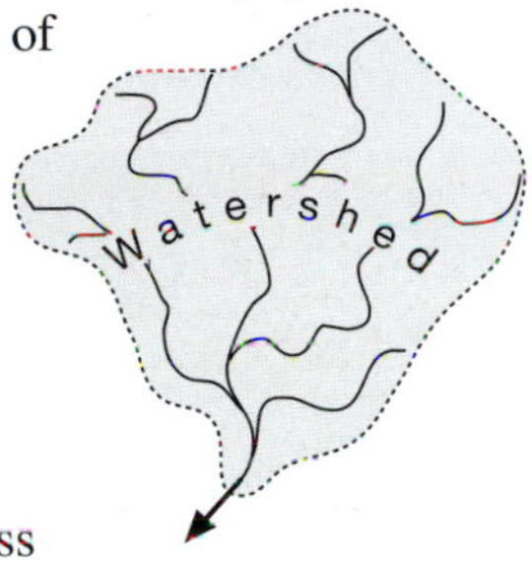

The five major watersheds, or **drainage basins**, of North America are separated by **continental divides**—each of which separates drainage into one ocean from that into another ocean (Fig. 9.25). But let's get back to your neighborhood and to assessing your position within a watershed relative to that of a prospective development.

What to do:

- Acquire a '7 1/2-minute quadrangle' map (printed as 7 1/2' Quadrangle on maps) of your area from the nearest library or government agency.

- Locate your home and the prospective development site on the map.

- Guided by streams and contours, sketch the boundaries of your watershed and the watershed of the development site.

- If both are within the same watershed, determine which is higher (in elevation) and which is lower.

- Is your home near a stream that might rise to the level of your home? Are access routes to your home threatened?

Q9.19 Turn to the Cody, Wyoming quadrangle on page 161. Imagine that you live in one of the houses along Sulphur Creek at coordinates D.5-8 (say, the house nearest the end of the word Creek). It has just been announced that a new development is to be constructed near the well at coordinates B-10. (A) Is there reason for you to be concerned? (B) How about if you own a home at the edge of the golf course at F-8?

Figure 9.25 Major watersheds (or drainage basins) of North America are separated by continental divides. The continental divide in the Rocky Mtns. is marked by lofty peaks and ridges, whereas the continental divide within the prairie lands of North Dakota is hardly perceptible to the eye.

I. River terraces

Where a river has a history of floodplain development alternating with downward erosion, **river terraces** can result (Fig. 9.26). Shoshone River has carved such terraces along its course through Cody, Wyoming (Fig. 9.27).

Q9.20 On facing page 161 is a portion of the Cody quadrangle shown at approximately the same scale as the satellite photograph in Figure 9.27. Using the Cody quadrangle (and its coordinates), draw a topographic profile from D-2.8 (near the Transmission Line) to Stampede Avenue at the marker 5088T, which is at H.5-7. This imaginary line approximates line A–B on the photograph. Label the highest terrace 'Powell' and the second-highest terrace 'Cody.'

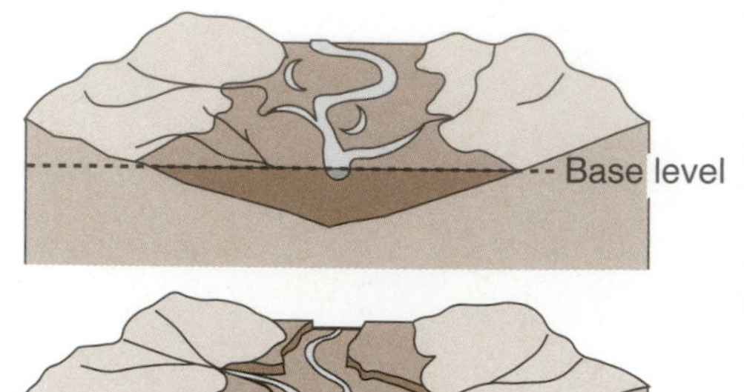

Figure 9.26 The abrupt lowering of base level can result in remnants of an earlier floodplain occurring as a river terrace.

River terraces
Cody, Wyoming

Cody, Wyoming, 7 1/2' quadrangle
N. 44° 31' 30", W. 109° 03' 18"

Figure 9.27 The Shoshone River has developed this series of river terraces in the town of Cody, Wyoming.

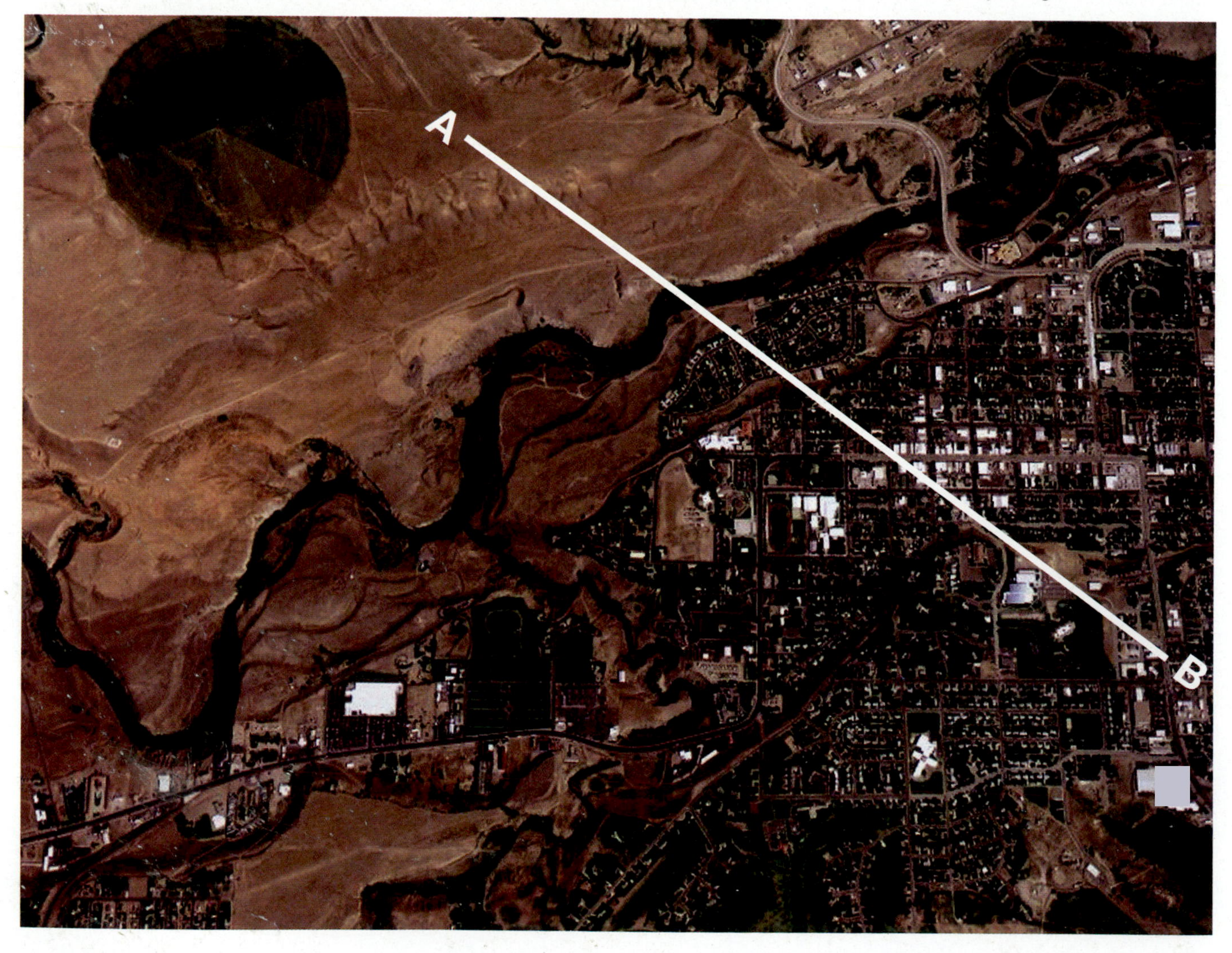

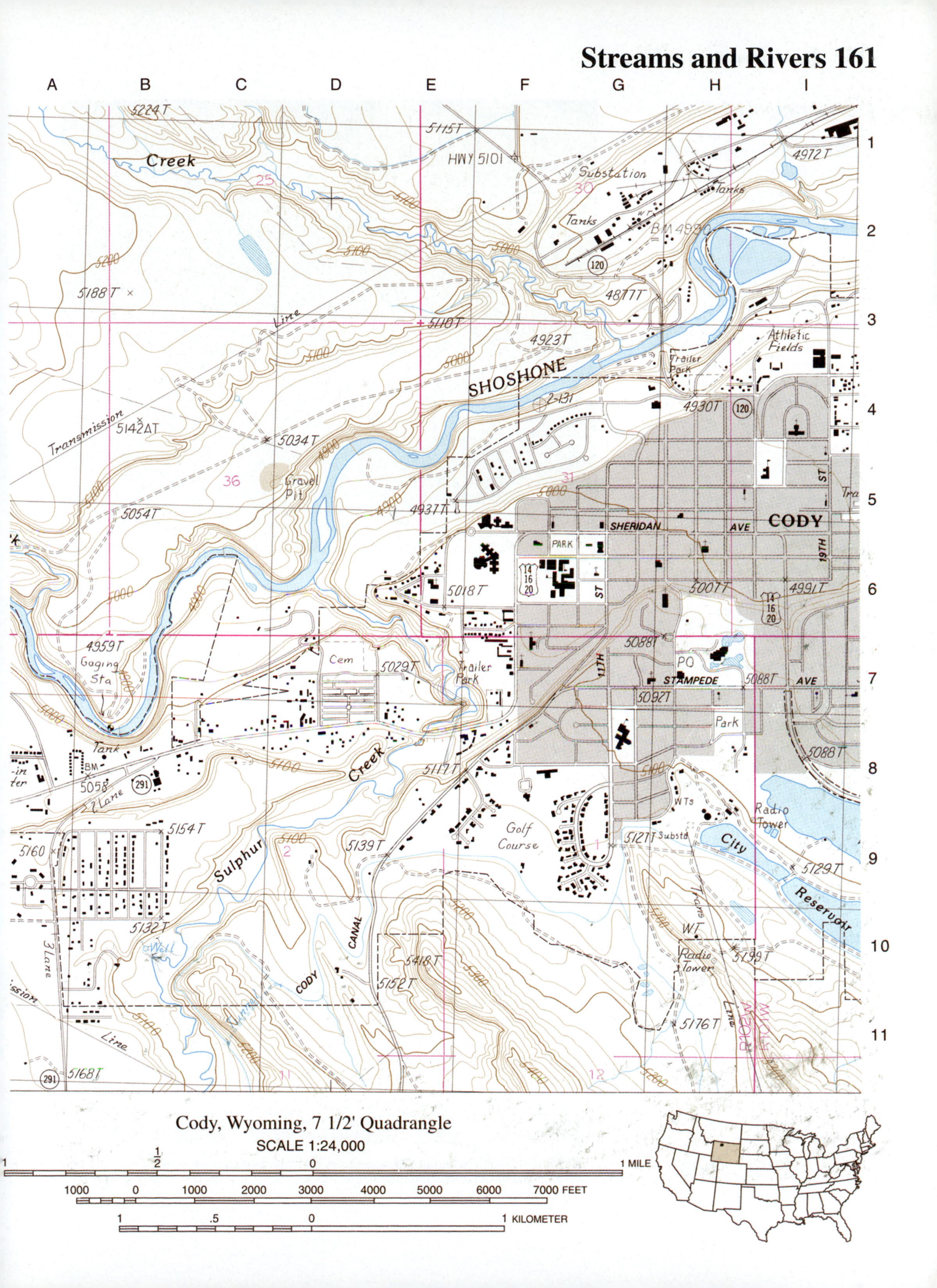
A B C D E F G H I
1 2 3 4 5 6 7 8 9 10 11
Creek
SHOSHONE
CODY
Substation
Tanks
Athletic Fields
Trailer Park
Transmission Line
Gravel Pit
Gaging Sta
SHERIDAN AVE
STAMPEDE AVE
PARK
PO
Park
Cem
Sulphur Creek
CODY CANAL
Golf Course
City Reservoir
Radio Tower
Cody, Wyoming, 7 1/2' Quadrangle
SCALE 1:24,000
1 MILE
7000 FEET
1 KILOMETER

J. Superimposed rivers

Superimposed river
Susquehanna River
Pennsylvania

Harrisburg West, Pennsylvania,
7 1/2' quadrangle
N. 40° 21' 54", W. 76° 58' 12"

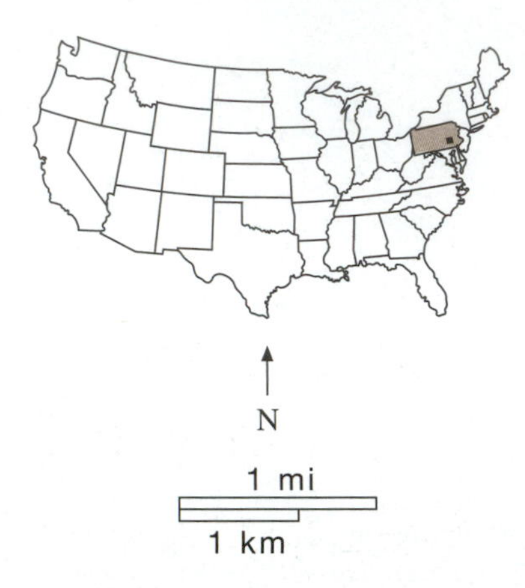

Figure 9.28 The superimposed Susquehanna River has carved water gaps through ridges.

How did the Susquehanna River manage to cut **water gaps** (aka gorges) through ridges shown in Figure 9.28? Surely the river didn't run up one side of a ridge and down the other until it sawed its way through. One idea is that an ancestral river flowing on a higher landscape worked its way downward, becoming **superimposed** on present-day topography (Fig. 9.29).

Q9.21 Incidentally, (A) Where in Figure 9.28 does the Susquehanna appear to be more shallow, at X or at Y? What's your evidence? (B) At which of these two places is the river probably more swift? Why do you say that?

Disclaimer—There are two additional processes that are also believed to form canyons: (1) **Antecedent streams**, an example of which is Columbia River Gorge. (2) **Headward erosion**, an example of which is Grand Canyon.

Figure 9.29 Three stages in the developmental history of a superimposed river.

A *First*—relatively hard rock (gray), flanked by softer rock (brown), stands up as a resistant ridge shaped by erosion in a humid environment.

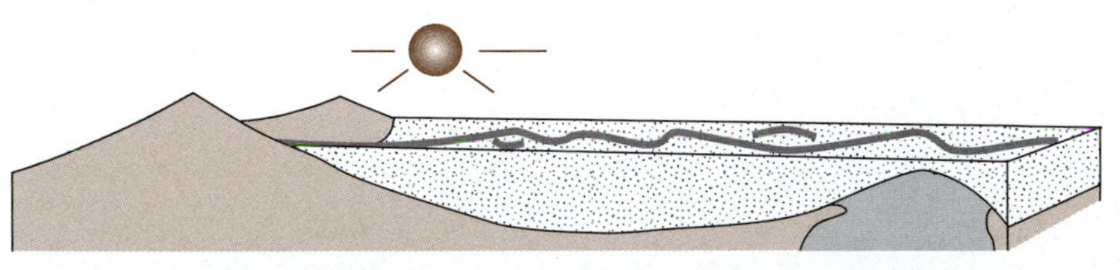

B *Second*—arid climate promotes stream deposition and the burial of the ridge by sediments. A river establishes its course on this sedimentary blanket.

C *Third*—a return to humid times promotes downward erosion while the river maintains its course through the hard rock.

K. Channelized rivers

The U.S. Corps of Engineers has straightened Chariton River at Novinger, Missouri, 'to prevent its meanders from wandering across farmland' (Fig. 9.30).

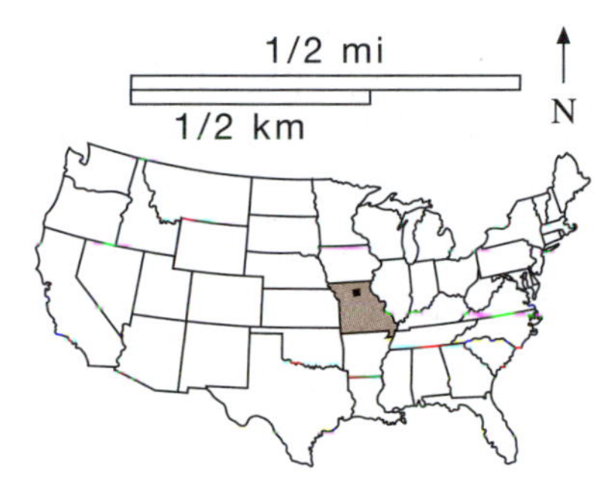

Figure 9.30 The relatively straight, channelized Chariton River (which is brown in this satellite photograph) is approximately marked by the white dashed line. A bridge is where the white dashed line crosses the east-west highway. The Chariton's floodplain is marked by a few segments of its earlier channel. Meander scars are apparent as well. Meanders and meander scars are more abundant on the west side of the river.

Recall (from Figure 9.2 on page 148) that increased gradient promotes increased velocity, which promotes increased erosion. Then address Q9.22.

Bell Park

Q9.22 Examine Figure 9.31. (A) What is the approximate gradient (in ft/mi) in the 'before' view? (It's not easy to measure gradient in the view, but do what you can.) (B) What is the approximate gradient in the 'after' view? (This is easier.) (C) In which case would you expect velocity—and therefore erosion—to be greater? *Hint:* Examine Figure 9.32.

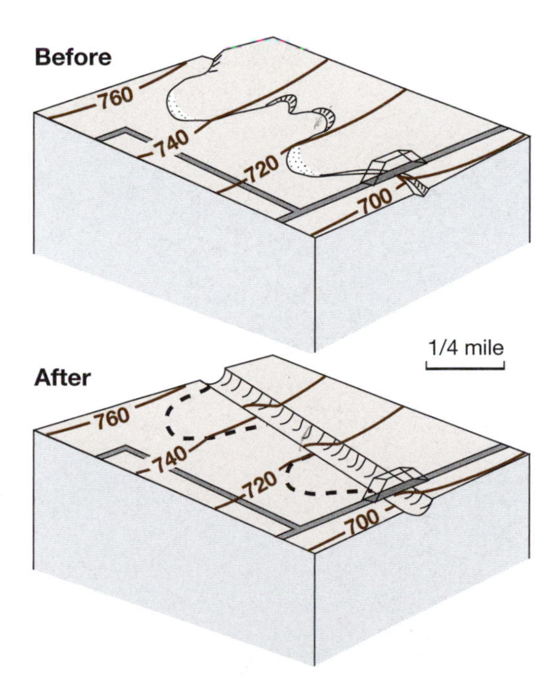

Figure 9.31 These are schematic sketches of a before-and-after portrayal of channelization. (The contour interval is 10 ft.)

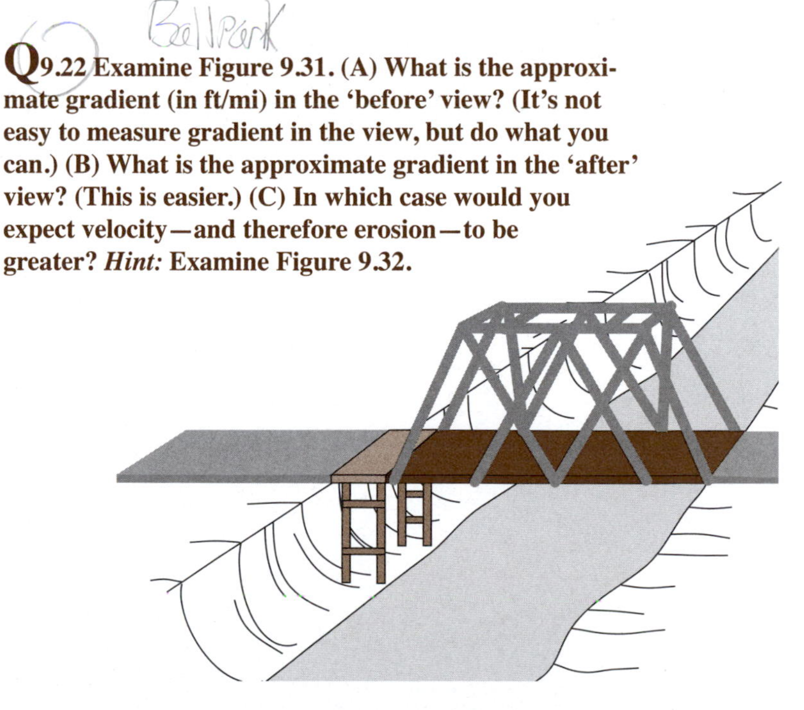

Figure 9.32 In some cases, bridges across channelized streams have had to be extended. Guess why!

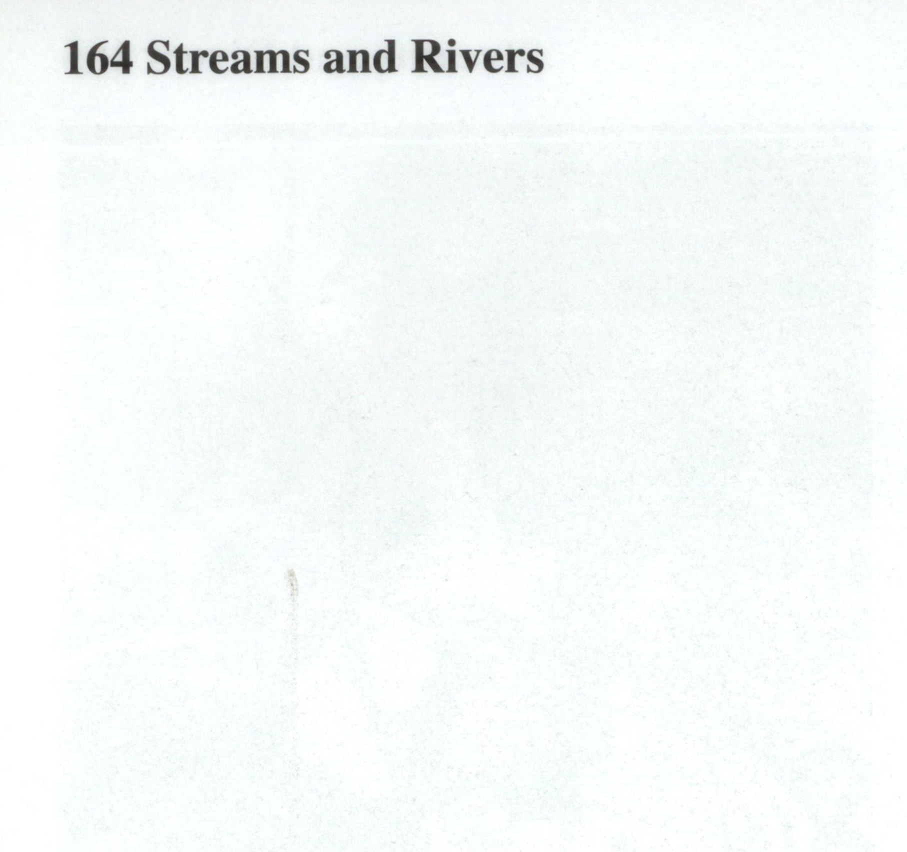

Intentionally Blank

10 Groundwater

Topics

A. What is the definition of groundwater? Why is the composition of geyser deposits variable within Yellowstone National Park? How does the total amount of groundwater on Earth compare with that of surface water?
B. How do the dynamics of groundwater in humid regions differ from those in arid regions?
C. What are two common problems associated with the water table described here? What is hydraulic gradient, and how does its measurement differ from the way in which geologists measure slope? What is hydraulic conductivity? What is Darcy's Law, and how does it apply in computing the rate at which groundwater flows through a saturated zone?
D. How can one map a water table from well data? How can one use such a map to decipher the direction and rate of flow of contaminants within the saturated zone?
E. What is a potentiometric surface, and what does it have to do with artesian conditions? What does a potentiometric surface have to do with flowing and non-flowing artesian wells?
F. What is a perched water table? What is the purpose of a reservoir liner?
G. What is karst topography, and what are its features? How are the direction and rate of flow of groundwater in Florida measured from a study of ponds within sinkholes?

A. Groundwater defined

In its broadest definition **groundwater** is all that water that occurs in otherwise open spaces within rocks and sediments. Groundwater that originates from the precipitation of rain and snow—the topic of this exercise—is called **meteoric water**.

In addition to meteoric water, there are two minor sources of subterranean water: *connate water* (aka *sediment water*), which is water that was entrapped within sediments at the time of their deposition in ancient seas; and juvenile water, water that was born of magmatic activity. Neither connate water nor juvenile water is a source of potable water, but connate water can locally be important as a high-salinity environmental contaminant associated with petroleum.

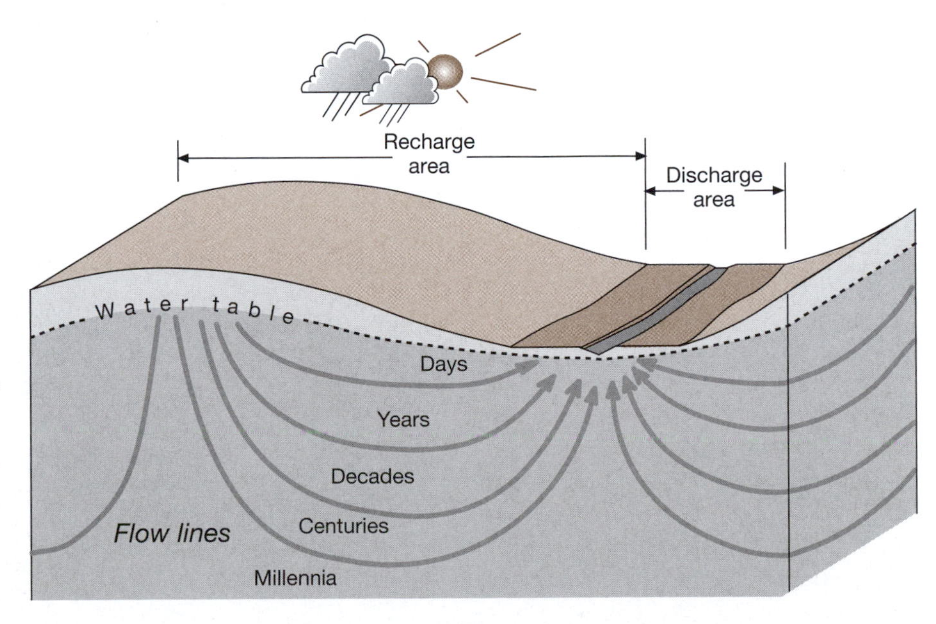

Figure 10.1 The vast majority of groundwater is meteoric in origin and is free to move with vagaries of climate. Rates of groundwater flow differ with depth, ranging from days to thousands of years to traverse an area the size of a county.

Groundwater—the great dissolver, the great precipitator Groundwater is physically and chemically *dynamic*. It is constantly on the move, constantly dissolving and/or precipitating a host of rocks and minerals—depending on the composition of the water and the composition of the rocks and sediments through which it moves (Fig. 10.2).

Dissolves some things	Precipitates many things
• Limestone, gypsum, salt (forms caves and landscapes in these rocks) • Few minerals from sandstones, shales, and igneous and metamorphic rocks (rarely forms caves in these rocks)	• Cave deposits (stalactites, stalagmites, etc.) • Cements that hold sedimentary rocks together (calcareous, siliceous, ferruginous) • Spring and geyser deposits • Concretions and geodes

Figure 10.2 Groundwater dissolves rocks and minerals, groundwater precipitates rocks and minerals—depending on the composition of the water and on the composition of the rocks and sediments through which it moves.

The variability in the composition of groundwater is illustrated by the variety of **geyserites** in Yellowstone National Park. (Geyserite is mineral material that is precipitated from groundwater as it emerges from the ground and evaporates, leaving behind elements that were in solution.)

Q10.1 In the southern part of Yellowstone National Park (e.g., the vicinity of Old Faithful), geyserite consists of varicolored siliceous material; whereas in the northern part of the park (e.g., in the vicinity of Mammoth Hot Spring), geyserite consists of snow-white calcareous material. Examine details in Figure 10.3 and try to explain why this difference in the mineral compositions of Yellowstone geyserites. *Hint:* What goes around, comes around. Whatever water dissolves, water might precipitate.

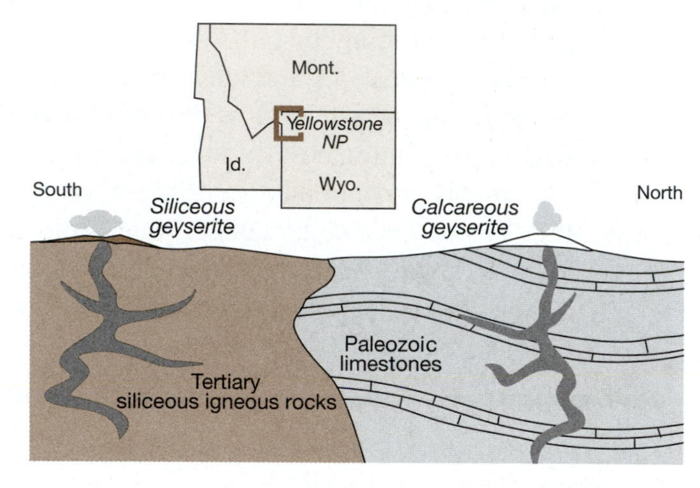

Figure 10.3 This is a schematic cross-section showing the variety of rocks in which the 'plumbing systems' of Yellowstone geysers occur.

Groundwater and geologic wonders—Minerals that grow within cavities in rocks are most commonly precipitated by groundwater (Fig. 10.4A and B). And, petrifaction of trees of Triassic age in the Petrified Forest of northern Arizona reflects the work of groundwater as well (Fig. 10.4C). In the world of geologic wonders, cases documenting the effects of groundwater abound.

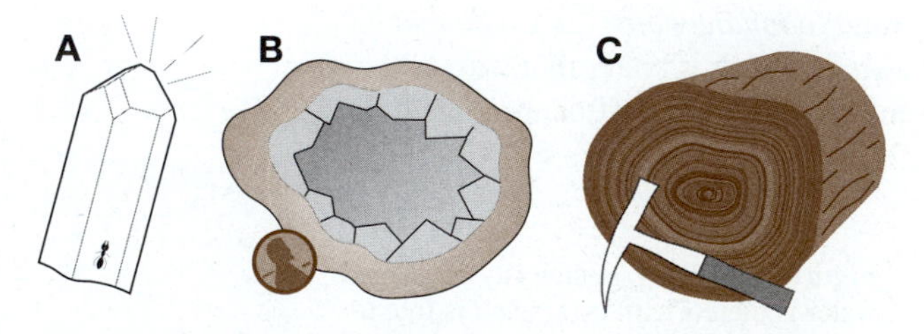

Figure 10.4 A This faceted quartz crystal was precipitated by groundwater within cavities in sandstone of the Ouachita Mountains of west-central Arkansas. **B** This geode was precipitated by groundwater within cavities in volcanic rocks of Brazil. C. This fossil tree was 'petrified' by groundwater in the Petrified Forest of Arizona.

The impending global water shortage not only requires that the world develop all available potable water resources in the near future, but we must also do a better job of minimizing the waste of water and guard it against pollution.

Groundwater will play a growing role in efforts to provide water for our growing global population, given the fact that it quantitatively competes with other fresh water resources (Fig. 10.5).

COLUMBIA DAILY TRIBUNE, MARCH 24, 2002

U.N. water report warns of impending shortages

Crisis will affect 5 billion worldwide by 2025.

VIENNA, Austria (AP)—Warning that 2.7 billion people face a critical shortage of drinkable water by 2025, the United Nations marked World Water Day on Friday with a call for a "blue revolution" to conserve and tap the seas for new supplies.

In fewer than 25 years, about 5 billion people will be living in areas where it will be difficult or impossible to meet all their needs for fresh water, creating "a looming crisis that overshadows nearly two-thirds of the Earth's population," a U.N. report said.

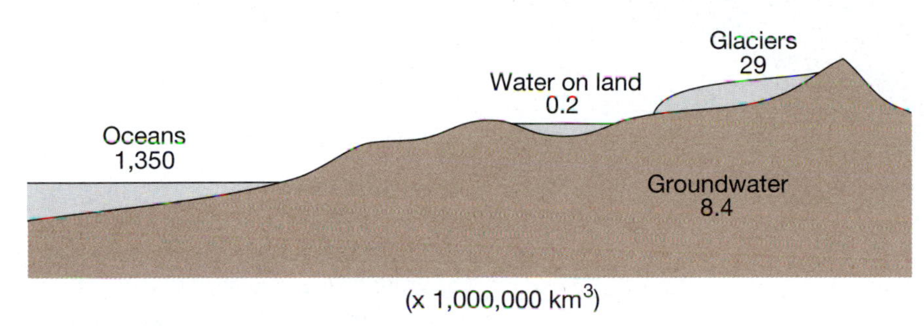

Figure 10.5 This is a comparison among the four vast reservoirs of accessible water on Earth. (Each unit is one million cubic kilometers.) Water on land consists of streams, rivers, lakes, and ponds.

Q10.2 (A) How many cubic kilometers of water reside within groundwater? (B) How many more times abundant is groundwater than water on land?

Groundwater has several advantages over surface water when it comes to providing for municipal needs.

Q10.3 Imagine that you are a member of a city council and your town is in need of a new and larger municipal water supply. Discussion has turned to the merits of well water compared to those of surface water. In what ways can you imagine that groundwater might be superior to surface water as concerns the following points?

(A) Paying the cost of drilling a well, compared to that of constructing a dam.
(B) Contending with the occasional drought in your semiarid region.
(C) Minimizing contamination from surface runoff and from the atmosphere.
(D) Protecting your water supply against the threat of terrorism.

B. Anatomy of water tables

Saturated and unsaturated zones—Within the subterranean realm of groundwater there are two main zones:

(1) The **saturated zone** (Fig. 10.6) is the zone in which open spaces in sediments and rocks are filled with water. The top of the saturated zone is the **water table**. The slow movement of groundwater—*toward streams in humid regions and away from streams in arid regions*—is impeded by friction, so water tables are rarely flat. The shape of a water table in a humid region mimics that of the land surface—i.e., high under hills and low under valleys, where it intersects perennial streams and lakes.

(2) The **unsaturated zone** is the zone in which intergranular spaces and fractures are filled with air and, at times, films of descending water.

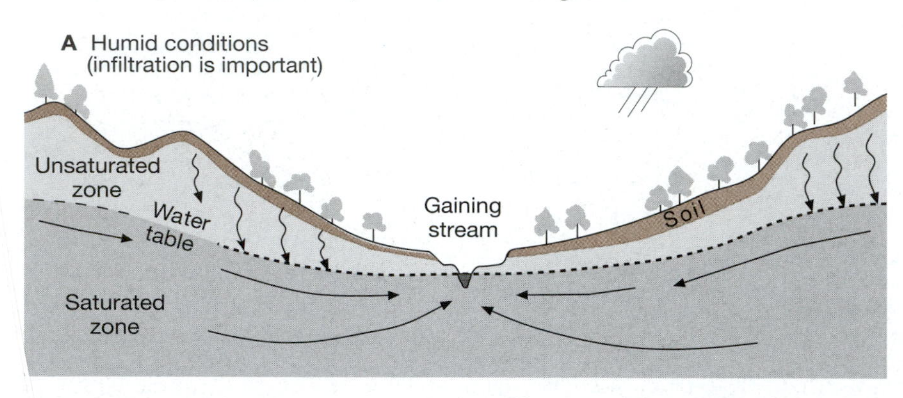

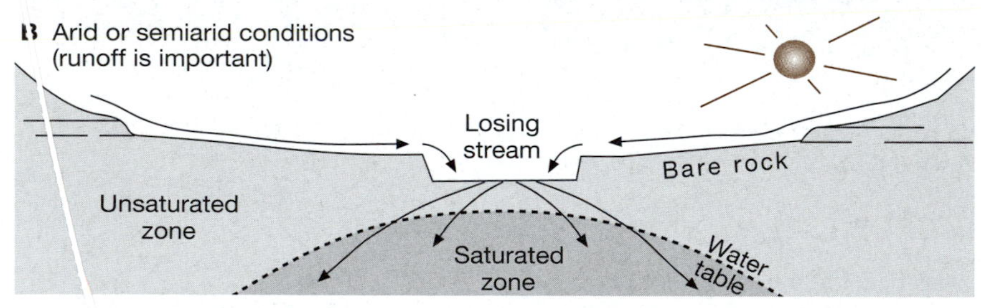

Figure 10.6 A In a humid region, water moves ('seeking its own level' as it were) in its tendency to develop a horizontal water table, and so the saturated zone feeds a *gaining stream*. **B** In an arid or semiarid region, the water table slopes downward from a *losing stream*, the source of water for the saturated zone.

Much of western United States lies within the Great American Desert (Fig. 10.7), a region in which the groundwater situation is like that in Figure 10.6B. In contrast, eastern United States is characterized by conditions shown in Figure 10.6A.

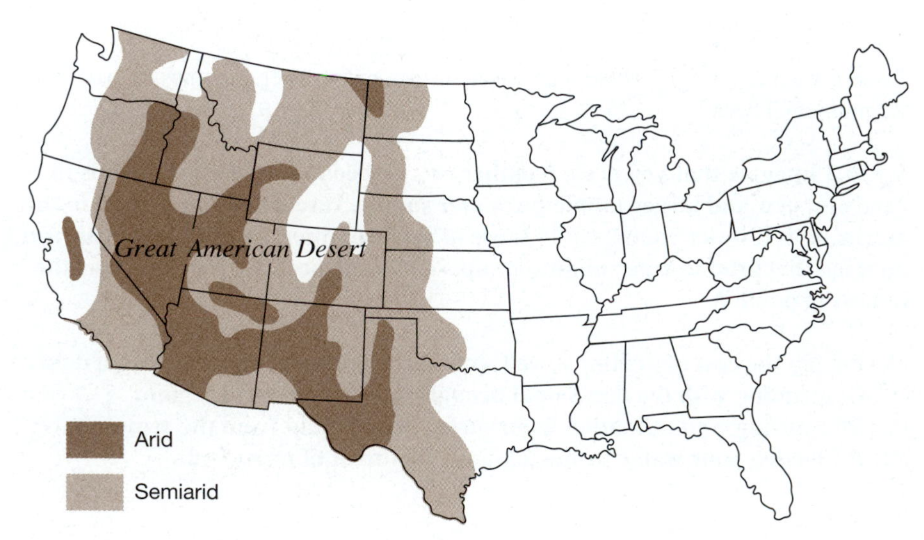

Figure 10.7 Because of arid to semiarid climate, approximately one-half of the conterminous 48 states is at risk as concerns the development and management of water resources.

A word about arid and semiarid streams

Q10.4 Judging from the information accompanying Figure 10.6, which kind of stream do you suspect would rise faster (though briefly) for a given amount of rain—that in a humid region or that in an arid or semiarid region?

If you answered the above question with, "in an arid or semiarid region," you were correct. Flat-bottomed losing streams like that in Figure 10.6B (Spanish *arroyos* or *barrancos*) can be dangerous. There is little rain in arid and semiarid regions, but when rain does come, it is typically *torrential*. Commonly, the entire annual amount of rainfall arrives in a single afternoon—often creating disastrous **flash floods**. In August 2003 this kind of flash flood swept automobiles and highway dividers from I-35 in Kansas, killing a number of people.

Aquifer defined

Before going further in our discussion of groundwater, we need to define the concept of **aquifer**—*a body of sediments or rocks that yields water sufficient to meet specific needs*. The saturated zone in Figure 10.6 might or might not be an aquifer. The definition of 'aquifer' is qualitative; e.g., an aquifer supplying a particular city might cease to be viewed as an aquifer were the population to grow beyond its capacity.

C. Dynamics of water tables

Common problems

In Los Angeles County, a 3.5-mile section of I-105 was constructed below ground level in an effort to minimize noise and visual pollution. Caltrans (Calif. Dept. of Transportation) believed the water table to be 30 feet below road level at the time of construction (Fig. 10.8). However, what Caltrans failed to learn was that the water table had been drawn down by over-pumping in the 1950s, and another state agency had recently mandated that the over-pumping cease. *(Ref: Calif. State Auditor rep't #99113, 1999.)*

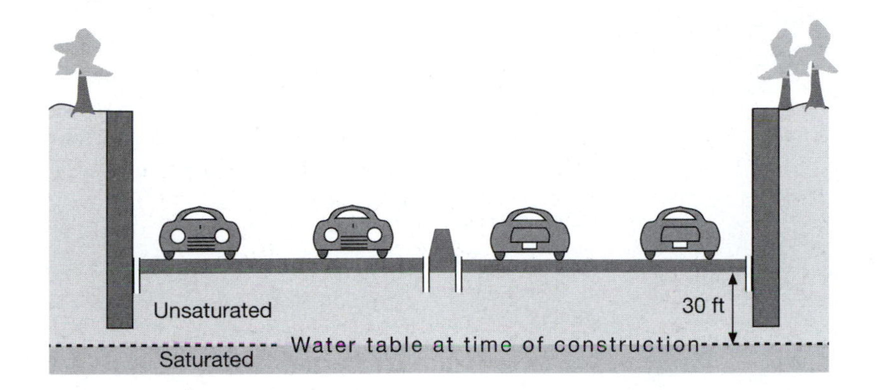

Figure 10.8 Highway engineers recessed a section of I-105 in Los Angeles County in an effort to mitigate noise and visual pollution. At that time, the water table was 30 ft below the highway.

Q10.5 So what do you suppose happened when over-pumping of the saturated zone was stopped by that other California state agency?

On more than one occasion a gas station in a low topographic setting has allowed the level of gasoline in its storage tanks to become too low (Fig. 10.9). Then came the rains, with runoff making its way into the saturated zone.

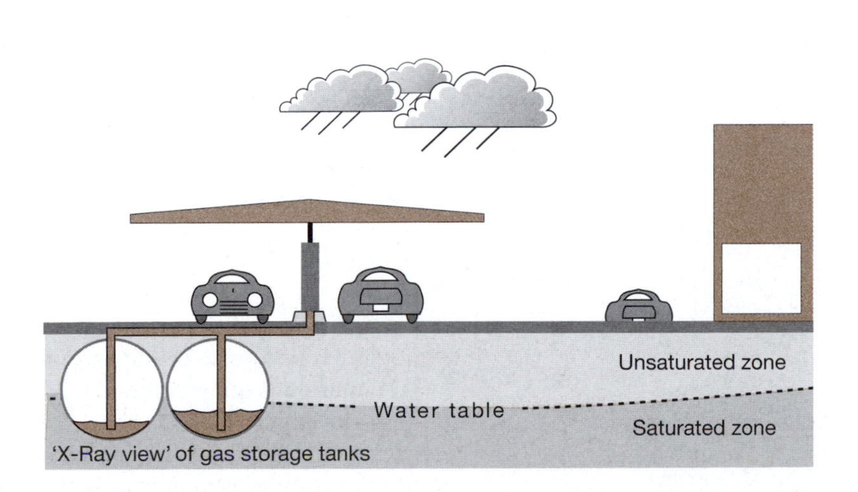

Figure 10.9 This gas station is very near the water table, which presents a threat to fuel storage tanks.

Q10.6 Can you imagine what happened when the water table rose? *Hint:* Asphalt and concrete are only so strong.

Hydraulic gradient

Geologists describe the magnitude of slope as the vertical angle between slope and the horizontal (Fig. 10.10).

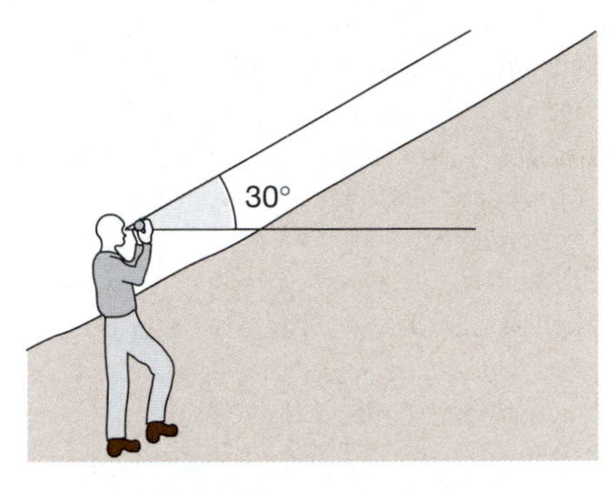

Figure 10.10. A Brunton compass, with its leveling bubble and sight-adjusted protractor, enables a geologist to measure the vertical angle between slope and the horizontal.

But engineers describe the magnitude of slope as the ratio of vertical drop to horizontal distance, aka the percent of grade. Thus, a gradient of 0.05, or 5 percent, designates a vertical drop of 5 feet per 100 feet of horizontal distance. This same convention is used in describing the hydraulic gradient of groundwater (Fig. 10.11).

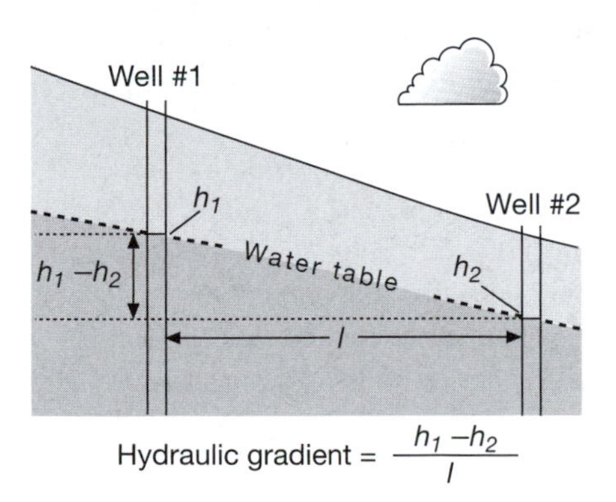

$$\text{Hydraulic gradient} = \frac{h_1 - h_2}{l}$$

Figure 10.11 h_1 is the elevation of the water table in well #1, h_2 is the elevation of the water table in well #2, and l (for length) is the horizontal distance between wells.

Q10.7 If, for the model in Figure 10.11, h_1 were 506 ft, h_2 were 497 ft, and l were 150 ft, what would be the hydraulic gradient (in percent) between well #1 and well #2?

Hydraulic conductivity

Just as surface water flows faster down steeper slopes, groundwater moves faster down steeper hydraulic gradients. But hydraulic gradient is not the only factor affecting the rate of groundwater movement. Equally important is hydraulic conductivity, which is the ease with which sediments or rocks transmit water. Hydraulic conductivity introduces the concepts of porosity and permeability.

Porosity is the percentage of a body of sediments or rocks that consists of open spaces, called pores. Porosity determines the amount of water that sediments or rocks can hold. There are many kinds of pores—ranging from pores among sedimented particles, to pores within volcanic rocks, to cavities within soluble rocks, to fractures in any kind of rock (Fig. 10.12). And any of these pores can be filled to differing degrees by cements.

Permeability is the ability of soil, sediment, or rock to transmit fluid. Material with low porosity is likely to have low permeability as well, but high porosity does not necessarily mean high permeability. In order for pores to contribute to permeability, they must be (a) interconnected, and (b) not so small that they restrict flow. For example, clay commonly has high porosity, but clay grains are so broad in proportion to their microscopic size (i.e., around 0.005 mm) that the molecular force between clay particles and water restricts flow. As concerns the potential of sediments and rocks to transmit water—a critical issue in the aquifers—permeability is paramount.

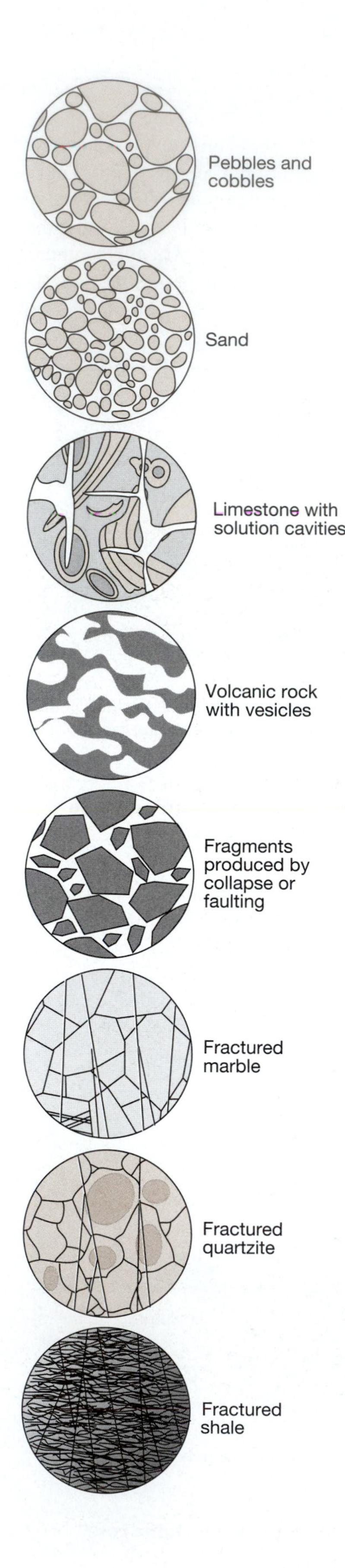

Figure 10.12 (White space is open space available to water. Open space along fractures is too thin to illustrate.) Porosity and permeability can result from sedimentation, volcanism, solution, collapse, faulting, and fracturing. Subsequent cementation can reduce the volumes of any of these pores. (Magnification is 5–10x.)

Darcy's Law

The most fundamental questions in targeting a prospective groundwater resource are, 'How much, and how often?' This *volume per time* issue is analogous to the discharge of a stream.

In 1856, **Henri Darcy**, a French engineer, attempted to determine whether a prospective aquifer could yield water sufficient for the city of Dijon. Darcy undertook a series of laboratory experiments in which he measured the rate of water flow through a variety of sediments in tubes tilted at various angles. Not surprising to us now, Darcy concluded:

(1) Groundwater flows faster through more permeable rocks.

(2) Groundwater flows faster where the water table is more steeply inclined.

Darcy identified the four key variables in groundwater flow (or discharge) as…

- Discharge *(Q)*
- Hydraulic conductivity *(K)*
- Hydraulic gradient $(h_1 - h_2 / l)$
- Area *(A)* (thickness x breadth of the aquifer)

…and crafted an algebraic expression ('Darcy's Law') of their relationship:

$$Q = (K)\,(h_1 - h_2 / l)\,(A)$$

Darcy's Law enables one to calculate the maximum amount of water that an aquifer might yield to an array of wells.

Example:

Q10.8 Hydraulic conductivity of an aquifer is known to be 8 ft/day, and its dimensions are estimated to be 40 ft thick and 18,000 ft wide. Two test wells drilled one mile apart in the direction of flow encountered the water table at elevations 5,030 ft and 5,050 ft. *Question*: How many gallons of water flow through the aquifer per day? (To convert ft^3 to gal, see page *ii* at the front of this manual.)

D. Mapping a water table

Map the direction of groundwater flow within your mapped area

Figure 10.23 on Answer Page 184 is a contour map on which 26 water wells have been plotted. Each well site shows the depth to the water table within that well as a negative value in feet below ground level.

Procedure:
For each well location:
(1) Estimate the *surface elevation* from the proximity of contour lines.

(2) Subtract the *depth to the water table* from surface elevation in order to determine the *elevation of the water table.* Record that elevation (of the water table) at the well site.

(3) After repeating the above two steps for each of the 26 wells, *contour the groundwater elevations* with a contour interval of 20 feet. (A 'getting started' example is framed with a gray rectangle in the lower-left corner of the map.)

Q10.9 Draw an arrow between Well A and Well B indicating the direction in which groundwater is likely to be moving. In which direction is the arrow pointing, northeastward or southwestward?

Q10.10 (A) At what map coordinates is the difference between the elevation of the ground and the elevation of the water table the greatest? (B) Give the coordinates of a place where you might expect to find a marsh or spring.

Q10.11 If contaminants were to find their way into groundwater at Acme Industries, in which well would those contaminants be more likely to appear—the well at the Smith farmhouse, or the well at the Jones farmhouse?

Rates of groundwater flow—applying Darcy's Law to your map of the water table

Q10.12 What is the difference in elevations of the water table at Well A and Well B?

Q10.13 What is the map distance (in feet) between Well A and Well B?

Q10.14 What is the hydraulic gradient $(h_1 - h_2 / l)$ between Well A and Well B?

Q10.15 If the hydraulic conductivity *(K)* of the aquifer is 10 ft/day, and the cross-sectional area *(A)* of the aquifer is 200 ft x 5,000 ft, what is the rate of flow or discharge, *(Q)*, through the aquifer in cubic feet per day?

E. Artesian conditions and confined aquifers

'Water seeks its own level,' which explains simple *artesian* conditions. To illustrate—if we were to add water to the glass tubing in Figure 10.13A, to a particular elevation in conduit A, that water would be driven by **hydrostatic pressure** to that same elevation in conduits B and C. However, were our tubing (a) filled with sand, and (b) open at its downstream end (Fig. 10.13B), water would not rise as high in B and C. Moreover, the farther from the recharge area, the less the **hydrostatic potential** for lifting water in a conduit. This reduction in potential away from the water's source describes a sloping **potentiometric surface**.

In the real world, an artesian aquifer is more like the situation in Figure 10.13B (rather than 10.13A) in that (a) an artesian aquifer is filled with sediments and/or rocks, and (b) water within an artesian aquifer is free to move within that aquifer.

Q10.16. In Figure 10.13B, what two things account for less hydrostatic pressure within Conduit C than within Conduit B? *Hint:* These two things are pretty much spelled out by labels in this figure, and one appears in the description of 'the saturated zone' (explaining why water tables are rarely flat) on page 170.

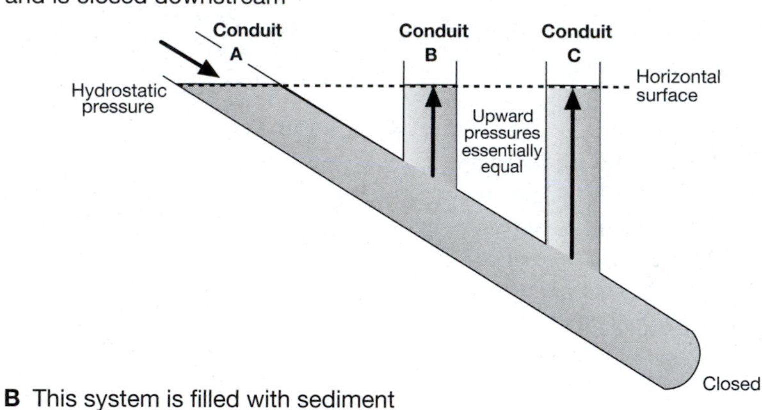

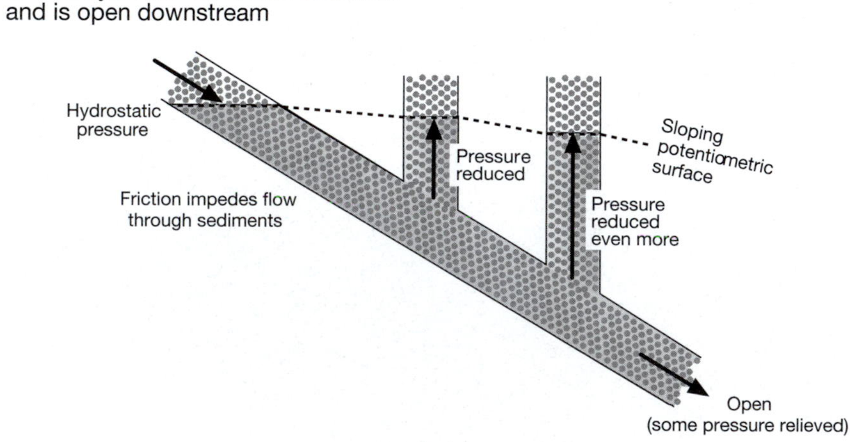

Figure 10.13 A This tubing illustrates the simple principle, 'water seeks its own level.' **B**. This tubing illustrates the variation of this principle that is applicable to the real world of artesian flow of groundwater. It's filled with sediment, and it's open downstream.

Notable artesian examples

When wells were first drilled in London circa 1900, fountains at Trafalgar Square were **flowing artesian wells** (Fig. 10.14). But hydrostatic pressure has since declined, so now water must be pumped to the surface.

Figure 10.14 Fountains at Trafalgar Square are a graphic reminder of the famous London Artesian Basin.

Artesian thermal waters at Hot Springs National Park, Arkansas, owe their heat to deep circulation (Fig. 10.15). Slow descent of rainwater is via a large collecting system, with rapid ascent via narrow passageways. (A bent funnel effect.) Thus, heat persists as waters make their quick escape to the surface.

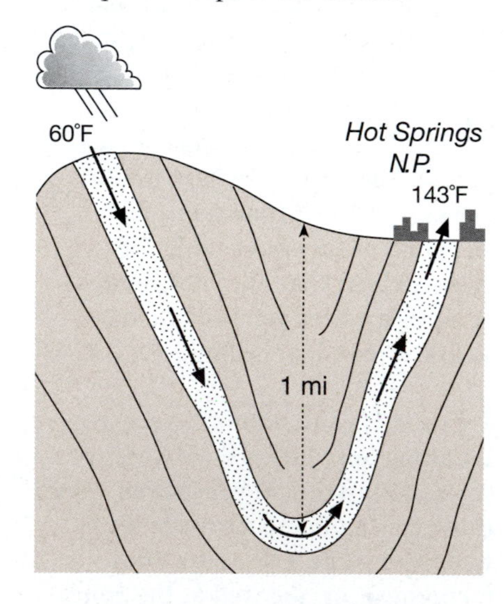

Figure 10.15 Within folded rocks of the Ouachita Mountains, rainwater enters an artesian aquifer at 60°F, descends to a depth of one mile, and emerges at 143°F (the temperature of hot coffee).

Details of confined aquifers—But first, a look back at **unconfined aquifers.** In Sections B and C (pages 170–171), which deal with the anatomy and dynamics of water tables, aquifers are *unconfined*; i.e., the top of the zone of saturation (i.e., the water table) is free to rise and fall with the vagaries of climate. But in a *confined* aquifer, such is not the case. A **confined aquifer** is one in which water is prevented from rising and falling by relatively impermeable intervals of rock called **aquitards** (from Latin, *retards water*). A visual metaphor: Swiss cheese (with its interconnected holes) between slices of dense bread. Aquitards typically consist of shale—the most common and the least permeable of all sedimentary rocks.

A confined aquifer receives its water in an area where it intersects the land surface, called the **recharge area** (Fig. 10.16). Where a well is drilled into a confined aquifer, water rises toward the elevation of the water table in the recharge area, a condition called **artesian** (recall the glass tubing model on facing page 174). Water will not rise quite as high as the water table in the recharge area because (a) friction is associated with water moving through the aquifer, and (b) the water is free to move laterally, thereby reducing hydrostatic pressure. The imaginary level to which water in a group of artesian wells tends to rise is (as defined on the facing page) the potentiometric surface (aka the piezometric surface). In Figure 10.16 the potentiometric surface is below ground level in the vicinity of Wells A and B, so they are **non-flowing artesian wells**. The potentiometric surface is above ground level in the vicinity of Well C, so that well is a **flowing artesian well**.

Q10.17 In Figure 10.24 on Answer Page 184, label each well with the correct letter as described in the '10.17 text' to the left of that figure.

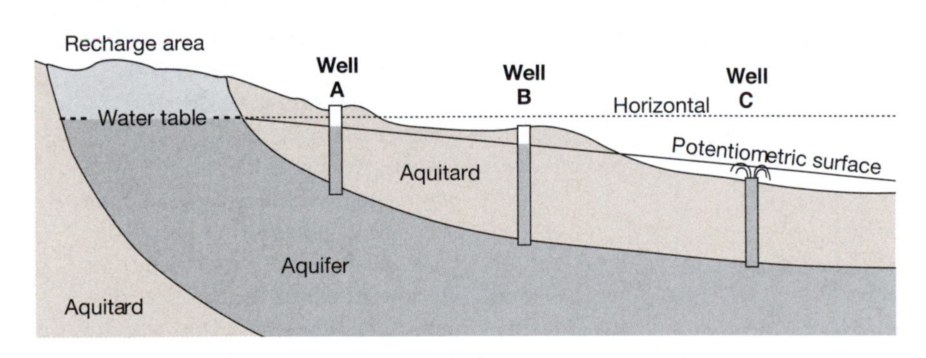

Figure 10.16 The dotted line is a horizontal projection of the water table in the recharge area. The solid line is the potentiometric surface. Well **A** is near the recharge area, so the water in it rises almost to the elevation of the water table. Well **B** is farther away, so friction over a greater distance accounts for the water's not rising as high as in Well **A**. The mouth of Well **C** is below the potentiometric surface, so it is a flowing artesian well.

Mapping a potentiometric surface—Figure 10.25 on Answer Page 185 shows six wells (#1–#6) on a ground elevation contour map in the area of a confined aquifer. All six wells penetrated the aquifer. The map includes a second set of contours (straight dashed lines) drawn on the aquifer's potentiometric surface. Flowing artesian wells should occur where the potentiometric surface is *higher* than ground elevation. Non-flowing artesian wells should occur where the potentiometric surface is *lower* than ground elevation.

Procedure:
At every place where a ground elevation contour line crosses a potentiometric contour line of the *same value*, place a small circle. Then connect the circles with a line. Ground elevations of wells on one side of that line are *lower* than the potentiometric surface, so the wells should be *flowing* artesian wells; whereas ground elevations of wells on the other side of that line are *higher* than the potentiometric surface, so the wells should be *non-flowing* artesian wells.

Q10.18 (A) Which of the six wells in Figure 10.25 on page 185 should be *flowing* artesian wells? (B) Darken the area of the map (like the swatch in the legend) in which the wells should be flowing artesian wells.

P.S. Such a map is useful in assessing land use values. Who wants to purchase a site for a home or a building in a swamp?

F. Perched water tables

Where descending surface water encounters an aquitard, it can accumulate as a local saturated zone with its own local **perched water table** (Fig. 10.17). This is one of the more common explanations for the occurrence of springs.

In areas of flat-lying sedimentary rocks where relatively impermeable shales alternate with permeable sandstones or limestones, perched water tables (and related springs) are likely.

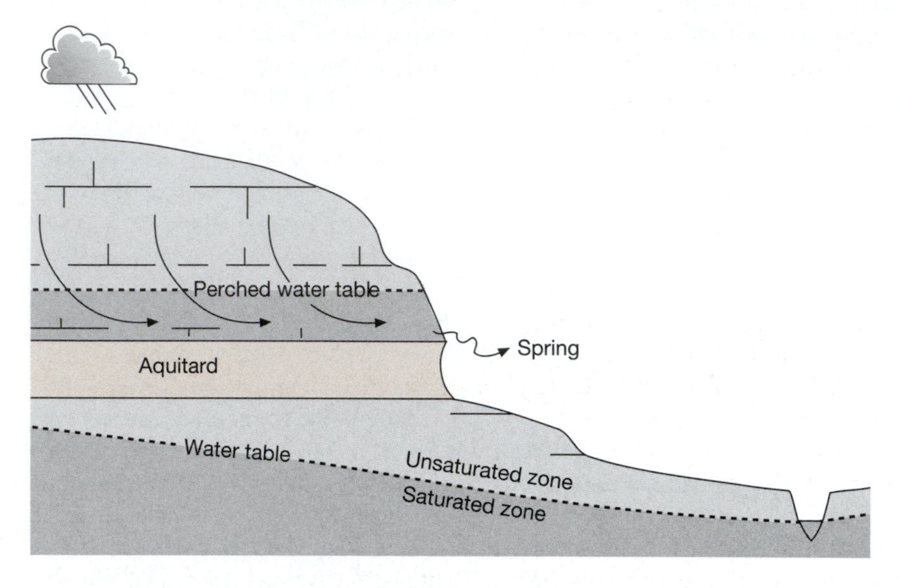

Figure 10.17 Here, descending water is deflected by an aquitard to emerge as a spring. The general water table is deeper.

Trouble for construction sites—Perched water tables can cause water problems in basements of homes and buildings. At Saint Louis University, excavation for two buildings breached a perched water table (Fig. 10.18). The only remedy was the installation of sump pumps to lift the water out and away to storm sewers.

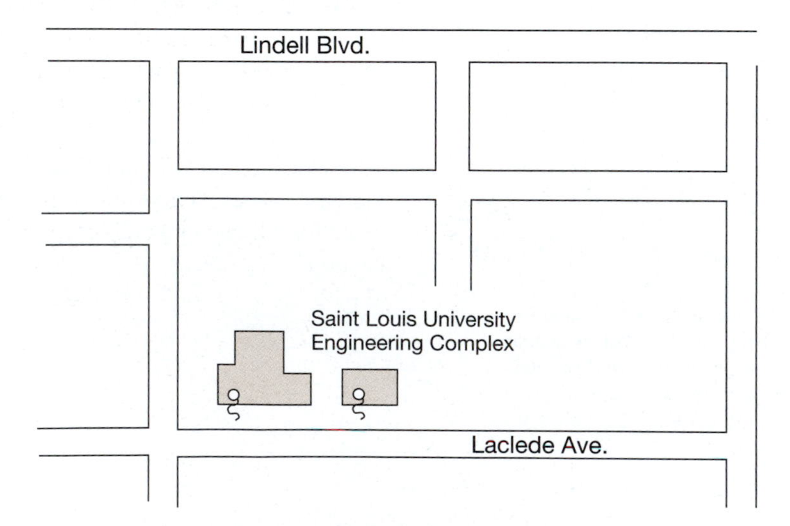

Figure 10.18 Perched water tables can cause springs to appear in basements.
The symbol for a spring is a small circle with a wiggly tail (see the last item under **RIVERS, LAKES, AND, CANALS** *on Topographic Map Symbols, page 17 of #1 Maps Exercise).*

Ephemeral ponds (i.e., short-lived) occur where water is temporarily prevented from sinking into the zone of saturation by relatively impermeable soil or sediments (Fig. 10.19). Such ponds disappear after a time by evaporating and/or sinking into the unsaturated zone.

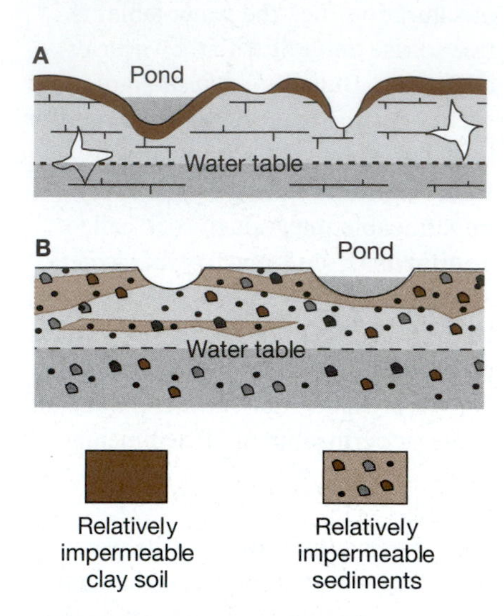

Figure 10.19 A This ephemeral pond occurs within a sinkhole formed by the collapse of a cave roof. **B** This ephemeral pond occurs within a depression typical of landscapes carpeted by glacial deposits.

Q10.19 In Figures 10.19A and B, two depressions are occupied by water, whereas others are dry. (A) Explain this presence and absence of ponds in these two figures. (B) If the two ponds were *perennial* (i.e., year-round) ponds, because of intersecting the water table, how would the presence or absence of water in the other depressions differ from that which is shown?

Reservoir liners and the Erin Brockovich case

A rancher can build a stock pond (with proper local permits) by simply bulldozing an earthen dam across a stream course. If stream flow is considerable, an added feature might be a metal tube or concrete spillway to handle excess water and prevent the water's topping the dam and washing it away. A second matter, especially common in cases of marginal water supply, is *leakage*, which can result in pond water descending into the unsaturated zone (Fig. 10.20).

Q10.20 Judging from what you learned from information in Figure 10.19A on facing page 176, how might one seal a leaking stock pond? *Hint:* We're talking three steps here, with steps #1 and #2 being the draining and restoring pond water.

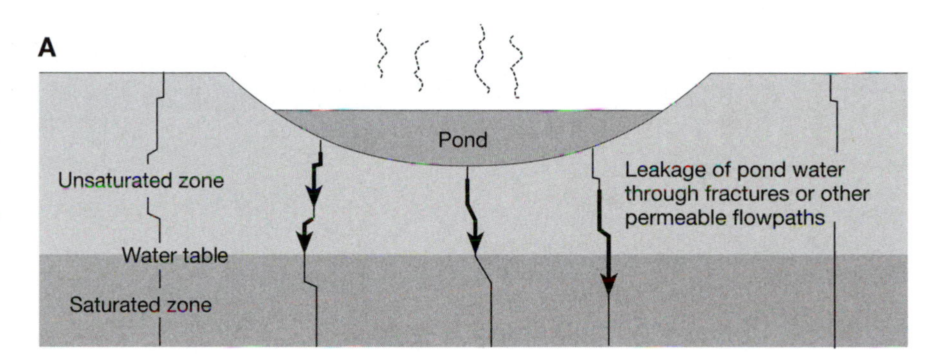

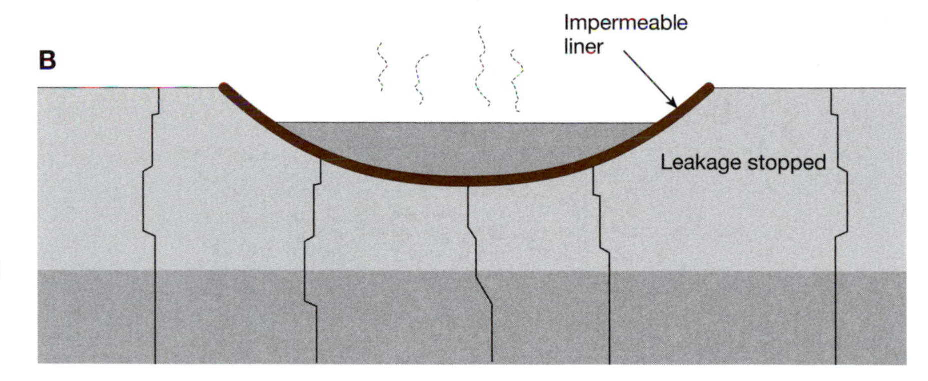

Figure 10.20 A A pond without an impermeable liner might leak water into the unsaturated zone. **B** The same pond with an impermeable liner loses water only through evaporation. Such liners are essential in the construction of best landfill sites.

The Erin Brockovich case (1994) grew out of the undisputed fact that Pacific Gas and Electric used water containing an alleged carcinogen, chromium-6, to cool pipes that were heated by the compression of gas at its pumping station near Hinkley, California. (Chromium-6 was added to the water to minimize corrosion of the cooling towers.) A group of plaintiffs alleged that chromium-rich water was placed in evaporating ponds that lacked proper reservoir liners. PG&E admitted that there was, in fact, seepage of pond water into the groundwater, but disputed the alleged health effects of such water. Hydrologic and geochemical data marshalled by Brockovich, legal assistant with the Masry law firm, which represented the plaintiffs, are not a matter of public record because the case was settled out of court. So any maps and cross-sections that were entered into evidence by either side are unavailable.

G. Karst topography

In humid areas where there is an abundance of limestones and/or dolostones (and, more rarely, gypsum and/or salt), the dissolving of these relatively soluble rocks accounts for the development of **caves** and cave-related features (Fig. 10.21). (The chemistry of *dissolution* appears on pages 100-101.)

Sinkholes commonly develop through the collapse of cave roofs, and **disappearing streams** descend into underground passages, only to reappear as **springs**.

Such a landscape is called **karst** topography, a name taken from a part of Slovenia that is characterized by such features. Karst landscapes are largely solution landscapes. Very little detrital sediment occurs in streams flowing from such regions.

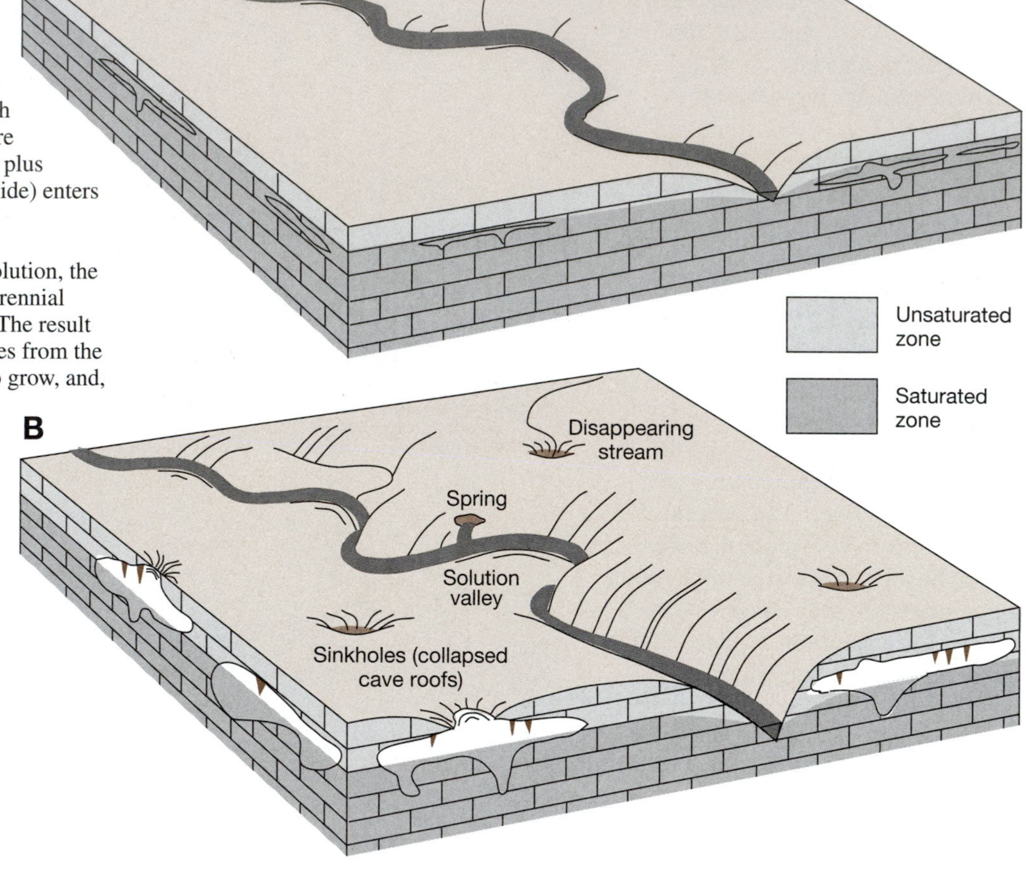

Figure 10.21

(A) Dissolution of limestone occurs within the uppermost part of the saturated zone (just beneath the water table), because it is there that carbonic acid (i.e., rainwater plus atmospheric and soil carbon dioxide) enters the saturated zone.

(B) As a valley is deepened by solution, the water table descends, tracking perennial streams with which it intersects. The result is that the cave effectively emerges from the saturated zone, the cave ceases to grow, and, instead, begins to be filled with an array of cave deposits (e.g., stalactites, stalagmites, columns, flowstone, curtains, and cave pearls).

Beware the karst

Karst is a proverbial red flag in almost every imaginable kind of land development. In karst terrain special care should be taken in an effort to contend with the following:

- Pollution of groundwater, resulting from the rapid flow of waste effluent from its source to water supplies.
- Catastrophic collapse of sinkholes, swallowing buildings and people.
- Failure of building foundations, causing irreparable damage to homes and commercial buildings.

Of these, pollution of groundwater is the most universal.

The movement of groundwater through soil and sediments magically cleanses water of ordinary waste such as that in municipal and home sewage systems—provided that the **residence time** within the soil and sediments is sufficiently long (i.e., measured in months and years). Not only does the movement through soil and sediments filter solid particles, but slow movement typical of fine-grained earth materials provides time for pathogenic bacterial and other troublesome micro-organisms to die.

In striking contrast to the tiny passageways in fine-grained soil and sediments, solution channels in limestone are so large that through-flow of liquids is virtually instantaneous.

Google Earth image of karst topography

An example of karst topography occurs at Rock Bridge State Park, Missouri (Fig. 10.22). This area is at the northern margin of the Ozark Plateau, the most spectacular karst region between Mammoth Cave, Kentucky and Carlsbad Caverns, New Mexico.

Q12-21 Examine the Google Earth image in Figure 10.22. Notice the many ponds. At a glance some of the ponds might be mistaken for stock ponds. However, there is good evidence indicating that the large muddy pond at letter 'A' in the southwest quarter of the photo is a sinkhole. What is the evidence indicating a sinkhole, rather than a stock pond? *Hint:* The evidence appears in the relationship of the pond to an adjacent construction feature.

Karst topography
Rock Bridge State Park
Missouri

Ashland, Missouri 7 1/2' quadrangle
N. 38° 52' 09", W. 92° 17' 10"

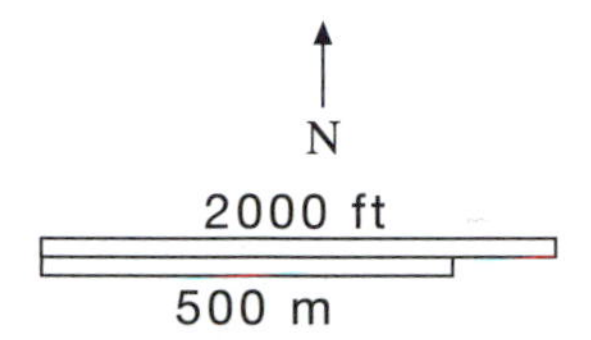

Figure 10.22 Sinkholes characterize karst topography at Rock Bridge State Park. Some sinkholes are naturally lined with relatively impermeable clay soil (see Figure 10.19A, page 176), so they can hold water long after the rains that filled them. In contrast, other sinkholes are without benefit of such natural liners, so they fail to retain rainwater for even short periods of time.

Karst topography on a map—Putnam Hall, Florida, quadrangle

The Putnam Hall, Florida quadrangle—on the facing—is in a region underlain by limestone of the Florida Aquifer. (The contour interval is10 feet.)

Countless sinkholes (some with ponds, some without) are marked by bundles of crudely concentric contours with hachures (tick-marks) indicating depressions. Some of these bundles of contours are 'pinched' on opposing sides, creating a bird's eye appearance (e.g., at C-6 and E-3) that suggests solution of the limestone along fractures, aka **joints**.

Q10.22 Do these probable joints appear to be oriented more nearly northeast–southwest, or more nearly northwest–southeast?

Q10.23 There appears to be no stream patterns indicating through-flowing surface streams on this map. Why do you suppose that is?

Q 10.24 What is the elevation of the surface of (A) Chipco Lake? (B) Mariner Lake? (C) Junior Lake?

Q 10.25 Do water levels in these three lakes (as well as others) appear to be governed by the vagaries of spotty rainfall and random surface drainage, or do they appear to mark systematic elevations on a water table? *Hint:* Notice the elevations of the bottoms of dry sinkholes relative to water levels in ponds.

Q 10.26 (A) What is the elevation of the surface of Grassy Lake? (B) What is the distance (in feet) between the center of Chipco Lake and the center of Grassy Lake?

Q 10.27 Blue biodegradable fluorescent dye was once added to Chipco Lake. Twenty-four hours later the dye appeared in Grassy Lake. What is the approximate velocity of groundwater flow between these two lakes in feet per day?

Q 10.28 Would you expect to detect that same fluorescent dye in Mariner Lake? Why, or why not?

For a comprehensive short course in groundwater, go to EPA site http://www.epa.gov/seahome/groundwater/src/geo.htm

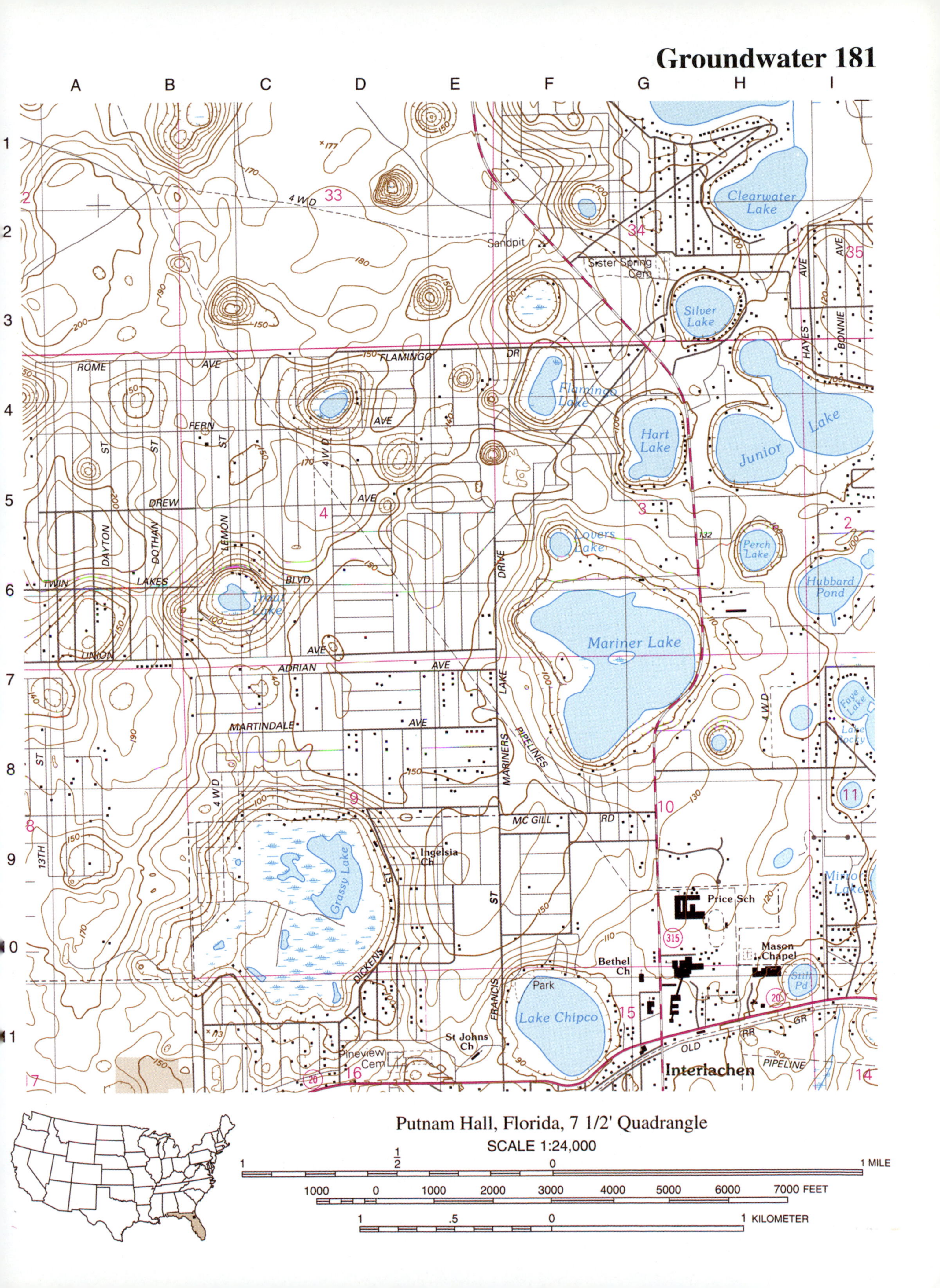
A
B
C
D
E
F
G
H
I
1
2
3
4
5
6
7
8
9
10
11
Clearwater Lake
Silver Lake
Sister Spring Cem
Sandpit
4 WD
33
34
35
Flamingo Lake
FLAMINGO DR
ROME AVE
FERN ST
Hart Lake
Junior Lake
HAYES AVE
BONNIE AVE
DREW AVE
DAYTON ST
DOTHAN ST
LEMON ST
Lovers Lake
Perch Lake
Hubbard Pond
TWIN LAKES BLVD
Trout Lake
Mariner Lake
UNION AVE
ADRIAN AVE
LAKE DRIVE
MARTINDALE AVE
MARINERS ST
PIPELINES
Faye Lake
Lake Rocky
MC GILL RD
13TH ST
Ingelsia Ch
Grassy Lake
DICKENS ST
Price Sch
Mirror Lake
Mason Chapel
Bethel Ch
Still Pd
Park
Lake Chipco
FRANCIS ST
St Johns Ch
Pineview Cem
OLD GR RR
PIPELINE
Interlachen
Putnam Hall, Florida, 7 1/2' Quadrangle
SCALE 1:24,000
1 MILE
1000 0 1000 2000 3000 4000 5000 6000 7000 FEET
1 KILOMETER

Intensionally Blank

(Student's name) (Day) (Hour)

(Lab instructor's name)

ANSWER PAGE

10.1

10.2 (A) (B)

10.3 (A)

(B)

(C)

(D)

10.4

10.5

10.6

10.7

10.8

10.9 ______________________________

10.10 (A) __________ (B) __________

10.11 ______________________________

____

10.12 ______________________________

10.13 ______________________________

10.14 ______________________________

10.15 ______________________________

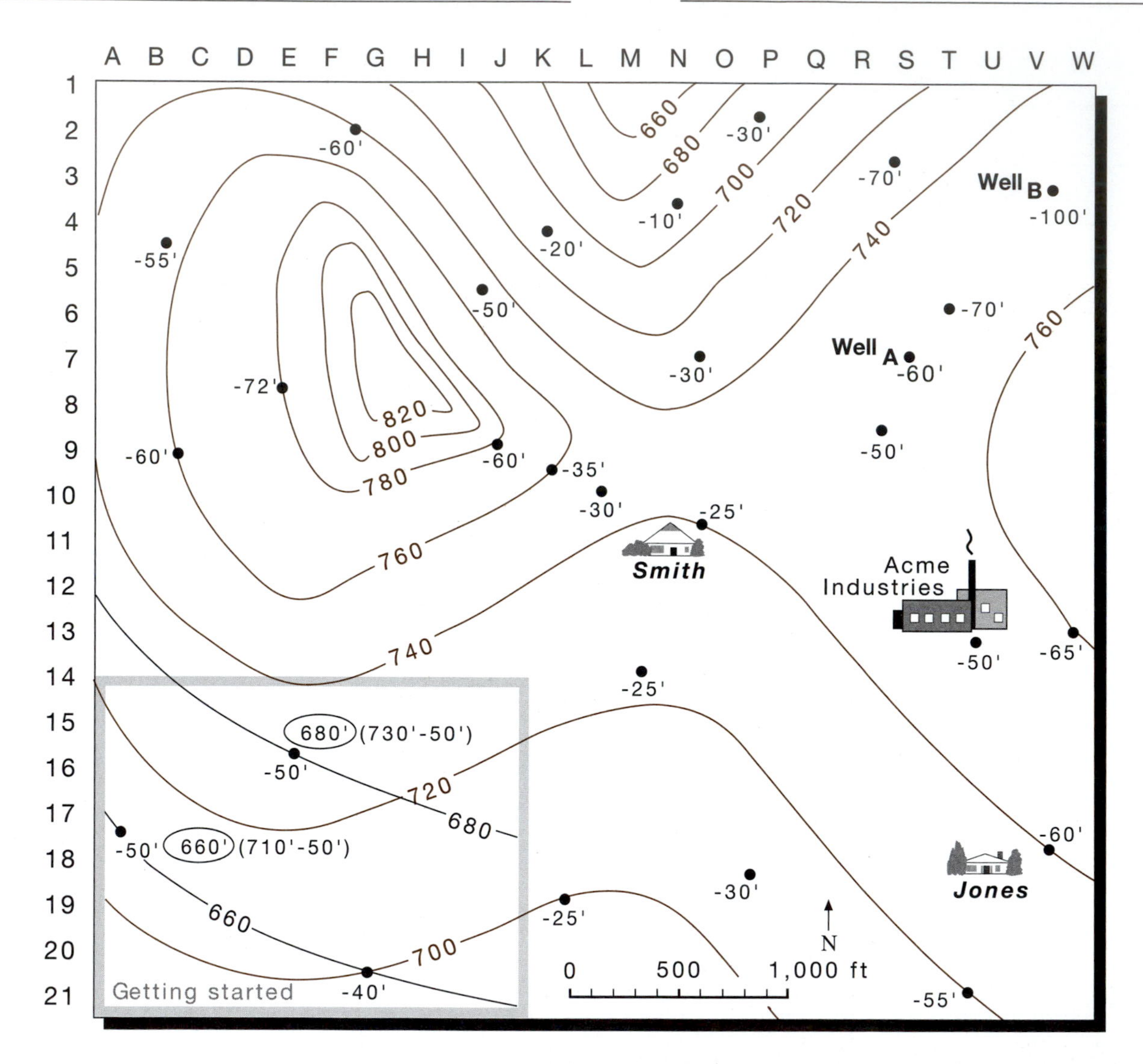

Figure 10.23. A contour map of a humid area. The contour interval is 20 ft. Each dot represents a water well with the depth to the water table indicated in feet with a negative value (e.g., -70').

10.16 ______________________________

____

____

10.17 (Label Fig. 10.24.) A—Possibly a flowing artesian well.
B—Possibly a nonflowing artesian well.
C—A well that might produce from an unconfined aquifer.

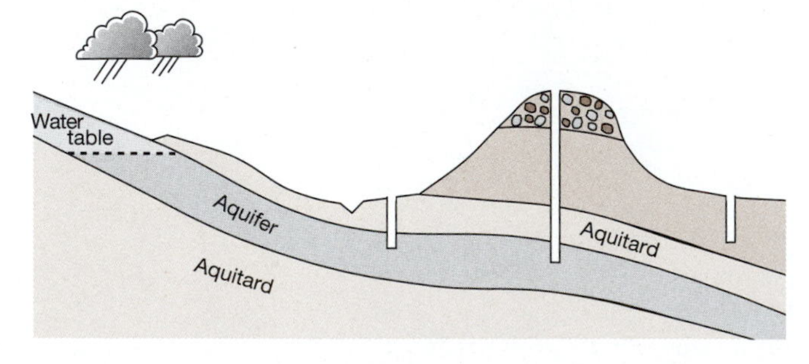

Figure 10.24. Three wells with different hydrologic settings.

10.18 (A) ______________________

10.19 (A) ______________________

____

____

____

(B) ______________________

____

____

10.20 ______________________

____

____

____

____

10.21 ______________________

____

____

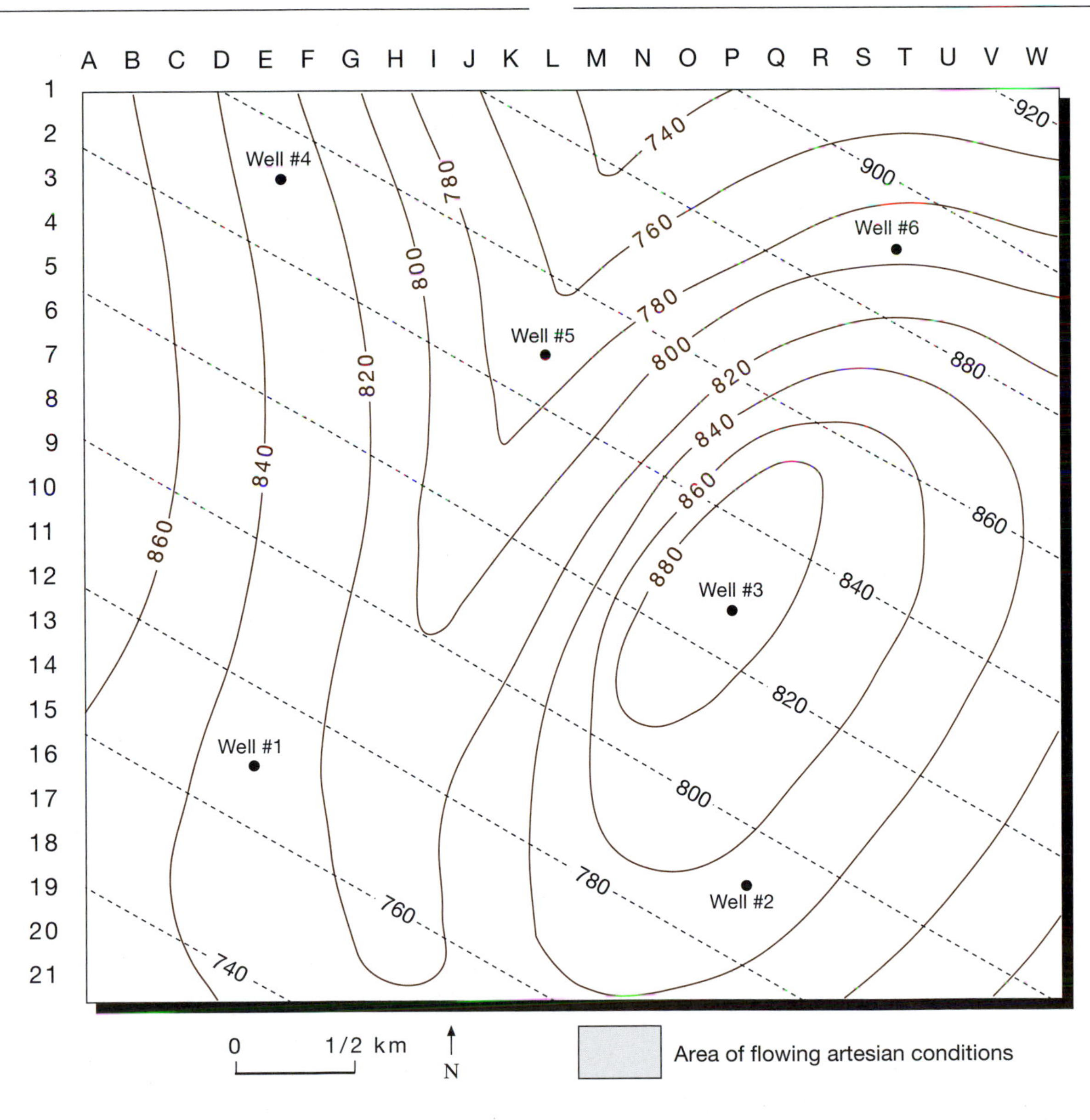

Figure 10.25 This is a contour map (with an interval of 20 ft) of an area underlain by a confined aquifer. The black dashed lines are drawn on the aquifer's potentiometric surface, also with a contour interval of 20 feet.

10.22 ____________________

10.23 ____________________

10.24 (A) (B) (C)

10.25 ____________________

10.26 (A) ____________________

(B) ____________________

10.27 ____________________

10.28 ____________________

11 Slopes and Subsidence

Topics

A. What is angle of repose? How do geologists and engineers differ in describing the magnitude of slope? What are lahars and what do they have to do with Mount Rainier?

B. What are five or six examples of mass wasting, and what are the features on which their classification is based?

C. What are the five common causes of slope failure? What is 'daylighting?' What are the particulars of the Madison River Canyon slide?

D. What are some of the factors that contribute to instability of Southern California hillsides? What evidence correlates groundwater levels with the Abalone Cove landslide? What are some of the ramifications of the Pacific Palisades landslide? What are some mitigating measures taken to stabilize slopes such as those in Southern California?

E. What four factors contributed to the collapse of rock at Hot Springs, Arkansas? Why can utility trenching be dangerous?

F. What is an example of natural subsidence? What two things account for most subsidence problems in the United States? What is the aquitard-drainage theory? What is the geologic setting and history of the Baldwin Hills Reservoir failure?

A. 'Liquid Earth'

Slope failure is just what the term implies—the downhill failure of earth on slope. Lots of things can contribute to landslides and avalanches, but the universal culprit is slope. The axiom: *View the Earth's surface as liquid if you wish to avoid disaster*. Some parts of that liquid are more viscous, some parts more fluid, but gravity tugs at all inclined earth surfaces.

Angle of repose is the *maximum slope angle at which material can remain stable*—for the moment. Angle of repose for dry sand is 34°. For rock fragments, 45° (Fig. 11.1A). Slope magnitude should be a concern when coexisting with the land. Pacifica, California requires studies before issuing building permits on slopes greater than 15 percent (Fig. 11.1B).

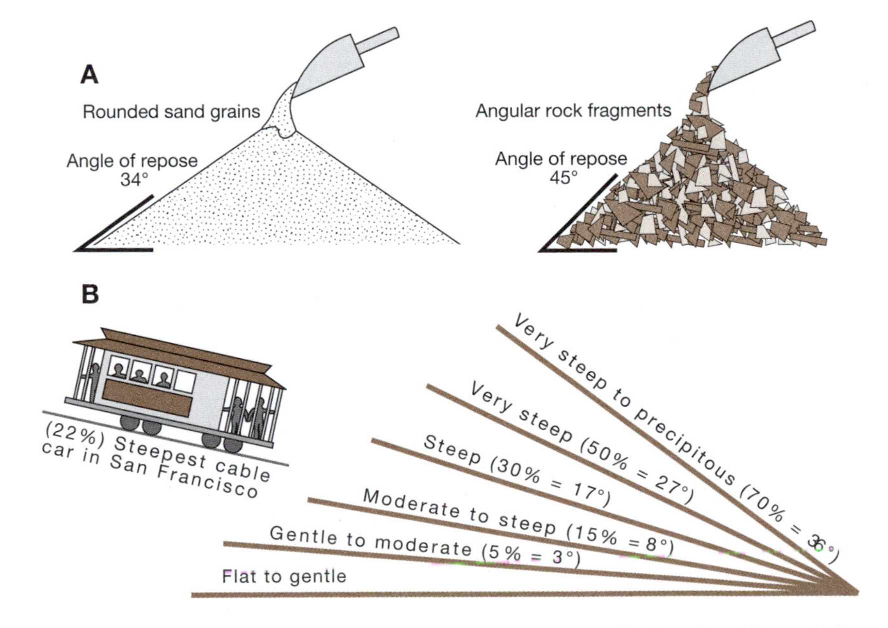

Figure 11.1 A Geologists describe the magnitude of slope as a vertical angle expressed in degrees. **B** Engineers describe the magnitude of slope as vertical drop in feet, divided by horizontal distance in feet, multiplied by 100 (i.e., drop/dist. x 100). Example: In the case of the 45° slope illustrated in Figure 11.1.A (above), vertical drop equals horizontal distance, so the grade is 100%.

Lahars of Mount Rainier

Loss of life and property—Slope failures in America cost more lives and more dollars than all other natural disasters combined. Even so, people continue to venture into places that promise to someday experience slope failure.

Actually, geologic disasters are the easiest of all disasters to predict. Why? Because geologic processes seldom if ever come to an abrupt halt. The simple rule: *If it has happened around here before, it's likely to happen around here again.*

Mount Rainier is an example of extrapolating past events into the future (Fig. 11.2). The flanks of this sleeping volcano in Washington state record a history of volcanic mudflows (called **lahars** in Indonesia), that have flowed down valleys now occupied by a dozen villages.

Lahars have killed >24,000 people in Central and South America, >11,650 in Japan, and >9,300 in Indonesia in recorded history. Lahars are usually triggered by volcanic eruptions onto snow and ice and by the saturation of volcanic debris during heavy rains. In the case of Mount Rainier, concern centers about the possibility of…

- a rise in thermal activity
- the melting of snow and ice
- meltwater liquefying volcanic ash

Present-day changes being monitored at the crater of Mount Rainier have prompted geologists to conclude that there are *"regions of hydrothermal alteration that are particularly susceptible to collapse, endangering population centers downstream from the volcano."* *(U.S. Geological Survey Circular 1172)*

Q11.1 Speeds of lahars differ as a function of slope and fluidity, but 25 mph is not unusual. At that speed, approximately how long would it take for a Mount Rainier lahar to reach Tacoma, Washington?

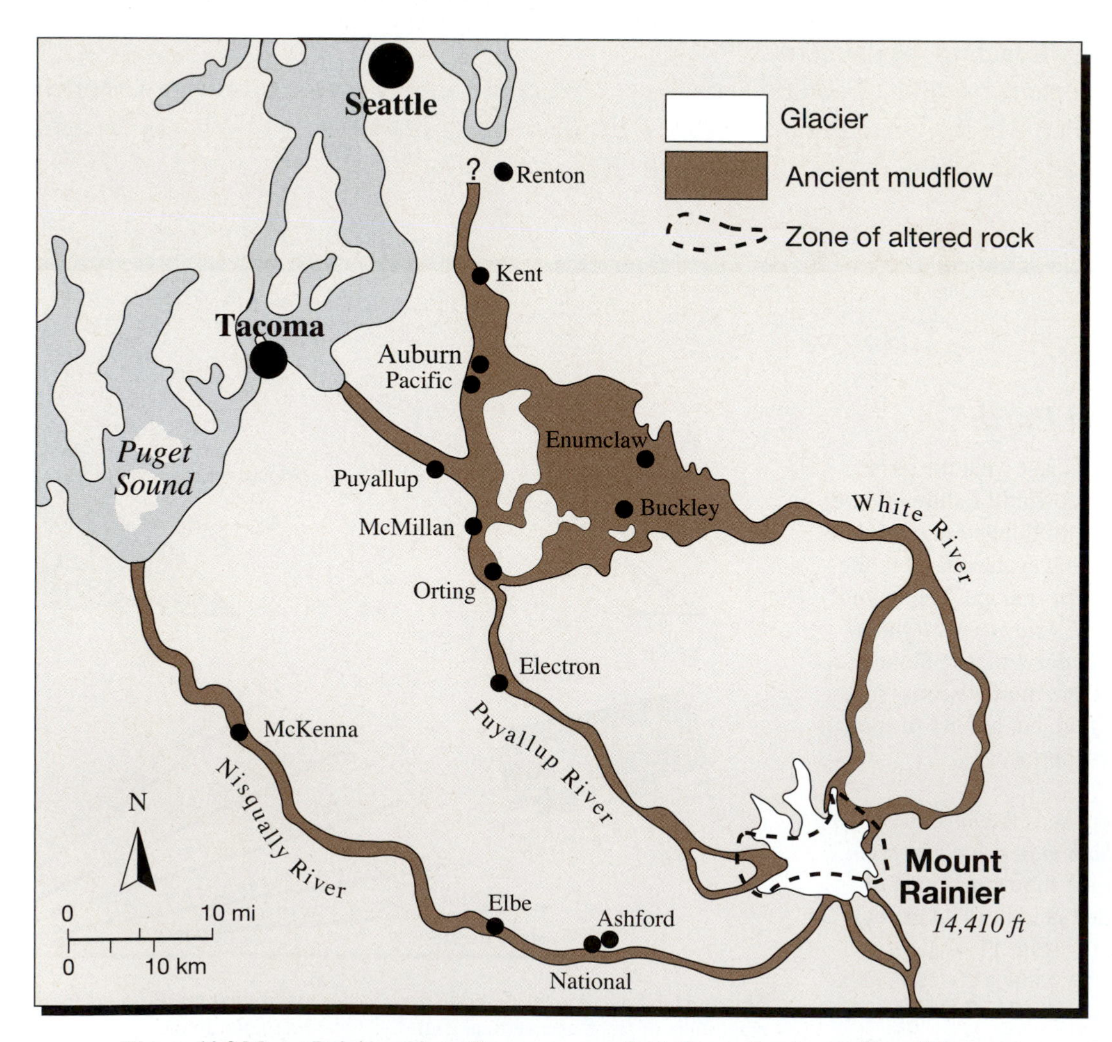

Figure 11.2 Mount Rainier and mudflows of the past 5,600 years. Mudflows have followed river valleys because they behave as a fluid. Towns develop within valleys because valleys are routes of transportation, power transmission, and communication.

B. Mass wasting

Mass wasting (aka *mass movement*) is the general term for gravity-induced downslope movement of earth material.

No states are immune from slope-related disasters, which prompted the U.S. Geological Survey to institute its *National Landslide-Loss-Reduction Program* in the 1980s to identify and map landslide-prone areas.

The classification of the kinds of mass wasting employed by engineering geologists and soil scientists consists of a matrix of the following features (Fig. 11.3):

- mechanism of movement (e.g., slide, flow, fall)
- type of material involved (e.g., rock, type of soil)
- velocity (e.g., creep, debris fall)

Figure 11.3 A classification of mass wasting based on mechanism, material, and velocity of movement.

MECHANISM		MATERIAL			VELOCITY
		Rock	Fine-grained soil	Coarse-grained soil	
Slide	 Earth slump	Slump	Earth slump *Stair-stepped slices with surfaces that slope downward in an uphill direction. Trees and shrubs within a slice have a similar orientation.*	Debris slump	Slow (days to weeks)
	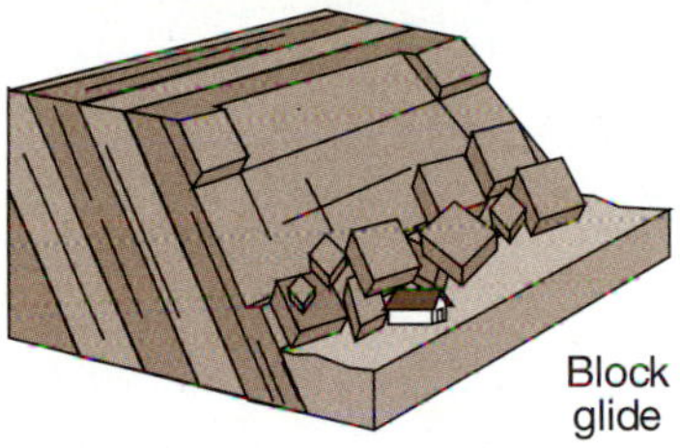 Block glide	Block glide *Down the plane of inclined rock layers where inclination is parallel to slope*	Earth slide 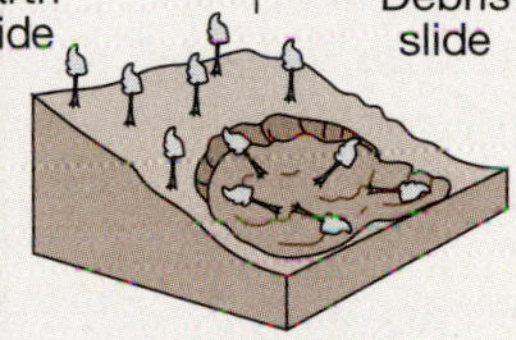 *Chaotic jumble of soil and debris; mixed orientations of trees and shrubs*	Debris slide	Variable (seconds for rock, minutes to days for debris/earth)
Flow	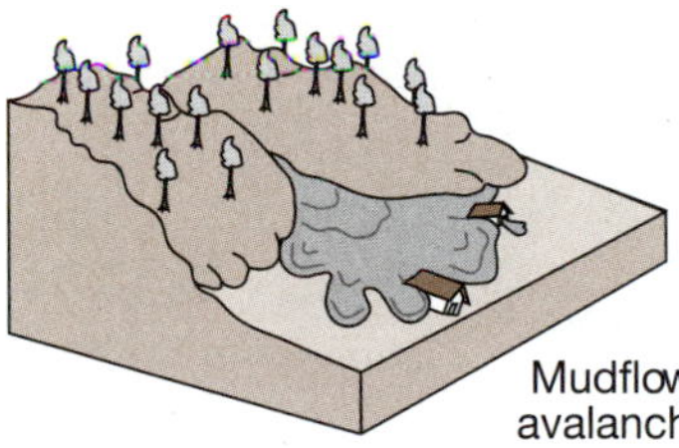 Mudflow, avalanche	Rock avalanche	Mudflow, avalanche *Behaves as a viscous fluid, follows valleys*	Debris flow, avalanche	Very rapid (minutes)
	 Creep	Creep	Creep *Marked by thin, weak layers of rock dragged downward by surface flow; trees bowed as result of tilting downslope, followed by upward growth*	Creep	Extremely slow (months)
Fall	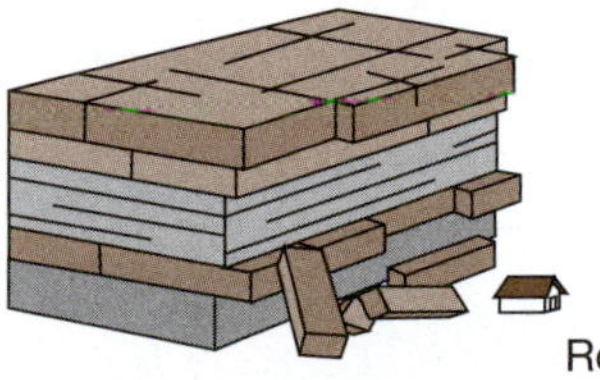 Rockfall	Rockfall	Earthfall *Undercut rock and debris fall vertically for lack of support*	Debris fall	Extremely rapid (seconds)

C. Causes of slope failure

- loading
- watering
- steepening
- undercutting
- earthquakes

Living on the edge—It should come as no surprise to anyone that loading a tenuous site with a home or building can trigger slope failure (Fig. 11.4). Watering by rainfall (or via a garden hose) is loading as well.

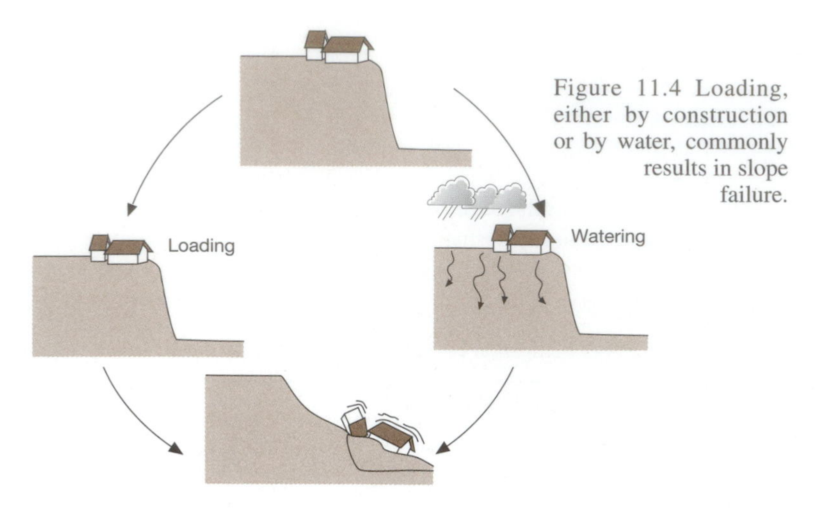

Figure 11.4 Loading, either by construction or by water, commonly results in slope failure.

Q11.2 Assume the following conditions:
(a) One cu ft of dirt weighs 60 lbs.
(b) The average density of the grains that comprise the dirt is 2.0.
(c) The porosity (i.e., open space) within the dirt is 50% of volume.
(d) Rainwater completely occludes porosity (i.e., fills the open space). Question: What is the weight of the saturated cu ft of dirt? *Hint:* In order to answer this question, you must know the density of water.

Q11.3 Beginning with dry dirt, there is a second effect resulting from watering—in addition to loading—that promotes slope failure. What do you suppose that second effect is? *Hint:* One word will do.

Roads commonly follow valleys, with roads carved along hillsides above stream courses (Fig. 11.5).

Q11.4 (A) Measure the natural slope in Figure 11.5A. (If you don't have a protractor, you can make a quick-copy transparency of the one on page *i* in the front of this manual.) What is the term in Figure 11.1 that most closely matches this magnitude of slope? (B) Measure the modified slope in Figure 11.5B. What is the term (in Figure 11.1) that most closely matches this magnitude of slope?

Loading and lubricating by water adds to the instability of ancient sea cliffs along California's coast. On January 10, 2005 a mudslide buried four blocks of the village of *La Conchita*. Ten people were killed and 14 injured. Of the 166 homes in the community, fifteen were destroyed and sixteen more were judged by the county to be uninhabitable. *(USGS photo)*

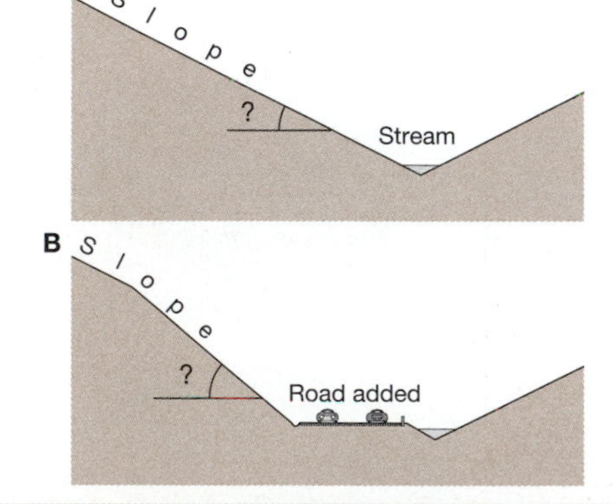

Figure 11.5 Carving a highway right-of-way from a valley wall commonly steepens an already perilous slope.

The National Landslide Information Center is at http://gldage.cr.usgs.gov

Problems with terraced tracts
Hillsides are commonly terraced—by *cutting-and-filling*—to create flat homesites. 'New dirt' is especially susceptible to subsidence and failure.

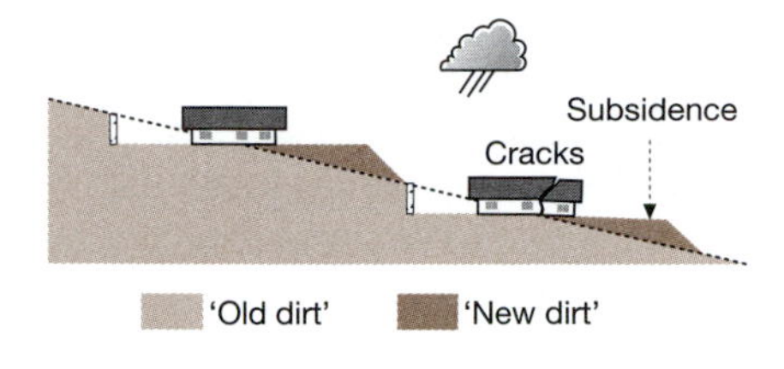

Excavation downslope from a property (i.e., undercutting) can create problems for an upslope homeowner, retaining wall notwithstanding (Fig. 11.6).

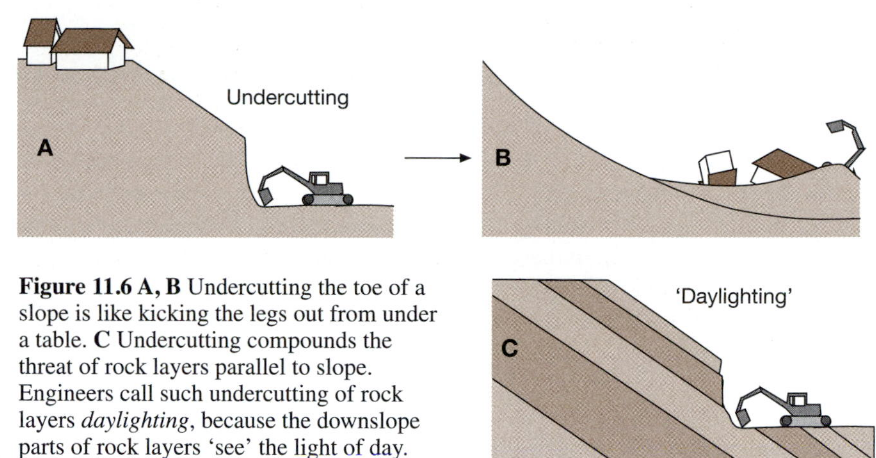

Figure 11.6 A, B Undercutting the toe of a slope is like kicking the legs out from under a table. **C** Undercutting compounds the threat of rock layers parallel to slope. Engineers call such undercutting of rock layers *daylighting*, because the downslope parts of rock layers 'see' the light of day.

Figure 11.7 Countless homes along seacoasts are in partial or total state of collapse, owing to undercutting by the adjacent surf.

Undercutting can be from natural causes, as well. Wave erosion gobbles up residential properties along innumerable seacliffs (Fig. 11.7).

Q11.5 What is another geologic process that commonly undercuts property, resulting in collapse? *Hint:* This process tends to be seasonal.

Earthquakes commonly trigger slope failures (Fig. 11.8). At midnight on August 17, 1959, a magnitude 7.1 earthquake dislodged a vast chunk of the south wall of the Madison River Canyon, burying 28 campers under 150 feet of rock. It has been estimated that some 27 million cubic meters of rock fell into the canyon at a rate sufficient to excavate the Panama Canal in 35 minutes. Wind produced by displaced air tumbled automobiles down the road along the north side of the canyon. The slide impounded the Madison River, forming Earthquake Lake.

Q11.6 Ref: Figure 11.8. Why did the *south* wall of Madison Canyon fail instead of the *north* wall? *Hint:* Look back at the *block glide* model in Figure 11.3 on page 189 and at Figure 11.6C on this page.

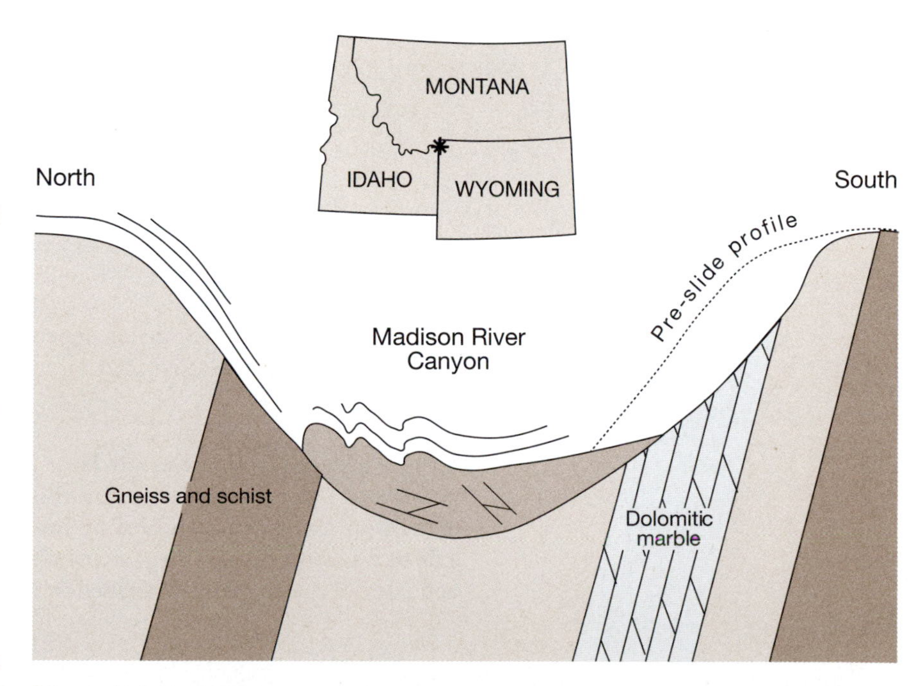

Figure 11.8 The Madison Canyon earthquake brought down the south wall of the canyon. Metamorphic foliation (a kind of layering) is inclined downward toward the north.

D. California case studies and mitigating measures

Debris slides and flows brought on by devastating rainstorms of 1977–1980 caused widespread damage to hillside residential properties in Southern California. Deaths attributed to these storms totaled 30, and 111 homes were destroyed. Damages were estimated at $400 million.

Natural slopes in Southern California, which are commonly proportioned 1.5 horizontal to 1.0 vertical, or steeper, and which are commonly exacerbated by human activities, are viewed by geologists as 'accidents waiting to happen.'

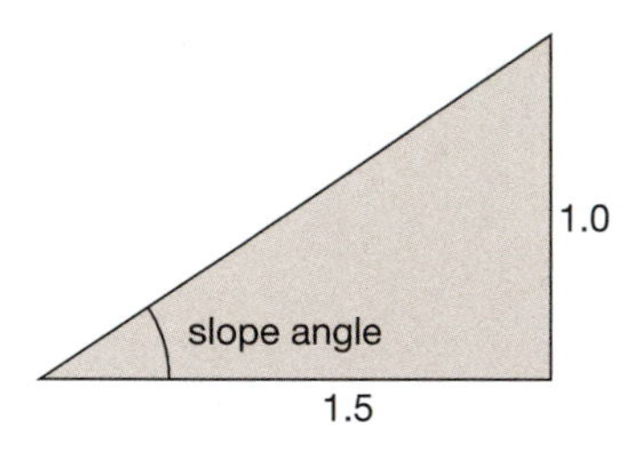

Q11.7 (A) If you have trig functions in your calculator, compute the slope angle in the above triangle. If not, use a protractor. (B) What label in Figure 11.1 on page 187 applies to this magnitude of slope angle?

Topanga Canyon and Angelo Drive

The role of rainstorms—Within the period early December 1979 through March 1980, a series of rainstorms occurred in the Los Angeles area (Fig. 11.9). Topanga Canyon, a residential community five miles west of Beverly Hills in the Santa Monica Mountains, received 17 inches of rain in less than 10 days.

Damage in Topanga Canyon was largely caused by *surface water*. Low bridges and unwise construction practices restricted water flow and flooded natural channels—washing out principal roads and isolating the canyon community for several days.

Q11.8 During what period of less than ten days in Figure 11.9 did Topanga Canyon receive 17 inches of rain?

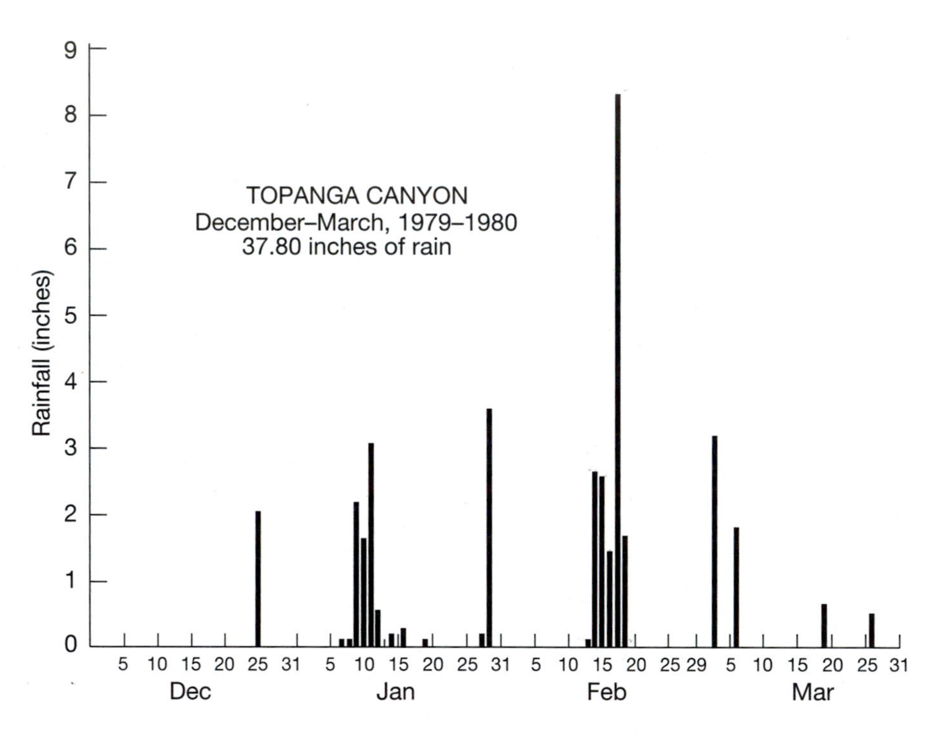

Figure 11.9 The total rainfall in the Topanga Canyon area during four months in 1979–1980 wasn't all that unusual. It was the *sustained* occurrence of rainfall during periods of a few days that proved disastrous.

While Topanga was spared major slope failures, the hillside downslope from Angelo Drive, which is also in the Santa Monica Mountains, was scarred by several debris flows.

Q11.9 Figure 11.10 on Answer Page 201 is a contour map of the hillside downslope from Angelo Drive. On this map there is sketched an outline of the debrisflow and its axis. Guided by the contours, try your hand at drawing the axis of a second equally large debrisflow on this map. ***Hint:*** **Look again at the behavior of debris flows described in Figure 11.3 on page 189.**

Abalone Cove

The role of groundwater—We saw on previous page 192 that rainstorms caused flooding and isolation of Topanga Canyon, and surface water is a big factor in mudflows and debrisflows, but *landslides* appear to be caused more by *groundwater*.

Figure 11.11 is a map of the Altamira watershed and Abalone Cove landslide area, including two monitoring wells.

Figure 11.12 is a plot of four variables, a combination of which provides clues to the cause of the landslide.

Q11.10 Study the trends in the four variables plotted in Figure 11.12, and see if you can make a case for groundwater's having been the cause of the Abalone Cove landslide. Incidentally, later dewatering of the slide area (via additional wells drilled on the slide) stopped the slide.

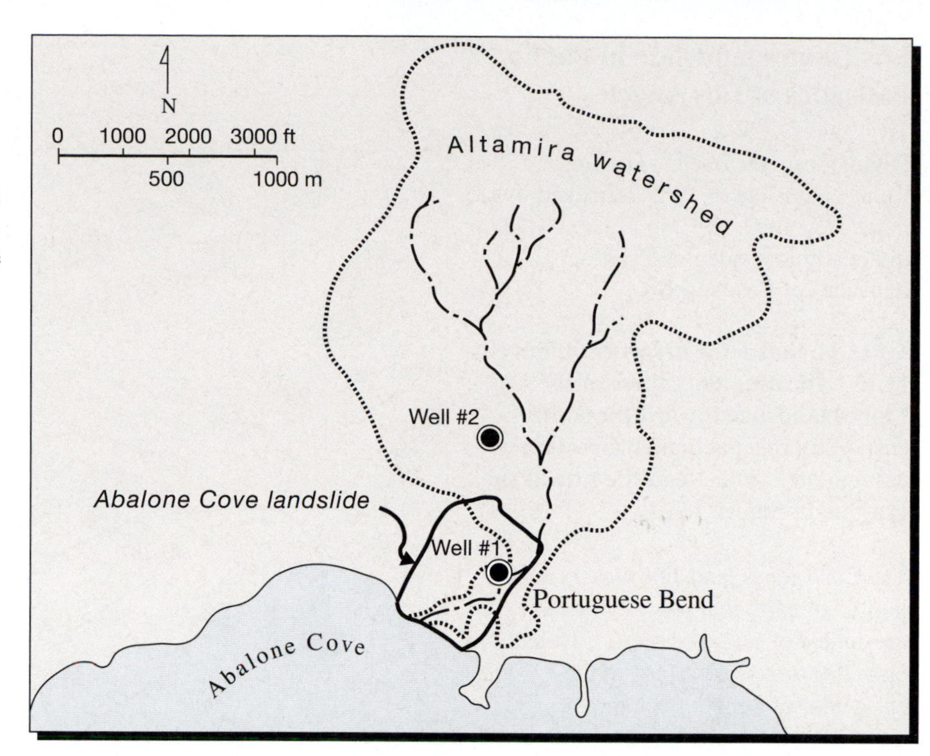

Figure 11.11 The Abalone Cove landslide area, in the Palos Verde Hills area south of Los Angeles, receives its groundwater from the Altamira watershed.

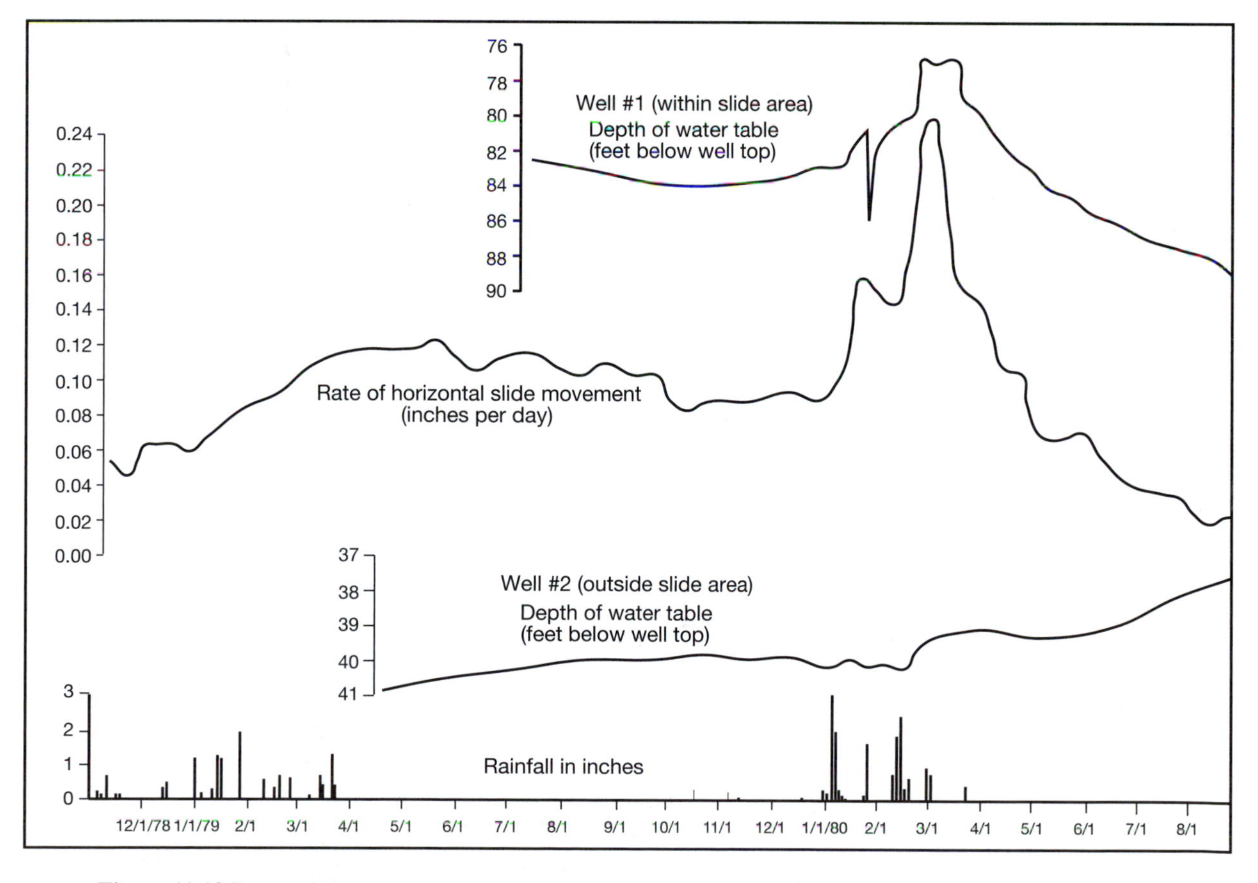

Figure 11.12 Four variables have been plotted through the period December 1, 1978–August 1, 1980.

Los Liones landslide in Pacific Palisades of Los Angeles

History repeats itself—The Los Liones landslide of 1965 destroyed five homes (A, B, C, D, E) in Figure 11.13) in the scenic residential Pacific Palisades of Los Angeles.

Q11.11 Notice the arrows in Figure 11.13 indicating the course of the Los Liones landslide toward the southeast. Why this particular direction of movement? *Hint:* Read the title of the graphic in Figure 13-11.

The Los Liones landslide was a chaotic jumble of earth that was mobile over the course of several days. It traveled only 100 ft or so, stopping at the back-door of building F, a motel.

Q11.12 Which of the kinds of slope failure illustrated in Figure 11.3 on page 189 best fits the Los Liones landslide?

Q11.13 When the earliest warning signs first appeared (e.g., incipient cracks in masonry), which home do you suppose exhibited the larger problems, A or B?

Q11.14 A dewatering attempt was made with a drill rig within the slide area along Los Liones Drive. The idea was to drill a hole and pump the water out—like with any water well. The effort was no sooner underway when the drill pipe met with technical failure. What do you suppose was the reason for the technical failure? *Hint:* Examine details in Figure 11.3 on page 189.

Q11.15 Incidentally, do you see any reason for the owner of home G to be concerned? Why or why not?

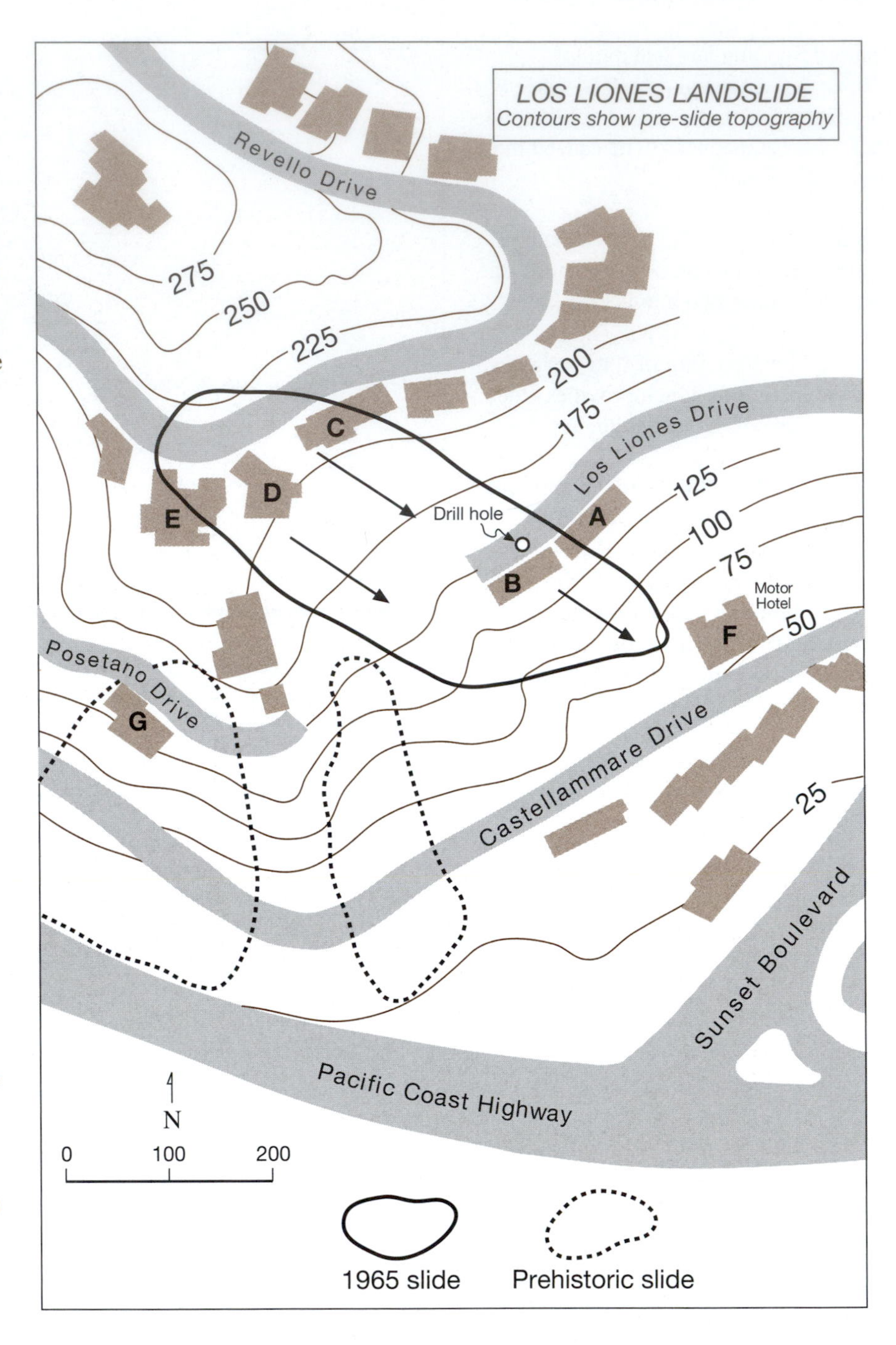

Figure 11.13 The Los Liones landslide was typical of hillside failures in Southern California, documenting the principle first stated on page 188: *If it has happened around here before, it's likely to happen around here again.*

Mitigating measures

Plants—In Southern California, and in many other semi-arid regions of the world, soil plus *regolith* (i.e., disintegrating bedrock), is commonly less than one meter thick, but increases to around three meters in ravines—near its angle of repose. So, small things, like plant cover, can make a difference.

In Southern California, vegetation consists largely of chaparral, which consists of shrubs that form a canopy over slopes. Chaparral protects slopes in two ways: (1) It shields soil from being dislodged by torrential rains, and (2) it binds soil with its root systems (Fig. 11.14). The proof: Mud and debris flows increase when heavy rains follow brush fires within three years.

But in many places, chaparral has been cleared from around homes in an effort to protect them from wildfire. So the dilemma commonly presents itself to homeowners: Protection from fire or protection from mudflows?

Q11.16 Some hillside homeowners imagine that a palatial lawn of manicured grass protects their home from mudflow and creep. But what do you suppose is the critical failing of grass, as compared with shrubs?

Q11.17 Bedrock geology, as well as topography, plays an important role in the effectiveness of plants like the California scrub oak. There are two features identified in Figure 11.14 having to do with the nature of bedrock. What are they? *Hint:* Contrast the description of bedrock in the caption of Figure 11.14 with a brief description of solid granite.

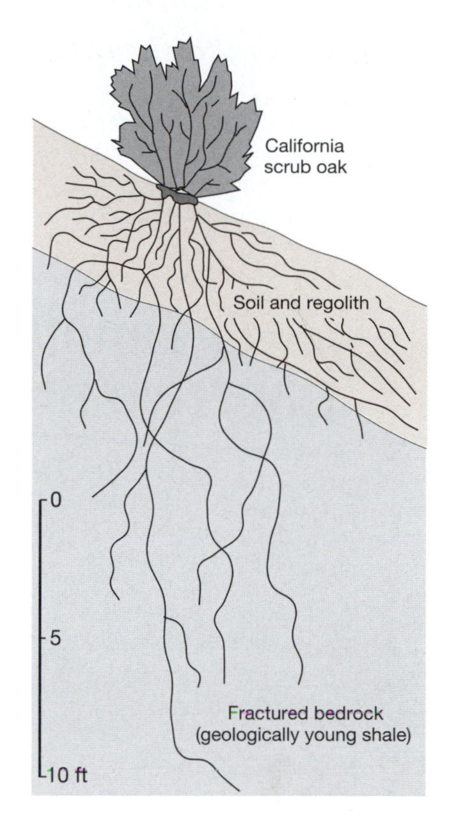

Figure 11.14 A stabilizing plant in Southern California is the California scrub oak (*Quercus dumosa*), with roots that can penetrate up to 20 ft. of rubbly fractured bedrock.

Rock bolts—Steepened slopes can be stabilized with rock bolts (as illustrated below).

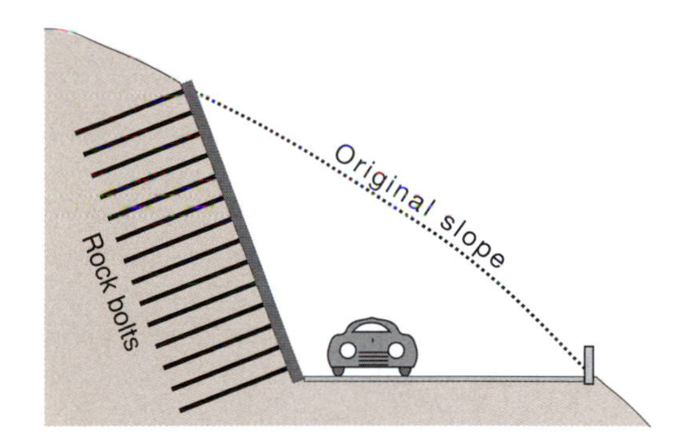

Retaining walls and pins—Vertical slopes can be contained with steel-reinforced concrete, gravel, and 'weepholes' for drainage.

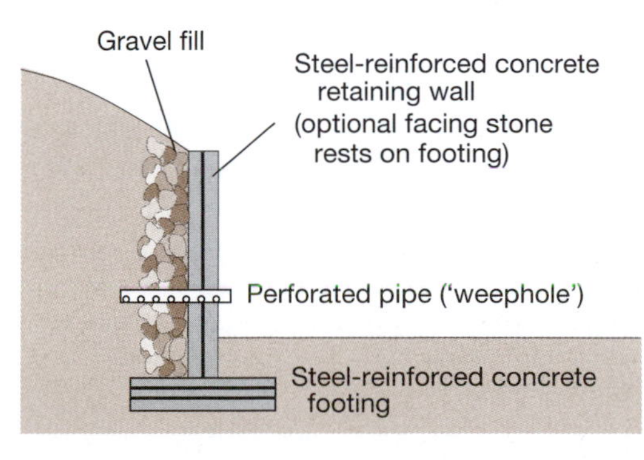

Dewatering—Groundwater can amount to double jeopardy as concerns slope stability. (A) Water adds weight. (B) Water acts as a lubricant, not only because it is a fluid, but also because under pressure of load (*hydrostatic pressure*) groundwater can have a heaving effect as well.

Dewatering of the Abalone Cove landslide eventually stopped the slide. Dewatering in that case was via shallow water wells, but elaborate proactive measures can also be taken (as illustrated below).

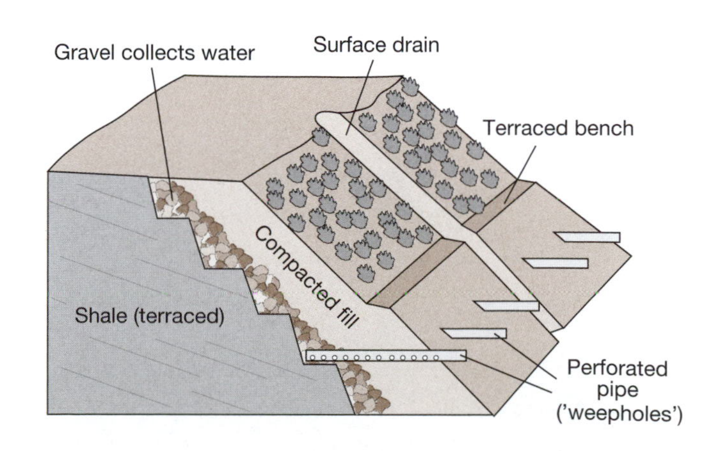

E. Problems in the heartland

Killer trenches—OSHA requires that supports be placed within utility trenches dug to a particular depth. But regulations are commonly ignored. Three men in eastern Missouri were buried alive in 1998 while digging a neighborhood trench without supports.

COLUMBIA DAILY TRIBUNE, JULY 27, 1998

Sewer repair cave-in kills two neighbors

IMPERIAL, MO. (AP)—Three neighbors trying to save their subdivision some money by repairing a faulty storm-sewer line were buried alive when the walls of a trench they were working in collapsed. Two died in the cave-in late Saturday night. A third, who remained hospitalized today, was rescued after being trapped for nearly six hours.

Concerns for children—Construction site developers commonly stockpile 'new dirt' at angles of repose. A bit of digging by kids could bring down dirt sufficient to take their lives. And one doesn't have to dig tunnels. One only has to create a steeper slope. Fences should be required around these 'attractive nuisances,' like those around neighborhood swimming pools.

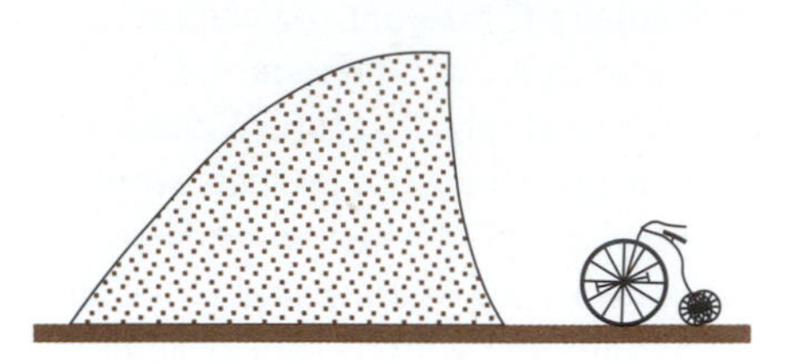

THE SENTINEL RECORD, NOVEMBER 12, 1995

"Mountain crumbles..."

HOT SPRINGS, ARK.—An estimated 200 tons of rock crashed into the back of three shops on Central Avenue Saturday afternoon, killing one shop keeper and doing extensive damage. "This has happened before along this street and so will probably happen again," said a geologist with the Arkansas Geological Commission in Little Rock.

Catastrophic block glide—According to local geologists, *four* conditions promoted the collapse of a wall of rock in downtown Hot Springs, Arkansas, in the autumn of 1995. Three of the conditions are natural, while a fourth was the undercutting of the slope by shop owners to create more space behind their stores. Undercutting did not produce the failure immediately, but when the three natural conditions coincided, the undercutting tipped the proverbial scale.

Q11.18 (A) Identify the three natural conditions itemized by the Arkansas Geological Commission. *Hint:* All three conditions are illustrated in Figure 11.15. One condition is also illustrated in Figure 6.4 on page 94. (B) How does the term 'daylighting' apply to shop owners' efforts to increase space behind stores? *Hint:* See Figure 11.6C on page 191.

Figure 11.15 This is looking southward behind shops on Central Avenue in Hot Springs, Arkansas. Fractures within the Arkansas Novaculite (a whetstone variety of chert, or flint) are inclined parallel to slope. Here we see collapsed Novaculite rubble impinging upon the back of a shop.

F. Subsidence

Natural subsidence

The load of the sediments that comprise the Mississippi Delta (or any other delta for that matter) depresses Earth's crust. For example, measurements taken along a traverse from Osyka, Mississippi, to New Orleans, Louisiana, show the greatest rates of subsidence during the period 1934–1986 in the area of sediment accumulation (Fig. 11.16).

Figure 11.16 The weight of sediments being deposited in the Mississippi River delta continues to depress earth's crust.

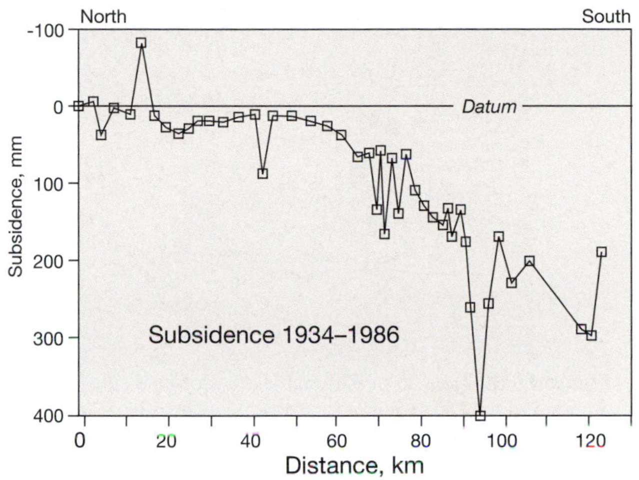

Q11.19 What is the greatest rate of subsidence (in millimeters per year) illustrated in Figure 11.16?

True, this doesn't sound like much subsidence, but given the fact that the French Quarter of New Orleans is 2 m below sea level, any degree of subsidence is foreboding. In case you're wondering, heroic levees hold out the Gulf of Mexico, and equally heroic pumps lift rainwater over the levees and into the Gulf. But Hurricane Katrina (August, 2005) toppled levees and New Orleans was flooded.

Subsidence caused by humans

Humans have caused ground subsidence largely through the extraction of groundwater and petroleum. A graphic record of subsidence caused by groundwater withdrawal is evident near Mendota, California, within the San Joaquin Valley (Fig. 11.17).

Q11.20 What was the average rate of subsidence in the Medota area in millimeters per year during the period 1925–1977?

Figure 11.17 Approximately 13,500 km^2 of the San Joaquin Valley have subsided because of the withdrawal of groundwater used for irrigation. This is said to be the largest human alteration of the Earth's surface.

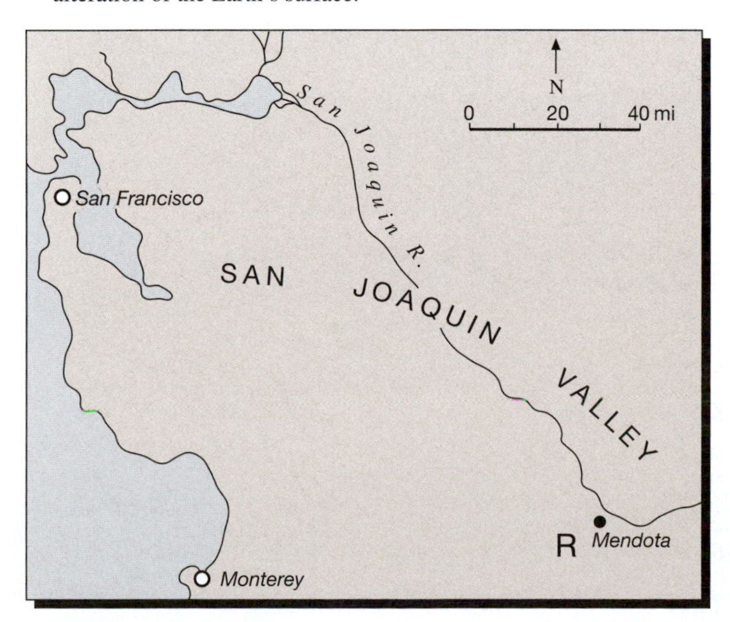

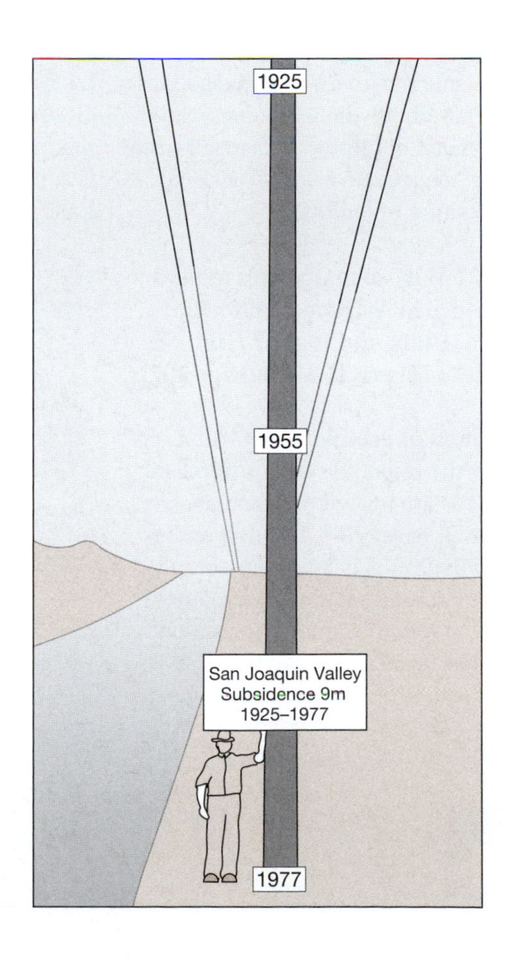

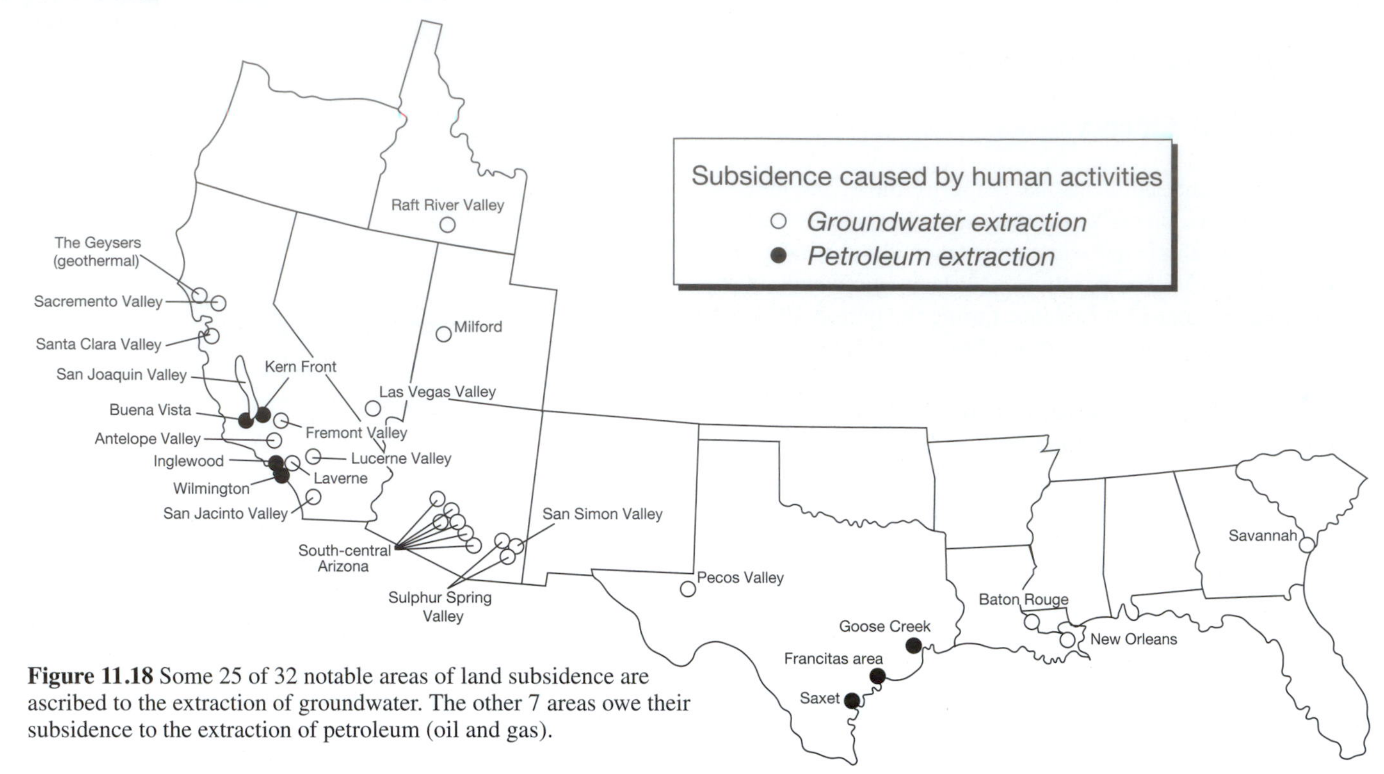

Figure 11.18 Some 25 of 32 notable areas of land subsidence are ascribed to the extraction of groundwater. The other 7 areas owe their subsidence to the extraction of petroleum (oil and gas).

The areas of subsidence shown above in Figure 11.18 are those that make the news. In addition, there are countless other cases of subsidence cursed by homeowners and small-town America. Some cases involve the collapse of cave roofs, while others relate to the collapse of ceilings above shallow mines. Still a greater number of problems result from erosion, inadequate foundations, and unwise siting of buildings.

Q11.21 Why so many sites of land subsidence in Arizona, California, Nevada, Utah, and Idaho? *Hint:* See caption to Figure 11.17 on page 197.

Extraction of groundwater—In Exercise 10, the principles of **aquifers** and **aquitards** are treated in an elementary manner on pages 174–175. But an understanding of land subsidence resulting from groundwater and petroleum extraction requires more information, which has been assembled by the U.S. Geological Survey in *Land Subsidence in the United States*.

We can begin our story with an undeveloped aquifer system with its balance between **recharge** and **discharge** and a relatively stable land surface (Fig. 11.19). The pumping of groundwater to meet urban and agricultural needs can upset this balance, resulting in land subsidence.

Actually, the fundamental principles now applied by USGS were first developed by W.E. Pratt and D.W. Johnson in 1926 while studying land subsidence at the Goose Creek oil field of east Texas. Their theory, which they labeled aquitard-**drainage**, posited that when oil is removed from sand, water from associated clay is squeezed—through compaction—into the sand to replace the oil. The result: Land subsidence.

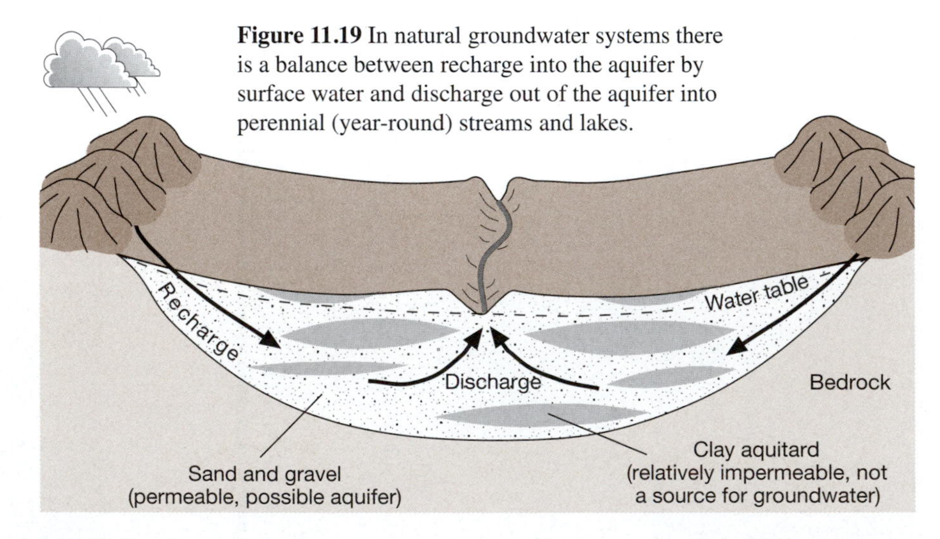

Figure 11.19 In natural groundwater systems there is a balance between recharge into the aquifer by surface water and discharge out of the aquifer into perennial (year-round) streams and lakes.

Principles of land subsidence can be reviewed at http://water.wr.usgs.gov/subsidence/sub_bib.html

The aquitard-drainage model

Dewatering/compaction/subsidence—Land subsidence associated with groundwater extraction from sedimentary basins is commonly caused by compaction of sediments in response to water moving out of aquitards and into aquifers—much like the *aquitard-drainage* model envisioned by Pratt and Johnson. But first, a metaphor illustrating the compressibility of clay grains.

In a vertical succession of lenses of clay suspended in sand and gravel (Fig. 11.20), clay grains are more subject to compaction than are sand and gravel because of the leaf-like shape of clay grains. A visual metaphor: A freshly raked pile of leaves is fluffy and porous, but if left unattended through one or two rainfalls, the pile of leaves takes on a higher degree of packing.

Porosity within a pile of leaves diminishes because of the flattening of the leaves. So it is with a lens of clay grains, which compacts following deposition, especially when pore water is withdrawn.

Q11.22 More from the world of Nature: Which can you compact more, a handful of snow flakes or a hand full of sleet granules?

Note: Down to depths of approximately 1,000 m sand and gravel aquifers compact to negligible degrees. So the process outlined here assumes that all compaction occurs within clay aquitards. In the real world, some percent of compaction would likely be assigned to the sand and gravel as well, especially at depths exceeding 1,000 m.

The process of subsidence owing to the extraction of water

- Before pumping (Fig. 11.20A) water pressure within the sand and gravel aquifer and water pressure within the clay aquitard are essentially equal. (Water can move across the interface separating the aquifer from the aquitard.)
- Pumping water out of the aquifer decreases its internal water pressure (Fig. 11.20B).
- With water pressure now greater in the aquitard than in the aquifer, water moves from the aquitard into the aquifer.
- Compaction of the aquitard accompanies the withdrawal of water—like the collapse of a toothpaste tube.

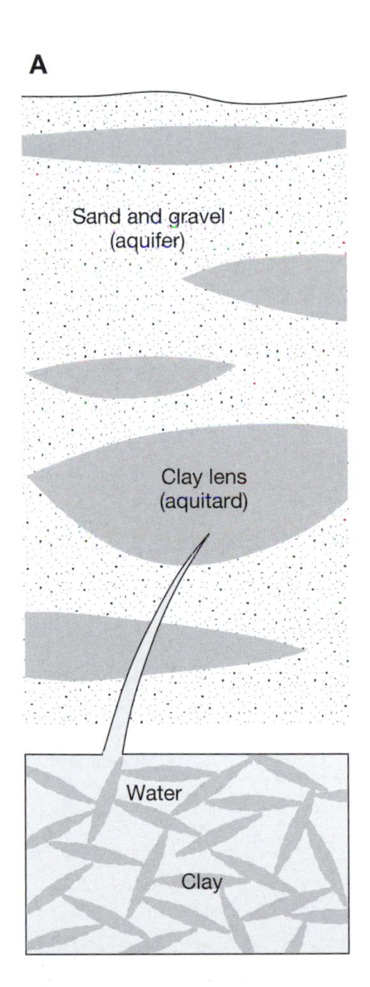

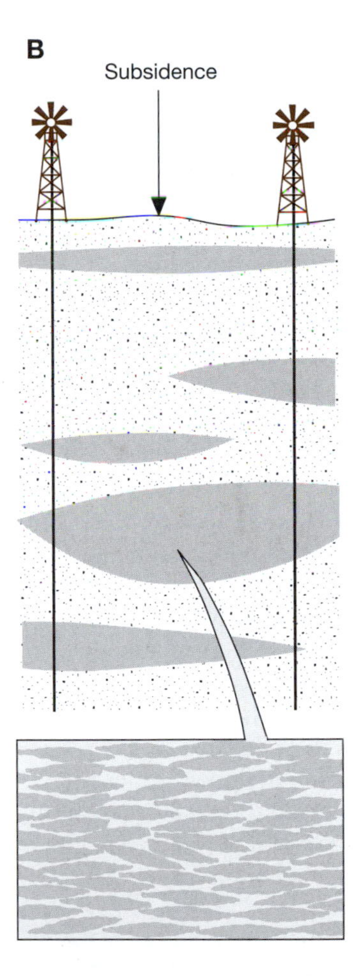

Figure 11.20 A Clay aquitard with various orientations of clay grains and porosity occluded by groundwater. **B** Withdrawal of groundwater from the sand and gravel aquifer reduces water pressure surrounding aquitards, causing partial dewatering and compaction of clay grains. Notice that compaction is limited to the clay aquitards. The volume of the sand and gravel aquifer remains about the same.

Q11.23 Given a 1,000-foot vertical succession of 70% sand and gravel and 30% lenses of clay. After dewatering for 50 years, lenses of clay have compacted 10%. (Assume negligible compaction within sand and gravel.) What should be the extent of land subsidence (in feet)?

The Baldwin Hills, California, reservoir disaster

The Baldwin Hills dam and reservoir were built in 1950 as a backup facility during peak hours of municipal water demand. The site chosen for this reservoir could hardly have been worse. The active Newport-Inglewood Fault is only a few hundred yards from the dam, and two faults with questionable activity run directly through the site (Fig. 11.21). The structural picture is compounded by subsidence caused by oil extraction at the adjacent Inglewood oil field. Measured subsidence during the period 1917–1962 is 9 ft.

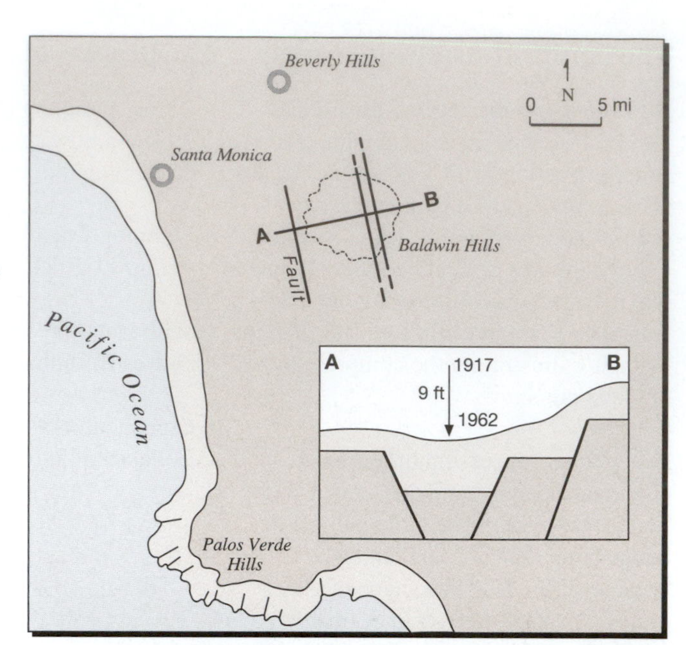

Figure 11.21 The site of the Baldwin Hills reservoir sits astride two faults and was actively subsiding at the time of its construction.

Planners were aware of the tenuous geologic setting, so they designed an elaborate reservoir liner, with drain pipes beneath to handle any leakage. This, they believed, could contend with ground motions (Fig. 11.22).

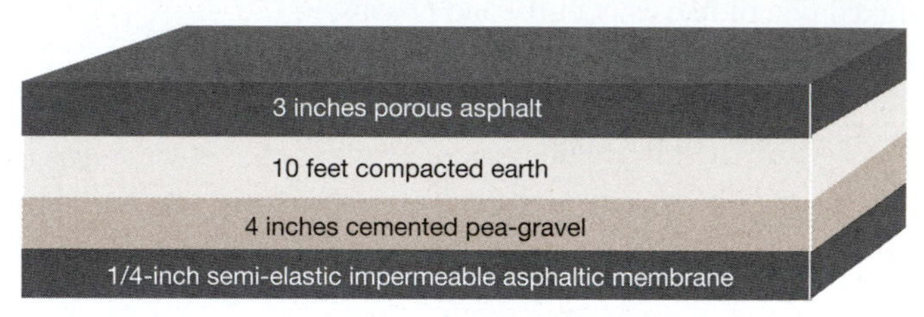

Figure 11.22 An elaborate four-layer reservoir liner (not to scale here) was designed, with leak-drains beneath it.

At 11:15 a.m. on Saturday, December 14, 1963 the liner failed along one of the faults (Fig. 11.23), releasing 65,000,000 gallons down the canyon to the north (Fig. 11.24). The wall of water swept away five people and destroyed scores of homes. Now only the shell of what once was marks that fateful spot.

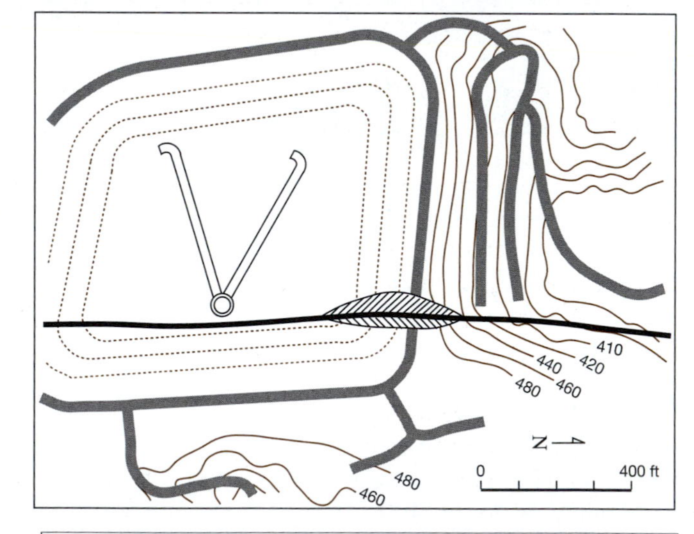

Figure 11.23 The empty reservoir. The rupture (shown with the hachured gash) was along one of the faults.

Q11.24 The Baldwin Hills reservoir was completed in 1950 and was immediately filled to capacity. It failed in 1963. Given this time line, which is more logical, (a) the reservoir was not designed sufficiently strong to hold 65,000,000 gallons of water, or (b) there was a geologic process going on here?

Q11.25 Review: What were the *two* features associated with the prospective site that should have been sufficient to dissuade thinking people from constructing a water reservoir on that site? *Hint:* Both features are presented in the first paragraph on this page and are illustrated in Figure 11.21.

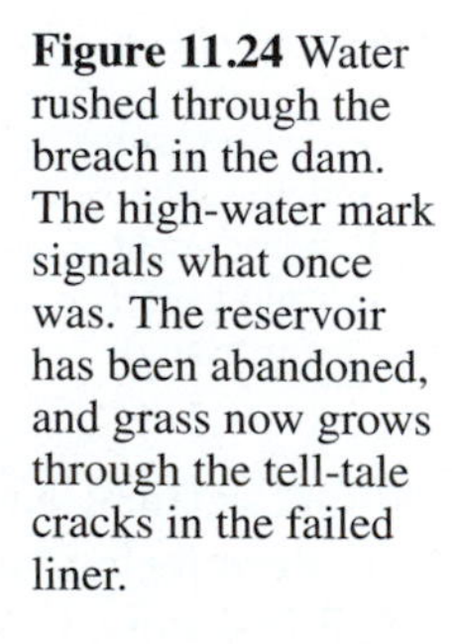

Figure 11.24 Water rushed through the breach in the dam. The high-water mark signals what once was. The reservoir has been abandoned, and grass now grows through the tell-tale cracks in the failed liner.

12 Coastal Processes and Problems

Topics

A. What percent of the U.S. population is expected to live within an hour's drive of a sea coast by the year 2025?

B. What physiographic coastal features occur along an active continental margin? What physiographic coastal features occur along a passive continental margin?

C. In what way does the profile of an ideal surface sea wave differ from that of a sine wave? What three variables determine the size of wind-generated waves?

D. What geometric form is described by the wave-generated motion of a floating object? Does a floating object actually move downwind? What circumstances produce a sea breeze? What is wave base, and how does it relate to wave length? How do marine terraces develop?

E. How are wave speed, wave length, and wave period related? In a breaking wave, how are wave length and wave height related? Why do headlands experience erosion whereas bays experience deposition? Why do seaports commonly occur within bays? How do longshore currents, rip currents, and spits develop?

F. What is a stack, a tombolo, a tied island? How does each develop?

G. What are two reasons for why coastal wetlands are valued by ecologists? Why are the Isles Dernieres—at the edge of LaFouche delta lobe—eroding so rapidly?

H. What are the five basic types of coastal installations, and what problems do they cause? How might troublesome groins be modified so as to become beneficial groins? What problems and solutions followed the construction of the harbor breakwater at Santa Barbara?

I. What is the where, when, and how of hurricane development? How does the Coriolis effect influence the behavior of wind and sea currents? Why is effective wind velocity different on the two sides (right, left) of an approaching hurricane?

A. Are we loving our coasts to death?

The 1990 Census reported that the greatest percent increases in U.S. population during the 1980s were in coastal regions. At present 50% of our population lives within an hour's drive of a coast, and that number is expected to grow to 75% by the year 2025.

- *The problem:* Exponential growth along our coasts is accompanied by building in high-risk localities.
- *The solution:* A better understanding of coastal processes.

We attempt to protect our coastal facilities with installations of stone and concrete—but such projects commonly either produce unexpected results, or fail entirely in the wake of incessant waves and currents and devastating weather.

Q12.1 Before going further, (A) can you name a *natural process* that seasonally presents big problems for coastal residents—namely the Gulf and Atlantic coasts? Try naming a notable 2005 example of that process. (B) In this context, can you name a *human activity* along coasts that commonly meets with this kind of disaster? Try naming a specific occurrence of such human activity. *Hint:* Think about a particular Texas 'island city' in 1900.

Q12.2 Can you name a kind of *inland* waterway management effort that also commonly meets with disaster?

B. Classification of coasts

North America is an example of a continent where the physiography of opposite coastal regions (both onshore and offshore) differs because of differences in proximity to global plate boundaries (Fig. 12.1).

Q12.3 (Fig. 12.1) What is the relationship between the proximity to a global plate boundary and the breadth of continental shelf?

Q12.4 (Fig. 12.1) What is the relationship between the breadth of continental shelf and the breadth of coastal plain?

Along our West Coast, relentless tectonic forces associated with the western boundary of the North American plate account for the abrupt boundary between land and sea—an example of an **active continental margin**. In contrast, our East Coast has long since departed from the eastern boundary of the North American plate, so land and sea have had more time to develop an equilibrium marked by the near seamless boundary between continental shelf and coastal plain—an example of a **passive continental margin**.

Classifying coasts is a bit difficult because—in addition to the broader effects of global plate boundaries—rivers, ocean currents, and climate exert their influences as well. But, as with all naturalists, geoscientists are moved to put names on things. Table 12.1 shows a classification of coasts based on **setting** and **process**, a classification that is independent of proximity to plate boundaries.

Q12.5 Name two coasts that are included within the same category of 'setting' (Table 12.1, column one), but which occur along different kinds of plate boundaries (Fig. 12.1).

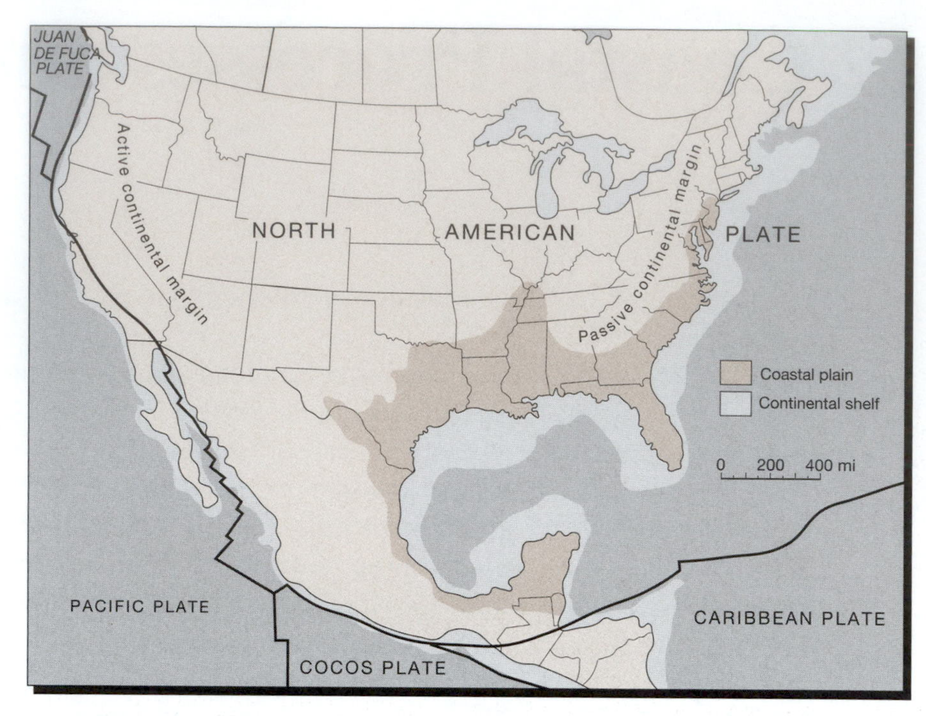

Figure 12.1 Differences in physiography between east and west coasts of our conterminous states reflect differences in proximity to global plate boundaries.

Table 12.1
A classification of coasts based on setting and process.

Setting	Process	U.S. example(s)
Subaerial (streams, glaciers, volcanism)	Stream erosion	Estuaries (northern U.S. Atlantic Coast)
	Stream deposition	Deltas (Mississippi Delta, Louisiana)
	Glacial erosion	Fjords (Alaska)
	Glacial deposition	Islands (Long Island, New York)
	Volcanic eruption	Hawaii (amoeboid-shaped islands)
Marine (waves and currents)	Erosion	Sea cliffs (U.S. Pacific Coast)
	Deposition	Beaches and barrier islands (southern U.S. Atlantic Coast and Gulf States)
Organic (corals, mangroves)	Reefs	Florida Keys
	Coastal wetlands	Florida Everglades

We will visit a number of these coasts later

C. Wave anatomy

The profile of an ideal surface sea wave is suggestive of a mathematical sine wave, but the two do differ (Fig. 12.2). Even so, physical oceanographers treat ocean waves as sine waves in certain calculations.

Q12.6 As shown in Figure 12.2, in what most obvious way does the profile of an ideal surface sea wave differ from that of a sine wave?

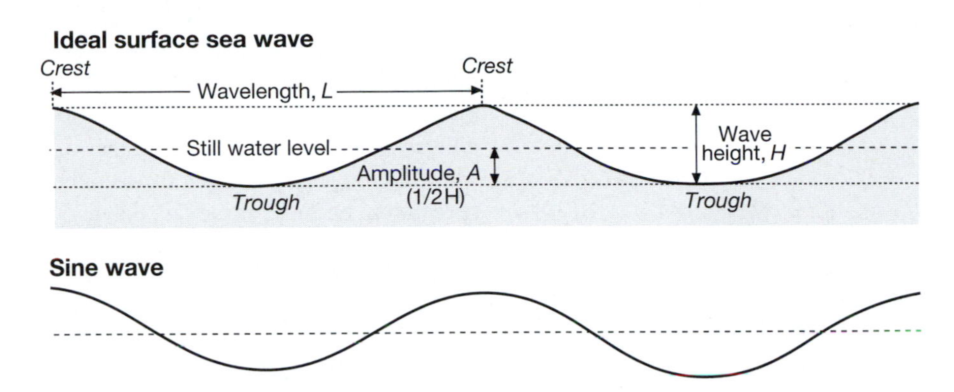

Figure 12.2 The profile of an ideal surface sea wave differs from that of a sine wave.

Earthquakes and submarine landslides locally produce ocean waves, but the universal wave-generating force is that of the wind. When air moves across the surface of water, friction roughens the surface into small rounded **ripples**, or **capillary waves**, with V-shaped troughs (Fig. 12.3). These tiny waves, which are characterized by wavelengths less than 1.74 cm (a precise number that is governed by the physics of wind and water), are produced by the interaction of wind and the surface tension of water molecules. Hence the term **capillary waves**. As more wind energy is added to the water surface, **gravity waves** develop with a form like that in Figure 12.2. Gravity waves take their name from the fact that gravity, rather than surface tension, is the leveling force when the wind abates.

As still more wind energy is transferred to surface waves, wave height increases more rapidly than does wave length, resulting in a steepness that causes waves to break as whitecaps (again, Fig. 12.3).

Three variables determine the size of wind-generated waves: (1) wind speed, (2) wind duration, and (3) fetch (the distance over which the wind travels)å.

Q12.7 Which of the above three variables do you suppose most likely explains why the largest waves occur at sea, rather than on lakes?

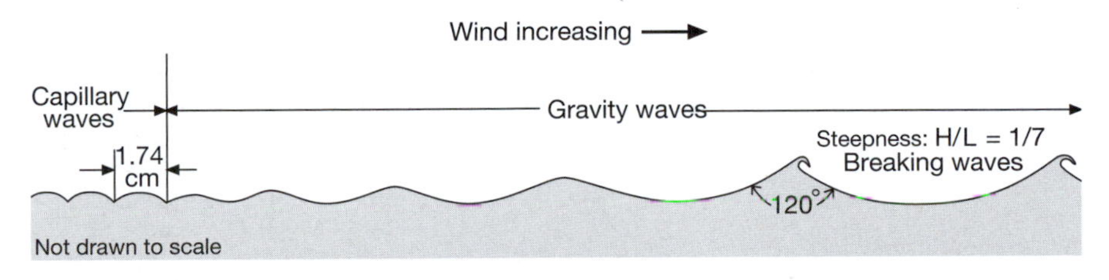

Figure 12.3 As surface waves gain energy, they increase in height and length. When wavelength exceeds 1.74 cm, waves assume the shape of the ideal surface sea wave shown in Figure 12.2. When wave steepness becomes 1/7 (height divided by length), waves become unstable and break as whitecaps.

D. Wave mechanics

Simple surface wave form

A floating object appears to bob up and down with passing waves. But, actually, the object's motion describes a *circle*, the diameter of which equals wave height (Fig. 12.4). Incidentally, this circular motion of water particles is analogous to the motion of earth particles affected by the *surface wave* of an earthquake. (An earthquake also produces two additional types of waves.)

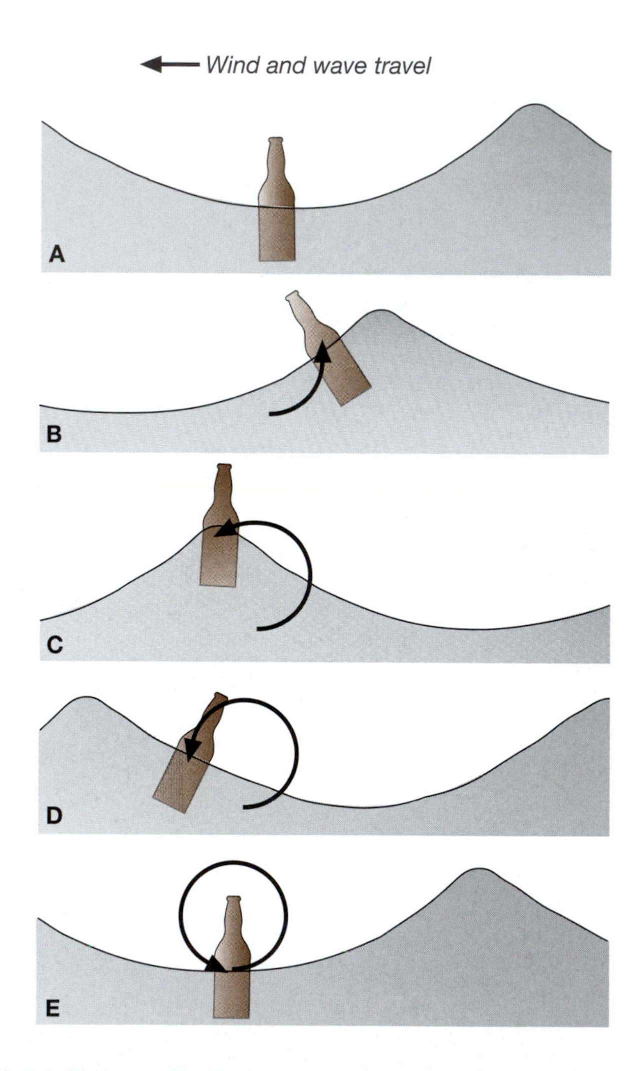

Figure 12.4 This illustrates the rise and fall of a floating bottle (beginning at **A** and ending at **E**) as the crest of a wave passes. So seasickness is caused not only by up-and-down motion, as in car-sickness, but by forward-and-backward motion as well.

Forward motion within the upper part of the orbit of a wave particle is a bit faster than backward motion within the lower part of that orbit. This is because of differences in resistance exerted by the water. A sensory metaphor: One might sense that the forward motion of a Ferris wheel is faster than its backward motion (Fig 12.5), but that is only perception, because of the fact that a sense of falling is more disquieting than a sense of rising.

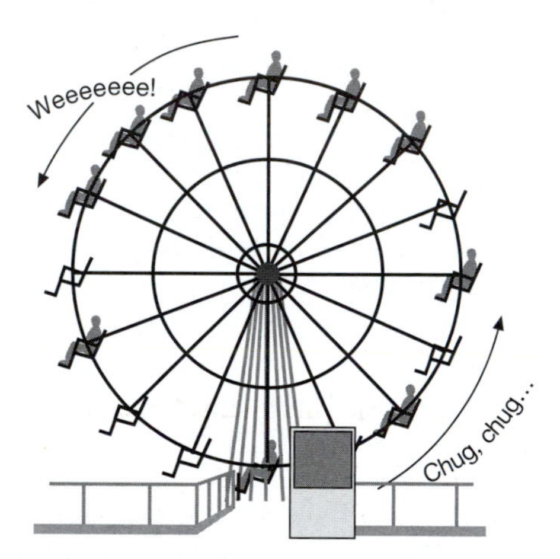

Figure 12.5 Perception: A Ferris wheel might produce the sensation of rotating faster where one is descending.

Q12.8 Does a floating object actually move downwind? On Figure 12.4, align a straight-edge along imaginary points marking the beginning (i.e., the 'base') of each partial circle. What does the orientation of the straight-edge suggest about the motion of the bottle—specifically, is it moving downwind or not?

Wave base

Because surface waves are produced by surface winds, the deeper the water particle, the farther it is from its energy source. Result: The agitation of water diminishes as a function of depth, as illustrated by the diminished orbits in Figure 12.6. The depth below which there is no excitation of water by surface wind is called **wave base**. It turns out that the distance between the still-water level and wave base equals one-half wavelength. To test this, do the following:

Q12.9 On Figure 12.6, measure the distance between the still-water level and wave base. Double that value and compare it with wavelength. How do the two compare?

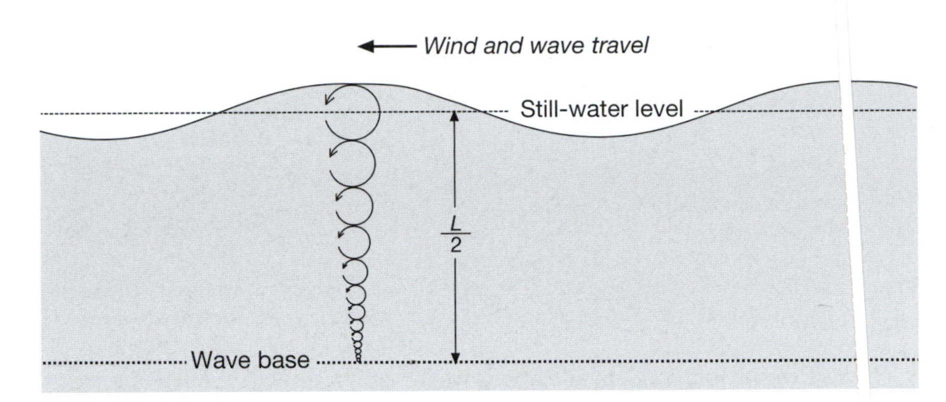

Figure 12.6 The depth below which there is little or no motion within the water is called *wave base*. Depth to wave base, from still-water level, is 1/2 wave length (L)

Landward ho—Waves are almost invariably driven landward by prevailing **sea breezes** (Fig. 12.7), which has to do with the fact that almost everywhere coastal land is warmer than adjacent ocean water. So…

Q12.10 Explain the mechanism that generates sea breezes. *Hint:* It is illustrated in Figure 12.7.

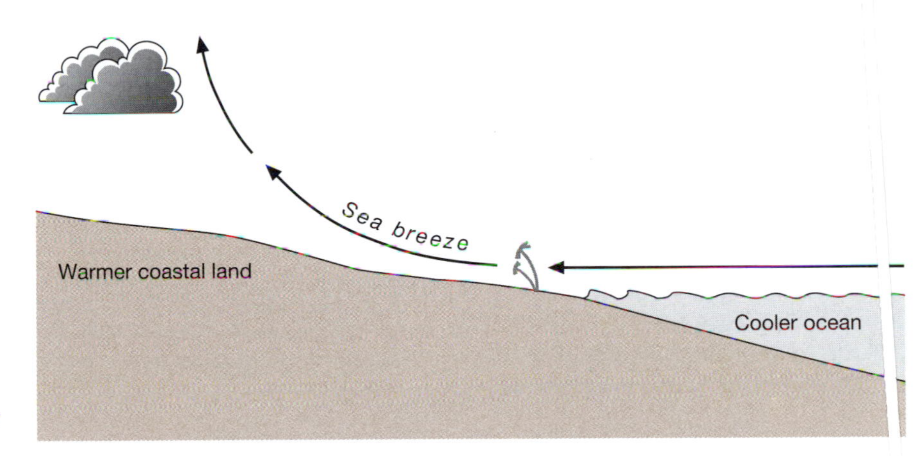

Figure 12.7 Sea breezes, which commonly flow at speeds of 13–19 km/hr (8–12 mi/hr) move from sea to land, reaching 15–50 km (9–30 miles) inland.

On rare occasions at night, when the air temperature over land declines, more gentle **land breezes** flow from land to sea—happily sweeping clouds of coastal mosquitoes with them.

Waves colliding with the sea floor—Figure 12.8 shows a shallow sea covering a seaward-sloping sea floor. Wind-driven waves are advancing toward the shoreline. Wave base, complete with schematic oscillating water particles, is also shown.

Q12.11 Sketch a projection of the wave base in Figure 12.8 horizontally to a point on the sea floor. How does the form of a wave approaching from point B to point A change in (a) general shape, (b) wavelength, and (c) wave height?

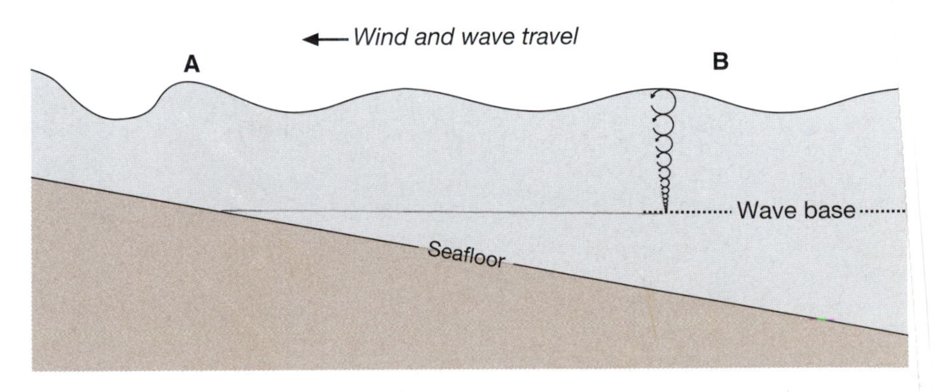

Figure 12.8 This is a cross-section that includes features in Figure 12.6, plus a sloping sea floor that rises landward to within wave base and beyond.

Wave forms reflect topography

Wave forms reflect sea floor topography (aka **bathymetry**). Shoaling conditions (the result of currents moving across shallow sea floors) are likely to produce breaking waves such as those that characterize shoreline surf zones (Fig. 12.9).

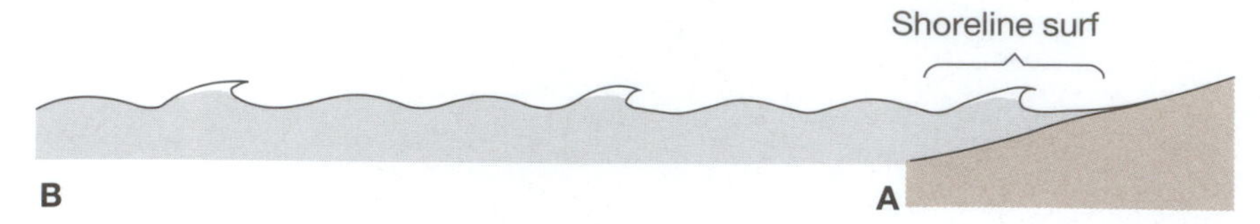

Figure 12.9 This is an incomplete sketch of sea floor and overlying wave forms. The shoreline surf zone—characterized by breaking waves white with foam—is shown as well.

Q12.12 From what you learned about waves colliding with the sea floor on page 207 (Fig. 12.8), and being mindful of wave forms, trace the sea floor from point A to point B in the reduced copy of Figure 12.9 on Answer Page 225. *Hint:* A breaking wave indicates that, *at that spot*, sea floor depth is within wave base; and, depth to wave base can be plotted from wave length.

Capsizing of the Taki-Tooo on June 14, 2003 (Fig. 12.10) was caused in part by the 32-foot fishing boat's venturing out across a sand bar topped by 12.foot-high breaking waves. Not so incidental is the fact that none of those who died were wearing life vests, whereas all eight survivors *were* wearing vests.

Q12.13 What *two* simple but valuable lessons should be learned from the Taki-Tooo tragedy? *Hint:* One has to do with pathways at sea, the other with on-board precautions.

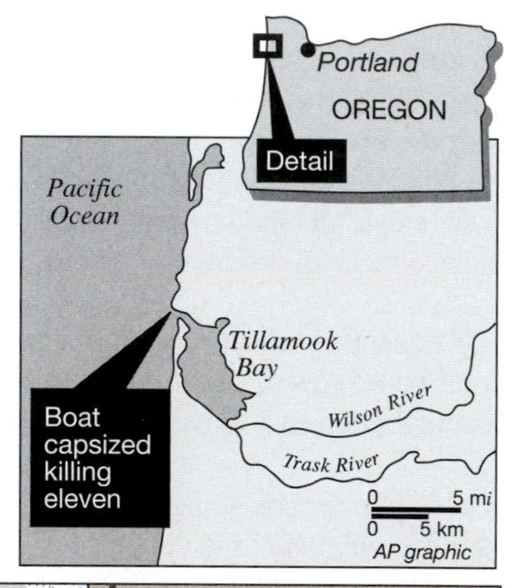

Figure 12.10 Rivers emptying into Tillamook Bay produce currents that keep the harbor channel swept free of mud and silt. But the mouth of the bay tends to become barred by sand drifting southward along the coast.

THE REGISTER-GUARD, JUNE 17, 2003

The Taki-Tooo tragedy

Charter boat disaster raises safety concerns

GARIBALDI, Ore.—The treacherous Tillamook sandbar—one factor in the capsizing of the Taki-Tooo and the loss of 11 people—hasn't been dredged by the U.S. Army Corps of Engineers since 1976. Decades of pounding surf, meanwhile, have eroded the rock jetties built to shelter boats from the swells and whitewater at the harbor's entrance. More than anything else, the Taki-Tooo tragedy serves as a haunting reminder of the power and unpredictability of the sea—and the vulnerability of all who sail upon it.

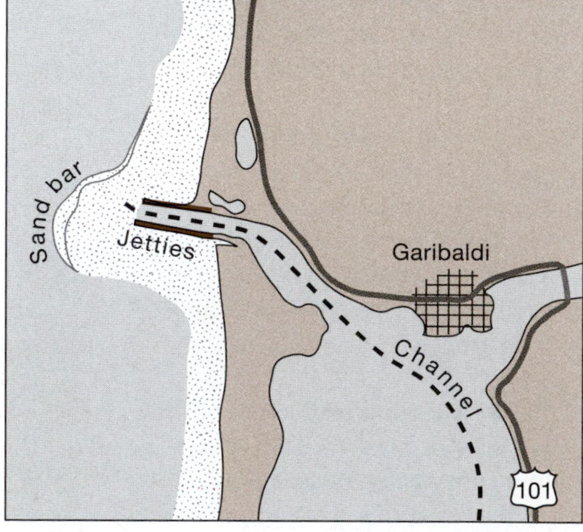

Taki-Tooo disaster: http://www.registerguard.com/news/2003/06/17/ed.edit.charterboat.phn.0617.html

Marine terraces

Given the turbulence of the surf zone, sediment suspended in the water provides grist for the proverbial mill—grinding away at rocks and installations at the water's edge, plus flattening the seafloor down to the depth of wave base (Fig. 12.11). The result: A **marine terrace** (aka wave-cut platform, aka wave-cut bench).

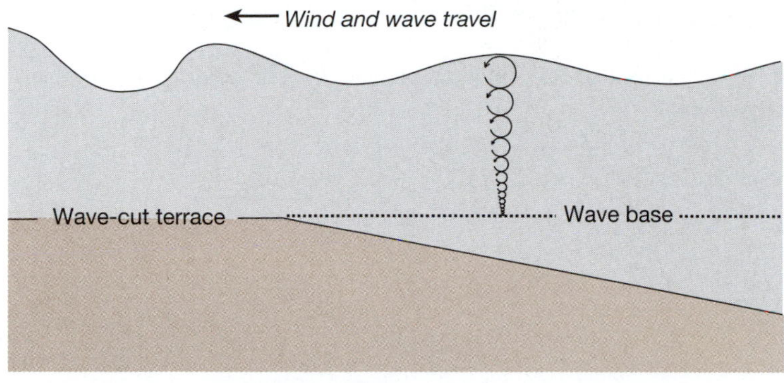

Figure 12.11 Where the elevation of the sea is stable—relative to that of the land—for a sufficient period of time, erosive action down to wave base can sculpt a marine terrace.

Marine terraces, hidden from view during their development, can be exposed through a rising of the land, a lowering of the sea, or some combination of the two. Because of its position along an active plate margin marked by vertical movements of continental crust, our west coast is replete with exposed marine terraces. A notable example is at Pebble Beach, California.

The area of Pebble Beach Golf Links, described by some as 'the greatest meeting of land, sea, and sky in the world,' exhibits innumerable exposed marine terraces. Figure 12.12C shows a view from hole 5 toward hole 6.

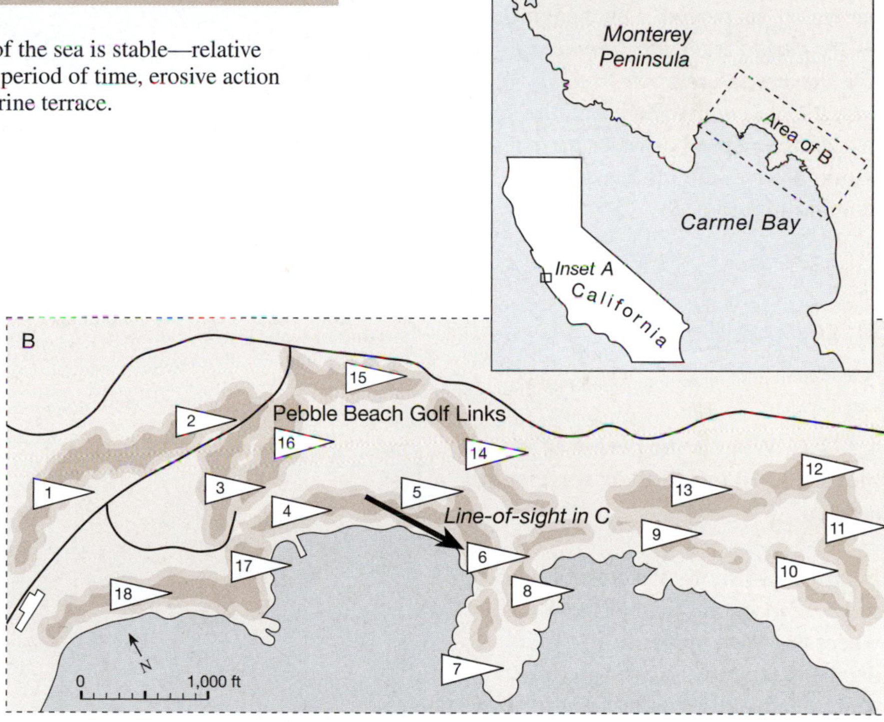

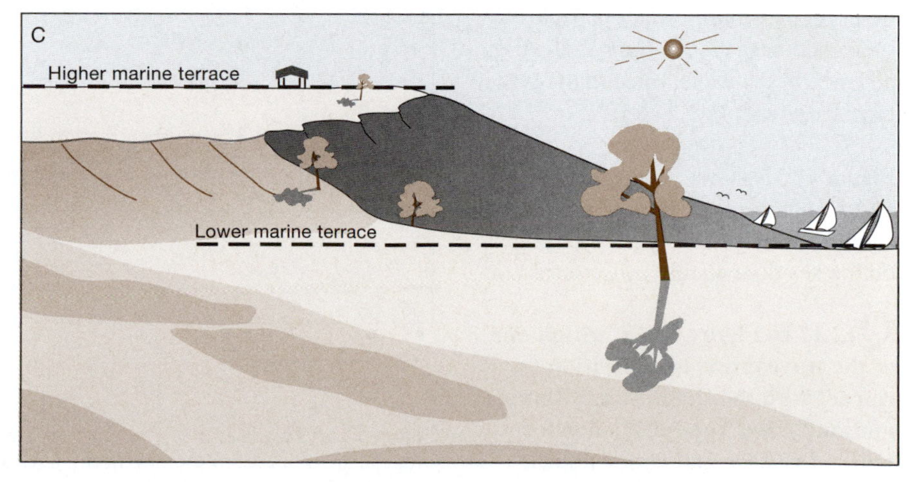

Figure 12.12 A–B. Famed Pebble Beach Golf Links. **C**. Looking southeastward from hole 5 toward hole 6 (as indicated by the arrow in B). *After web site pebblebeach.com*

Q12.14 From the information in Figure 12.12, on page 226 of the Answer Page draw a profile (i.e., a side-view) from NW to SE along the line-of-sight indicated by the arrow in Figure 12.12B. Show the relative elevations of each of the two marine terraces and that of the Pacific Ocean.

E. Wave refraction

Where 'stacks' of orbits of downward-diminishing wave motion encounter the seafloor (at wave base), frictional drag distorts the stacks, (a) flattening orbits, (b) decreasing wavelength, and (c) increasing wave height (Fig. 12.13).

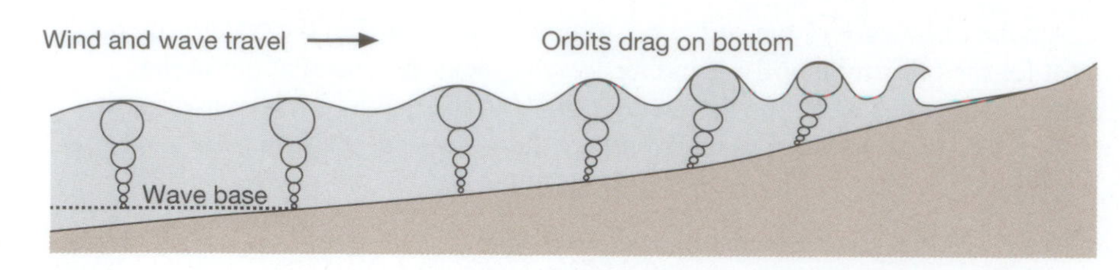

Figure 12.13 Wave-generated orbits of water particles become distorted and flattened where they meet the sea floor.

Q12.15 What does the simultaneous decrease in wavelength and increase in wave height have to do with the *conservation of energy*? *Hint:* A water particle within a wave crest has both *kinetic energy* (energy of motion) and *potential energy* (energy of elevation).

Curiously, once a wave has been generated, its **period** is unchanging. (*Wave period is the time required for two successive crests, or troughs, to pass a fixed point.*) Therefore, the speed of a wave (= speed of water particles within a wave) diminishes as wave length diminishes.

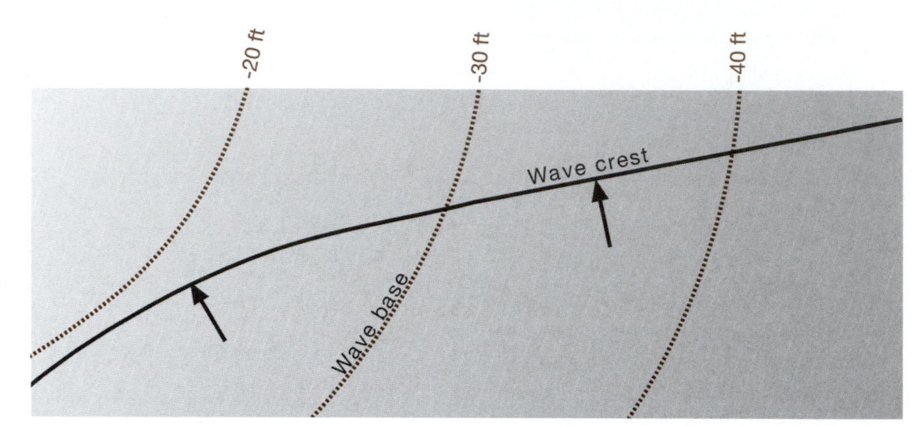

Figure 12.14 A contour map of the sea floor showing a solitary wave crest being refracted as it moves from water in which depth exceeds wave base into water in which depth is less than that of wave base.

Q12.16 What is the period of waves moving at 10 m/s, if their wavelength is 64 m?

Figure 12.14 shows the crest of a single wave on a contour map of the sea floor where the wave moves from water depths greater than wave base into water depths less than wave base. The dragging of orbiting water particles as they encounter the sea floor slows the advance of the wave, causing the wave crest to **refract** (i.e., bend*)*.

Figure 12.15 shows wave crests X, Y, and Z approaching shore. Crests are shown as straight lines, but topography on the sea floor should cause refraction.

Q12.17 In Figure 12.15, which one of the wave crests is in position to refract with the slightest additional motion? *Hint:* You must solve for both wave base and water depth.

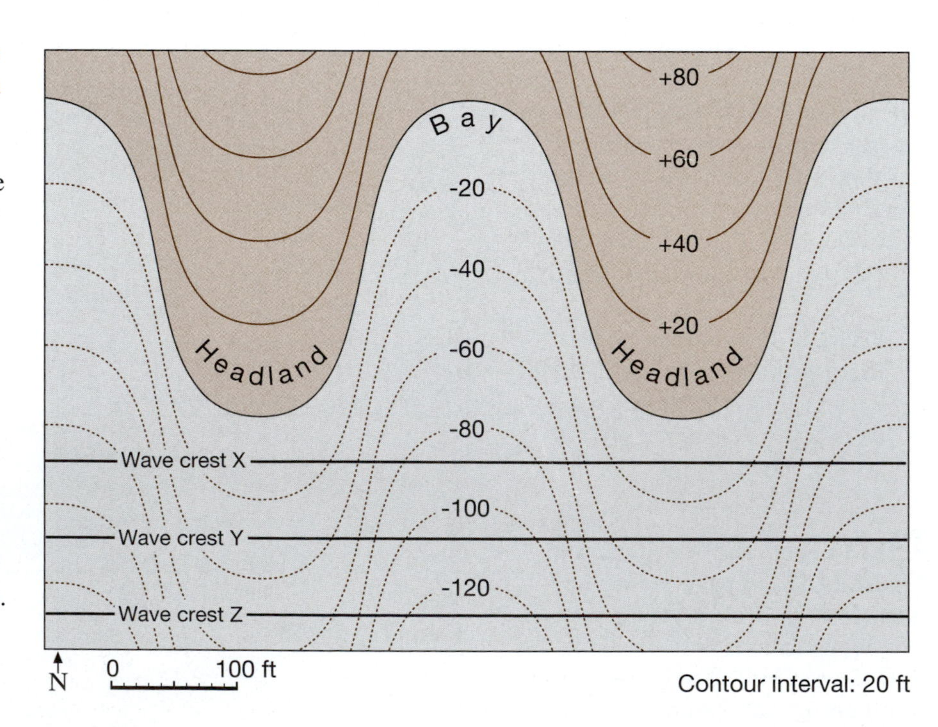

Figure 12.15 This is an artificial representation of straight wave crests approaching an irregular shoreline from south to north. Contours show land elevations and water depths.

Erosion of headlands, deposition within bays

Headlands (i.e., peninsulas) are sites of erosion, which produces sea cliffs, whereas bays are sites of deposition, which produces sand beaches. This is because wave energy (the grist mill mentioned earlier) is concentrated on headlands for reasons illustrated in Figure 12.16 and tabulated as follows:

- Lines of force are perpendicular to wave crests.
- When a wave crest refracts (bends), lines of force are redirected so as to continue to be perpendicular to the wave crest.
- Lines of force converge (thereby concentrating force) on headlands.
- Lines of force diverge (thereby dissipating force) within bays.

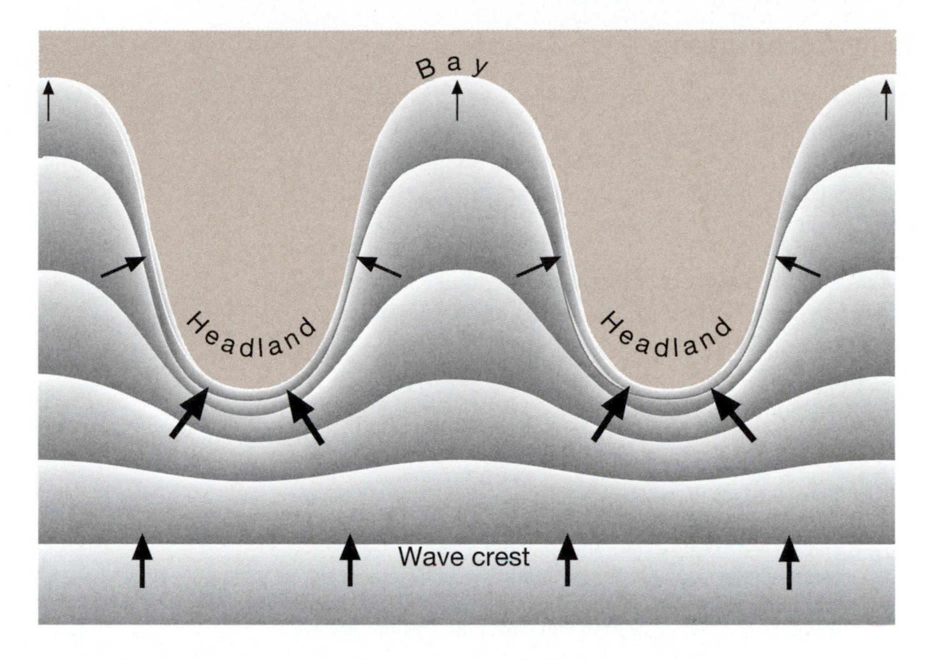

Figure 12.16 Lines of force (arrows) associated with advancing waves are perpendicular to wave crests. So as a wave refracts (bends), lines of force either converge or diverge. These arrows are vectors of sorts, i.e., an arrow points in the direction of applied force, and the weight of an arrow indicates the amount of relative force.

Q12.18 Seaports are typically constructed within bays. Why is that? (Seaports also commonly develop within estuaries—i.e., *drowned river mouths*—to facilitate inland commerce.)

Famed sand beaches of the world, e.g., Waikiki (Hawaii), Lloret de Mar (Spain), Puerto Vallarta (Mexico)—the list goes on—are all within bays (Fig. 12.17).

Q12.19 What is there about the physical geography of the Mexican coastline shown in Figure 12.17 that sets Puerto Vallarta apart from neighboring coastal communities?

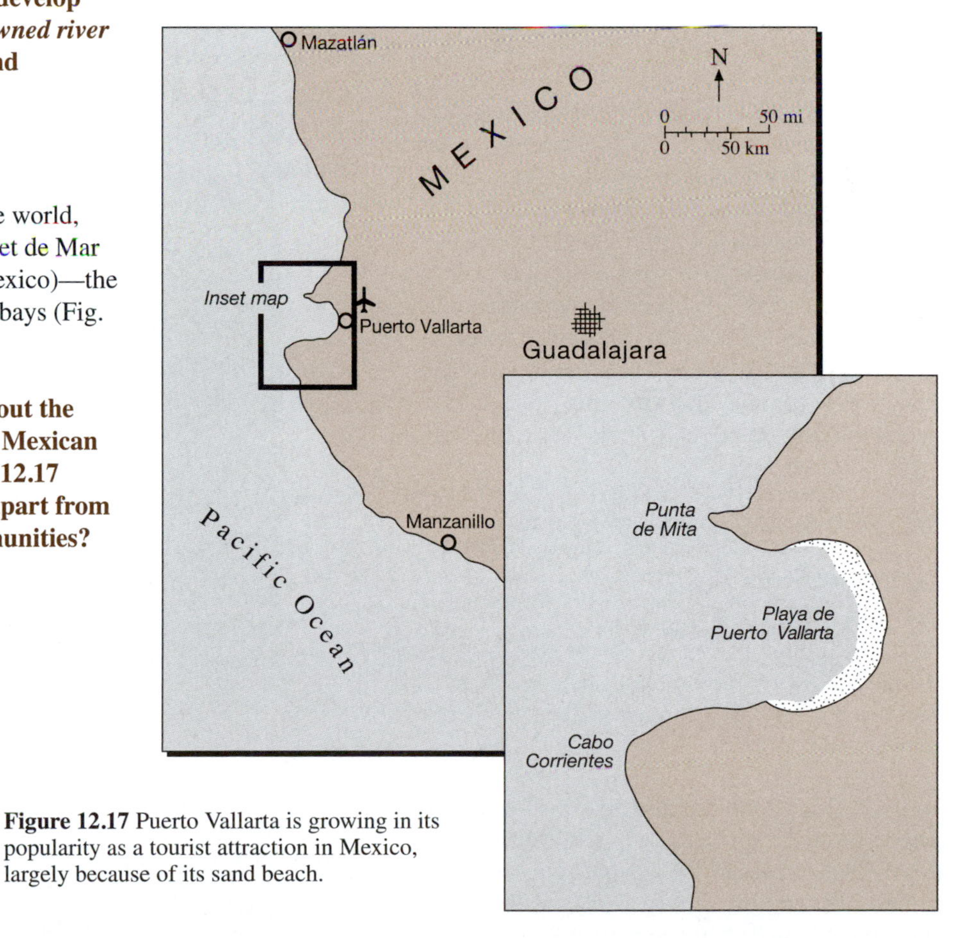

Figure 12.17 Puerto Vallarta is growing in its popularity as a tourist attraction in Mexico, largely because of its sand beach.

Longshore currents

Waves striking a coast at an angle—the usual condition—produce a **longshore current** (Fig. 12.18). Longshore currents can, in turn, effect **longshore transport** of coastal sediments, prompting the aging elementary classroom film title: "*Beach, A River of Sand.*"

Q12.20 Longshore currents and longshore transport are critical concepts in the business of coastal management. *Question:* Where—earlier in this exercise—did it appear that a longshore current and transport specifically led to fatalities?

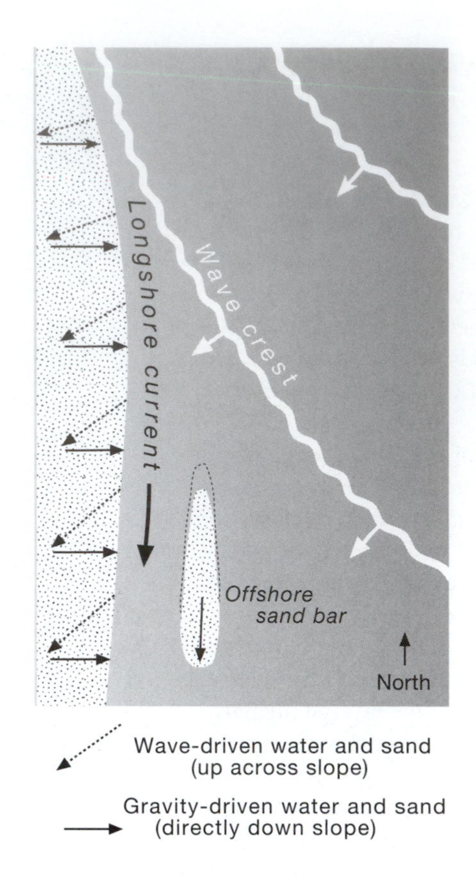

Figure 12.18 Waves striking a north–south coast from the northeast produce a current moving southward, dragging coastal sediments with it. (Notice the refraction of the wave crests as they enter shallow water.) On the beach, wave crests push water and sand up the beach in the direction of wave motion, while, within wave troughs, gravity drives water and sand directly downslope. This zigzag motion results in the longshore transport of beach sands.

Rip currents

In cases where the sea floor is relatively smooth, backwash from the surf zone returns to the sea as thin sheet-flow along the sea floor. Standing in waste-deep water, one can sense that seaward flow at ankle level. It is commonly called **undertow**. But where the sea floor is scalloped into channels and ridges at some angle to the surf zone, backwash can become concentrated within the channels, much like surface water finding its way into gullies. This channeled water, which is difficult to detect from the shore, can be swift and strong—giving rise to the name **rip currents** (Fig. 12.19). Of course, undertow and rip currents grade into each other.

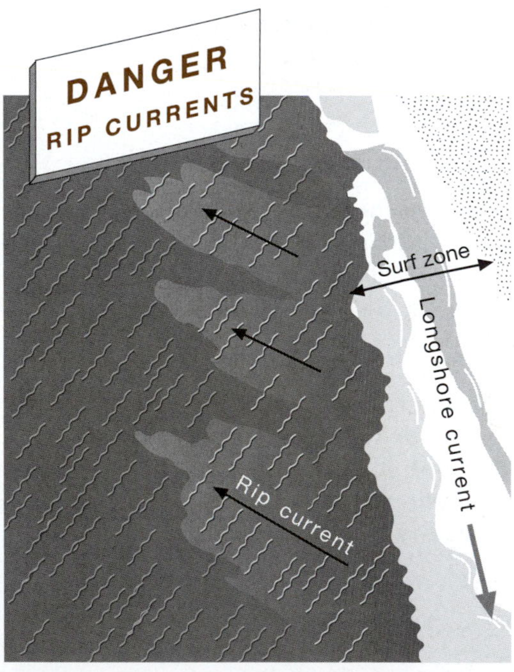

Figure 12.19 This is a sketch of an actual low-altitude photograph taken along the coast near La Jolla, California. Rip currents and their direction of flow are indicated with arrows. In this case, rip currents run directly opposite of the motion of onshore waves.

Rip currents move slowly by automobile standards, e.g., 8 km/hr (5 mi/hr), but that's fast enough to get swimmers into big trouble.

Q12.21 In the Sydney 2000 Olympics, Americans Gary Hall, Jr. and Anthony Ervin tied (yes, tied!) in the 50 m freestyle event. Their time: 21.98 seconds. *Question:* could Gary and Anthony make it to shore swimming against a rip current moving at the rate of 8 km/hr?

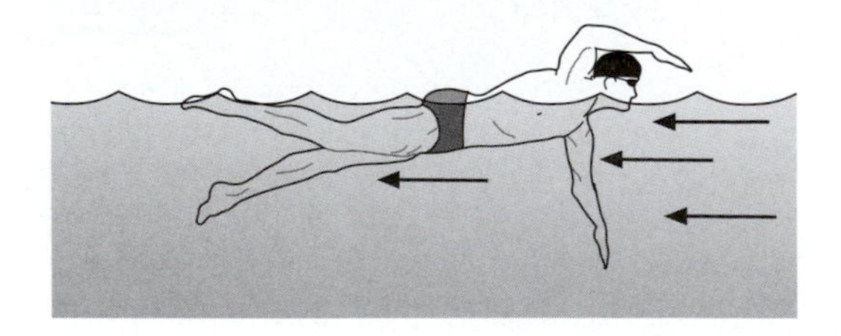

Land forms shaped by currents—Cape Cod, Massachusetts

The complexity of currents along our Atlantic coast presents special challenges to coastal developers and managers. The converging, diverging, and swirling of currents shape myriad land forms in this region, including curious **sand spits** (Fig. 12.20).

A sand spit is basically an extension of a beach or bar into deeper water of a bay mouth. Currents entering the bay commonly shape a spit into a hook-like feature. In fact, a number of spits bear the name 'hook,' e.g., Sandy Hook, New Jersey—a picturesque part of Gateway National Recreation Area.

A disclaimer: Larger islands in Figure 12.20 are neither spits nor sand bars, but gravel bodies left by retreating glaciers. But most of the delicately narrow peninsulas, along with the hook-shaped features, consist of sand strewn about by present-day currents.

Q12.22 On the map in Figure 12.20, try to find peninsulas or islands that appear to be spits that were shaped by currents. Place an arrow along each spit indicating the direction of the current that shaped it, and record its map coordinates (letter-number) on page 226 of the Answer Page. *Hint:* Do not be dismayed by currents that appear to converge or diverge over short distances. Such irregularities are typical of the sea.

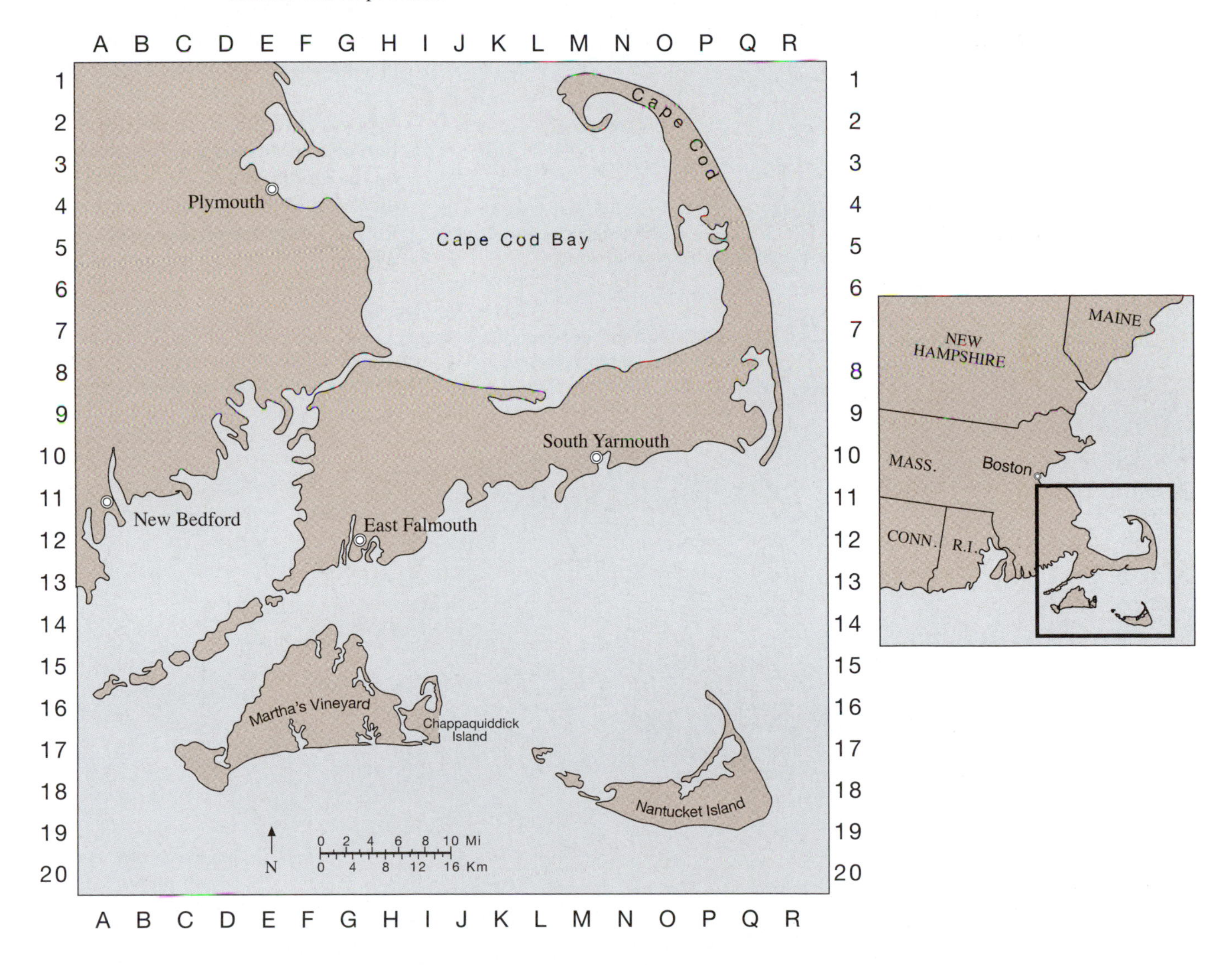

Figure 12.20 Sand spits of various shapes and sizes along our Atlantic coast are graphic testimonies to the complexity of currents that shaped them.

F. A summary of shoreline landforms

Figure 12.21 illustrates a few of the processes and products, already introduced, that occur along rocky coasts such as our own Pacific coast:

- erosion of headlands—producing wave cut cliffs
- deposition within bays—producing sand beaches
- sculpting of a marine terrace—with erosional remnants called stacks

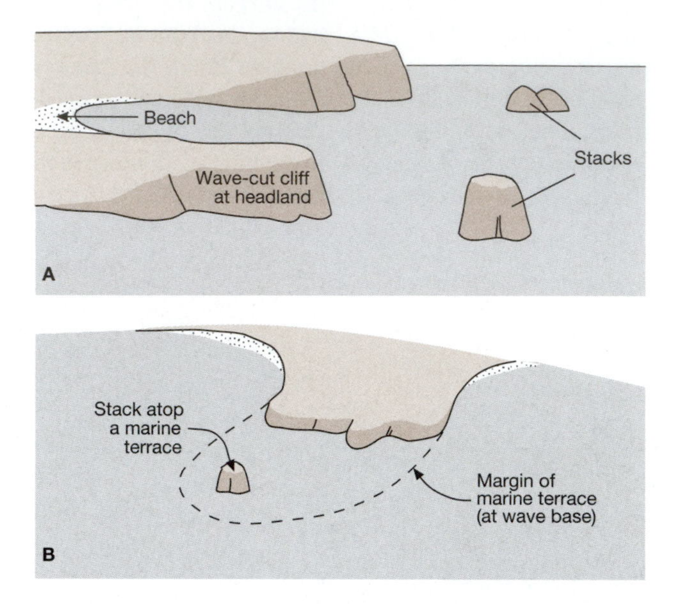

Figure 12.21 A View of partially eroded headlands, with wave-cut cliffs and deposition of the eroded material as a sand beach within a bay. Stacks (erosional remnants of headlands) occur offshore. **B** An 'X-ray view' of a wave-cut platform below water level.

Shoreline features—La Push, Washington quadrangle

Drawing on the features illustrated in Figure 12.22, the following three questions apply to the La Push, Washington, 15' quadrangle on facing page 215.

Q12.23 (A) On Answer Page 226 draw a topographic profile between coordinates F.9-7.7 (shoreline) and I.2-7.5 (highest point in vicinity). (B) There are indications of marine terraces in this transect. At what approximate elevations are they?

Q12.24 (A) Give the coordinates of the tied island with the largest tombolo. (B) What is the name of the beach that includes the seaward side of the attached tombolo?

Q12.25 Judging from the distribution of sand brought to the ocean by the Quillayute River, what is the direction of the coastal current at La Push—southeastward or northwestward?

Figure 12.22 Erosional and depositional features of a rocky coast. The spit reflects the generation of sand through erosion of a headland and transport of that sand by a westward-flowing current. The shape of the tombolo reflects deposition of sand by a river (forming a delta) and transport of that sand by a current.

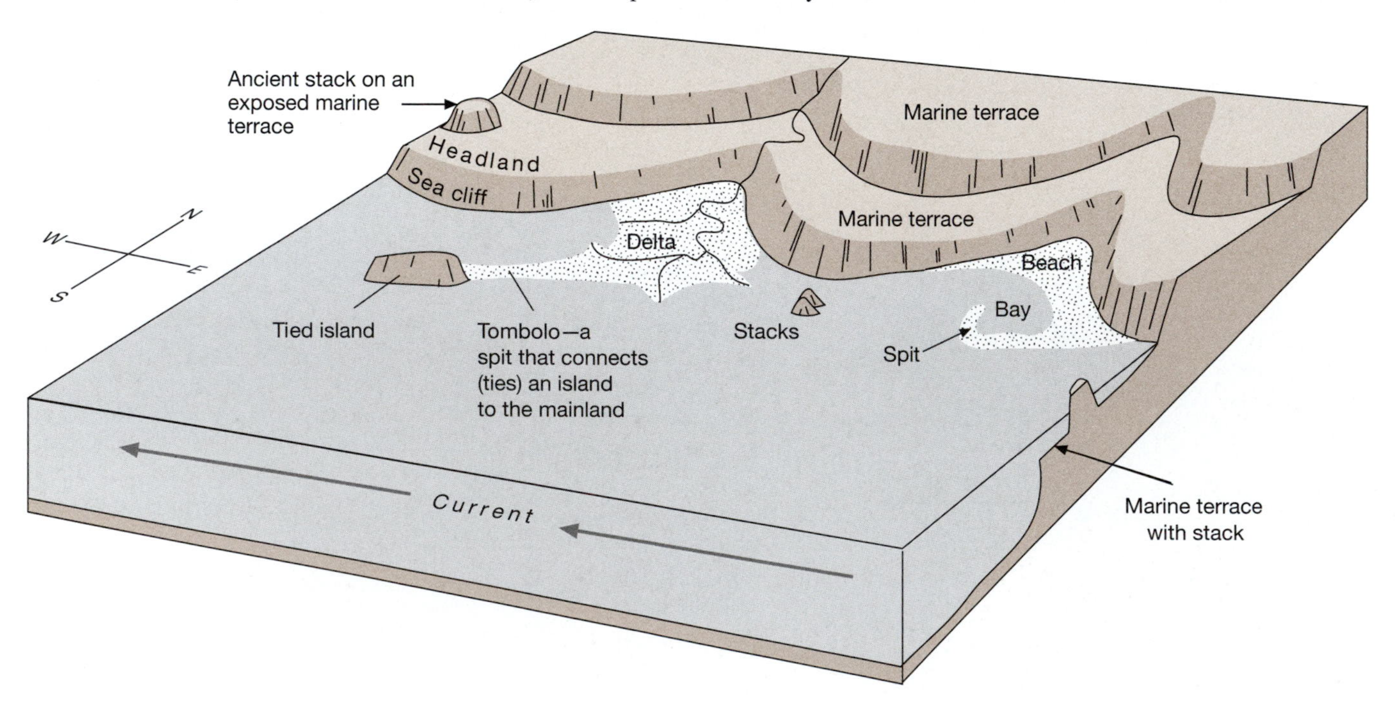

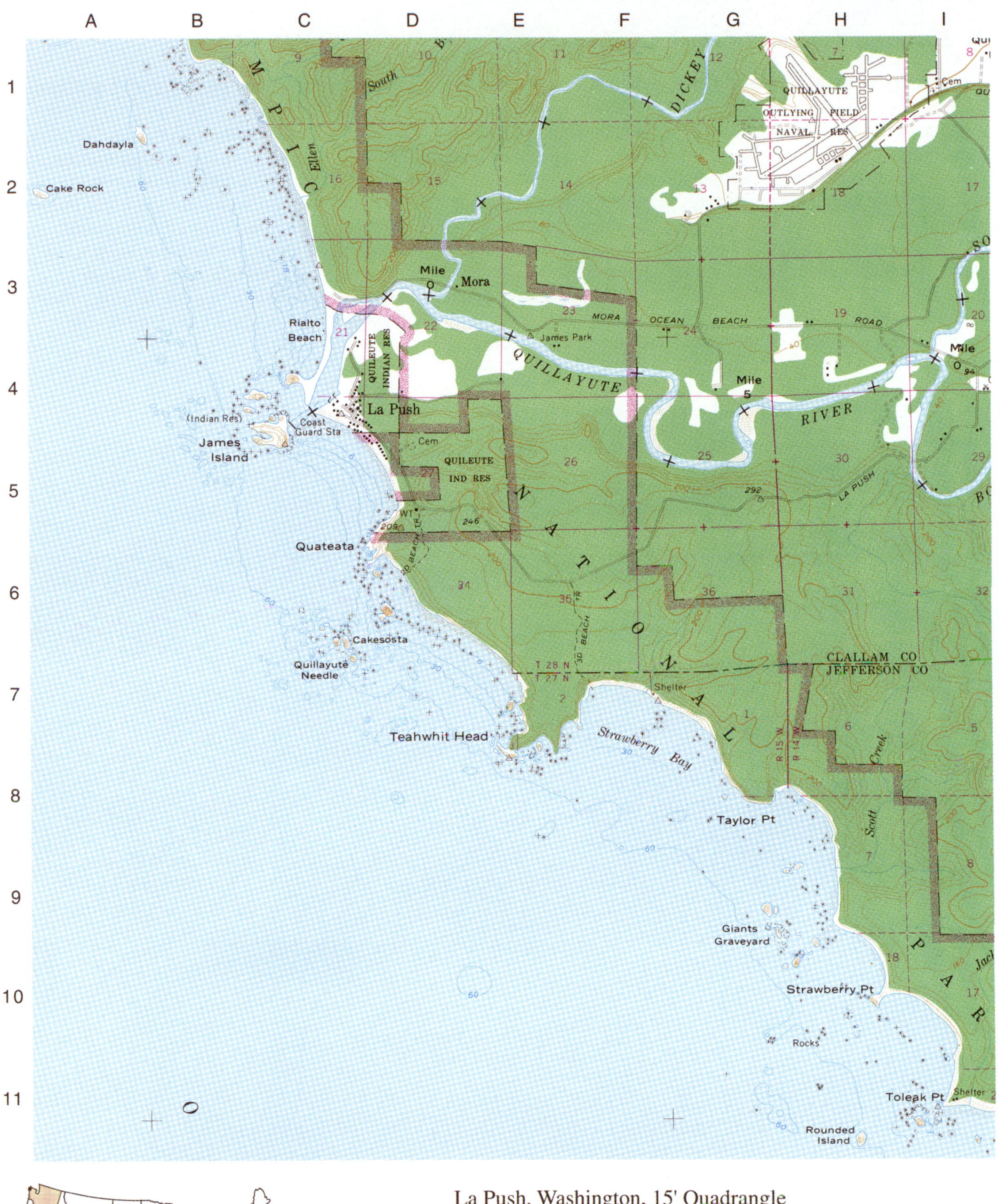

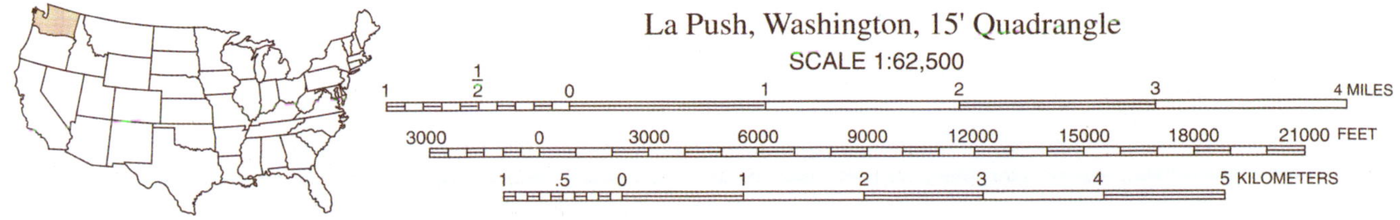

La Push, Washington, 15' Quadrangle

SCALE 1:62,500

G. Unwanted effects of waves and currents

Threats to coastal wetlands (see cover story on facing page)—example: Isles Dernieres, Louisiana

A **delta** such as the Mississippi River delta (Fig. 12.23) develops where a river emptying into the sea dumps sand and mud faster than ocean waves and currents can sweep them away. It's sort of like trash trucks getting ahead of bulldozers at your local landfill.

In the case of our Mississippi delta, *Old Man River* has been winning the battle with the Gulf of Mexico for 7,000 years. That is, the delta is still very much there, despite the fact that early subdelta lobes, e.g., the St. Bernard and LaFouche, aren't looking so good these days (again, Fig. 12.23). The reason: When a delta lobe is abandoned by a river, because the river has changed its course, waves and currents begin to eat away at the older lobe.

Coastal wetlands (e.g., marshes and swamps) are home to essential links in food chains that—although based on land—extend to the sea where they play vital roles in coastal fishing industries. Additionally, wetlands are home to microbial organisms that cleanse polluted surface waters as they make their way to the Gulf of Mexico.

Figure 12.23 Mississippi River delta consists of a succession of subdelta lobes, each of which reflects an earlier course of the river. Isles Dernieres is at the margin of the abandoned LaFouche lobe (bold rectangle), whereas Mississippi River in its present course is building the Balize lobe. (The B.P. values—i.e., years **before present**—are based on radiometric carbon measurements.)

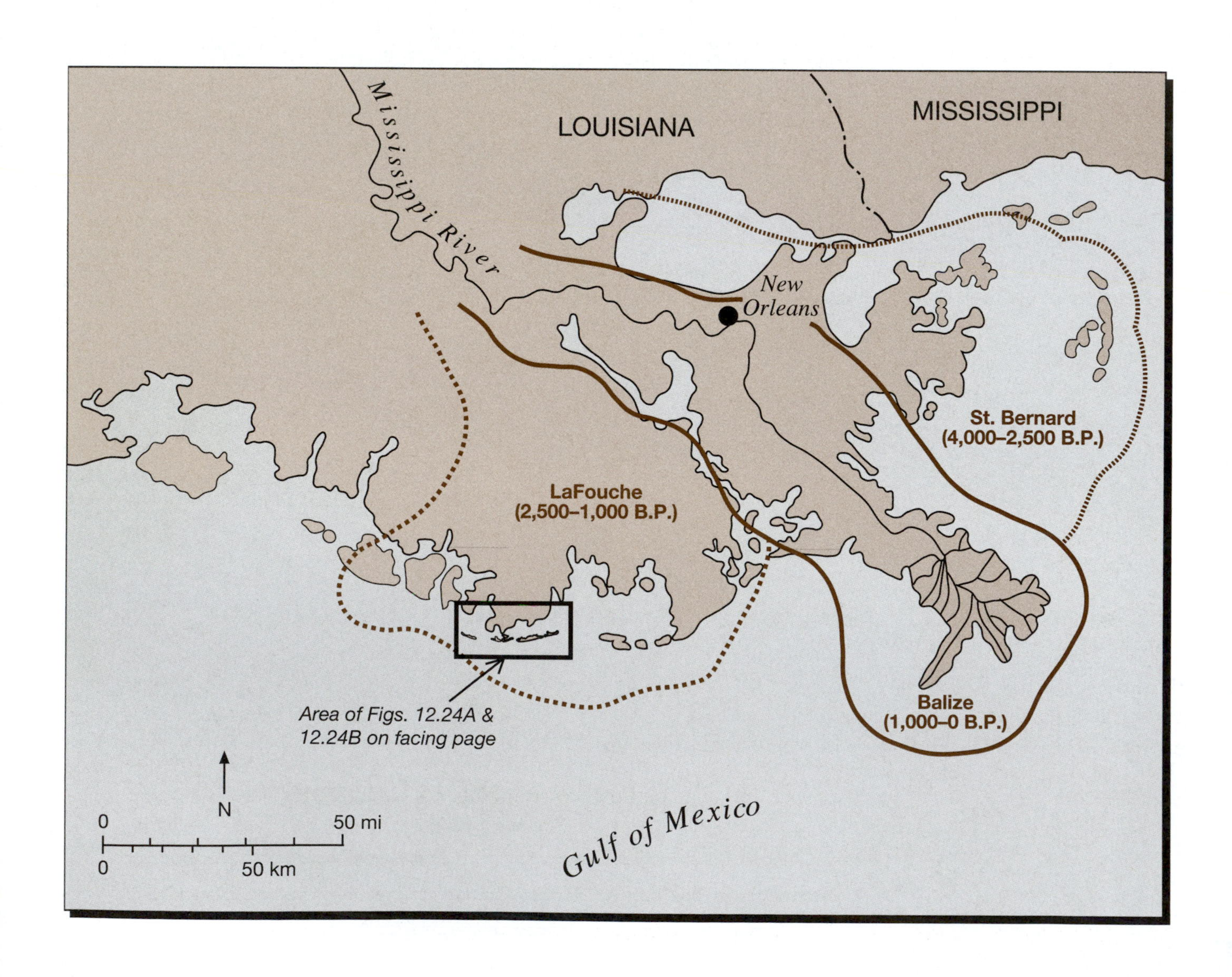

Isles Dernieres—a 32-kilometer-long stretch of southern Louisiana **coastal wetlands**—is being eroded at a rate of 20 meters per year, placing the Isles among the most rapidly deteriorating shorelines in the world (Figs. 12.24A and 12.24B). Understandably, ecologists are alarmed. Some argue that coastal wetlands act as 'speed bumps' in lessening storm surges produced by hurricanes.

Not to question the importance of coastal wetlands, but let us for a moment place the 'demise' of Isles Dernieres in its regional geologic context.

Q12.26 The deterioration of Isles Dernieres is bad news, to be sure, but given the proverbial balance of nature, isn't there some good news as well? If so, what is it? *Hint:* Look back again at Figure 12.23.

But the answer to Q12.26 is not to say that the Mississippi River is behaving as always before. The amount of delta-building sediment delivered to the Gulf of Mexico by the Mississippi now is judged to be only 50% of that delivered a century ago.

Q12.27 Can you think of a couple of reasons why there has been such a huge decrease in the amount of sediment within the lower Mississippi? One reason is improved tillage practices, minimizing erosion of croplands. *Hint:* The other reason has to do with projects within the river and its tributaries.

Cover story—The photo in Figure 12.25 illustrates destruction unleashed by the tragic oil spill of 2010 in the Gulf of Mexico. Green wetland and marsh foliage of the Mississippi Delta is subject to being killed by such oil and thereby rendered less resistant to erosion by waves and currents. Moreover, this area is astride the gigantic Mississippi flyway for countless migratory waterfowl. Damage to wetlands and marshes such as this is believed to be more fatal and long lasting than damage to open water and beaches.

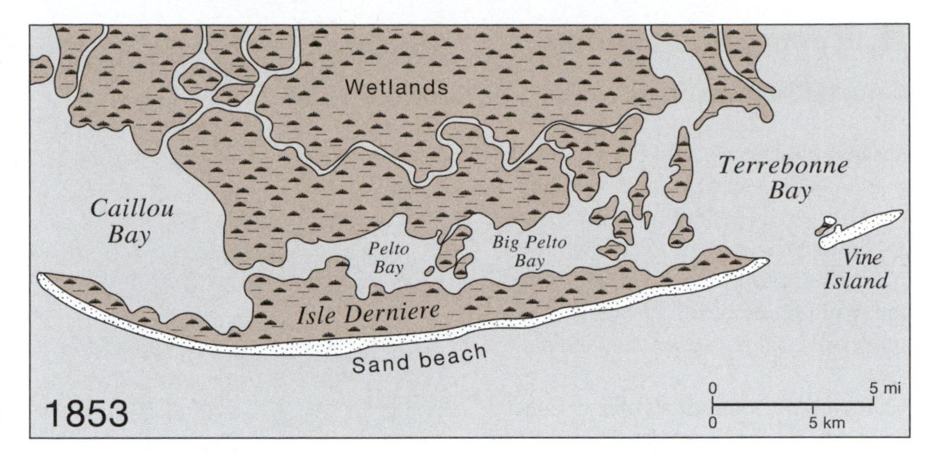

Figure 12.24A This is the area outlined by the bold rectangle in Figure 12.23, showing Isles Dernieres area as it appeared in 1853. There was a single Isle Derniere at that time.

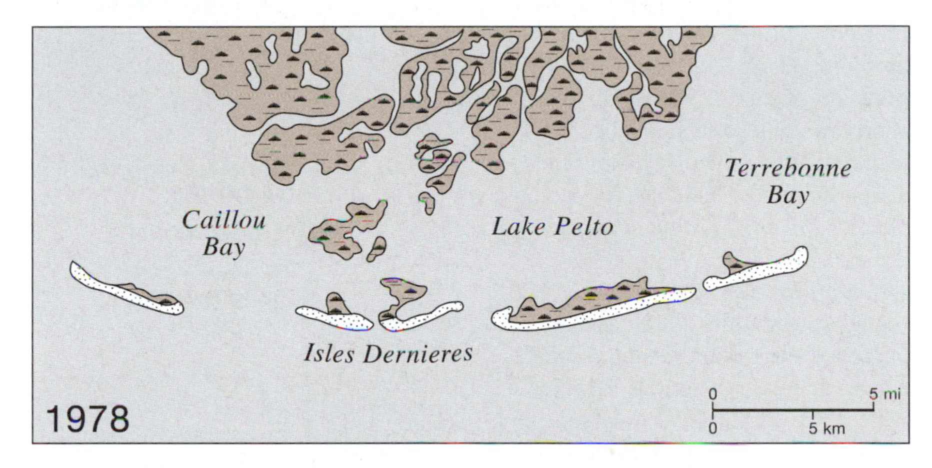

Figure 12.24B This is the same area as that shown in Figure 12.24A, but as it appeared after 125 years of erosion by waves and currents of the Gulf of Mexico.

Figure 12.25 This photo shows oil invading wetlands and marshes of the Mississippi flyway for migratory waterfowl. May 23, 2010. Mississippi Delta on May 23, 2010. (See front cover and title page)

H. Human intervention

Coastal installations

Valued beaches—Sand beaches are not only valued as playgrounds, but they are also valued as buffers to waves that might otherwise gnaw away at coastal properties. For these and other reasons, the maintaining of sand beaches is a universal goal in coastal management.

There are five common types of coastal engineering emplacements (Fig. 12.26), three of which, **groins**, **jetties**, and **harbor breakwaters** (Figs. 12.26A, B, C) are designed to impede longshore transport of sand. But there is one huge problem: All three disrupt downdrift transport of sand—causing downdrift shorelines to lose sand and become susceptible to erosion. This reminds us again of a fundamental law of ecology: *You can't change just one thing!*

Offshore breakwaters (Fig. 12.26D), though intending to simply shelter beaches from the sea, result in sand accumulation in the relatively quiet water landward of breakwaters—forming a dam of sorts that checks downdrift transport of sand much like groins, jetties, and harbor breakwaters. An example has developed at Santa Monica, California (Fig. 12.27).

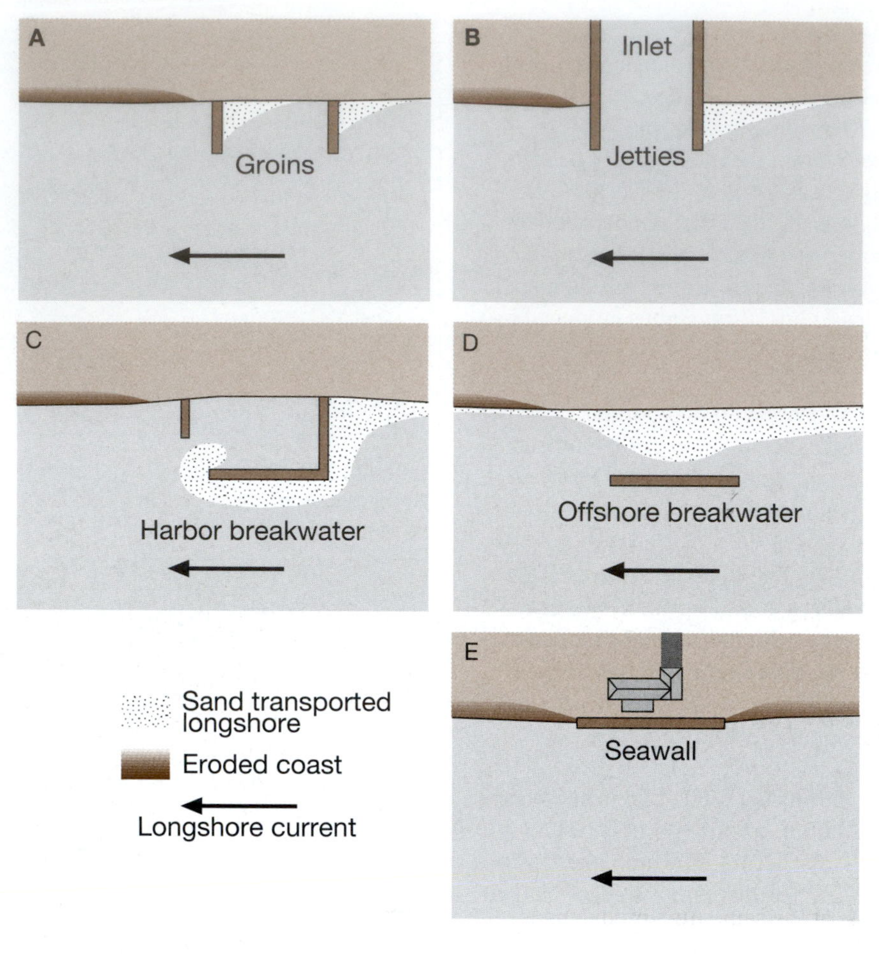

Figure 12.26 Three of the five common kinds of engineering emplacements (A–C) interfere with longshore transport of sand, converting downdrift beaches into sites of wave erosion. Offshore breakwaters (D) result in much the same problem, and modest sea walls (E) are commonly undermined and destroyed by storm waves.

Q12.28 In Figure 12.27, in what direction does the longshore current appear to be moving—northwestward or southeastward?

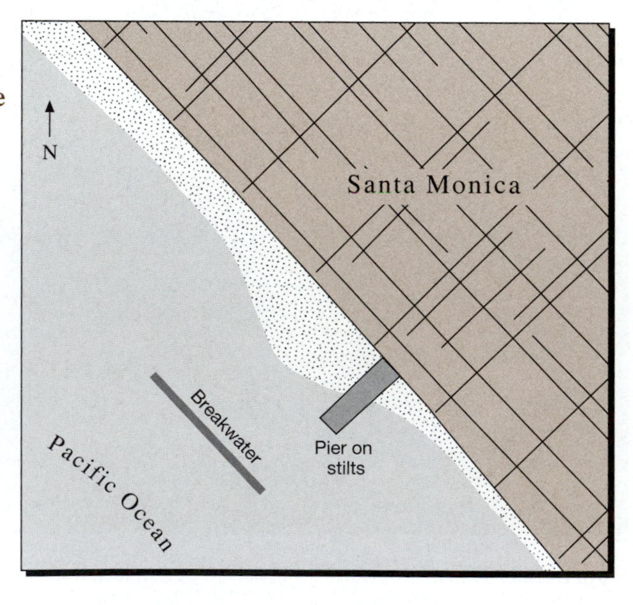

Figure 12.27 The breakwater at Santa Monica was built in 1933. Sand promptly began to accumulate landward of the breakwater, and erosion has become a problem downdrift.

Sea walls or revetments (Fig. 12.26E). On September 8, 1900, Galveston, Texas, was demolished by a hurricane with a storm tide of 4.6 m (15 ft.). Loss of life: 6,000, which continues to be the record for deaths caused by an American natural disaster.

To contend with future hurricanes, Galveston residents constructed a 4.9 m (16 ft.) high sea wall in 1904. Then, in 1915, that seawall was tested by a similar hurricane. Death toll: 12.

But there's a problem with sea walls constructed by homeowners with fewer resources. Such dubious seawalls are commonly destroyed by storm waves.

Mitigating unwanted effects of groins

Year One—Homeowners at a fictitious seaside community built a sea wall to protect their properties from wave erosion. But, alas, their sand beaches promptly washed away.

Year Two—Homeowners built sturdy groins to protect their beaches. But, alas, while beaches at some properties grew in size, neighboring beaches shrank to nothing (Fig. 12.28).

Year Three—After considerable study, homeowners decided on yet another measure—a project designed to slow the longshore transport of beach sand, while at the same time producing a more equitable distribution of sand among home sites (Fig. 12.29). Eureka! Their final project met with the success they had hoped for.

Q12.29 From what you can see in before-and-after Figures 12.28 and 12.29, what appears to have been the rather simple measure taken by residents, and why did that measure result in a more equitable distribution of sand among their homesites? *Hint:* Study details of Figures 12.28 and 12.29, *including insets in each.*

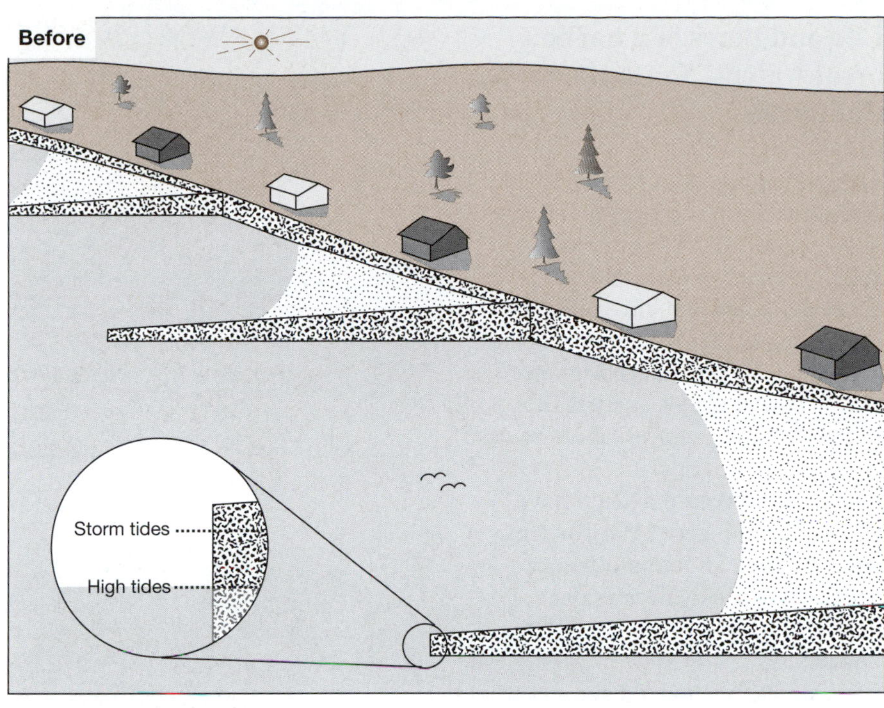

Figure 12.28 Homeowners in this seaside community, who had first built a seawall to stop erosion of the shore, constructed groins in an effort to preserve sand beaches along the seawall. But sand became unequally distributed among home sites.

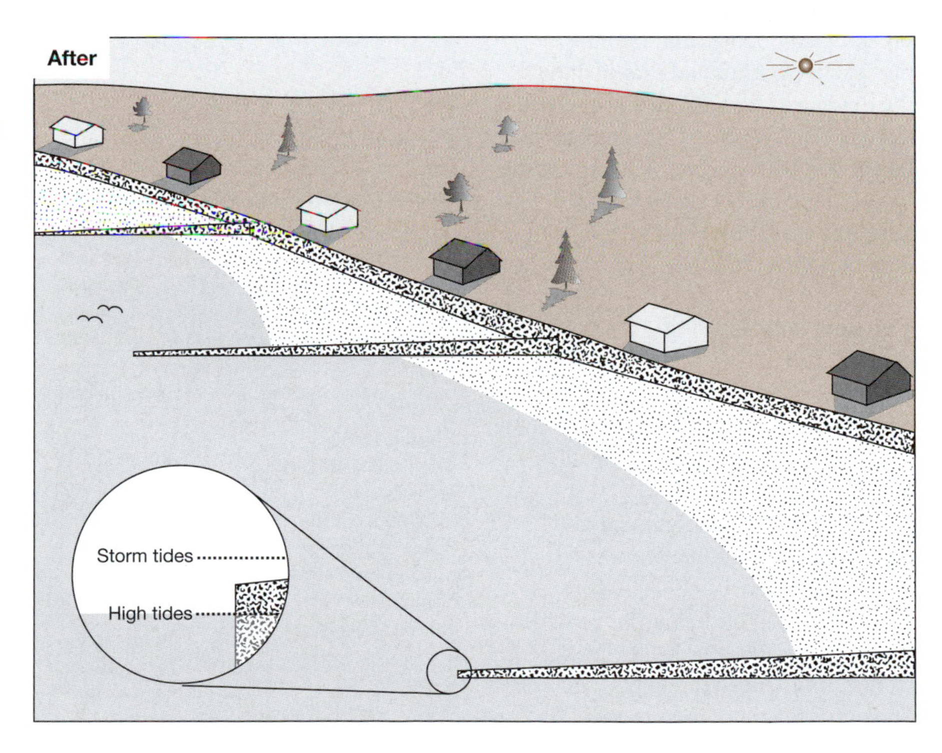

Figure 12.29 Homeowners undertook a measure that resulted in more equitable distribution of sand beaches among home sites.

Life and times of a harbor breakwater—Santa Barbara, California

The challenge—Santa Barbara had a delightful beach running the full length of a delightful city (Fig. 12.30) until 1927, when the powers that were decided that Santa Barbara needed a safe anchorage for recreational boats. Managing that anchorage has since proven to be a classic study in the effects of blind engineering of beaches.

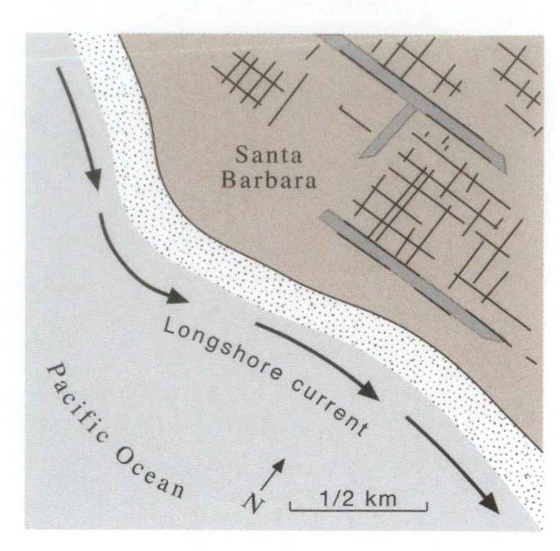

Figure 12.30. Santa Barbara shoreline in 1927, before commencing its harbor project.

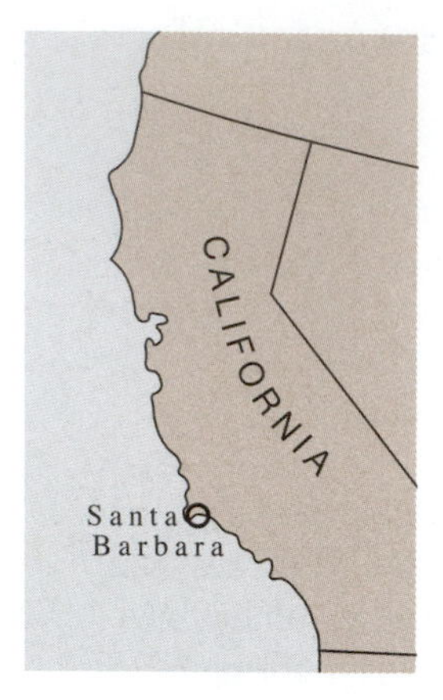

Q12.30 (A) Would a groin have been advisable here? Why or why not? (B) How about an offshore breakwater? Why or why not?

The solution?—In 1927 it was believed that a simple offshore breakwater would magically shelter boats, while causing little or no problems with the shoreline. Wrong! No sooner had the breakwater been finished than sedimentation began forming a tombolo between beach and breakwater, oozing sand into the prospective harbor and diminishing the downdrift flow of sand to the beach beyond (Fig. 12.31). Once the downdrift beach buffer disappeared, undercutting of shoreline cliffs caused houses to topple into the sea for reasons not understood at the time.

Figure 12.31 The initial offshore breakwater resulted in the development of a tombolo dam between shore and breakwater and the starving of downdrift beaches.

The final solution (Fig. 12.32) consisted of…

- enclosing the harbor
- vacuuming sediments that find their way into the harbor via spit-like currents
- disgorging those sediments downdrift from the harbor

This bypassing of the harbor provides two benefits: (1) It prevents the harbor from filling with sediments, and (2) it prevents erosion of the downdrift shoreline.

Q12.31 (A) There is a continuing expense associated with this solution. What is it? (B) Who do you believe should pay for this continuing expense?

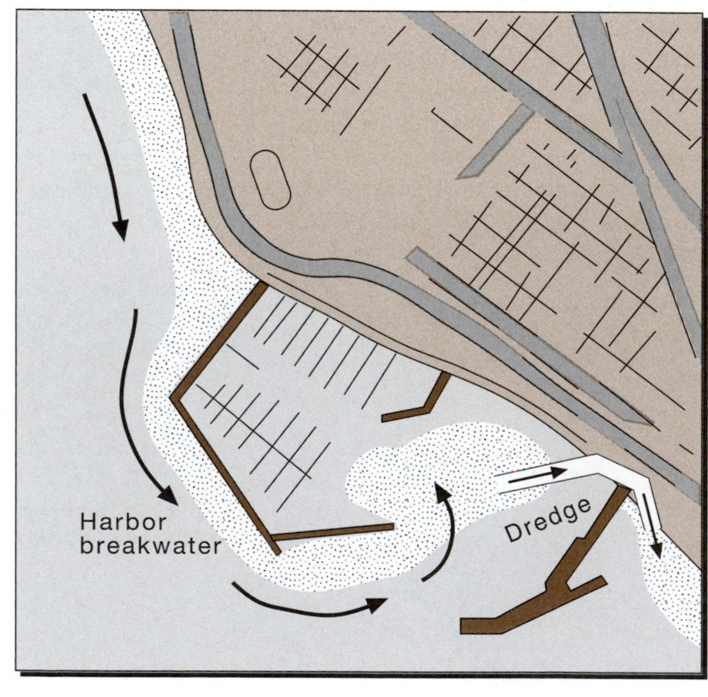

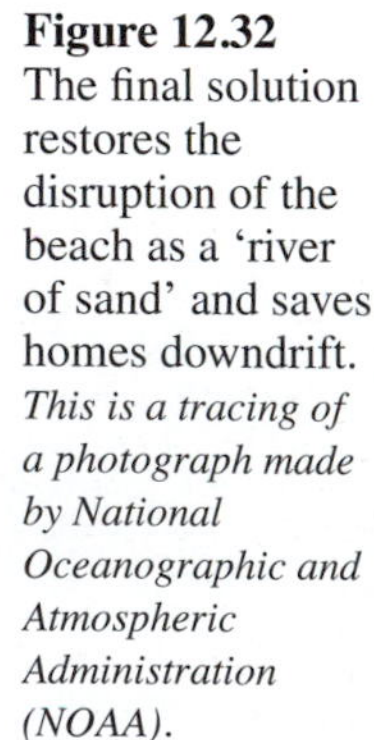

Figure 12.32 The final solution restores the disruption of the beach as a 'river of sand' and saves homes downdrift. *This is a tracing of a photograph made by National Oceanographic and Atmospheric Administration (NOAA).*

I. Hurricane Dynamics—Katrina (August 2005)

The where, when, and how of hurricane development—A single hurricane can unleash energy sufficient to provide the electrical needs of the United States and Canada for a whole year. Such huge storms are 'born-and-raised' of *heat*—the source of which is the release of **latent heat** (defined below) from vapor that is produced through the evaporation of sea water. For this reason, hurricanes develop where there is a surface layer of sea water tens of meters thick at temperatures above 80 °F. These conditions occur within a global belt roughly delineated by latitudes 20° N and 20° S. The region of western North Atlantic hurricanes is explained in the caption of Figure 12.33.

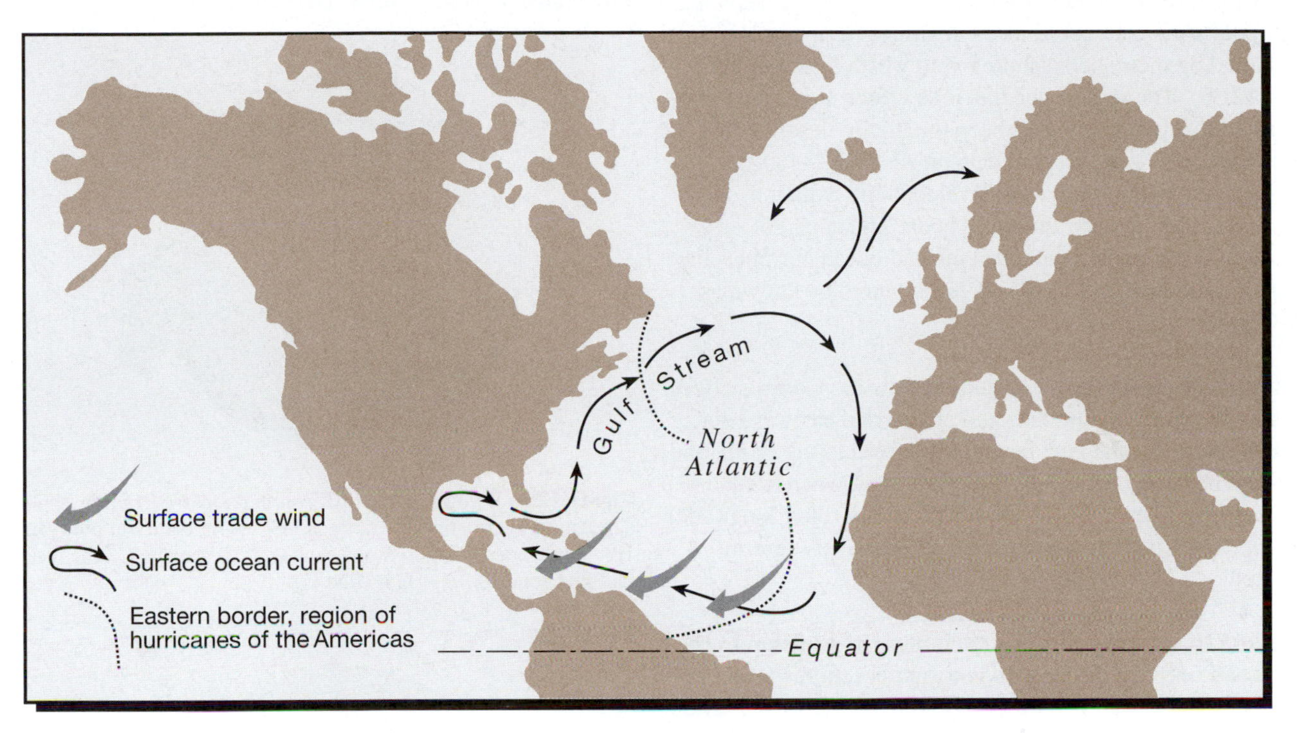

Figure 12.33 Hurricanes of the Americas are driven by the **Gulf Stream**—a clockwise cell of circulation within the North Atlantic that is driven by the westward trade winds. The southern, westward-flowing part of the Gulf Stream is heated by the tropical sun during late summer and early fall in its journey from Africa to the West Indies and the Gulf of Mexico.

In order to understand *how* hurricanes form we need to define a few concepts that you might already know about from courses in chemistry and/or physics:

Latent heat—Energy associated with the change of *phase* (as in solid, liquid, and gas) of a substance.

Latent heat of fusion (aka latent heat of melting)—Energy *released* when a substance freezes, *consumed* when a substance melts.

Latent heat of vaporization (aka latent heat of condensation)—Energy *consumed* when a substance evaporates, *released* when a substance condenses.

In the case of water-air interactions—the stuff of hurricanes—we're dealing with *latent heat of vaporization*. (Fig. 12.34).

Q12.32 When air temperatures in Florida dip slightly below freezing—thereby threatening an orange crop—owners spray their trees with water in an effort to raise the temperature of the trees a few degrees. Explain the reasoning behind this curious practice.

(c) Warm moist air cools aloft, condensatioin occurs, releasing latent heat

(b) Warm moist air rises

(a) Evaporation produces vapor with latent heat

Warm water

Figure 12.34 The phenomenon of latent heat can be viewed as a transport system. Evaporation of sea water stores heat within water vapor. As the water vapor rises, it is cooled down to the point where condensation (precipitation) occurs, which releases latent heat to the atmosphere.

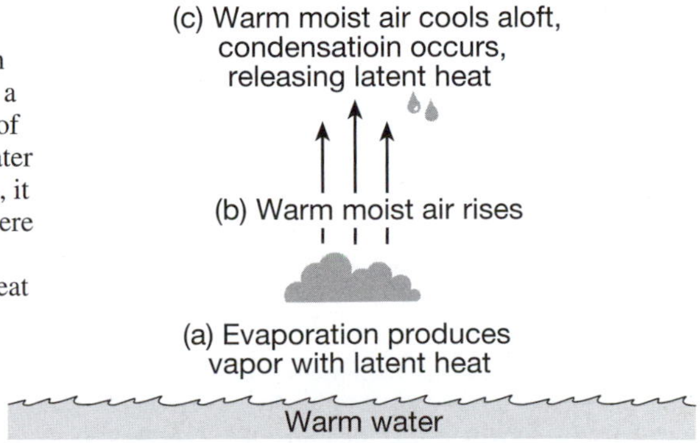

The Coriolis effect—The discussion on the previous page dealt with the geography and physics of hurricane development. The understanding of the circulation of hurricane winds requires an understanding of the **Coriolis effect**.

During World War I, the Germans introduced huge artillery pieces called 'Big Berthas,' mammoth guns that could hurl shells 15 miles high at targets 75 miles away. But they had a lesson to learn when they fired southward on Paris. Having been guided by map coordinates in aiming, gun crews were befuddled by their shells landing well west of their targets (Fig. 12.35). The reason: the **Coriolis effect**. It is ironic that it was a Frenchman who had been the first to describe this phenomenon. It was Gaspard de Coriolis (1792–1843), a French mathematician, who developed the equation describing motion over a rotating body. The apparent diversion in the paths of air currents and ocean currents relative to latitude and longitude has come to be known as the Coriolis effect.

Figure 12.35 shows the mechanics of how the Coriolis effect works. Both the gun and the shell are carried eastward via Earth's rotation—the gun, because it's fixed to the Earth, and the shell because of its eastward component when fired. But points on the ground along the meridian (their longitude) move progressively faster, and, so, progressively 'outrun' the shell.

Q12.33 In this model, what variables (in addition to the speed of Earth's rotation) do you suspect might affect the distance that a shell fired directly southward would land west of the gun's meridian? *Hint:* Think of things that affect the path of a thrown object (e.g., a ball).

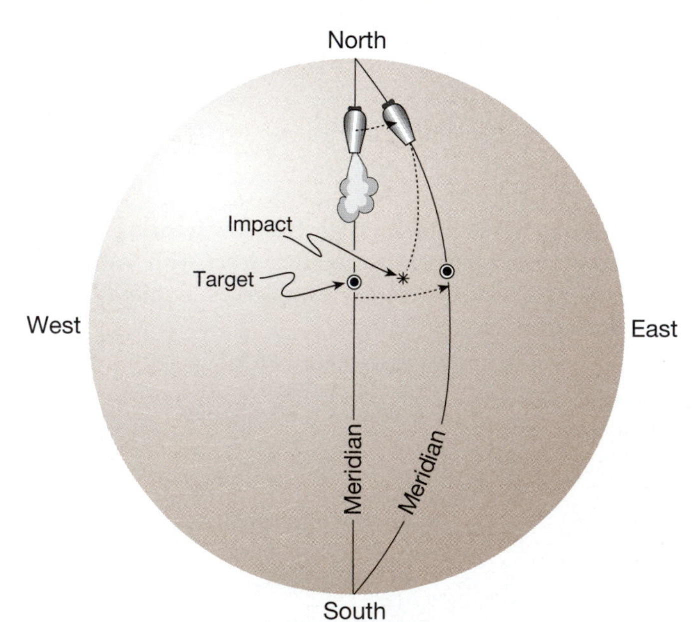

Figure 12.35 This is a schematic representation of how the Coriolis effect causes a shell to be diverted from its intended path, relative to a meridian (aka line of longitude).

If you find it difficult to envision the Coriolis effect, you can demonstrate it to your satisfaction with a disk of cardboard tacked through its center to a surface below. Then Follow instructions in captions to Figures 12.36 and 12.37.

You and a classmate can demonstrate the Coriolis effect at your desk. One of you can rotate a sheet of paper about its center while the other tries to draw a straight line from the center outward. Voilá.

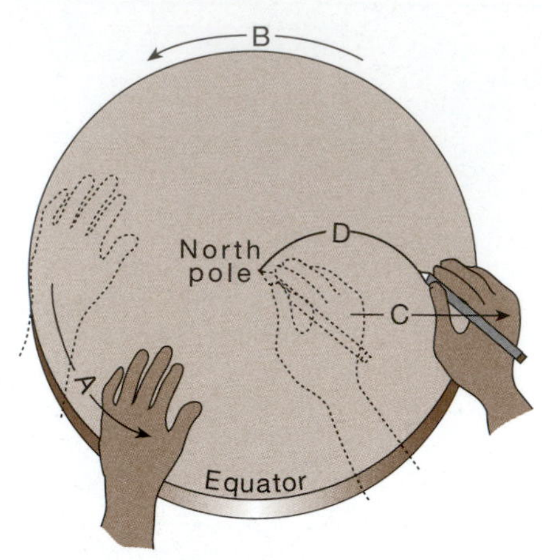

Figure 12.36 If you rotate a disk *counterclockwise* (arrow A)—simulating the rotation of Earth as in viewing the north pole (arrow B)—while attempting to draw a straight line (arrow C), the resultant line curves to the *right* (line D).

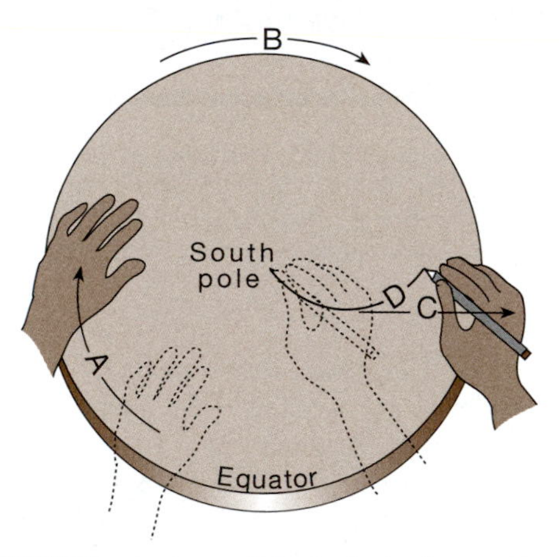

Figure 12.37 If you rotate a disk *clockwise* (arrow A)—simulating the rotation of Earth as in viewing the south pole (arrow B)—while attempting to draw a straight line (arrow C), the resultant line curves to the *left* (line D).

The Coriolis rule—In the Northern Hemisphere, air currents and ocean currents tend to turn to the right (line D in Figure 12.36), whereas in the Southern Hemisphere, currents tend to turn to the left (line D in Figure 12.37).

Q12.34 It doesn't happen in the real world, but imagine a current of air or water flowing *exactly* along the equator (or along any other line of latitude). What do you suppose would be the Coriolis effect, if any?

The Coriolis effect and hurricane motion—The familiar symbol of a hurricane that appears on weather maps is a pinwheel that graphically indicates counter-clockwise rotation.

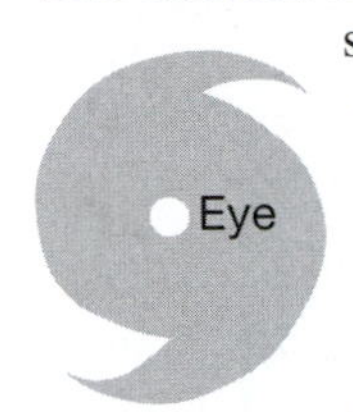

So why the counter-clockwise rotation? Answer: Because (a) a hurricane develops within a low-pressure system (Fig. 12.38A), (b) a low-pressure system draws air inward, and (c) in the Northern Hemisphere converging air veers to the right, which (d) results in a cell rotating counter-clockwise (Fig. 12.38B).

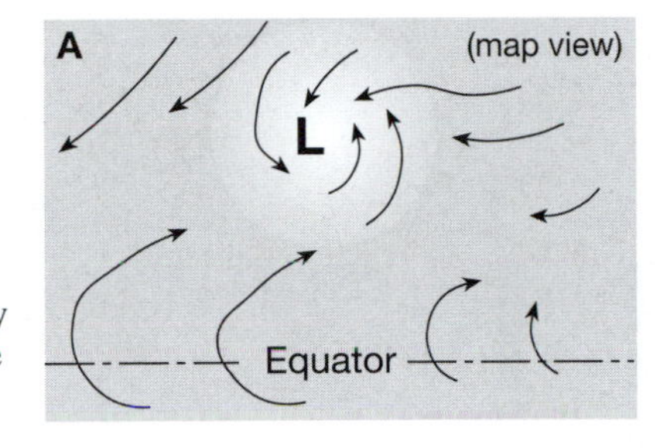

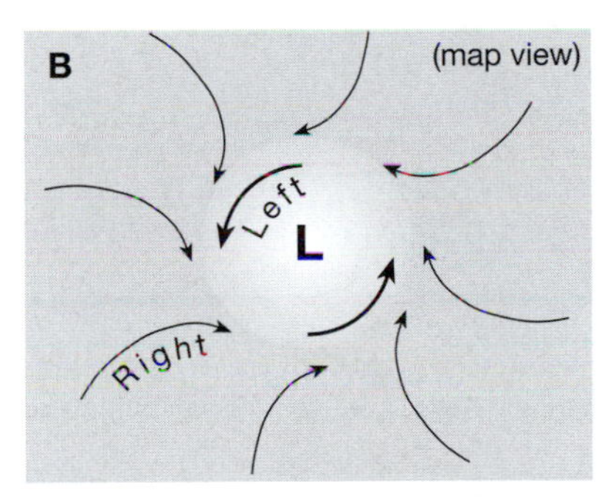

Figure 12.38 A. Opposing trade winds cause air to swirl, thereby producing a low-pressure system that draws air inward. **B**. In the Northern Hemisphere, converging air veers to the right (light arrows), which results in a cyclonic cell rotating counter-clockwise (heavy arrows).

Q12.35 Sketch a hurricane (or cyclone) symbol as it would appear on a map of offshore weather at Sydney, Australia.

Q12.36 So why the familiar 'eye' in the center of a hurricane—the volume of air that is practically devoid of wind and rain? *Hint:* No doubt you have heard of this 'slinging' force. The name by which it's known?

Timeline of Hurricane Katrina at…
http://en.wikipedia.org/wiki/hurricane_katrina

Outline of hurricane development:

• Cyclonic motion commonly begins as opposing trade winds swirl around one another, creating a low-pressure system (Fig. 12.38A).

• The rotating low-pressure system pulls air toward its center (Fig. 12.38B). causing hot moist air to rise Fig. 12.39A).

• At higher elevations the moisture condenses. Torrential rain ensues.

• The latent heat released through condensation further warms the rotating air, causing it to become even lighter and rise even faster.

• As more moist tropical air moves in to replace the rising air, more condensation occurs. So air inside the storm rises higher, faster, and more broadly (Fig. 12.39B).

Q12.37 Quite simply, why don't hurricanes develop over land?

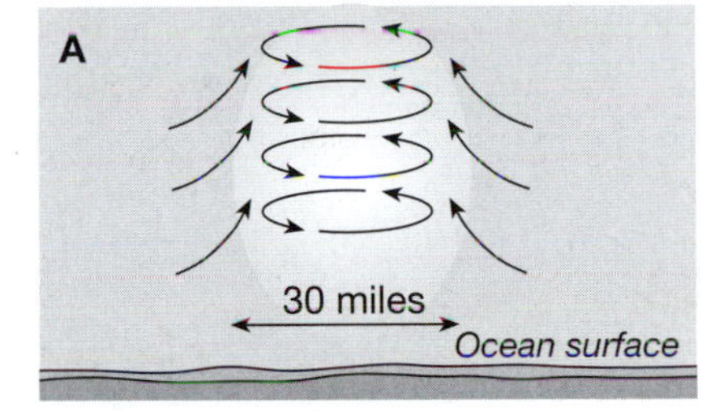

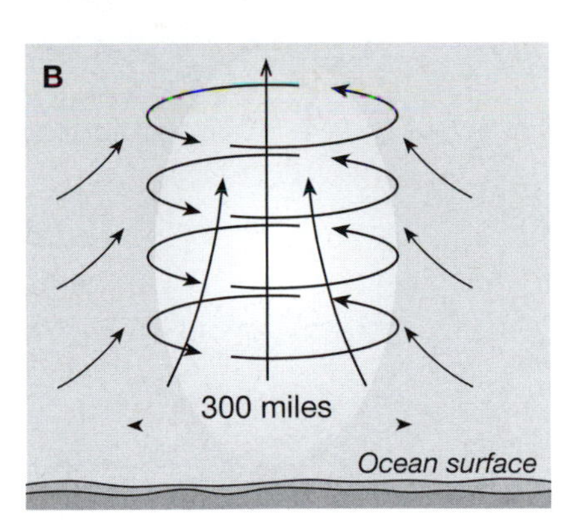

Figure 12.39 This figure illustrates the development of a hurricane—which is invariably over an ocean.

Hurricanes present a **triple-threat** to coastal areas through (a) wind speed, (b) storm surge (high water), and (c) flooding (through high water and levee failure).

Storm surge is driven by wind and low atmospheric pressure that accompany a hurricane.

Q12.38 Some scientists argue that there is another factor that contributed to the storm surge of Hurricane Katrina. What is that factor? *Hint:* See loss of 'speed bumps' in first paragraph on page 217.

Enlarging on the Coriolis effect—Figure 12.40 shows a hypothetical hurricane—on which wind speeds and the speed of the hurricane are plotted—at a location within the path taken by Hurricane Katrina.

Q12.39. During Katrina, parts of Mississippi suffered greater winds than did New Orleans. Why? Compute net wind speeds east and west of the eye in Figure 12.40, and refer to those net wind speeds in your answer.

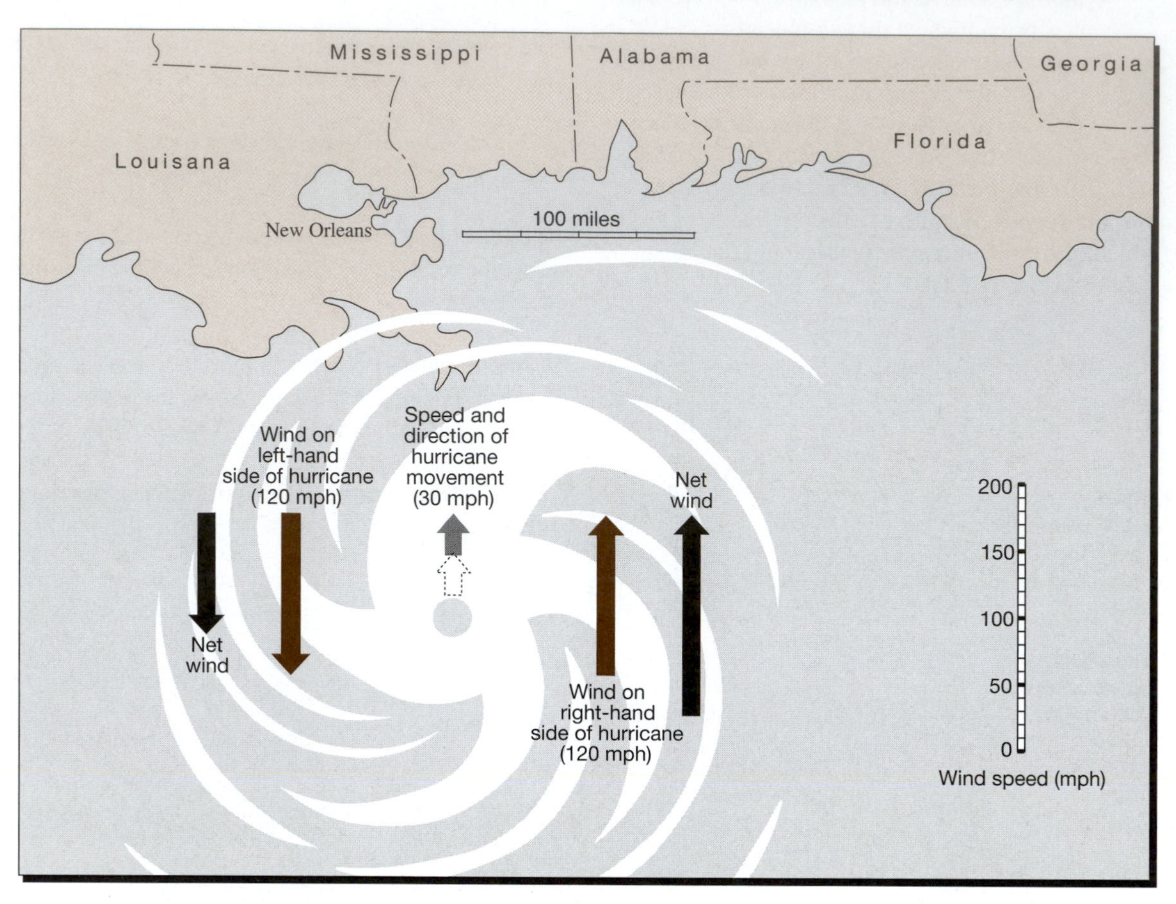

Figure 12.40 This shows a hypothetical hurricane approaching the Gulf Coast. Direction and speed of the wind are indicated by the orientations and lengths of arrows—as with mathematical vectors.

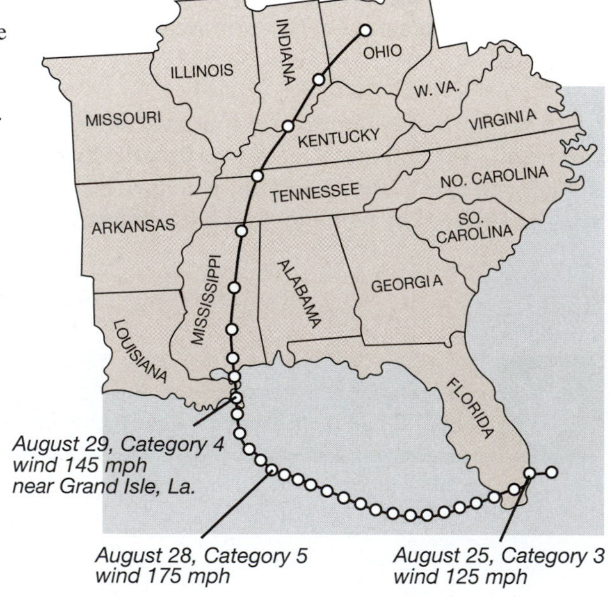

Figure 12.41 This figure shows the actual path of Hurricane Katrina. Notice the greater density of dots (symbolizing intensity) over water.

Figure 12.41 shows the path of Hurricane Katrina and wind speeds from August 25 to August 29. Greater wind speeds have slammed our Gulf Coast at other places, other times. It was levee failure and flooding that created the tragedy of 2004 in New Orleans.

Q12.40 Notice in Figure 12.41 that—like all other hurricanes—Katrina diminished over land. Again, what is the missing essential ingredient on land?

______________________________ (Student's name)

________ (Day) ________ (Hour)

______________________________ (Lab instructor's name)

ANSWER PAGE

12.34

12.35

12.36

12.37

12.38

12.39

12.40

Intentionally Blank

13 Atmosphere and Oceans

Topics

A. What is mass? What is Newton's first law of motion? What is density, (a) expressed in terms of mass and volume, and (b) expressed in terms of the mass of a substance relative to that of an equal volume of pure water at 4 °C?

B. How does temperature affect the moisture-holding capacity of air? So why does condensation occur at higher elevations?

C. What is adiabatic heating and cooling? How does an air conditioner apply adiabatic principles? How does adiabatic heating explain chinook winds and Santa Ana winds?

D. What is a Hadley cell? Ferrel cell? Why do great deserts occur in the vicinity of 30° latitude?

E. What are the factors that promote vertical circulation of seawater? How did one of these factors assist WW II German submarines in their Mediterranean campaign?

F. What is the pycnocline? What two variables in seawater affect the pycnocline? What process explains the tendency toward oxygen depletion in deep ocean water? By what process in high latitudes does atmospheric oxygen find its way into deep ocean water?

G. What is the Coriolis effect? How does the Coriolis effect explain Ekman transport? How does Ekman transport explain upwelling and downwelling along seacoasts? What is the photic zone, and what does it have to do with organic recycling? How do these phenomena account for the fertilizer industry in coastal Peru?

H. What is El Niño? Why does it bring flooding and erosion to western parts of the Americas? How did El Niño contribute to the near demise of the Peruvian anchovy industry?

A. Earth's four spheres

Lithosphere, hydrosphere, atmosphere, and biosphere. These four systems comprise the all-inclusive system—*Earth*. Interesting things occur at interfaces among these 'spheres.' You can probably imagine what some of these things are from a cursory examination of Figure 13.1.

Q13.1 List a substance essential to human life that is provided—either directly or indirectly—by each of these four spheres.

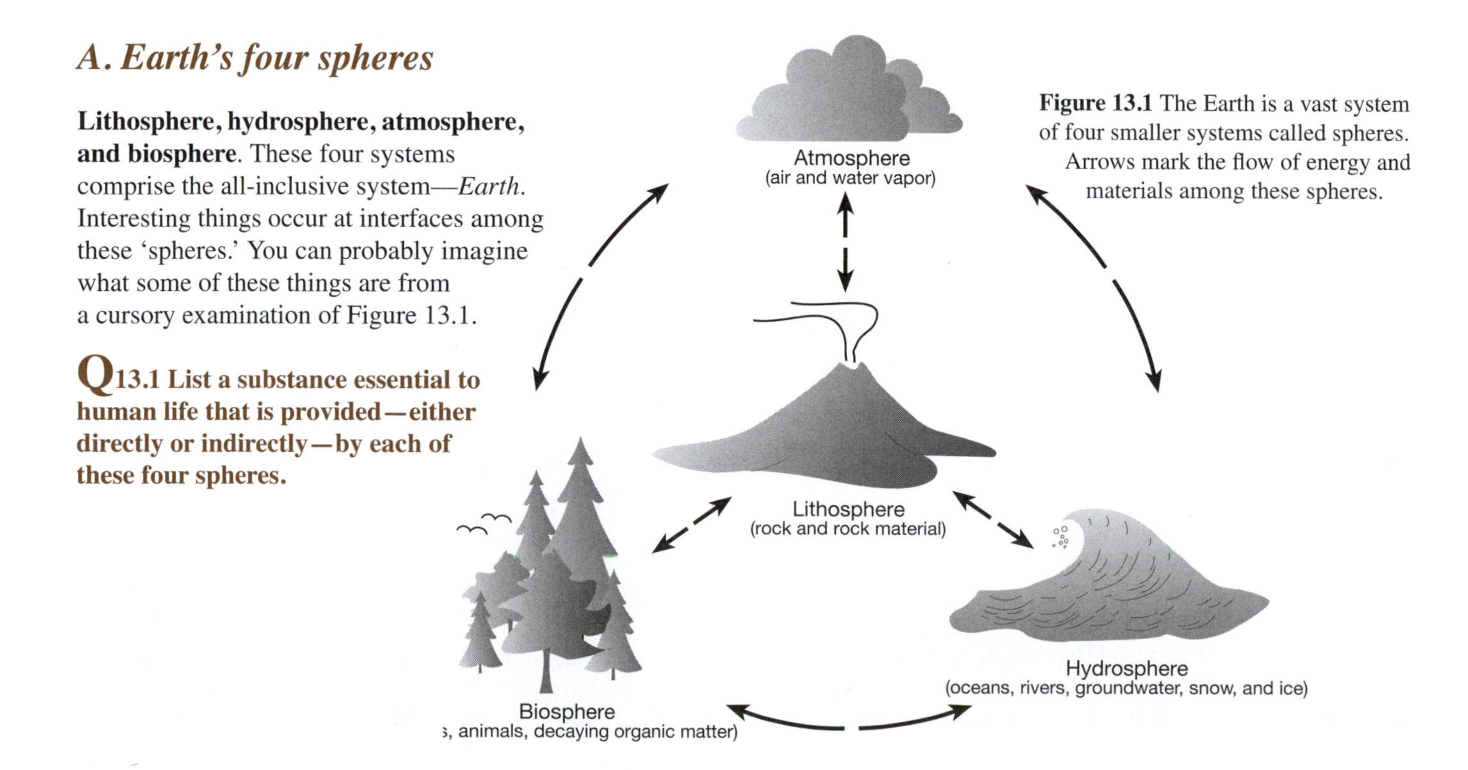

Figure 13.1 The Earth is a vast system of four smaller systems called spheres. Arrows mark the flow of energy and materials among these spheres.

B. Importance of density

Atmosphere and oceans—Although all four spheres shown in Figure 13.1 interact with one another to some degree, clearly the most dynamic coupling is that between the atmosphere and the oceanic part of the hydrosphere. There are few phenomena associated with physical oceanography that are not affected by the atmosphere, and there are few atmospheric phenomena in which the ocean doesn't play an important role (Fig. 13.2).

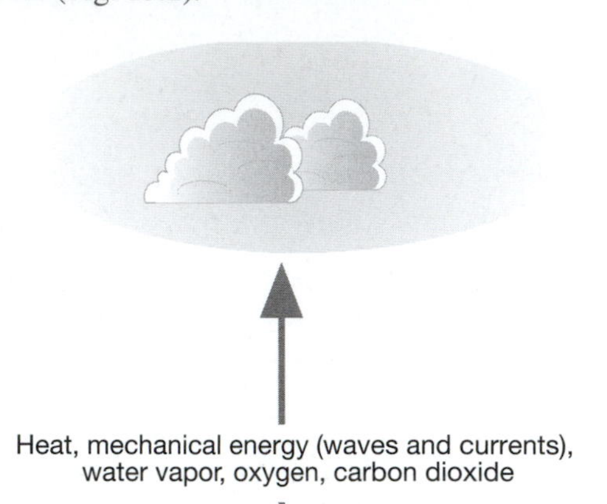

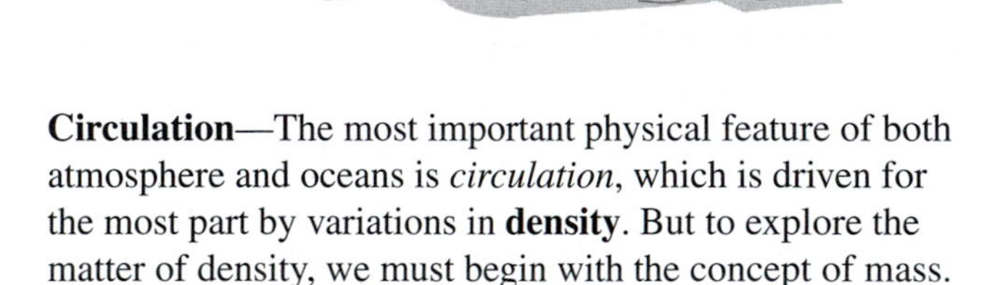

Figure 13.2 Important exchanges between the atmosphere and the hydrosphere include heat, mechanical energy, water vapor, and gases.

Circulation—The most important physical feature of both atmosphere and oceans is *circulation*, which is driven for the most part by variations in **density**. But to explore the matter of density, we must begin with the concept of mass.

Mass is the amount of matter (i.e., the total of electrons, protons, neutrons, etc.) within an object (be it solid, liquid, or gas). Mass is that quality that determines the **inertia** of an object, and is the subject of Sir Isaac Newton's first law of motion, which states: (a) *a motionless object will remain motionless until a force is applied to it; and (b) a moving object will continue moving at the same speed and in the same direction until a force is applied to it.* Objects are stubborn. They tend to continue doing what they're doing.

Q13.2 Power brakes and power steering assist motorists in dealing with inertia. Explain this statement in the context of Newton's first law.

Q13.3 (A) Do you suppose that an object aboard a space shuttle exhibits 'weight' to the same degree that it does on Earth? (B) How about the object's inertia?

The mass of an object can be expressed in grams (g), and its **volume** (i.e., the amount of space the object occupies) can be expressed in cubic centimeters (cc). A gram is arbitrarily defined as the mass of one cubic centimeter of pure water at 4 °C. (Why at 4 °C? Because that's the temperature at which water is most dense, so that is taken as the standard.)

Density can be expressed as *mass (grams)* per volume *(cubic centimeters):*

$$Density = \frac{Mass\ (g)}{Volume\ (cc)}$$

So one can compute density (D) by dividing mass (M) by volume (V):

$$D = \frac{M}{V}$$

Q13.4 (A) Consider an object with a mass of 100 grams and a volume of 20 cubic centimeters. What is its density? (The number you compute is the mass of 1 cubic centimeter of the substance that comprises this object.) (B) How much greater is the density of this substance than that of pure water at 4 °C?

Under the influence of gravity, low-density material tends to rise, and high-density material tends to sink. Inasmuch as the *heating* of an air mass causes it to expand, thereby *decreasing* its density, warm air tends to rise.

In Figure 13.3, air masses A and B are equal in total mass particles, but mass B is greater in volume, so is less dense. Therefore, it rises. The boundary between the warmer less-dense air and the cooler more-dense air behaves like an optical lens, bending light as it passes from air of one density into air of a different density. This optical distortion accounts for mirages reported from desert landscapes.

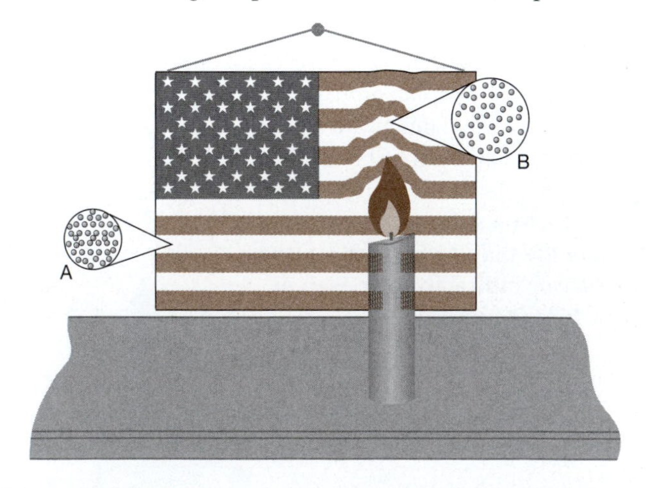

Figure 13.3 Here, stripes viewed through hot air rising from a candle appear to be distorted by the lens effect. Turbulence within the rising air causes the image to wiggle.

C. Atmosphere, temperature, and pressure

But first, a counterintuitive fact about density and air:

Q13.5 Which do you think is heavier, dry air or humid air? Record your answer on the Answer Page, and then read ahead to see if your answer is correct.

Intuitively, one might think that humid air is heavier (i.e., more dense) than dry air, but such is not the case. Molecular oxygen (O_2) and nitrogen (N_2) make up 98% of the mass of the atmosphere. The molecular weights of oxygen and nitrogen are 28.01 and 32.00 respectively, with the average weight of dry air being 28.50. Water vapor has a molecular weight of only 18.10. Therefore, as water vapor is added to air, the percent of other gases is diminished proportionately, so the molecular weight of the humid air is lessened.

Q13.6 What element do you suppose accounts for such a low molecular weight for water vapor?

Moisture-holding capacity of air—Warm air can hold more water vapor than can cool air. And, as you know, people go to the mountains in summer to escape the heat of the lowlands. So temperature must decrease with increasing elevation.

Q13.7 So what happens to water vapor held within warm air when that air climbs in elevation? *Hint:* See graphic details in Figure 13.4.

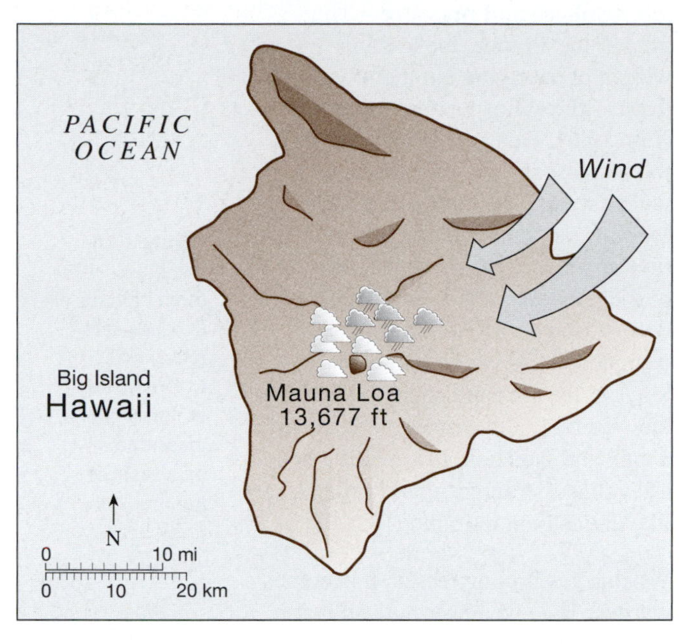

Figure 13.4 Wind drives warm moist air from the Pacific Ocean upward thousands of feet in its traverse across Mauna Loa volcano.

Atmosphere's temperature—You are in flight and you hear over the intercom, "Good afternoon, ladies and gentlemen. This is your captain speaking. We are flying at an altitude of 20,000 feet, and the outside air temperature is __ °C."

Q13.8 Judging from the graph in Figure 13.5, which temperature did the captain report, -40 or -60 (°C)?

Earth's atmosphere is divisible into intervals, the lowest of which are the **troposphere** (Gr. *tropos*, turning) and the **stratosphere** (L. *stratum*, layer) (Fig. 13.5).

Note—**Altitude** signifies the height of a point within the atmosphere above sea level, whereas **elevation** signifies the vertical distance between sea level and a land surface.

Notice in Figure 13.5 the curious reversal in the graph of temperature as a function of altitude. The reason for this reversal is that the principal source of heat for one of the intervals is the Earth, whereas the principal source of heat for the other is the sun.

Q13.9 (A) Which interval appears to derive its heat from the Earth?

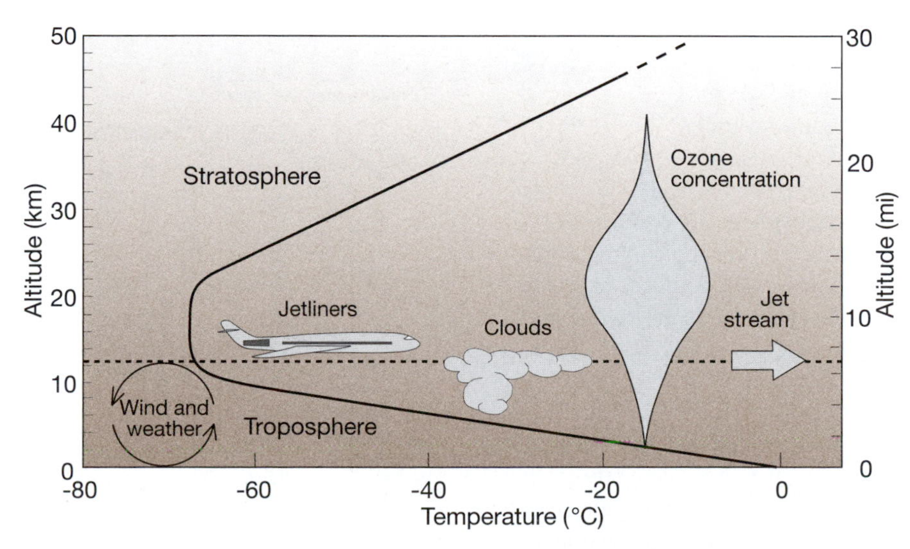

Figure 13.5 The atmosphere is divisible into intervals. Turbulence and weather occur within the lowest interval—the troposphere, which contains 80% of the atmosphere's mass.

Atmosphere and pressure—The greater the altitude, the less the weight of overlying air, therefore the less the atmospheric pressure (Fig. 13.6). This is analogous to the increase in pressure with depth within water. The increase in *psi (pounds per square inch)* as a function of water depth presents one of the greatest hazards associated with SCUBA diving. A SCUBA tank is designed to deliver air to a diver's lungs at the pressure equal to that of the water. Should a diver inhale from a tank and then hold her breath while ascending, her lungs would expand like the balloon in Figure 13.6.

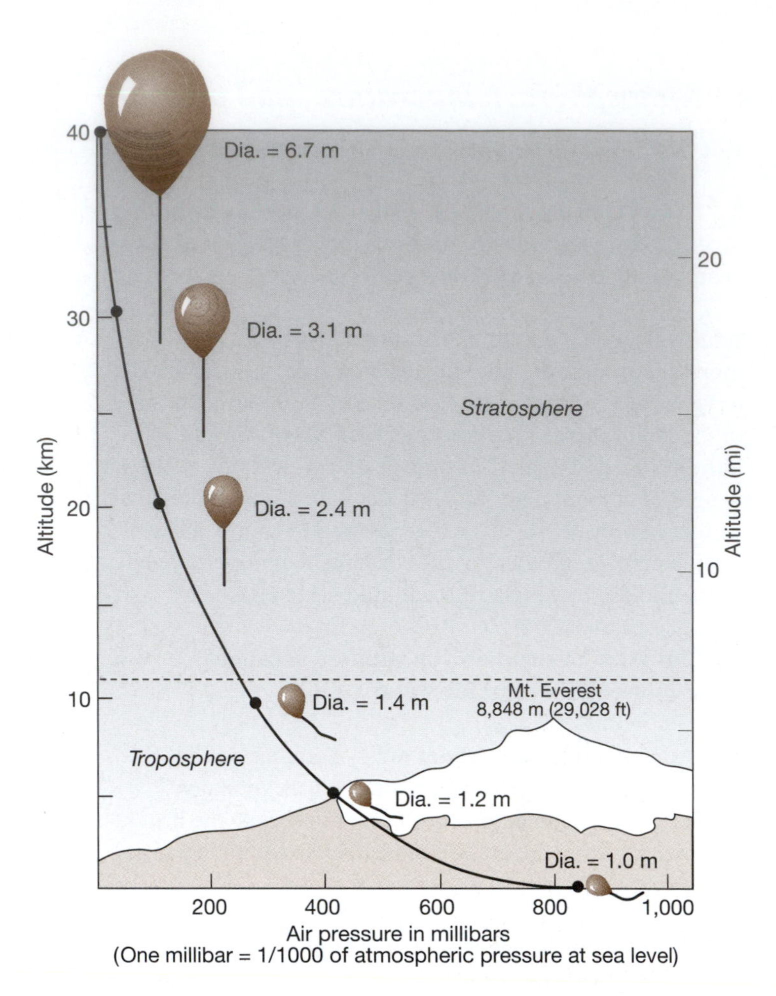

Figure 13.6 A helium-filled balloon expands as it rises, owing to a reduction in pressure measured on a scale of hundreds of millibars.

When a gas is compressed, it becomes warmer. Have you ever noticed that a bicycle pump grows warm as you pump, compressing the air within it?

Q13.10 So what do you suppose happens to the temperature of gas when it expands? *Hint:* You might have sensed this when releasing pressurized spray/foam from a can.

This phenomenon is called **adiabatic heating** and **adiabatic cooling** (Gr. *adiabatos*, impassable; i.e., heat does not pass in or out of the gas.)

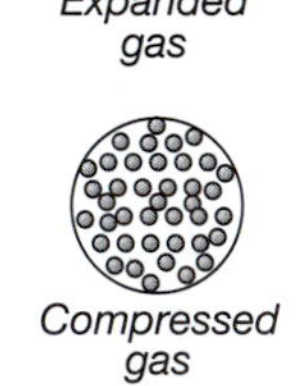

A graphic metaphor: Compression results in more frequent collisions among molecules per unit volume, so the gas becomes warmer.

The troposphere air conditioner—Adiabatic heating and cooling might seem new to you, but these processes are what make refrigerators and air conditioners work. Consider a common home air conditioner (Fig. 13.7):

1. Pressure produced by a compressor outside the house converts gas into liquid. The surrounding air absorbs the heat produced by this conversion, and a fan is required to blow the heated air out through an exhaust system.

2. The compressor pumps liquid into the house via a pipe that includes a nozzle, which—like an atomizer—converts liquid back into gas.

Q13.11 What happens to the temperature of the liquid when it is converted from liquid to gas? *Hint:* This is the opposite of compression.

3. A fan pulls warm room air across the chilled coils—thereby cooling the air and driving it into ducts that convey it to other parts of the house.

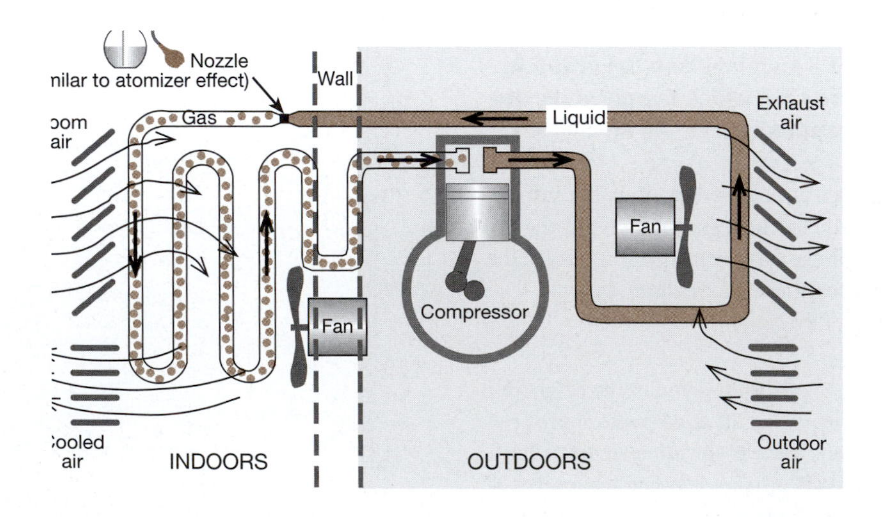

Figure 13.7 The inner-workings and hidden mechanisms of a common air conditioner convert warm outside air to cool inside air with a compressor and fans.

Natural adiabatics—Dry air generally cools at a rate of 1.0 °C per 100 m (5.5 °F/1,000 ft) of increased elevation.

Q13.12 Does this rate agree with that shown in Figure 13.8?

In the case of dry air, the vertical *decrease* in temperature on the *windward* side of a mountain range is the exact reciprocal of the *increase* on the *leeward* side. So, in Figure 13.8, the temperature of air is 74.5 °F at elevation 1,000 ft on both sides of the mountain range. The model is symmetrical. But if that's the rule, why are valleys on the west (windward) sides of mountain ranges in California pleasant, whereas valleys on the east (leeward) sides of those ranges are unbearably hot? The answer: condensation.

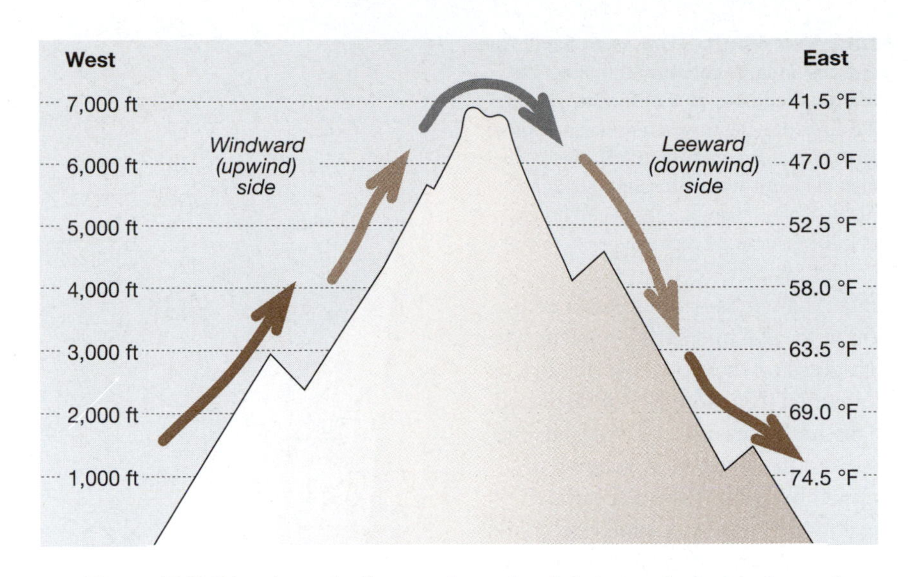

Figure 13.8 This schematic diagram shows dry air being pushed up-and-over the Rocky Mountains (for example) by prevailing winds from the northwest.

Condensation of water vapor complicates adiabatic cooling as follows:

Note—Water vapor is water in its gaseous phase; not to be confused with small water droplets that comprise clouds.

You probably know that evaporation is an **endothermic** reaction, i.e., boiling water absorbs heat from its surroundings.

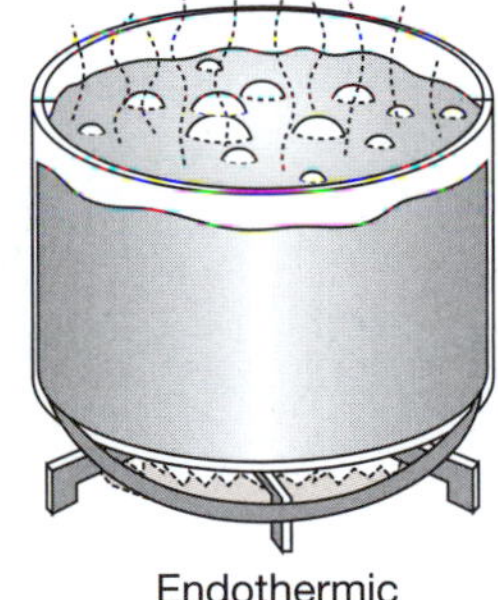

Figure 13.9 Evaporation is an *endothermic* reaction.

But do you know that the reverse process—*condensation*—is an **exothermic** reaction, i.e., condensing water releases heat to its surroundings?

Figure 13.10 Condensation is an *exothermic* reaction.

Rising humid air within which condensation occurs is warmed by the process, lessening the effect of adiabatic cooling. The result: significantly warmer temperatures at lower elevations on the leeward sides of mountains. For example, imagine the situation illustrated in Figure 13.11.

On the windward side of this range…
(1) Air at 60 °F rises from sea level on the windward side.

(2) From sea level to 10,000 ft, adiabatic cooling *decreases* temperature 5.5 °F/1,000 ft.

(3) From 4,000–10,000 ft, condensation *increases* temperature 2.3 °F/1,000 ft.

On the leeward side of the range…
(4) From 10,000 ft to sea level, adiabatic warming increases temperature 5.5 °F/1,000 ft.

Q13.13. (A) What is the temperature of the air at sea level on the leeward side? (B) How much warmer is this than the air's beginning temperature on the windward side?

Figure 13.11. Again, air being pushed up-and-over a mountain range, but in this case the air is humid and condensation begins to occur on the windward side at elevation 4,000 ft.

Chinook winds—These extraordinarily warm winds descending from mountain ranges in the West are called *chinook* winds, which can be a blessing to ranchers in the Great Plains, where they melt snow and expose grasses for livestock. But there can be a downside, too. During the 1988 Winter Olympics in Calgary, Alberta, chinook winds melted snow in the midst of the Games, forcing postponement of ski competition until new snowfalls replenished the slopes.

Santa Ana winds—Dry, swift Santa Ana winds are in large measure to blame for disastrous wildfires in California, e.g., the 1993 fires that destroyed a thousand homes from Laguna Beach to Santa Barbara. These winds also swept fires through Oakland Hills in 1991, burning a thousand homes and killing some 24 residents.

Santa Ana winds—which most common in the fall and spring—occur when high pressure develops over the Rocky Mountains (Fig. 13.12). Air displaced by the high pressure cell descends the slopes to the southwest, where it is warmed by compression at lower elevations. Santa Ana winds are similar to chinooks, but differ in that they are driven laterally by high pressure, rather than by prevailing winds.

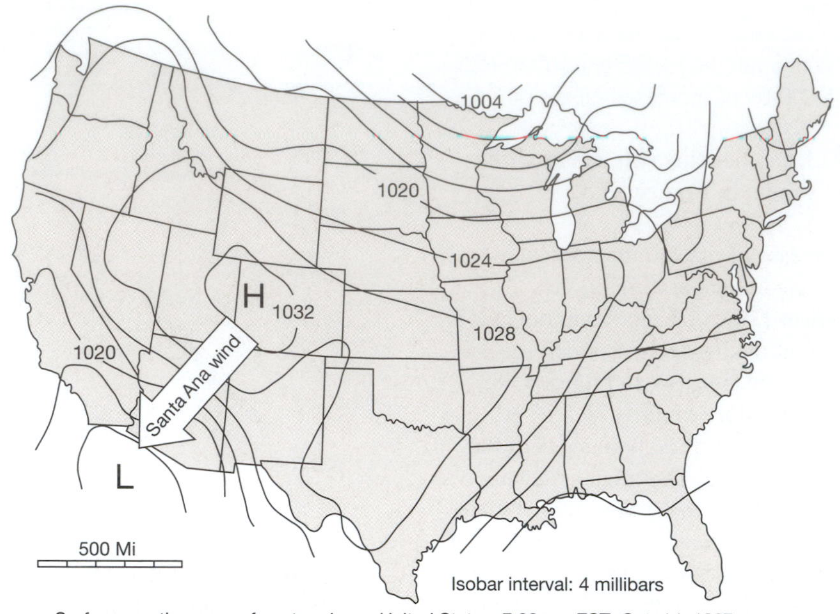

Figure 13.12 Air pressures plotted on surface weather maps are adjusted to sea level, so they exhibit relatively weak *horizontal pressure gradients*, with isobars differing on a scale of *tens* of millibars. Recall from Figure 13.6 (page 232) that *vertical pressure gradients* are much stronger, with isobars differing on a scale of *hundreds* of millibars.

Q13.14 Imagine that a Santa Ana wind originates in October on a 8,000-foot high plateau in southern Utah. Air temperature at this particular time is 60 °F. That air warms at a rate of 5.5 °F/1,000 ft as it descends to sea level along the California coast. What is the temperature of the wind reaching coastal California?

Southern California surfers rush to beaches during Santa Ana winds to enjoy unseasonably warm weather, only to find surface water surprisingly cold.

Q13.15. What! The arrival of a hot Santa Ana wind produces a chilling of coastal Pacific water. How can this be? *Hint:* The answer to this question is provided by the last of the four models for vertical circulation of seawater shown in section E on page 236: *Global ocean circulation.*

Note—Although there are other theories, most agree that the term *Santa Ana wind* originated when an Orange County, California newspaper reported such a wind emerging from nearby Santa Ana Canyon.

D. Global atmospheric circulation

As we have already seen, the heating of air causes it to expand, thereby making it less dense, which causes it to rise—displacing cooler, more dense air. This phenomenon, known as **convection**, lifts hot-air balloons and accounts for the common problem of uneven heating within multistory homes.

The sun's heat is most effective in warming the atmosphere in equatorial regions, so air tends to rise at low latitudes and sink in polar regions (Fig. 13.13). This simple circulation is called a **Hadley cell**, named for George Hadley (1685–1768), the British physicist who first envisioned this mechanism.

Q13.16. On Answer Page 245, sketch Hadley cells for the other three quadrants on this two-dimensional map.

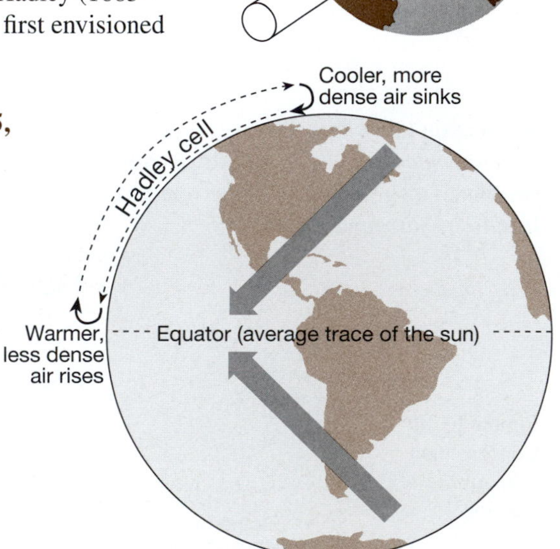

Figure 13.13 Hadley cells would be the simple consequence of solar heating, were it not for the fact that Earth is rotating. (The bold arrows will be explained later.)

Hadley did not think that winds would simply move in north-south directions, however. He believed that Earth's rotation from west to east deflects air to the right in the Northern Hemisphere and to the left in the Southern Hemisphere (re: the large arrows in Figure 13.13). An elaboration of Hadley's model, which took Earth's rotation into account, was proposed by U.S. meteorologist William Ferrel (1817–1891) in 1865 (Fig. 13.14). Ferrel's model accounts for the world's great deserts in the vicinity of 30° latitude (e.g., North American and Sahara deserts at 30° N and the Australian desert at 30° S). Wind directions with east-west components (e.g., Northeast trades, Southeast trades, and Westerlies) reflect Earth's rotation.

Ferrel's model derives in large measure from Earth's 24-hour period of rotation. Other planets have different *angular velocities* (e.g., Jupiter turns on its axis once every 11 hours).

Q13.17 How many atmospheric cells do you suppose Jupiter has in each hemisphere—fewer, equal, or more than that of Earth? *Hint:* Think of cells other than the simple Hadley cell as by-products of Earth's rotation.

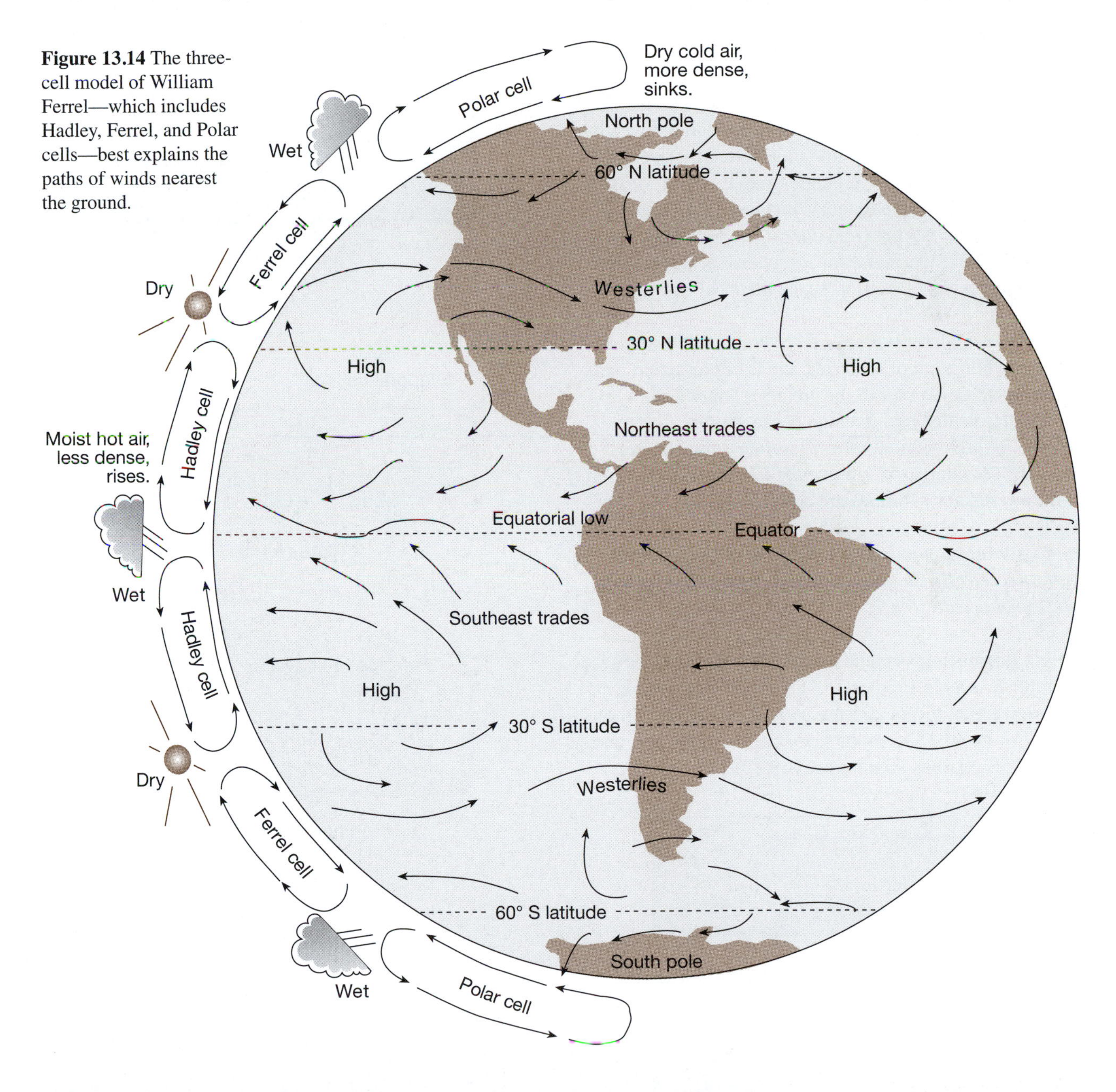

Figure 13.14 The three-cell model of William Ferrel—which includes Hadley, Ferrel, and Polar cells—best explains the paths of winds nearest the ground.

National Oceanic and Atomospheric Administration home page is at www.noaa.org

E. Global ocean circulation

Density revisited—Recall that global atmospheric circulation is in large measure driven by a *single factor*: latitudinal differences in air density resulting from differences in solar heat energy—complicated by Earth's rotation. But global ocean circulation is driven by *three factors*, two of which affect density:

Internal (*affecting density*)
- Temperature (warmer water is less dense).
- Salinity (more saline water is more dense).

External
- Surface winds drag surface waters.

Salinity is the percent by weight of dissolved elements in seawater—elements that were dissolved through the weathering of rocks of Earth's crust and transported by rivers to the ocean during eons of geologic time. Evaporation of seawater during this same time has increased salinities to their present levels. Sodium and chlorine, which make up common salt (NaCl), account for some 85% of the major dissolved elements in seawater, hence the label *salinity*.

Salinity of ocean water is measured in grams of dissolved elements per kilogram of seawater, and is expressed in parts per thousand (‰). To repeat, the greater a water's salinity, the greater its density. Average ocean salinity is 35‰. In latitudes of high-rainfall tropics, surface salinities are around 34.5‰, whereas in latitudes of low rainfall and high evaporation, surface salinities are around 36.7‰.

Note—Evaporation and freezing separate pure H_2O from seawater, increasing the salinity/density of residual water, thereby causing it to sink.

Processes promoting vertical circulation of seawater—Each of the vessels in Figures 13.15A and 15B contains water in vertical motion indicated by arrows. As viewed on the page, Figure 13.15A exhibits *clockwise* circulation, and Figure 13.15B exhibits *counterclockwise* circulation.

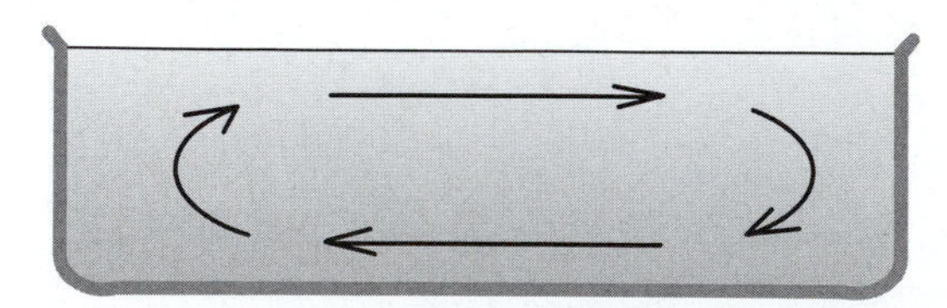

Figure 13.15A Clockwise vertical circulation of water.

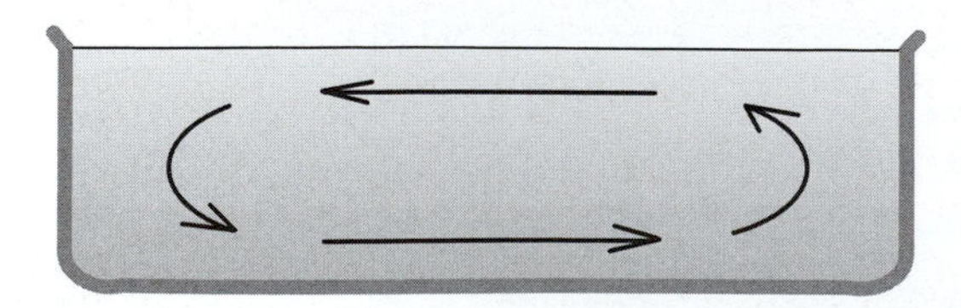

Figure 13.15B Counterclockwise vertical circulation of water.

The following four vessels (Figs. 13.15C through 13.15F) schematically represent processes that promote the vertical circulation of water within ocean basins.

Q13.18. (A) Sketch a vertical circulation cell (like that in either Figure 13.15A or 13.15B) within each of the four vessels below. (B) Then, on Answer Page 245, record the label *clockwise* or *counterclockwise* for each of the Figures 13.15C through 13.15F.

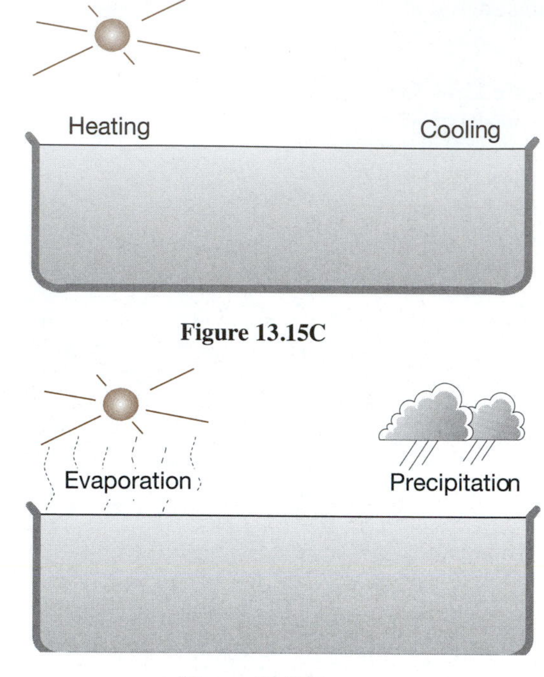

Figure 13.15C

Figure 13.15D

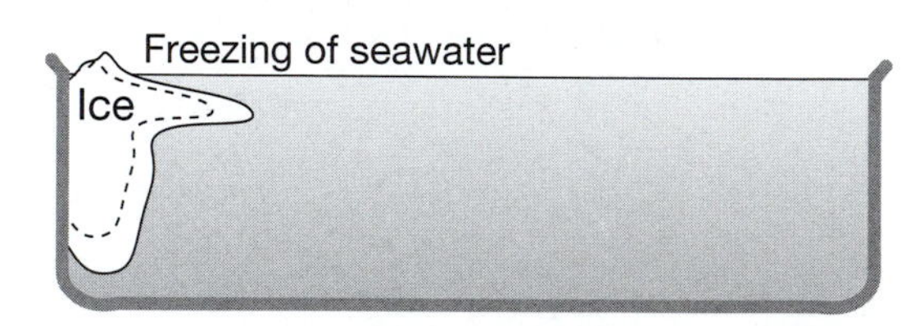

Figure 13.15E

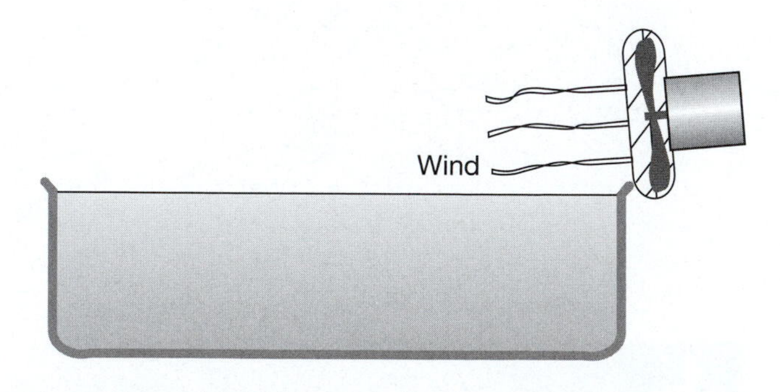

Figure 13.15F This model explains the surprisingly cold surface coastal waters accompanying Santa Ana winds of California.

A story that emerged from World War II claimed that German *unterseebootes*, or U-boats—in an effort to escape detection by the British military presence at Gibraltar—were able to quiet their engines and stealthily ride a current of Atlantic Ocean water into the Mediterranean Sea. Then, after completing their missions, they could ride a current of Mediterranean water back into the Atlantic. What could be the explanation for these two opposing currents?

Because of its mid-latitude warm and dry climate, the Mediterranean Sea is a huge evaporating basin (Fig. 13.16). The loss of water through evaporation exceeds the total of fresh water entering via rainfall and rivers, so the salinity increases to about 38.6‰, i.e., 10% higher than that of normal seawater.

Q13.19 (A) So what happens to the surface water made hypersaline through evaporation? *Hint:* Recall how salinity affects density and the resultant circulation. (B) Why doesn't the Mediterranean Sea become a lifeless body of water like the Dead Sea (Fig. 13.17)? *Hint:* Evaporation causes the water level in the eastern Mediterranean to be generally 15 cm (6 in) lower than that at the Strait of Gibraltar. So why isn't the water level in the eastern Mediterranean even lower?

Q13.20 Try modeling this complexity of currents in the diagram on Answer Page 245 by sketching the relative depths of these two currents where they enter and leave the Mediterranean at the Strait of Gibraltar.

Note—The rate of flow of high-saline water out through the Strait of Gibraltar is approximately equal to that of the lower Mississippi River.

Figure 13.16 The Strait of Gibraltar is not only a bottleneck as viewed on a map, but a bottleneck as viewed in cross-section as well—owing to the presence of a shallow sill.

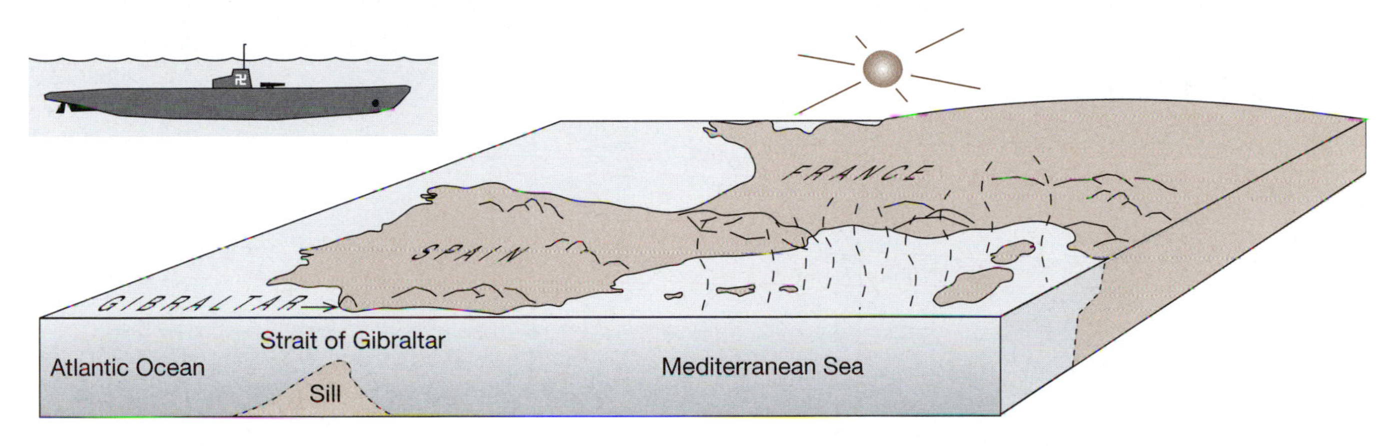

An effect of high rates of evaporation in the Mediterranean region is apparent in the excessive salinity of the Dead Sea (Fig. 13.17). Can you think of a body of water in the lower 48 states with similar salinity?

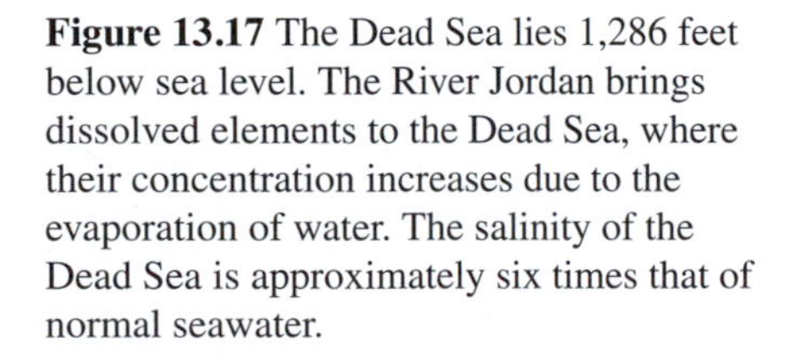
Figure 13.17 The Dead Sea lies 1,286 feet below sea level. The River Jordan brings dissolved elements to the Dead Sea, where their concentration increases due to the evaporation of water. The salinity of the Dead Sea is approximately six times that of normal seawater.

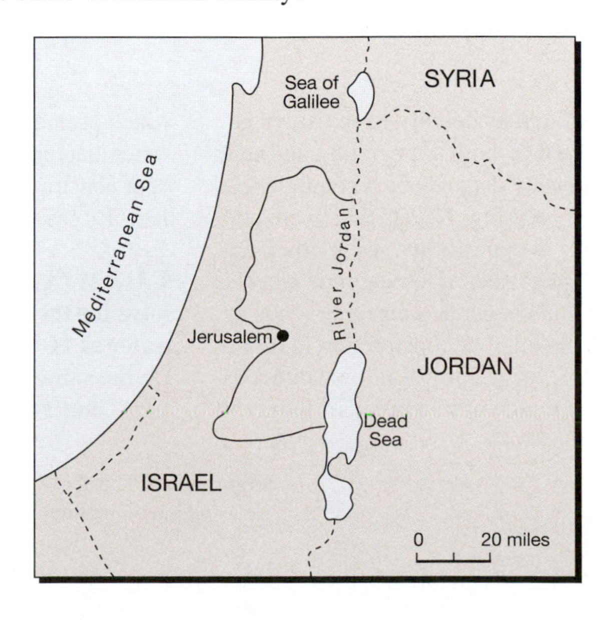

F. Density of seawater as a function of temperature and salinity

Ocean density stratification (layering)—Relative to density, the ocean is rather homogeneous down to a depth of around 100 m (330 ft), which reflects mixing by storm waves. From 100 m to around 1,000 m (3,300 ft) density increases abruptly with depth—an interval called the **pycnocline** (Fig. 13.18). Below 1,000 m, water again appears relatively homogeneous relative to density.

Layering is also characteristic of each of the two variables that affect density—*temperature and salinity* (Fig. 13.19). And, each of these two parameters, in general, changes along with density as a function of depth; i.e., temperature *decreases* with depth (causing water to become more dense), and salinity *increases* with depth (similarly causing water to become more dense).

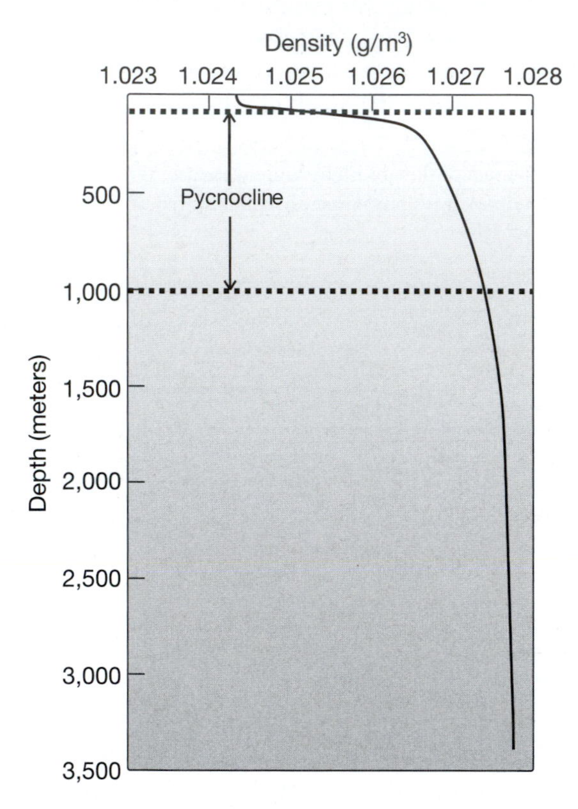

Figure 13.18 Relative to *density*, ocean water is divided into three intervals. The intermediate interval, where density increases abruptly with depth, is called the pycnocline.

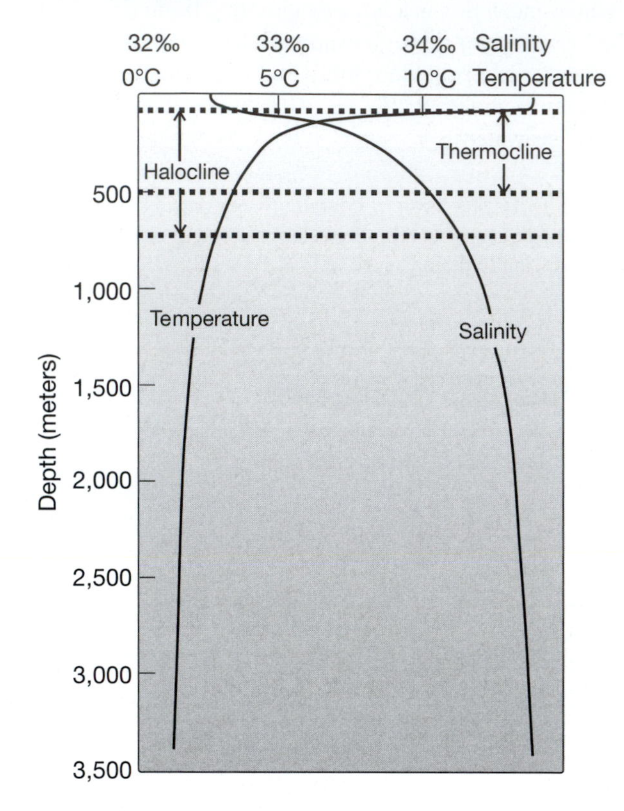

Figure 13.19 Relative to *salinity*, the intermediate interval is called the halocline. Relative to *temperature*, the intermediate interval is called the thermocline.

Inasmuch as density of seawater is affected by both temperature and salinity, density depends on a combination of the two (Fig. 13.20), which are commonly in opposition to each other. For example, water at low latitude receives maximum solar heating, so it tends to *rise* (recall your construction in Figure 13.15C, page 236). But low latitude is also a region of maximum evaporation, which promotes high-salinity water that tends to *sink* (recall your construction in Figure 13.15D, page 236).

Q13.21 (A) On Figure 13.20, solve for the density of seawater at 11 °C and 34.0‰. (B) Do the same for seawater at 14 °C and 34.9‰.

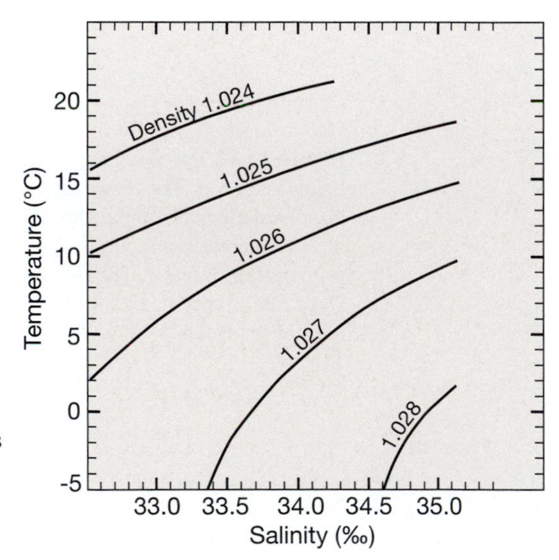

Figure 13.20 The density of seawater depends on a combination of temperature and salinity.

There is still another variable that affects both temperature and salinity—*latitude*. Examine Figure 13.21, which graphs density at three different latitudes.

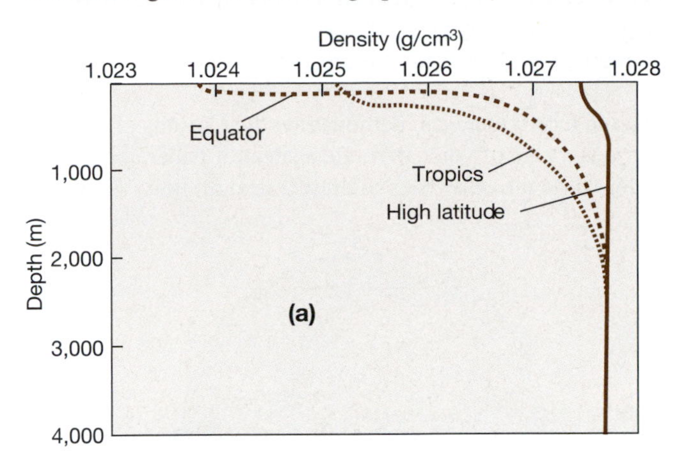

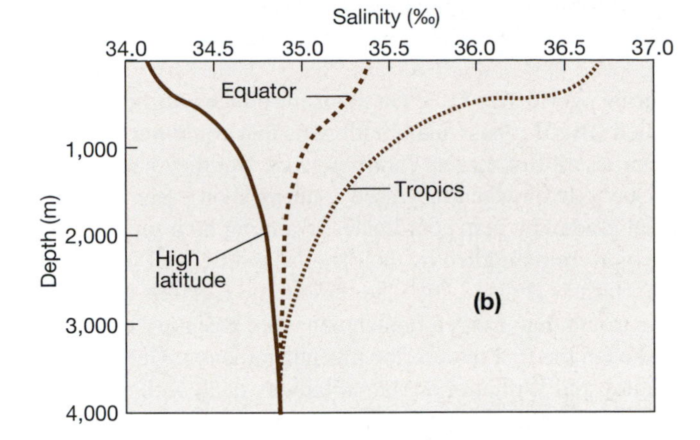

Figure 13.21 (a) Density, (b) salinity, and (c) temperature within the world ocean differ as a function of both depth and latitude.

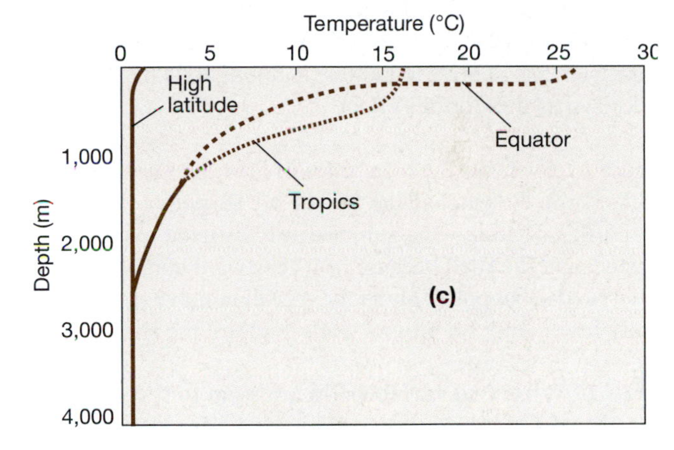

Q13.22 Which of the two factors that affect density appears to be dominant at tropical latitudes—salinity or temperature?

Q13.23 Which of the two factors that affect density appears to be dominant at the equator—salinity or temperature?

Q13.24 At which of the three latitudes do salinity and temperature appear to act in concert—rather than in opposition—in increasing density with depth?

Oxygen within the deep sea—Surface seawater absorbs oxygen from the atmosphere, so shallow marine waters contain oxygen sufficient for marine animals to thrive. But, in time, shallow-water organic matter sinks toward the ocean floor, where, en route, it is degraded by bacteria that consume dissolved oxygen (Fig. 13.22). So, deep ocean water tends to be depleted in dissolved oxygen. Yet deep ocean water has its share of animals that are alive and well. How do those animals obtain essential oxygen?

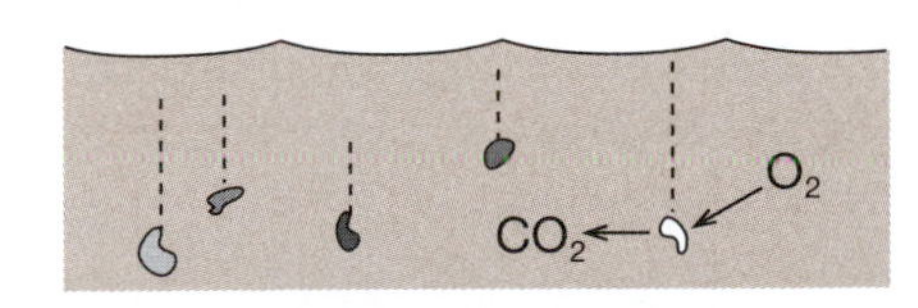

Figure 13.22 Bacterial decomposition of organic carbon compounds consumes oxygen and produces carbon dioxide.

Q13.25 Notice the vertical line A–B in Figure 13.23. Along this line, (A) what is the trend in oxygen content from the surface to a depth of 1 km? (B) What is the trend in oxygen content from a depth of 1 km to a depth of 2 km?

Q13.26. Explain the reversal in the downward trend along line A–B. *Hint:* Notice the geography of the line of cross-section in the figure. Also, it might be helpful to once again examine Figures 13.15C and 13.15E on page 236.

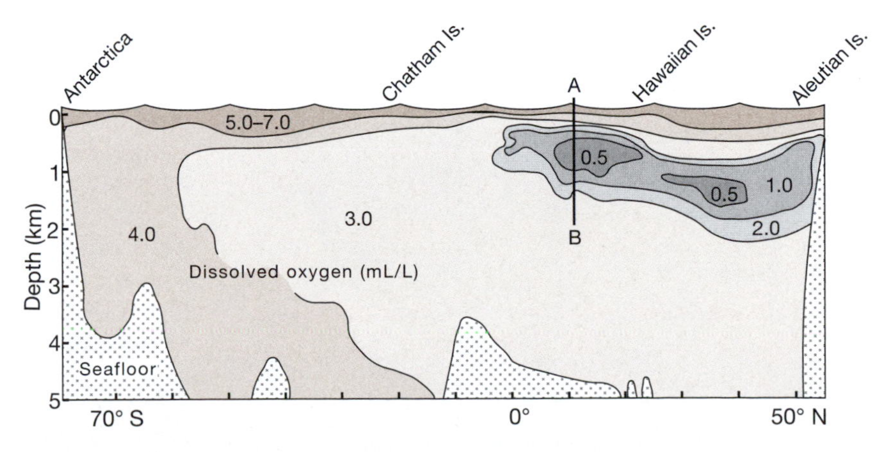

Figure 13.23 This cross-section through the Pacific Ocean shows dissolved oxygen in milliliters per liter. The line of section is approximately along meridian 170° W.

G. The Coriolis effect

An understanding of the Coriolis effect is critical to the understanding of details of atmosphere and ocean circulation.

During World War I, the Germans introduced artillery pieces called 'Big Berthas,' mammoth guns that could hurl shells 15 miles aloft at targets 75 miles away. But they had a lesson to learn when they fired southward on Paris. Having been guided by map coordinates in aiming their guns, gun crews were befuddled by their shells landing well west of their targets (Fig. 13.24). The reason: the **Coriolis effect**. It is ironic that it was a Frenchman who, a century earlier, had been the first to describe this phenomenon. George Hadley and William Ferrel qualitatively dealt with atmospheric circulation on a rotating globe, but it was Gaspard de Coriolis (1792–1843), a French mathematician, who developed the equation describing motion over a rotating body. The apparent diversion in the paths of air currents and ocean currents relative to latitude and longitude has come to be known as the Coriolis effect.

Figure 13.24 shows the mechanics of how the Coriolis effect works. Both the gun and the missile are carried eastward via Earth's rotation—the gun, because it's fixed to the surface, and the shell because of its eastward component when fired—but points along the meridian move progressively faster, and, so, progressively 'outrun' the shell.

Q13.27 What two variables (in addition to the speed of rotation of Earth) do you suspect determine the distance that the shell lands west of its target?

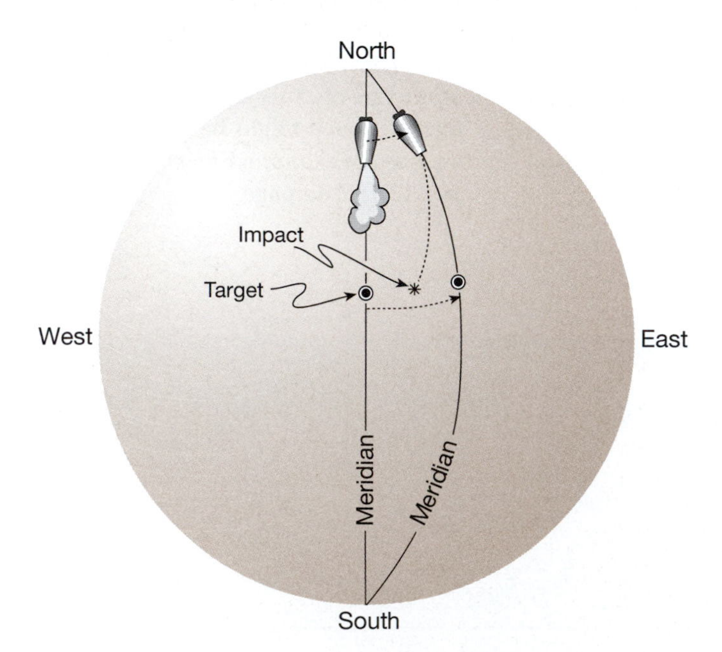

Figure 13.24 This is a schematic representation of how the Coriolis effect causes a shell or missile to be diverted from its intended path, relative to a meridian (aka line of longitude).

If you find it difficult to envision the Coriolis effect, you can demonstrate it to your satisfaction with a disk of cardboard tacked through its center to a surface below. Then Follow instructions in captions to Figures 13.25 and 13.26.

You and a classmate can demonstrate the Coriolis effect at your desk. One of you can rotate a sheet of paper about its center while the other tries to draw a straight line. Voilá.

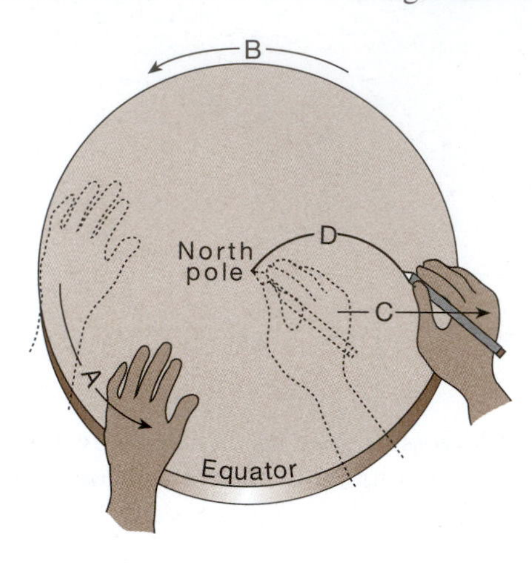

Figure 13.25 If you rotate a disk *counterclockwise* (arrow A)—simulating the rotation of Earth as in viewing the north pole (arrow B)—while attempting to draw a straight line (arrow C), the resultant line curves to the *right* (line D).

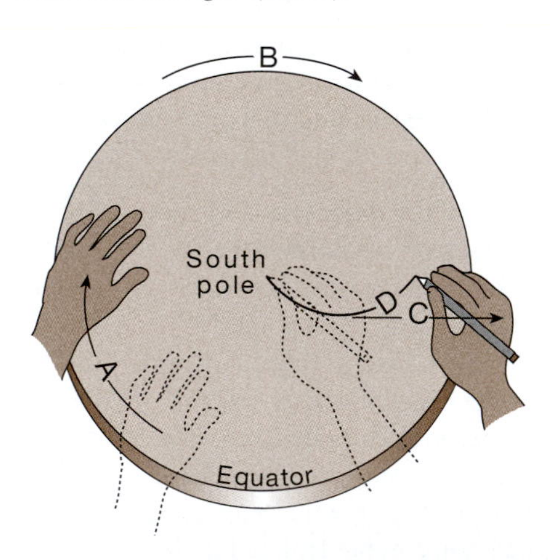

Figure 13.26 If you rotate a disk *clockwise* (arrow A)—simulating the rotation of Earth as in viewing the south pole (arrow B)—while attempting to draw a straight line (arrow C), the resultant line curves to the *left* (line D).

The Coriolis rule—In the Northern Hemisphere, air currents and ocean currents tend to turn to the right (line D in Figure 13.25), whereas in the Southern Hemisphere, currents tend to turn to the left (line D in Figure 13.26).

Q13.28 It doesn't happen in the real world, but imagine a current of air or water flowing *exactly* along the equator (or along any other line of latitude). What do you suppose would be the Coriolis effect, if any?

Deep-water currents—as we have seen in Figure 13.23—are important over time in exchanging atmospheric oxygen and deep sea carbon dioxide, but it is surface currents that are of most immediate interest to humans. Surface currents aid in navigation and promote good fishing in certain areas.

Surface currents are produced by the wind blowing over the ocean, driving global-scale currents in constant patterns. Surface currents flow more in response to perennial atmospheric circulation than to daily weather changes.

Coriolis effect and surface currents—In 1893, Norwegian explorer Fridtjof Nansen set out on the epic voyage of the Fram across the Arctic Ocean. While negotiating shifting pack ice Nansen noticed that icebergs drift at high angles to prevailing winds. He reported this curious feature to physicist V. Walfrid Ekman, who developed a mathematical model explaining Nansen's observation, a model that is now known as **Ekman transport** (or the Ekman spiral).

Ekman transport works as follows: The upper 100 m of ocean water can be viewed as consisting of a series of laminae or sheets. Surface wind exerts a frictional drag on the surface sheet of water, which (in the Northern Hemisphere) veers to the right some 45° owing to the Coriolis effect (Fig. 13.27). The surface sheet, in turn, exerts a frictional drag on the underlying sheet, which, like the surface sheet, veers to the right. This downward transfer of frictional energy diminishes to negligible effects at a depth of around 100 m, depending on wind velocity. The net motion of this bundle of sheets—which comprises a surface current—is in a direction approximately 90° to that of the surface wind.

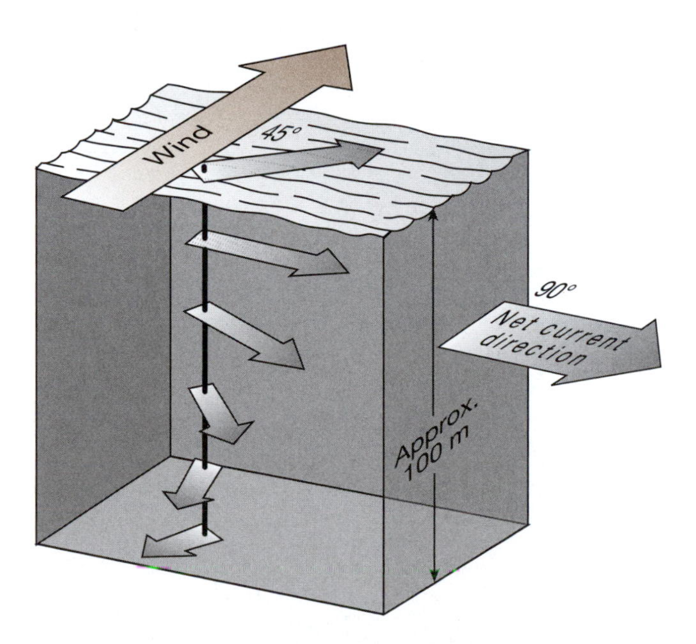

Figure 13.27 In Ekman transport in the Northern Hemisphere, frictional drag extends downward, veering to the right and decreasing in strength. The net direction of the 100-meter-thick surface current is 90° to the right of the prevailing wind.

Coastal upwelling and downwelling—A coastal wind blowing parallel to shore produces Ekman transport of surface water perpendicular to shore. Where perpendicular transport is in an *offshore* direction (i.e., away from the coast) a surface depression tends to develop, and deeper water rises to fill it—a process called **upwelling**. On the other hand, where the perpendicular transport is in an *onshore* direction (i.e., toward the coast) it sweeps water toward the coast, tending to create a bulge that causes water to sink—a process called **downwelling**.

Figures 13.28 and 13.29 depict upwelling and downwelling. Both illustrations are for coastal settings in the Northern Hemisphere. In each figure, two possibilities are sketched to indicate wind direction (labeled A and B).

Q13.29. Assume that Figure 13.28 illustrates a setting in the Northern Hemisphere. Which arrow indicating wind direction is consistent with upwelling, A or B?

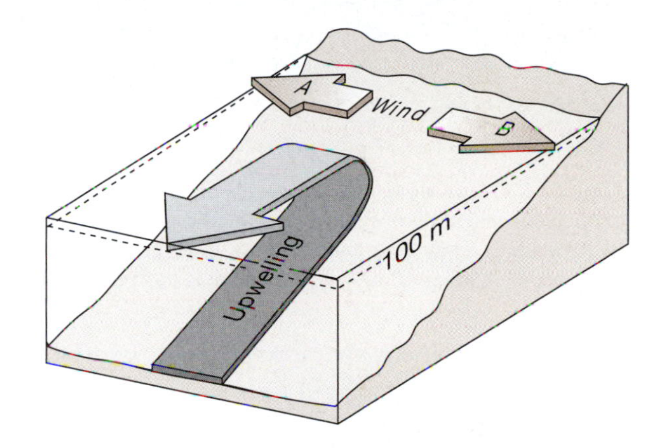

Figure 13.28 A surface current moving *offshore* (i.e., away from a coast) causes *upwelling* of water.

Q13.30 Assume that Figure 13.29 illustrates a setting in the Southern Hemisphere. Which arrow indicating wind direction is consistent with downwelling, A or B?

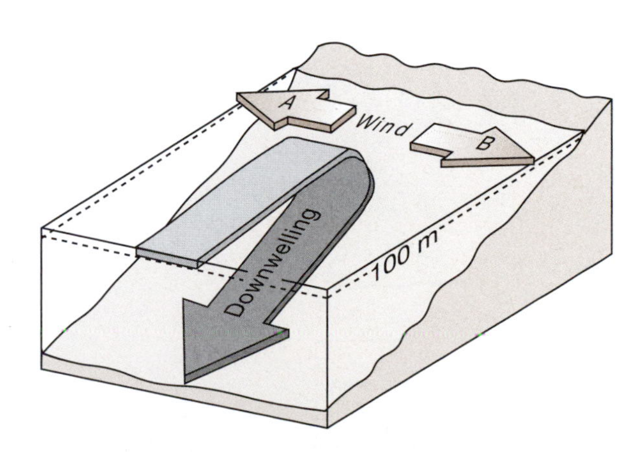

Figure 13.29 A surface current moving *onshore* (i.e., toward a coast) causes *downwelling* of water.

Coriolis effect and organic recycling. Bacterial decomposition of sinking organic matter (Fig. 13.22 on page 239) not only consumes oxygen and produces carbon dioxide, but it also converts organic compounds into organic nutrients (e.g., phosphates and nitrates) that can serve as fertilizer in organic recycling. But there must be a mechanism for bringing these nutrients up into the **photic zone**, where there is light sufficient for photosynthesis. Only there can nutrients be synthesized by plants into food for animals (Fig. 13.30). *That mechanism is upwelling.*

Where there is upwelling, fishing and related industries thrive; but where there is downwelling, fishing and related industries are poor indeed.

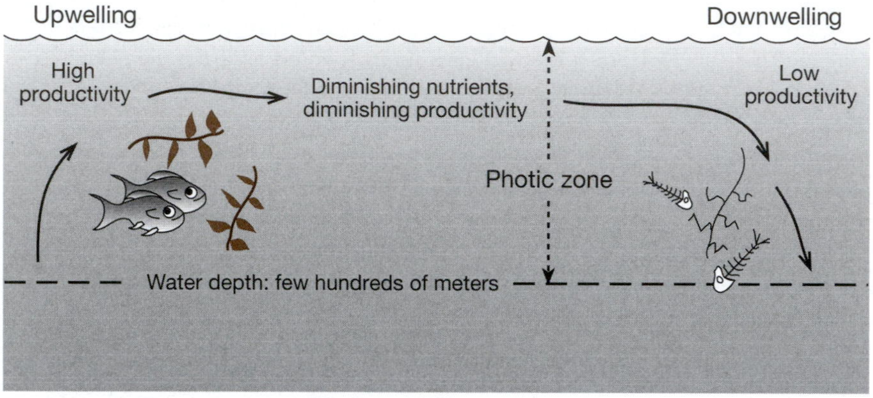

Upwelling
Downwelling
High productivity
Diminishing nutrients, diminishing productivity
Low productivity
Photic zone
Water depth: few hundreds of meters

Figure 13.30 Upwelling brings essential organic nutrients up within the shallow photic zone where photosynthetic productivity by plants promotes productivity by animals. Down-current from upwelling these nutrients are progressively depleted, so in areas of downwelling there is low productivity.

Coastal productivity—Applying what you have learned about Coriolis effect, Ekman transport, upwelling, and downwelling, label (Answer Page 246) 'good fishing' or 'poor fishing' for each of the coastal profiles on the global map of currents in Figure 13.31.

Q13.31 (A) a–b? (B) c–d? (C) e–f? (D) g–h?

Q13.32 On islands along the Peruvian coast (Fig. 13.32), rich deposits of guano (i.e., sea bird droppings) are the source of a global fertilizer industry. Why here?

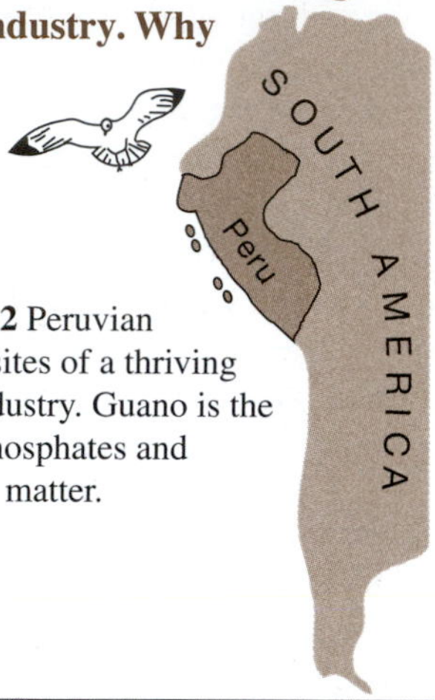

SOUTH AMERICA
Peru

Figure 13.32 Peruvian islands are sites of a thriving fertilizer industry. Guano is the source of phosphates and nitrogenous matter.

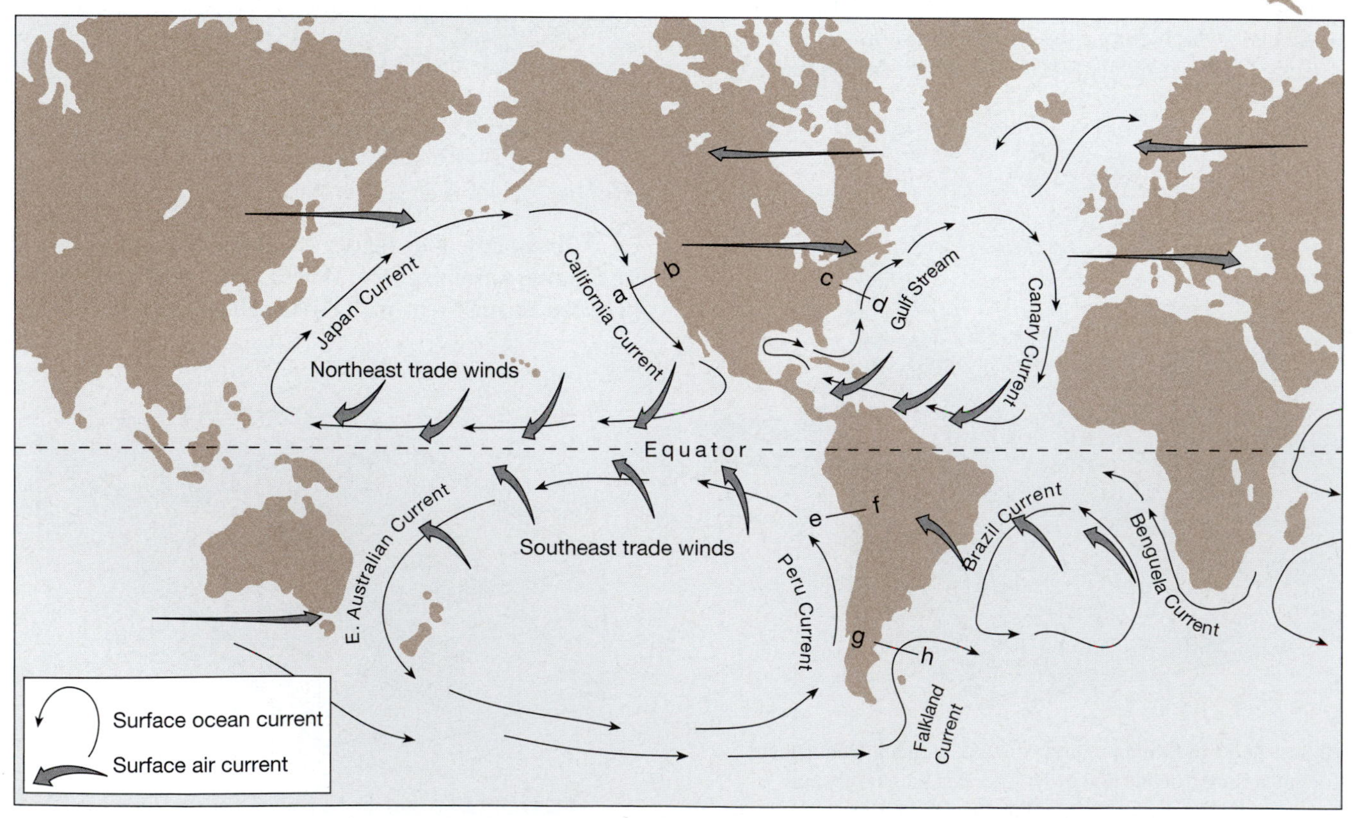

Figure 13.31 The angular relationship between the directions of surface winds and surface currents reflects Ekman transport.

H. El Niño

- Floods and landslides in Southern California, Peru, Ecuador, and east Africa
- Drought and fires in Indonesia, Brazil, and Australia
- Tornadoes in Florida
- Hurricanes in Baja and Mexico

Such is the litany of disasters blamed on **El Niño**, an atmospheric-oceanic event that occurs at two- to five-year intervals around the Christmas season—prompting Peruvian fishermen to call it **El Niño**, a reference to the Christ child. (El Niño is also known as the **Southern Oscillation**, or, collectively, **ENSO.**)

The cause of El Niño—The link between atmosphere and ocean resides in an unusual warming of surface waters in the eastern Pacific Ocean. During normal years, trade winds blow shallow equatorial Pacific waters westward (Fig. 13.31), warming it en route and piling it up some 400 cm higher than waters offshore of South America (Fig. 13.33). This westward sweep of Pacific water enhances upwelling along the Peruvian coast; and warm water in the western Pacific promotes evaporation and rainstorms in Malaysia.

Q13.33 Reflecting on Figure 13.4 on page 231 and its accompanying text, explain in more detail the rainfall in Malaysia during normal years.

El Niño begins with a slacking of trade winds, followed by western Pacific surface waters flowing back 'down-hill' toward the east, suppressing or reversing upwelling along the Peruvian coast (Fig. 13.34).

Q13.34 What has happened to the normal-year 400-cm difference in elevation between western Pacific and eastern Pacific waters, i.e., what is the difference in elevation at the time of El Niño?

Q13.35 Why are Peruvian coastal waters warmer during El Niño than during normal years?

Although we are looking here at only a narrow belt of atmospheric/oceanic activity, its consequences can be global. We see a microcosm of similar interrelated activity on a daily basis as weather fronts shift areas of rainfall and drought across North America. El Niño is simply a global-scale example of the dynamics of continental atmospheric activity.

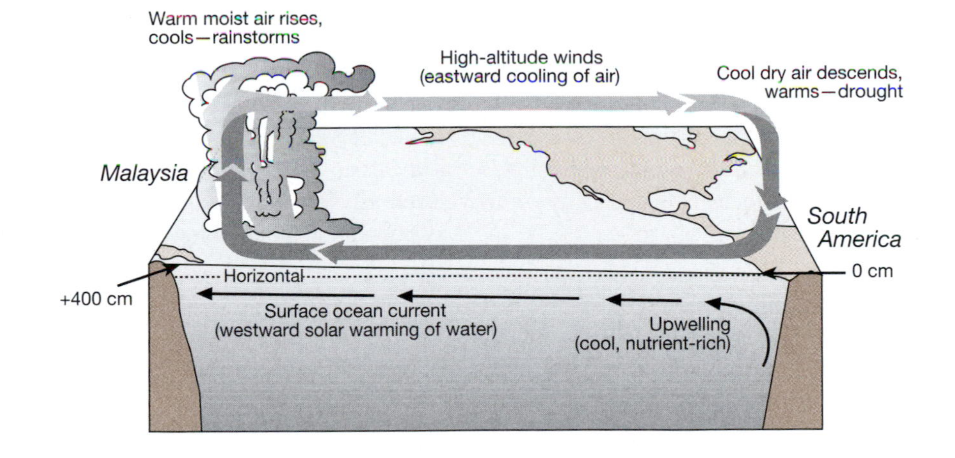

Figure 13.33 Normally, trade winds blow shallow Pacific waters westward, piling them up some 400 cm higher than waters offshore of South America. At the same time a low-pressure cell, with its excessive rainfall, occurs over Malaysia.

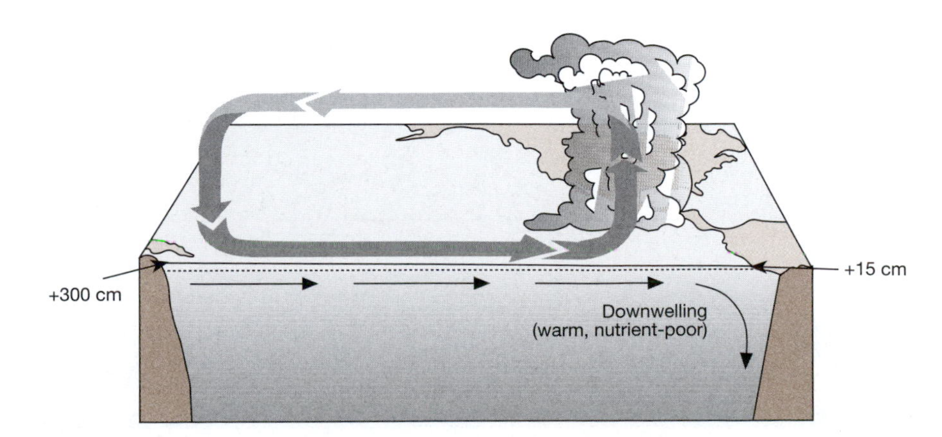

Figure 13.34 El Niño begins when trade winds slacken, high water from the western Pacific flows eastward, and the low pressure cell and rainfall migrate eastward, tracking the flow of warm water. It is the reverse of normal years.

History of El Niño—Although as early as the late 1960s the link between El Niño oceanic and atmospheric conditions had been made, it was not until its disastrous appearance in 1982–83 that it caught the attention of the general public. But El Niño as we know it surely has a history going back as far as the present configuration of continents. Its record over the past half-century has been compiled from studies of equatorial Pacific corals that can live hundreds of years (Fig. 13.35). Coral skeletons record changes in both the chemistry and temperature of seawater, so they are virtual tape recorders of past oceanic conditions—much like tree rings are for past terrestrial conditions.

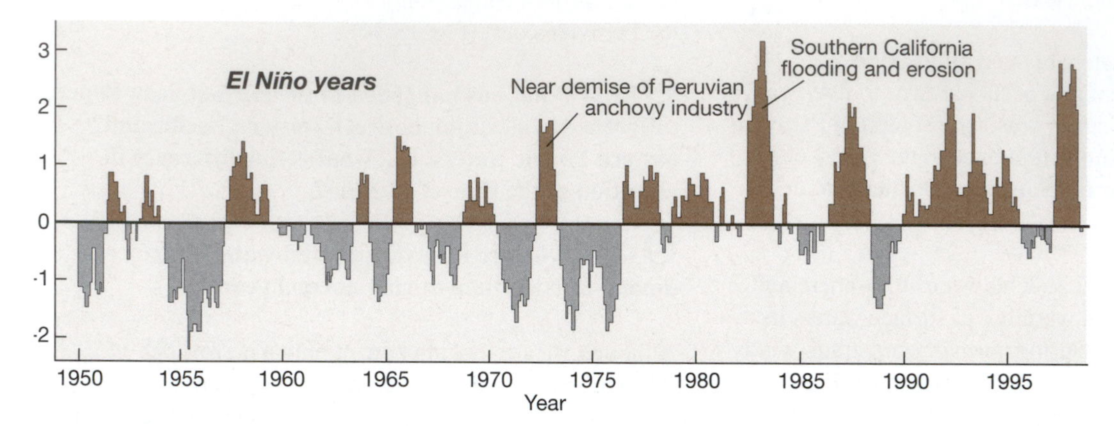

Figure 13.35 The record of the qualitative severity of El Niño events over the past half-century has been compiled from studies of coral skeletons.

From NOAA–CIRES Climate Diagnostics Center, University of Colorado

The near demise of the Peruvian anchovy industry—During the early 1950s Peruvian fishermen discovered a market in more developed countries for anchovies (Sp. *anchovetas*), which are processed into fish meal and used as a source of protein in livestock feed. By 1971 Peruvian anchovies comprised 20% of the world's commercial fish catch, but the ocean bonanza was not to be sustained. Biologists with the United Nations Food and Agriculture Organization (UNFAO) warned that the sustainable yield of Peruvian anchovies, based on what they knew of the population and life cycle, was around 9 million metric tons per year. Within 20 short years, annual harvests exceeded that amount —*a classic case of overfishing*. A severe El Niño in 1972 dealt the anchovy population a body blow from which it has yet to recover (Fig. 13.36). By placing short-term profits ahead of long-term planning, Peru lost an important source of jobs and national income. Its national debt has since grown.

Q13.36 Examine Figures 13.35 and 13.36 and briefly review El Niño's role in the near-demise of Peru's anchovy industry.

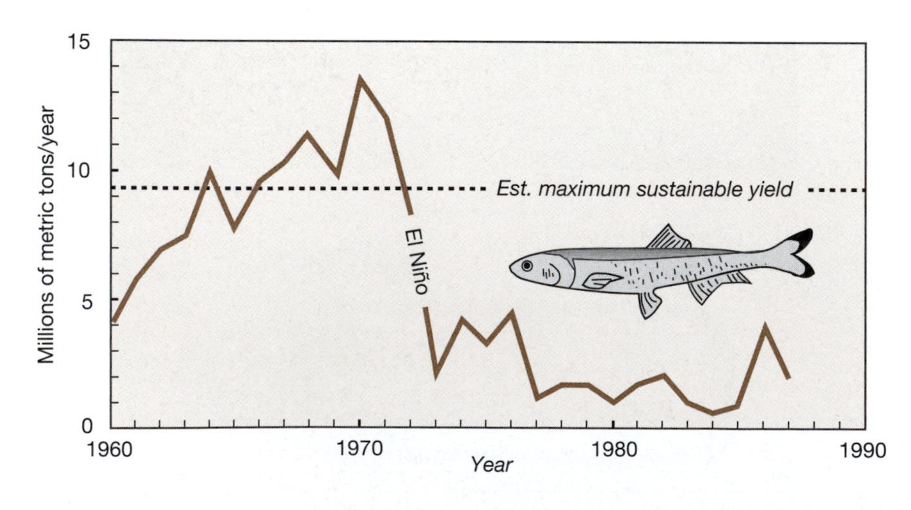

Figure 13.36 The near-demise of the Peruvian anchovy industry since 1972 resulted from a brief history of overfishing compounded by El Niño.

For Worldwide Weather and Climate Events see http://www.ncdc.noaa.gov/oa/reports/weather-events.html

(Student's name) (Day) (Hour)

(Lab instructor's name)

ANSWER PAGE

13.1

13.2

13.3 (A) (B)

13.4 (A) (B)

13.5

13.6

13.7

13.8

13.9

13.10

13.11

13.12

13.13 (A) (B)

13.14

13.15

13.16

13.17

13.18 (13.15C)

(13.15D)

(13.15E)

(13.15F)

13.19 (A)

(B)

13.20

13.21 (A) ______ (B) ______

13.22 ______

13.23 ______

13.24 ______

13.25 (A) ______

(B) ______

13.26 ______

13.27 ______

13.28 ______

13.29 ______

13.30 ______

13.31 (A) ______

(B) ______

(C) ______

(D) ______

13.32 ______

13.33 ______

13.34 ______

13.35 ______

13.36 ______

14 Waste and Water

Topics

A. What are two ways in which water can complicate waste management?
B. What is porosity, and how can it be quantitatively measured in the laboratory? What is permeability, and how can it be qualitatively measured in the laboratory?
C. Of what two main components does a home septic system consist, and what are their functions? Within the leach field of a septic system, why should the odor of rotten eggs and the presence of cattails be matters of concern for a buyer? What complication does slope present in designing a septic drain field, and how is that complication mitigated? What is the preferred grain-size of soil in a septic drain field, and why? How is a soil percolation test conducted?
D. What are the three phases of a municipal sewage-treatment plant, and what goes on within each?
E. What is the anatomy of a simple landfill? What is the anatomy of a sanitary landfill?
F. What is thought to cause the Gulf of Mexico *dead zone* (not to be confused with the 2010 oil spill)? How might it be mitigated?

A. Biggest player in waste management

The biggest player in waste management is *water*.

For example, flood waters can inundate waste repositories and scatter wastes from treatment plants—thereby disrupting the operations of such facilities.

Water is the universal solvent that mobilizes toxic chemicals within all types of waste repositories.

Water operates within two broad domains—both the *surface* and the *subsurface*.

Simple, but commonly ignored: Surface water runs downhill. The carrying capacity of streams and rivers can be readily observed and measured.

But the dynamics of subsurface water (aka **groundwater**) are more enigmatic because of the web of earth materials through which groundwater moves—a web characterized by **porosity** and **permeability** and their spatial distribution.

B. The nature of earth materials

Porosity and permeability

Porosity is the ratio—*stated as a percentage*—of void space in a sample of sediment or rock to the total volume of that sample. The volume of void space plus the volume of solid material = 100% of the total volume of the sample.

Porosity can be either of two broad types.

Primary porosity develops at the time sediment and/or rock accumulates. Two common types of *primary* porosity:

(1) Pore spaces that develop among grains of clay, silt, sand, and pebbles at the time those grains are deposited.

(2) Pore spaces that develop from gas bubbles within lava, producing irregular cavities before the rock crystallizes.

Secondary porosity develops long after a rock originates. Two common types of *secondary* porosity:

(1) Pore spaces that develop as open space along fractures, which are common in all rocks.

(2) Pores that develop as cavities produced by the dissolving of soluble rocks (e.g., limestone) by groundwater.

Q14.1 Review: (A) Which of the drawings in Figure 14.1 illustrate primary porosity? (B) Which of the drawings illustrate secondary porosity?

A. Fractures (any kind of rocks)

B. Sedimented grains (pebbles, sand, silt, clay)

C. Volcanic rocks (frozen gas bubbles)

D. Solution cavities (limestone, gypsum, salt)

Figure 14.1 Common examples of primary and secondary porosity.

Intergranular porosity (Fig., 14.1B) is among the most common types of porosity that affect the behavior of groundwater. So let's examine some of its aspects.

The maximum theoretical intergranular porosity possible among spherical grains of equal size occurs in *cubic* packing, where imaginary lines connecting the centers of four contiguous grains form a cube (Fig. 14.2A).

Q14.2 Compute the porosity (i.e., percent of intergranular open-space volume) shown in Figure 14.2A. ***Hint:*** **4/3 • pi • r³. (With Net connection, go to…)**

http://www.1728.com/diam.htm

Figure 14.2 These two views of equal diameter spheres show cubic and rhombic packing. Think 3-dimensionally.

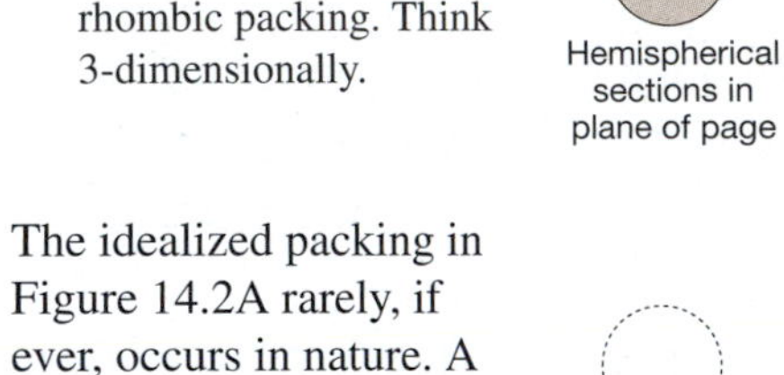

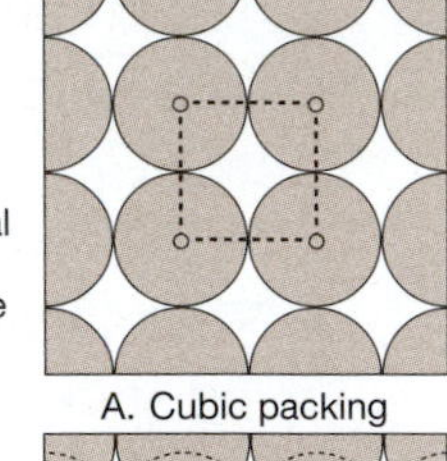

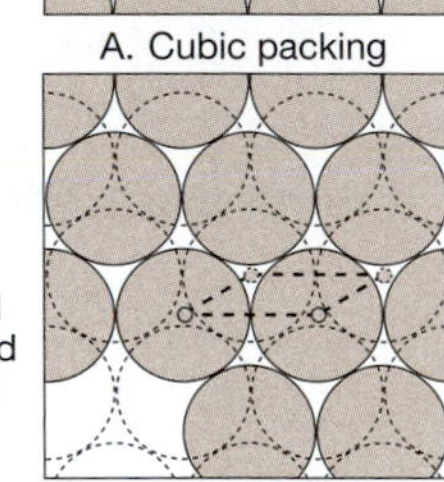

The idealized packing in Figure 14.2A rarely, if ever, occurs in nature. A more likely packing is the *rhombic* packing in Figure 14.2B (with its porosity of 25.95%).

In Figure 14.3 imaginary spheres of equal size have been poured into a box (so they are constrained by the bottom and four sides of the box), frozen in space, and sawed.

Q14.3 Which cross-section in Figure 14.3 is most likely to result from a random slice through a box of frozen spheres—A, B, or C?

Figure 14.3 Imagine sawing a random plane through a box of frozen spheres of equal size—producing one of the cross-sections: A, B, or C.

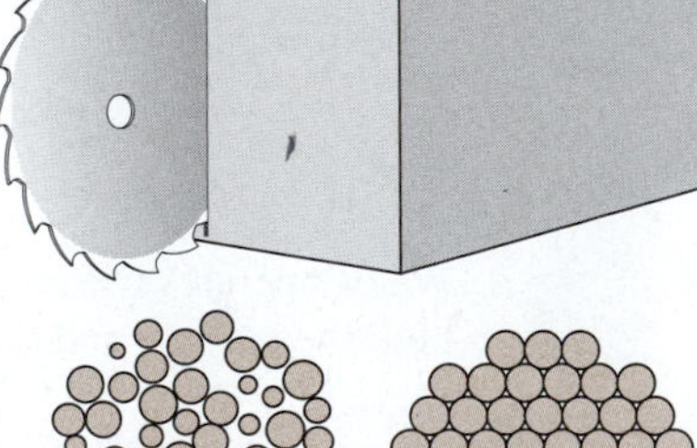

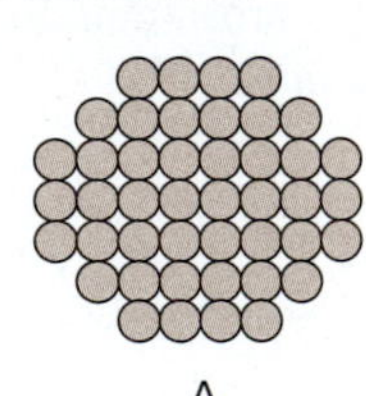

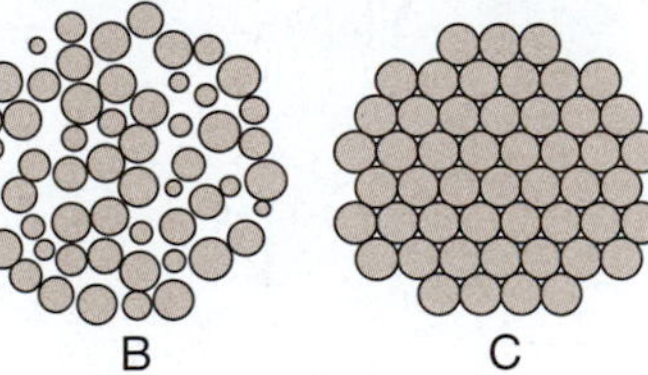

Porosity and permeability (cont.)

Does grain size affect porosity? Figure 14.4 shows cross-sections of two samples, each consisting of spheres of equal diameter in cubic packing.

Q14.4 How does porosity in Figure 14.4A compare with porosity in Figure 14.4B?

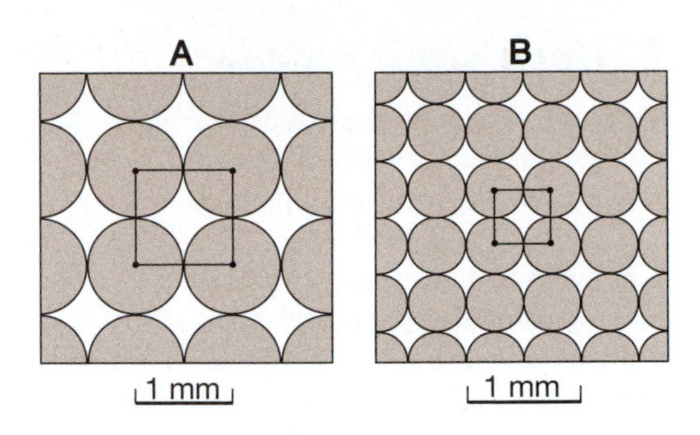

Figure 14.4 Illustration A consists of grains 1 mm in diameter, and illustration B consists of grains 1/2 mm in diameter. Each exhibits cubic packing.

Quantitatively measuring porosity

Q14.5 Refer to Figure 14.5. Write an outline of 7 or so steps you would take in quantitatively measuring the porosity in a sample of, say, the coarse spheres. *Hint:* For this mental exercise, you will need all of the items in the figure except the fine spheres, the fine screen, and the timer.

Qualitatively measuring permeability

Permeability is the fluid transmissivity of soil, sediment, or rock—i.e., the ease with which water moves through such material.

Q14.6 (A) Write an outline of the steps you would take in qualitatively measuring the permeability of each of two samples of spheres. (B) Which would you expect to find to be more permeable, the coarse spheres or the fine spheres? (C) Explain your expectation. *Hint:* You might include the terms *surface-area* and *friction* in your explanation.

Q14.7 In conclusion, (A) how do grain sizes of coarser spheres and finer spheres affect porosities? (B) How do grain sizes affect permeabilities?

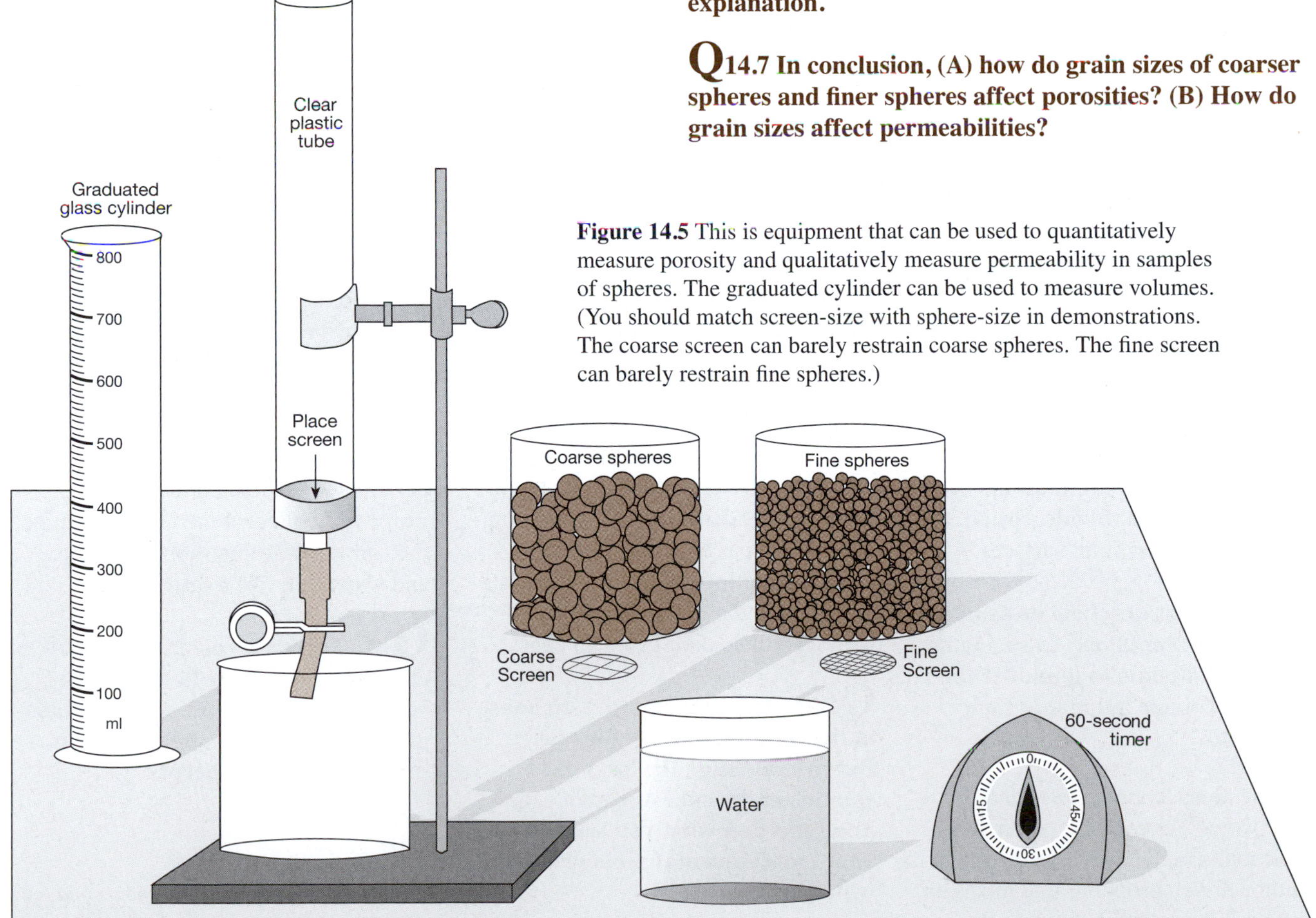

Figure 14.5 This is equipment that can be used to quantitatively measure porosity and qualitatively measure permeability in samples of spheres. The graduated cylinder can be used to measure volumes. (You should match screen-size with sphere-size in demonstrations. The coarse screen can barely restrain coarse spheres. The fine screen can barely restrain fine spheres.)

C. Home septic systems

There are two broad categories of liquid waste: (a) *domestic and municipal sewage*, and (b) *industrial effluents*.

Q14.8 Which do you imagine is more commonly characterized as 'hazardous waste?'

Sewage includes all waste that exits a house, hotel, or apartment building. All things considered, on average, each person in the U.S. accounts for 550 liters of sewage per day.

Home septic systems

Approximately 25 percent of the U.S. population lives in areas where a home **septic system** is the only option for sewage disposal.

Q14.9 Migration from the family farm to the city reduced the number of rural septic systems, but a more recent demographic trend is adding septic systems. The total is now increasing. What is the colloquial term for this reverse migration? *Hint:* Growing cities are examples.

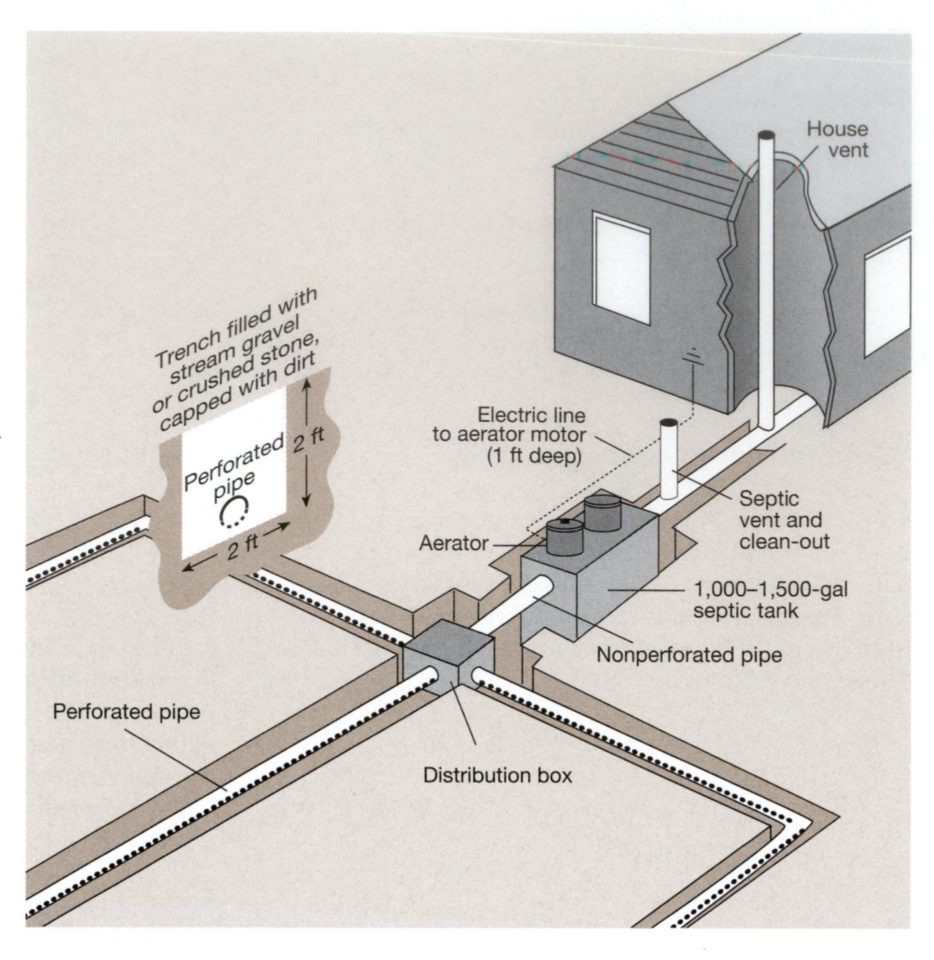

Figure 14.6 This is an example of a septic system with an aerator. The size of the tank depends upon the size of the house—as defined by the number of bedrooms. The total length of drain pipes is dictated by the size of the house and the condition of the soil.

A septic system consists of two main components (Fig. 14.6):

(1) A tank in which solids settle out of the sewage, and which requires a pump-out every few years.

(2) A leach field that consists of perforated pipes that carry liquids away from the tank and into the soil. In passing through the soil, the liquid is purified by (a) physical filtration and bacterial processes; and (b) **adsorption**, the clinging of ions to solid surfaces.

Q14.10 What practical measure can be taken (or *not* taken) by residents of a modern home to minimize the amount of waste going into their septic tank?

In some areas, county codes and/or homeowners' covenants require that septic tanks include an aerator, which is a motor-driven propeller that mixes air with the liquid, thereby adding free oxygen that facilitates the activity of **aerobic** bacteria. Certain aerobic bacteria make their living through **oxidation** of organic waste, a process that is efficient in cleansing septic system liquids that otherwise can clog drain fields. In contrast, **anaerobic** bacteria, which are less effective in processing liquid waste, derive oxygen from compounds such as sulfates—a process called **reduction**. A product of sulfate reduction is the gas hydrogen sulfide, with its telltale odor of rotten eggs.

Q14.11 You find your dream house in the country, and it's for sale. But in examining the leach field you detect the odor of rotten eggs. You reflect on what you learned in your environmental geology course, which is…?

There are strict specifications for the installation of a septic system, so the home builder should consult with county authorities early in order to assure an eventual occupancy permit.

A critical regulation requires that drain pipes be inclined not more than 1/4 inch per 10 feet. If inclined more steeply, water will collect at the end of the pipe, the soil will become saturated there, and water will rise to ground level.

Q14.12 You are considering purchasing a particular house with a septic system, but you notice cattails growing in the leach field. What is probably going on here?

Tip: Direct water from downspouts and lawn sprinklers away from your drain field.

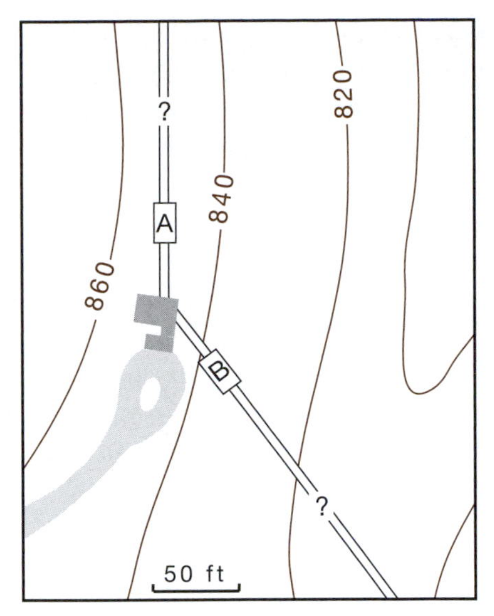

Figure 14.7 Two plans for drain pipes (A and B) appear on contoured topography.

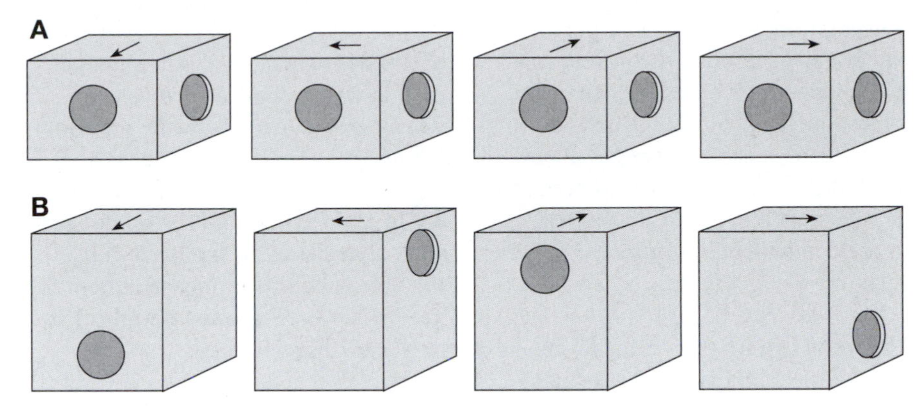

Figure 14.9 These are two different designs of boxes used in septic drain fields. Each design is rotated (arrows) so as to show all four sides. Box A is called a distribution box. Box B is called a stepdown or drop-box.

Q14.13 At a glance, which drain pipe in Figure 14.7 (above) is more advisable, A or B, and why?

Check out the two different designs of distribution boxes in Figure 14.9. One can be is a safe design for running a septic drain pipe *downslope*.

Q14.14 (A) How can a septic drain pipe be designed to safely run downslope as in Figure 14.8? (B) Sketch a cross-section running the length of the pipe in your design.

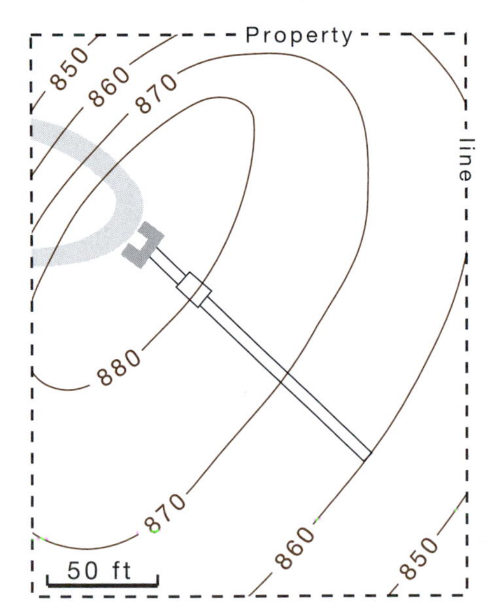

Figure 14.8 In this case, a drain pipe runs directly downslope, as shown by contours.

Q14.15 At a glance, which soil type in Table 14.1 would you intuitively expect to be the most efficient material for a septic drain field?

Table 14.1
Sediments and their porosity and permeability.

Soil type (Decreasing grain size ↓)	Absorption area needed (square meters per bedroom)
Gravel and coarse sand	6.5
Fine sand	8.3
Sandy loam	10.6
Clayey loam	13.9
Sandy clay	16.2
Clay with sand or gravel	23.1
Clay	Prohibited

After T. Dunne and L.B. Leopold, 1978

Loam: Soil that consists of a mixture of clay, silt, sand, and organic matter. Absorption *area* will be explained later.

The best soil for effective cleansing of sewage effluent is intermediate in its grain size. Reasons: **A**. Bacteria require *time* to cleanse effluent liquids, and coarse sediments transmit effluent too fast. **B**. Bacteria work better if they have *solid surfaces* (e.g., clay, silt, sand, pebbles) on which to attach themselves.

Q14.16 Consider two samples of soil—one fine-grained silt, the other coarse-grained sand. In which sample is there greater solid surface area per volume—the fine-grained silt or the coarse-grained sand? *Hint:* Recall your laboratory investigation of relative permeabilities on page 249.

But if grains are exceedingly small—as in clay—permeability is practically nil. So intermediate grain size provides an effective mix of permeability and surface area.

Testing soil percolation rate (yes, this *can* be done at home)

Percolation is the process whereby water moves through permeable soil. Percolation tests differ from place to place, but the goal is universal—to *quantify* permeability using standard procedures. The customary **three steps** are illustrated in Figure 14.10. (Inches is the common unit of measure in perk tests.)

(**Step 1**) Dig a hole 6–8 inches in diameter approximately 2 feet deep, then drive a nail near the top of hole.

(**Step 2**) Fill the hole with water (*dashed arrow upward*) to the level of the nail and record the 'start time.'

(**Step 3**) From the nail, measure the drop in water level (*dashed arrow downward*) with a tape to the nearest 1/8 inch at 10-minute intervals.

Q14.17 Why the nail? Why not just stick a yardstick to the bottom of the hole each time a measurement is needed? *Hint:* Examine details of the hole's bottom.

Frequent measurements will detect any changes within the 2-foot interval.

Q14.18 Examine Figure 14.10. What is there about this illustration that indicates vertical differences in permeability within this 2-foot interval of soil?

You can now calculate the rate of *drawdown in inches per hour*, which is the **percolation rate of the soil**.

Perk-testing soil is most reliable during a wet period, so that true percolation is measured, rather than *wicking*. Several holes within the proposed drain field should be tested and then averaged together to derive a reliable perk rate.

Figure 14.10 Three steps (**A-C**) in determining the percolation rate of soil. Darker shade of brown illustrates the penetraton (i.e., the soaking in) of the water.

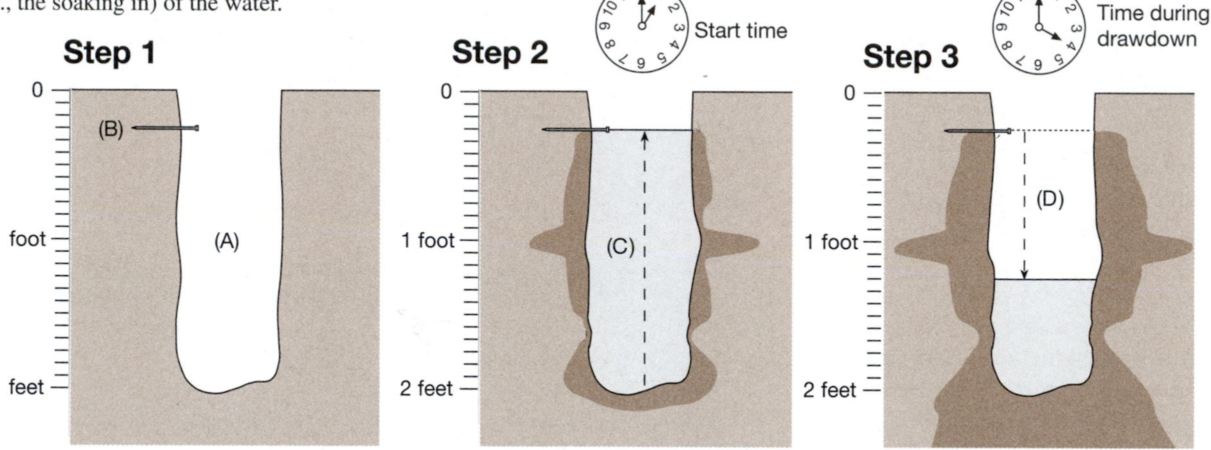

Calculating trench and pipe requirements

Figure 14.11 is a typical graph showing percolation rate in inches per hour vs. area required per bedroom. But the cost basis for a trench and perforated pipe is length. So the question arises: How does one convert area to lineal feet?

As shown in Figure 14.6, a trench is typically two feet deep and two feet wide. By convention, "area" is that of trench bottom only, excluding areas of the sides of the trench. So the procedure for converting area (Fig. 14.11) to lineal feet is to simply divide the number of square feet by two.

The example shown in Figure 14.11 reveals the following:
A soil percolation rate of 5
= 180 sq ft, divided by 2
= 90 lineal feet of trench and pipe (required per bedroom)

Q14.19 How many total lineal feet of trench and pipe would be required if percolation were as illustrated in Figure 14.10, and the number of bedrooms were four?

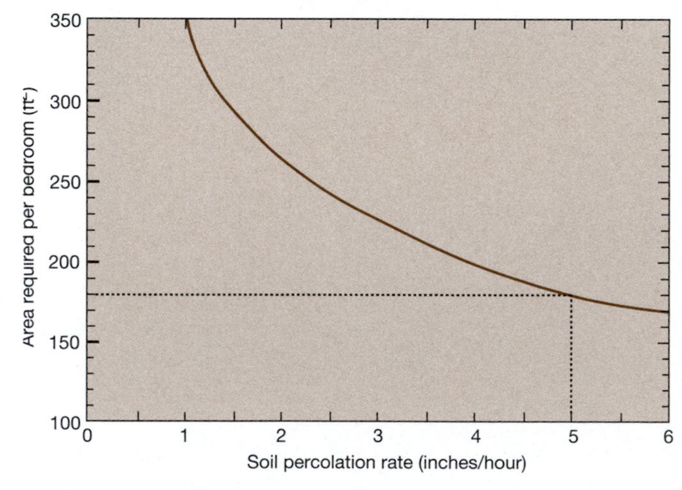

Figure 14.11 Graphs used in calculating the extent of a septic drain field typically plot percolation rate vs. area.

Incidentally, in some regions county regulators now require that a soil scientist provide a description of *soil morphology*, from which the length of trench and pipe is calculated.

D. Municipal sewage systems

The U.S. Clean Water Act of 1977 set regulations dealing with discharges of contaminants into U.S. streams and rivers and established the Environmental Protection Agency (EPA) as the authority charged with setting standards for such discharges. This led to the development of the modern sewage-treatment plant, which operates in three phases (Fig. 14.12):

- *Primary phase*—Solids are screened and settle, then are physically removed.

- *Secondary phase*—Microbes break down biodegradable materials.

- *Tertiary phase*—Additional impurities are removed through *carbon filtration, evaporation, precipitation, and wetland microbial extraction of excess nitrogen.

Q14.20 Notice the aeration tank in the secondary phase of sewage treatment in Figure 14.12. Judging from what you have already learned in this exercise, (A) what's going on here? (B) What kind of bacteria are at work here?

*A handful of finely ground carbon can contain the surface area of one acre.

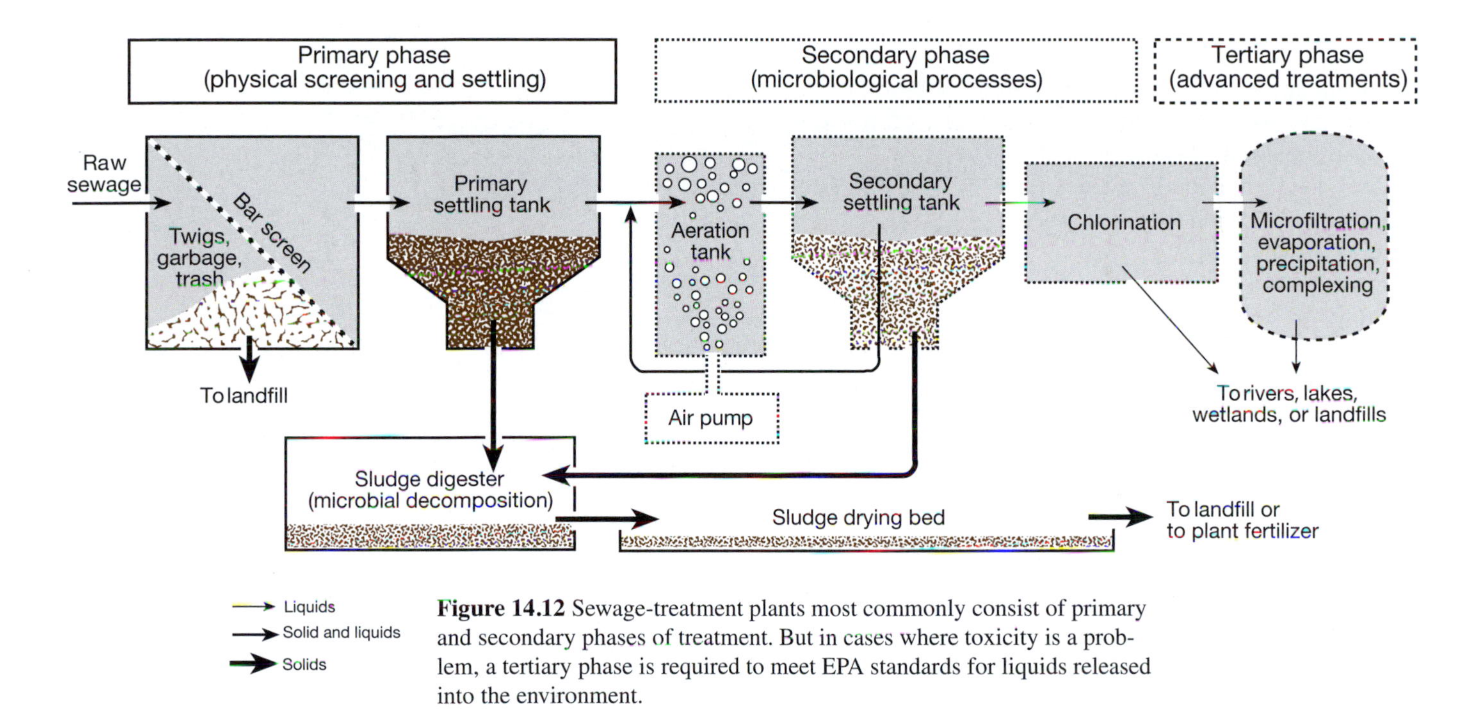

Figure 14.12 Sewage-treatment plants most commonly consist of primary and secondary phases of treatment. But in cases where toxicity is a problem, a tertiary phase is required to meet EPA standards for liquids released into the environment.

Sewage sludge (aka 'Metrogro') can in some cases be superior to commercial fertilizers. Sludge can be heat-treated to kill harmful bacteria. And, it can be treated to remove toxic metals and chemicals, but that is costly. Untreated sludge is acceptable as a fertilizer where plant nutrients will not directly find their way into a food chain.

Q14.21 (A) Given a situation of untreated toxic sludge, name a couple of soil applications where toxins might *directly* (i.e., in one step) find their way to our dinner table. (B) Name a couple of soil applications where toxins might only *indirectly* (i.e., in a number of steps) find their way to our dinner table.

E. Municipal effluents

Sanitary landfills

The minimal objective of a *simple landfill* (Fig. 14.13) is out-of-sight, out-of-mind. But *sanitary landfill* practices (Fig. 14.14) include the management of **leachate** and **gases**—effluents that are invariably produced within landfills.

Actually, it isn't the solid waste per se that presents problems in waste management. It's rainwater and groundwater that find their way in and out of landfills, mobilizing hazardous chemicals en route. Water can dissolve just about anything, given enough time, and it's this garbage juice, called *leachate*, that is troublesome. Leachate should be collected via a plumbing system within landfills and then treated either on-site or at a remote facility.

Landfill gases—produced by bacterial decomposition of organic compounds—consist largely of **methane** (CH_4) and **carbon dioxide** (CO_2). These greenhouse gases are colorless and odorless, but there is sufficient hydrogen sulfide (H_2S) present to impart the stench of sewer gas.

Q14.22 In the world of aerobic and anaerobic bacteria, (A) which do you suppose produces methane? (B) Which produces carbon dioxide?

Methane, aka swamp gas, is the gas typically associated with oil. And it is methane, when mixed with oxygen, that accounts for coal mine explosions. Landfill methane can be bled off via pipes and burned in gas turbines. At the Fresh Kills landfill on Staten Island, NY, the burning of methane provides electricity for thousands of homes.

Q14.23 Before burning landfill gas, a gaseous component should be extracted and discarded. Its name?

Selecting a landfill site

When selecting a landfill site, the most essential requisite is that of **dryness** (i.e., minimal surface water and minimal groundwater).

Q14.24 How about using inexpensive flat land adjacent to soccer fields and a babbling brook?

Q14.25 Why should test holes be drilled before choosing an appealing site in a rolling country setting?

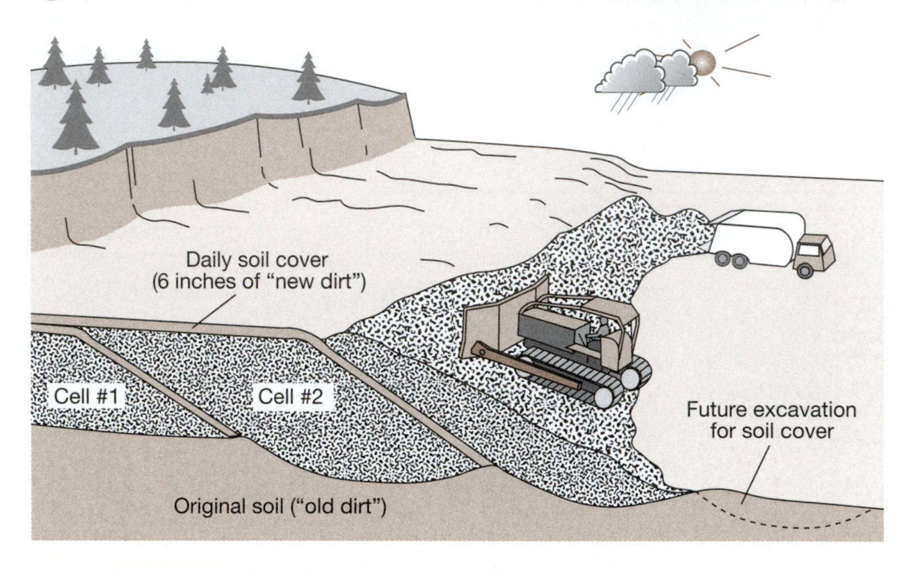

Figure 14.13 A simple landfill, designed to bury waste and recover (replant) the site. There are no provisions for managing leachate and gases.

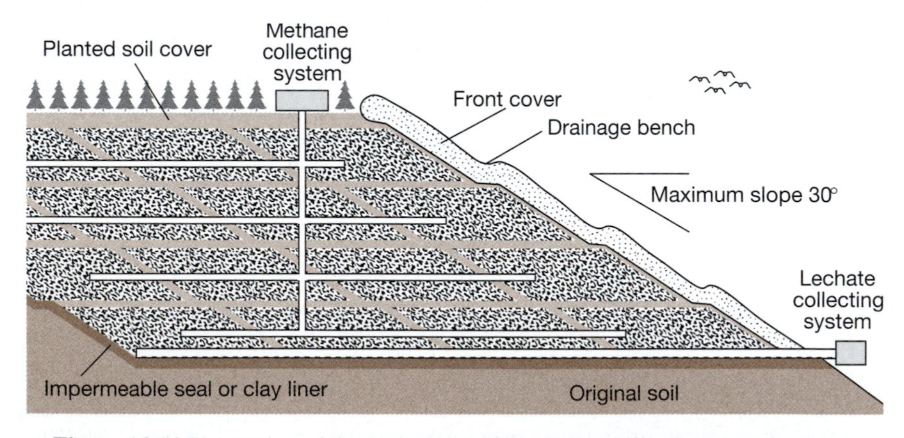

Figure 14.14 Elaboration of the sanitary landfill at Puente Hills, Los Angeles County with liner, leachate collection system, and methane collection system.

(After B.W. Pipkin and D.D. Trent, 2001)

Q14.26 Why do you suppose that there is a maximum slope angle dictated for the face of the landfill in Figure 14.14? *Hint:* This hazard is compounded in Southern California by earthquakes.

Information about landfill gases is a http://www.epa.gov/outreach/lmop/products/factsheet.htm

F. Gulf of Mexico "Dead Zone"

The New York Times, JANUARY 20, 1998

A 'Dead Zone' Grows in the Gulf of Mexico

A worried White House dispatches teams of scientists.

It can stretch for 7,000 square miles off the coast of Louisiana, a vast expanse of ocean devoid of the region's usual rich bounty of fish and shrimp, its bottom littered with the remains of crabs and worms unable to flee its suffocating grasp. This is the gulf of Mexico's "dead zone," which last summer reached the size of the state of New Jersey.

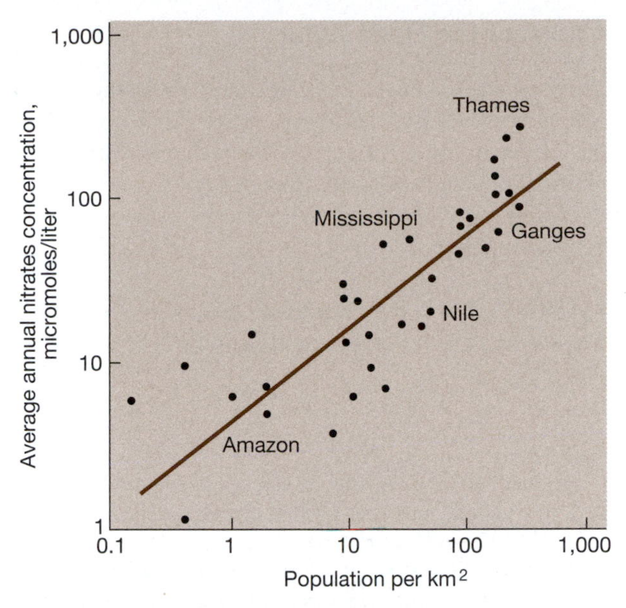

Figure 14.15 Human population density correlates with levels of pollutants in river waters. (*After B. Peierls, et al., 1991*)

Mobile organisms flee, immobile organisms die. Such is the case within water in a state of **hypoxia** (Gr. *hypo*, under or less than—in this case, less than sufficient oxygen). Water is considered hypoxic where dissolved oxygen is less than 2 mg/L, which is about 20% of shallow sea water's capacity at mid-latitude salinity and temperature.

A number of aquatic environments can be naturally hypoxic, but coastal waters at the mouths of rivers are especially at risk owing to **anthropogenic** (human) activities (Fig. 14.15). The culprits appear to be nitrates and phosphates from developed lands. Perhaps you have noticed unsightly algae choking a suburban pond—algae that are nourished by runoff from fertilized lawns and shrubbery (a process called **cultural eutrophication**). Where such nutrients find their way into oceans, they promote similar overabundance of algae, which in turn promotes an overabundance of microscopic marine animals. Decay of such abundant organisms can deprive water of its dissolved oxygen. Such is the case in the Gulf of Mexico hypoxic **dead zone**, which has been growing since its discovery in the early 1980s (Fig. 14.16).

Q14.27 What is the apparent direction of flow of the longshore current along the Louisiana coast if it's fed by the Mississippi River—east to west or west to east?

Scientific sampling of water offshore of Louisiana indicates a surface layer of relatively fresh river water floating on a layer of hypoxic seawater.

Q14.28 What is there about the chemical compositions of river water and seawater that explains this stratifcation (i.e., layering) within water offshore of Louisiana (or offshore of any river, for that matter)?

Q14.29 (A) Why does hypoxia tend to develop in deeper water? *Hint:* What depletes deeper water of its oxygen? (See Figure 13-22 on page 239) (B) Why does the hypoxic condition of deeper water tend to persist? *Hint:* What is the source of oxygen in seawater?

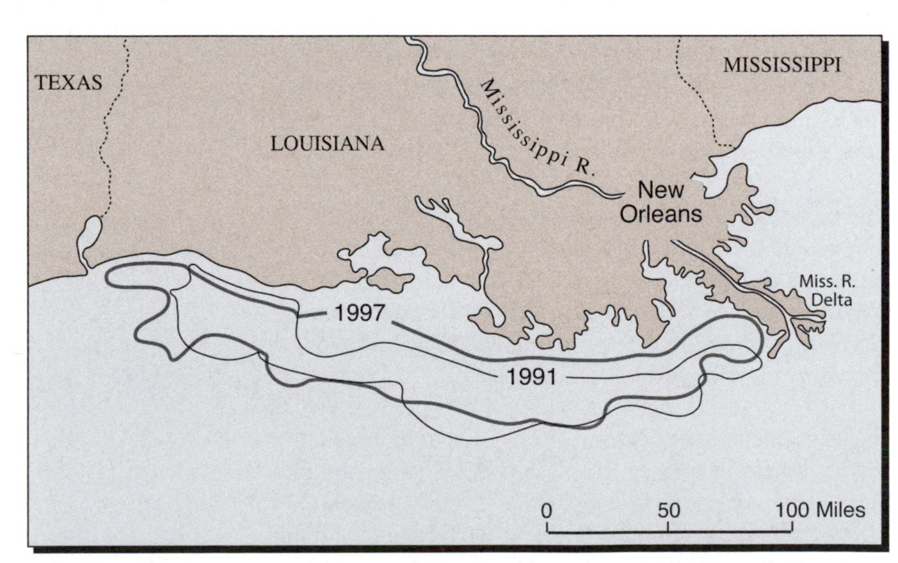

Figure 14.16 This map shows growth in the area of the hypoxic zone from 1991 to 1997. (*Based on data from Nancy Rabalais and her team at Louisiana Universities Marine Consortium*)

Cause of the Dead Zone

There is debate between marine scientists on the one hand and agricultural and fertilizer industries on the other over the source of nitrates and phosphates that account for the Gulf of Mexico dead zone. However, evidence pointing to the Mississippi River and its tributaries as the source area is compelling indeed.

Q14.30 What is the circumstantial evidence shown in Figure 14.16 (previous page) that points to the interior of the American Midwest as the source of excessive nitrates in this dead zone?

Q14.31 Examine the histogram in Figure 14.17, which is a plot of the annual variation in the size of the dead zone over a recent 15-year period. What evidence lies within this graphic that points to the Mississippi River and its tributaries as the source of troublesome nitrates and phosphates?

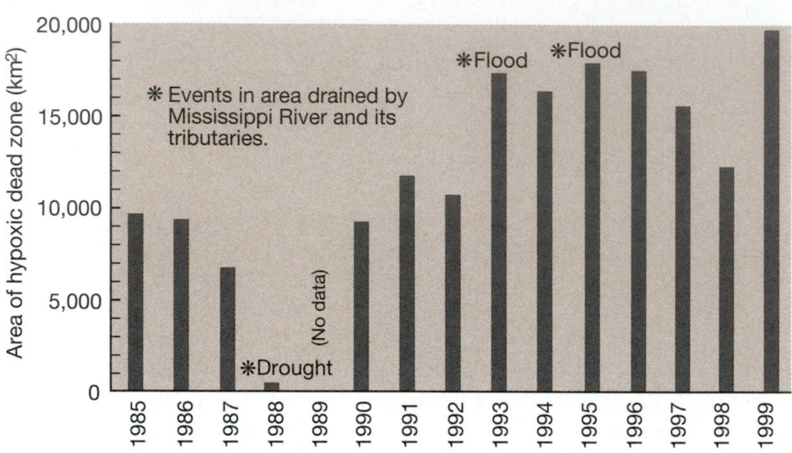

Figure 14.17 This histogram shows growth in the hypoxic dead zone year-by-year during the period 1985–1999. Added are three events that occurred within the Mississippi River drainage basin within this interval of time. (*From Nancy Rabalais and her team at Louisiana Universities Marine Consortium*)

Mitigation of the problem—Conventional wisdom is that the source of upstream pollutants is largely agricultural (i.e., from fertilizers and manure), but scientists with the *U.S. Geological Survey* also list…

- Sewage treatment waste water
- Industrial waste (e.g., pulp mill effluents)
- Atmospheric pollutants from the burning of fossil fuels

The most promising suggestion for the mitigation of the Gulf of Mexico dead zone is wetland restoration and creation within the Mississippi River drainage basin. Microbiological processes within wetland surface water and groundwater effectively **denitrify** the water—that is, they make nitrogen unavailable as a nutrient for plant uptake by removing it from the water as nitrogen gas.

New threat to oxygen in the Gulf of Mexico

The BP 2010 Deepwater Horizon blowout on the mile-deep-seafloor of the Gulf of Mexico prompted this from Professor John Kessler of Texas A&M University: "This is the most vigorous methane eruption in modern human history."

The National Science Foundation has funded research on methane in the Gulf amid concerns about the depths of the oil plume and questions about what role natural gas is playing in keeping the oil below the surface, said David Garrison, a program director in the federal agency who specializes in biological oceanography.

In early June, a research team led by Samantha Joye of the Institute of Undersea Research and Technology at the University of Georgia investigated a 15-mile long plume drifting southwest from the leak site. They said they found methane concentrations up to 10,000 times higher than normal and oxygen levels depleted by 40 percent or more. In an e-mail, Joye called her findings "the must bizarre-looking oxygen profiles" she has ever seen anywhere.

Columbia Daily Tribune, June 18, 2010

Methane becomes new Gulf concern

High gas levels threaten wildlife

NEW ORLEANS (AP)—It is an overlooked danger in the oil spill crisis: The crude gushing from the well contains vast amounts of natural gas that could pose a serious threat in the Gulf of Mexico's fragile ecosystem.

The oil emanating from the seafloor contains about 40 percent methane, compared with about 5 percent found in typical oil deposits, said John Kessler, a Texas A&M oceanographer who is studying the impact of methane from the spill.

That means huge quantities of methane have entered the Gulf, scientists say, potentially suffocating marine life and creating "dead zones" where ocygen is so depleted that nothing lives.

F. Gulf of Mexico "Dead Zone"

The New York Times, JANUARY 20, 1998

A 'Dead Zone' Grows in the Gulf of Mexico

A worried White House dispatches teams of scientists.

It can stretch for 7,000 square miles off the coast of Louisiana, a vast expanse of ocean devoid of the region's usual rich bounty of fish and shrimp, its bottom littered with the remains of crabs and worms unable to flee its suffocating grasp. This is the gulf of Mexico's "dead zone," which last summer reached the size of the state of New Jersey.

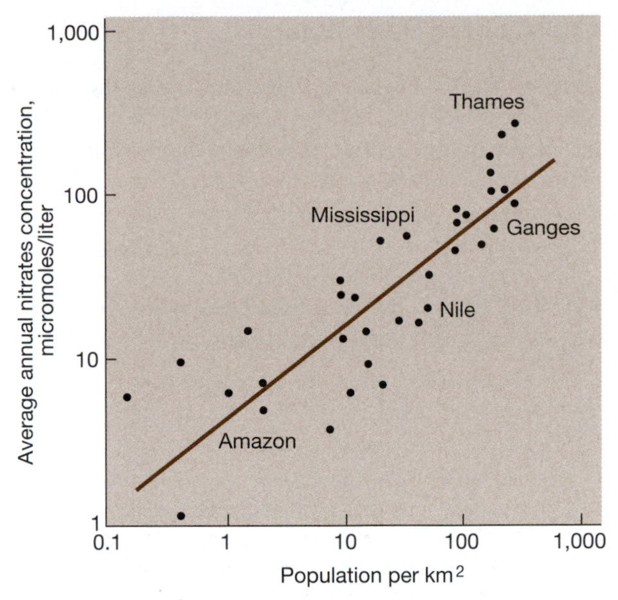

Figure 14.15 Human population density correlates with levels of pollutants in river waters. (*After B. Peierls, et al., 1991*)

Mobile organisms flee, immobile organisms die. Such is the case within water in a state of **hypoxia** (Gr. *hypo*, under or less than—in this case, less than sufficient oxygen). Water is considered hypoxic where dissolved oxygen is less than 2 mg/L, which is about 20% of shallow sea water's capacity at mid-latitude salinity and temperature.

A number of aquatic environments can be naturally hypoxic, but coastal waters at the mouths of rivers are especially at risk owing to **anthropogenic** (human) activities (Fig. 14.15). The culprits appear to be nitrates and phosphates from developed lands. Perhaps you have noticed unsightly algae choking a suburban pond—algae that are nourished by runoff from fertilized lawns and shrubbery (a process called **cultural eutrophication**). Where such nutrients find their way into oceans, they promote similar overabundance of algae, which in turn promotes an overabundance of microscopic marine animals. Decay of such abundant organisms can deprive water of its dissolved oxygen. Such is the case in the Gulf of Mexico hypoxic **dead zone**, which has been growing since its discovery in the early 1980s (Fig. 14.16).

Q14.27 What is the apparent direction of flow of the longshore current along the Louisiana coast if it's fed by the Mississippi River—east to west or west to east?

Scientific sampling of water offshore of Louisiana indicates a surface layer of relatively fresh river water floating on a layer of hypoxic seawater.

Q14.28 What is there about the chemical compositions of river water and seawater that explains this stratifcation (i.e., layering) within water offshore of Louisiana (or offshore of any river, for that matter)?

Q14.29 (A) Why does hypoxia tend to develop in deeper water? *Hint:* What depletes deeper water of its oxygen? (See Figure 13-22 on page 239) (B) Why does the hypoxic condition of deeper water tend to persist? *Hint:* What is the source of oxygen in seawater?

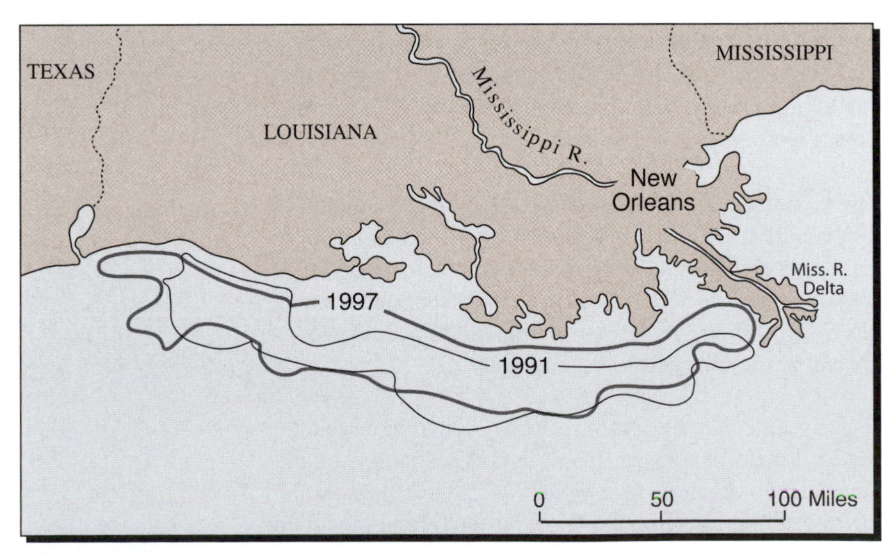

Figure 14.16 This map shows growth in the area of the hypoxic zone from 1991 to 1997. (*Based on data from Nancy Rabalais and her team at Louisiana Universities Marine Consortium*)

Cause of the Dead Zone

There is debate between marine scientists on the one hand and agricultural and fertilizer industries on the other over the source of nitrates and phosphates that account for the Gulf of Mexico dead zone. However, evidence pointing to the Mississippi River and its tributaries as the source area is compelling indeed.

Q14.30 What is the circumstantial evidence shown in Figure 14.16 (previous page) that points to the interior of the American Midwest as the source of excessive nitrates in this dead zone?

Q14.31 Examine the histogram in Figure 14.17, which is a plot of the annual variation in the size of the dead zone over a recent 15-year period. What evidence lies within this graphic that points to the Mississippi River and its tributaries as the source of troublesome nitrates and phosphates?

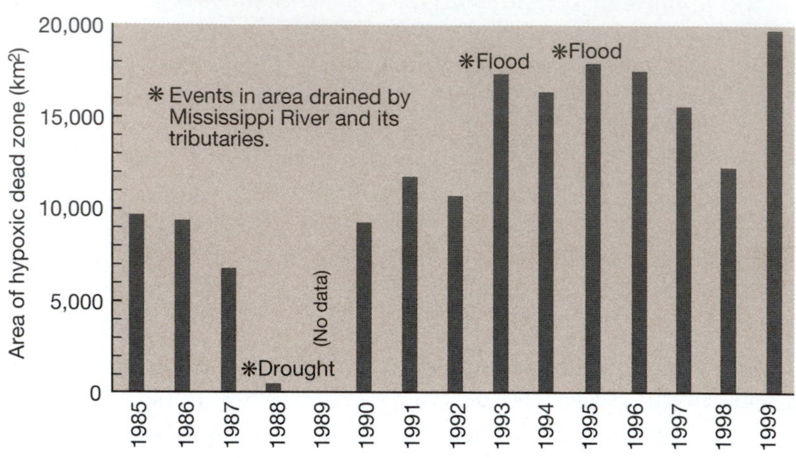

Figure 14.17 This histogram shows growth in the hypoxic dead zone year-by-year during the period 1985–1999. Added are three events that occurred within the Mississippi River drainage basin within this interval of time. (*From Nancy Rabalais and her team at Louisiana Universities Marine Consortium*)

Mitigation of the problem—Conventional wisdom is that the source of upstream pollutants is largely agricultural (i.e., from fertilizers and manure), but scientists with the *U.S. Geological Survey* also list…

- Sewage treatment waste water
- Industrial waste (e.g., pulp mill effluents)
- Atmospheric pollutants from the burning of fossil fuels

The most promising suggestion for the mitigation of the Gulf of Mexico dead zone is wetland restoration and creation within the Mississippi River drainage basin. Microbiological processes within wetland surface water and groundwater effectively **denitrify** the water—that is, they make nitrogen unavailable as a nutrient for plant uptake by removing it from the water as nitrogen gas.

New threat to oxygen in the Gulf of Mexico

The BP 2010 Deepwater Horizon blowout on the mile-deep-seafloor of the Gulf of Mexico prompted this from Professor John Kessler of Texas A&M University: "This is the most vigorous methane eruption in modern human history."

The National Science Foundation has funded research on methane in the Gulf amid concerns about the depths of the oil plume and questions about what role natural gas is playing in keeping the oil below the surface, said David Garrison, a program director in the federal agency who specializes in biological oceanography.

In early June, a research team led by Samantha Joye of the Institute of Undersea Research and Technology at the University of Georgia investigated a 15-mile long plume drifting southwest from the leak site. They said they found methane concentrations up to 10,000 times higher than normal and oxygen levels depleted by 40 percent or more. In an e-mail, Joye called her findings "the must bizarre-looking oxygen profiles" she has ever seen anywhere.

Columbia Daily Tribune, June 18, 2010

Methane becomes new Gulf concern

High gas levels threaten wildlife

NEW ORLEANS (AP)—It is an overlooked danger in the oil spill crisis: The crude gushing from the well contains vast amounts of natural gas that could pose a serious threat in the Gulf of Mexico's fragile ecosystem.

The oil emanating from the seafloor contains about 40 percent methane, compared with about 5 percent found in typical oil deposits, said John Kessler, a Texas A&M oceanographer who is studying the impact of methane from the spill.

That means huge quantities of methane have entered the Gulf, scientists say, potentially suffocating marine life and creating "dead zones" where ocygen is so depleted that nothing lives.

15 Hazardous Waste

Topics

A. What are some of the hazardous wastes associated with (a) commerce, services, and agriculture; (b) small-scale industry; and (c) large-scale industry?
B. What is the anatomy of a secure landfill? What is the bathtub effect, and what causes it? What is the history of the contaminant groundwater plume on Cape Cod?
C. What is the history of contamination at Love Canal, and how has it been mitigated?
D. What can go wrong with deep-well disposal of hazardous waste?
E. What is the present-day example of hazardous waste incineration, and what are the concerns?
F. What are the four laws of ecology developed by Barry Commoner in 1971? What are the three levels of radioactive waste, and what is their general distribution in the United States? Where and what are the five sites that were considered as possible deep-earth repositories? What are the nine items in the wish list of secure geologic features? What is the general geologic setting of the Yucca Mountain site? What is the projected timeline for its development? What now?
G. What is biomagnification of hazardous waste, and what are the details of the Clear Lake, California, example?

A. Hazardous wastes

Hazardous wastes are those that pose a present or potential threat to the health and well-being of humans and wildlife (Table 15.1). People fight the building of hazardous waste facilities in their neighborhoods. Objections range from diminished property values to adverse health effects. A poster in the office of the Evironmental Protection Agency (*EPA*) in Washington, D.C. lists a few of the objections raised by citizens and politicians.

The ABCs of Waste Disposal

NIMBY...Not in my back yard
NIMFYE...Not in my front yard either
PIITBY...Put it in their back yard
NIMEY...Not in my election year
NIMTOO...Not in my term of office
LULU...Locally unavailable land use
NOPE...Not on planet earth

Q15.1 Which, if any, of the hazardous wastes in Table 15.1 are produced near *your home?*

Table 15.1
Wastes identified by the *EPA* as being potentially carcinogenic, tumor-causing, and/or mutagenic.

Sector	Source	Hazardous waste
Commerce, services, agriculture	Vehicle servicing	Oil and gasoline
	Airports	Oil, hydraulic fluid
	Dry cleaning	Solvents
	Electrical transformers	PCBs
	Hospitals	Pathogenic/infectious wastes
	Farms and parks	Pesticides, containers
Small-scale industry	Metal treating	Acids, heavy metals
	Photofinishing	Solvents, acids, silver
	Textile processing	Cadmium, mineral acids
	Printing	Solvents, inks, dyes
	Leather tanning	Solvents, chromium
Large-scale industry	Oil refining	Spent catalysts, oily wastes
	Chemical industry	Tar residues, solvents
	Chlorine production	Mercury

From The World Environment, 1972–1992

B. Secure landfills

Hazardous wastes require **secure landfills**, which are elaborate systems fully enclosed by water-tight synthetic and/or clay liners (Fig. 15.1). Collecting systems keep the landfill free of leachate (aka garbage juice) and gas (a mix of methane and carbon dioxide), which are removed and treated in separate facilities. Special wells monitor the condition of the groundwater. But if certain of the features of a secure landfill are omitted, a phenomenon called the 'bathtub effect' can develop.

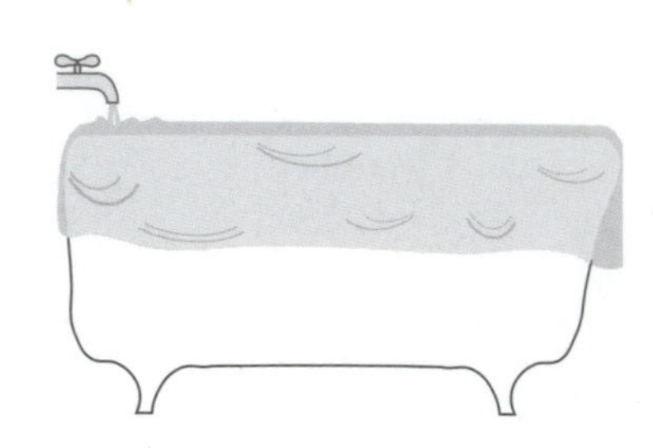

Q15.2 (A) What do you suppose the bathtub effect is, and (B) which of the features in Figure 15.1 is essential for the prevention of the bathtub effect?

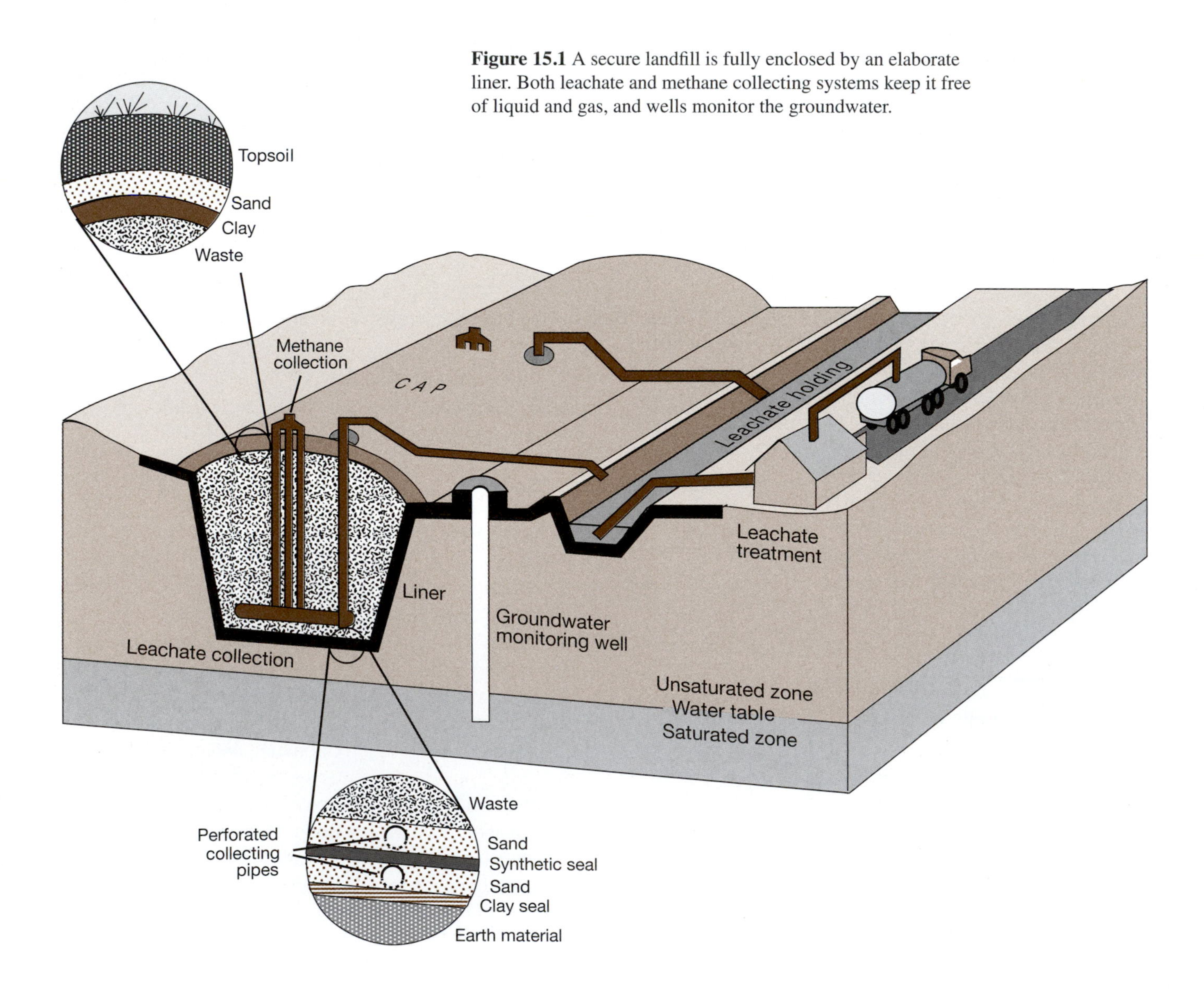

Figure 15.1 A secure landfill is fully enclosed by an elaborate liner. Both leachate and methane collecting systems keep it free of liquid and gas, and wells monitor the groundwater.

Lack of a secure landfill—Cape Cod, Mass.

A number of contaminated areas in northeastern United States have resulted from disposal of waste water into landfills without benefit of proper liners. Moreover, some sites are in sediments ill-suited for such disposal. An example is on Cape Cod, Massachusetts (Fig. 15.2), an area that is underlain by coarse sand and gravel deposited by retreating glaciers 10,000 years ago. Porosity within these sediments is around 35%, and permeability is high.

During the period 1936 to 1986 Otis Air Base discharged 2.5-billion gallons of sewage from a secondary-treatment plant, plus waste from a fuel dump, into insecure pits. Contaminants included the carcinogens tetrachloroethene and trichloroethene, along with nonbiodegradable detergents that were allowed before 1964.

Movement of the contaminant plume has been tracked by injecting dye into water wells and monitoring them over time. Conclusion: The plume is moving along with the groundwater at a rate of approximately one foot per day.

Q15.3 Refer to the map in Figure 15.2B. (A) In what direction is the groundwater moving, as indicated by contours on the water table? (B) Is the shape of the contaminant plume consistent with your answer to A? Explain.

Q15.4 Given the rate of movement of one foot per day, and from information in Figure 15.2, compute the approximate number of years before the contaminant plume reaches Nantucket Sound.

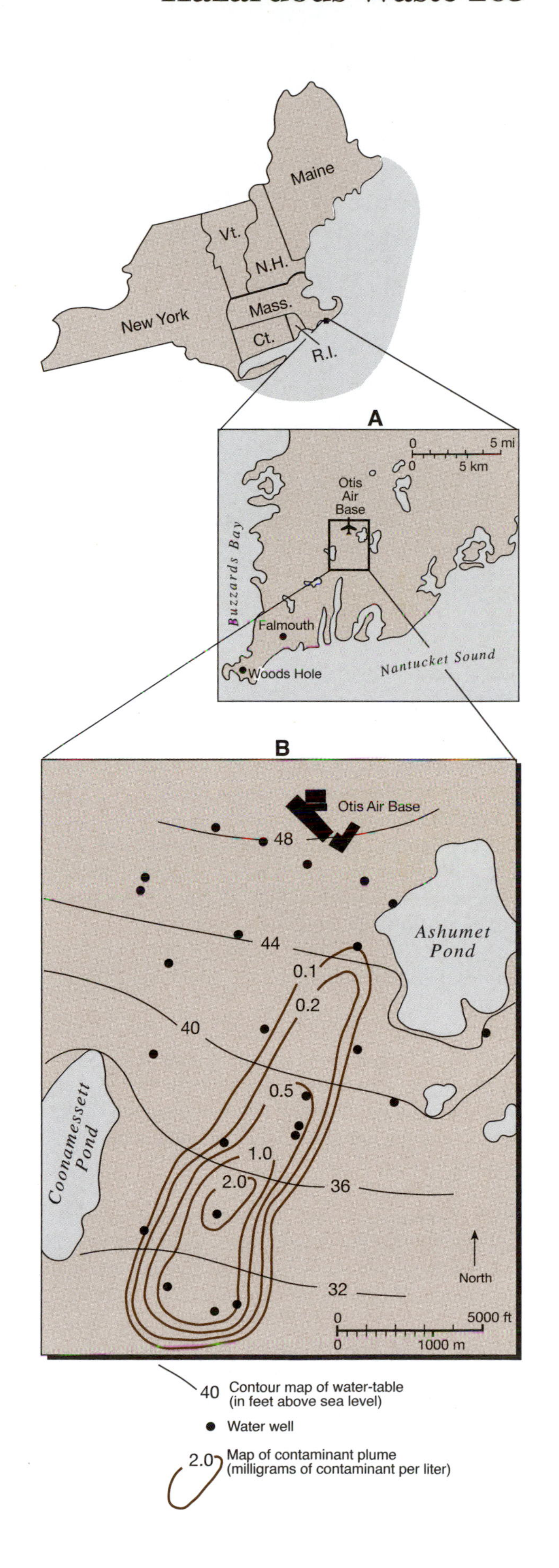

Figure 15.2 The plume of contaminated groundwater on Cape Cod is 3,000–5,000 feet wide and three miles long. Contours on the water table delineate hydraulic gradient. *(From U.S. Geological Survey)*

The U.S. government estimates that it would cost $250 million over 20 years to eliminate the contaminated groundwater plume on Cape Cod by pumping it to the surface, treating it with tertiary processes, and pumping it back into the subsurface.

Q15.5 If, instead of coarse sand and gravel, sediments on Cape Cod consisted of fine sand, silt, and some clay, (A) how would the rate of movement of the plume be affected? (B) How might this difference in sediment particle size affect the efficiency with which microbial processes mitigate the contaminants? *Hint:* Microbes attach themselves to solid surfaces.

C. Love Canal

EPA press release: June 1, 1989

Occidental Chemical signs consent order for storage and destruction of Love Canal wastes

Calling the occasion a "landmark event," New York and federal officials today announced that Occidental Chemical Corporation has signed a consent order with the state and federal governments obligating the company to take over storage and destruction of wastes which originated at Love Canal. The agreement represents Occidental's first acknowledgment and assumption of responsibility for cleanup work associated with Love Canal in the ten years since the toxic waste dump was discovered.

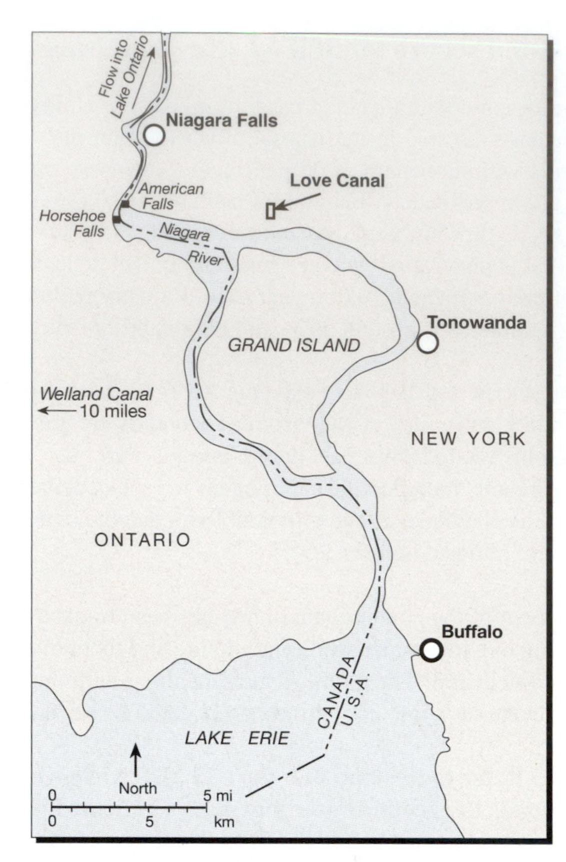

Figure 15.3 Love Canal became our nation's very first environmental 'federal disaster area.'

Origin

Love Canal, an 18-acre tract near the town of Niagara Falls, New York (Fig. 15.3), was planned by William T. Love in the 1890s as an American shipping canal on the U.S. side of Niagara Falls. His idea was to compete with Welland Canal in Canada—which connects Lake Ontario with Lake Erie. After excavating a trench 1,000 meters long, 25 meters wide, and 35 meters deep (Figs. 15.3 and 15.4), Love abandoned the project for lack of resources.

Love Canal timeline

1930s—Hooker Electrochemical Corp. purchases Love Canal from William Love for use as a chemical waste landfill. Hooker disposes of thousands of tons of drummed wastes—including chemical residues, process sludge, and fly ash. The site is also used by the City of Niagara Falls for disposal of solid waste.

1953—Hooker covers the site with an 'impermeable' clay cap and sells it to the Board of Education, City of Niagara Falls for one dollar. An elementary school is built on top of the canal, and hundreds of tract homes soon follow.

1970s—Following a period of extraordinary rainfall, water breaks through the clay cap, and the community begins to suffer anomalous miscarriages, birth defects, liver cancer, and seizures among children.

Q15.6 Recall the bathtub effect. Generally speaking, how does that model apply in the case of Love Canal?

1978—EPA detects benzene, chlorinated hydrocarbons, and dioxin in home basements. President Jimmy Carter declares Love Canal a federal disaster area—the first polluted area to be so designated. One thousand families are relocated, and two hundred homes are razed.

1980—Passage of the law providing for remediation of polluted groundwater—the Comprehensive Environmental Response, Compensation, and Liability Act (aka Superfund), which authorized the Environmental Protection Agency to define the cleanup that polluters must perform. If polluters refuse, the EPA performs the cleanup and bills the polluters. Costs can amount to millions or hundreds of millions of dollars. In some cases, the identity of the polluters is lost in antiquity. When this occurs, the present landowners are held responsible.

Q15.7 You are considering the purchase of a lot in the center of a village on which to build a gift shop. But you hear that it was once the site of a gas station. Why should you visit the office of records and deeds?

Finale. Occidental Chemnical Corp., which purchased Hooker in 1968, paid Superfund $102 million, the Federal Emergency Management Agency (FEMA) paid $27 million, and the state of New York paid $98 million. The approximate total for the cleanup of Love Canal: *$275 million*.

The original anatomy of the Love Canal landfill is shown in Figures 15.4 and 15.5. (*Produced by Fred C. Hart Associates, Inc. in 1978, preparatory to the cleanup*)

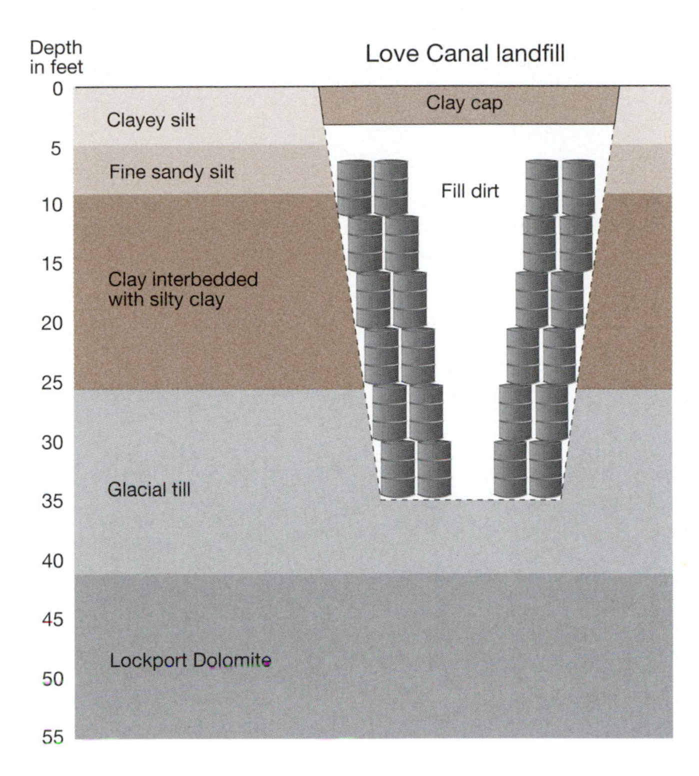

Figure 15.4 Geologic formations of the Love Canal landfill.

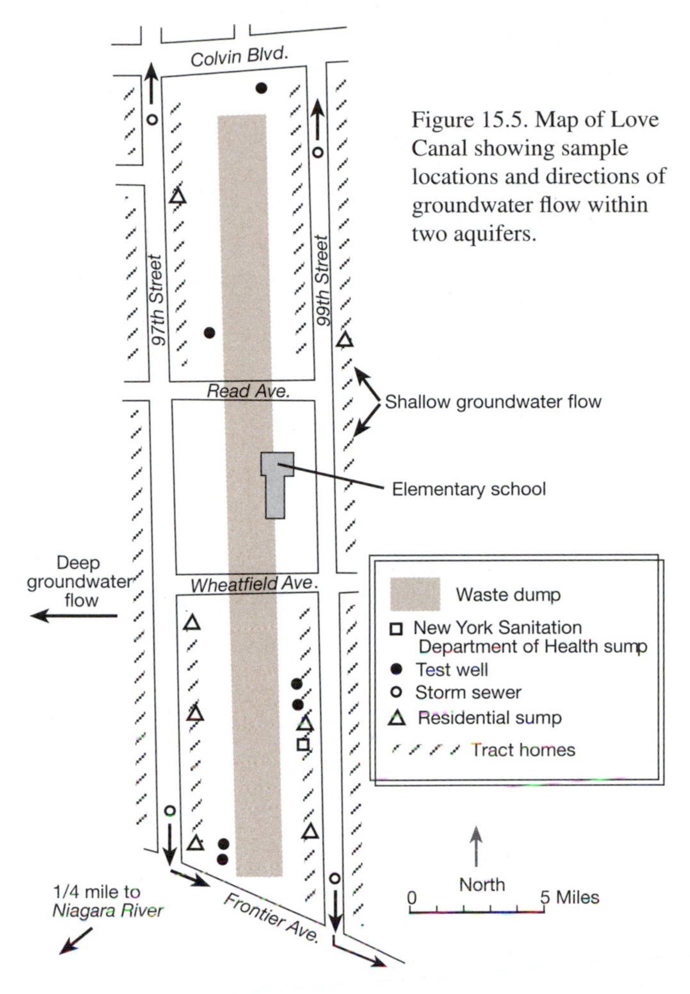

Figure 15.5. Map of Love Canal showing sample locations and directions of groundwater flow within two aquifers.

Remediation mandates from EPA (Fig. 15.6):

1. Cover the landfill with an arched clay cap.
2. Install barriers to prevent leachate from flowing out of the landfill, and groundwater from flowing into the landfill.
3. Install drain pipes at the base of the clay barriers to collect leachate and groundwater, and pipe both to the city processing facility.
4. Collect and dispose of surface runoff.

Q15.8 Encircle and number three of the four preceding remedial measures on the diagram at 15.8 on the Answer Page.

Q15.9 Figure 15.5 shows shallow groundwater flowing toward the landfill. Again, what measure was taken to stop that flow into the landfill?

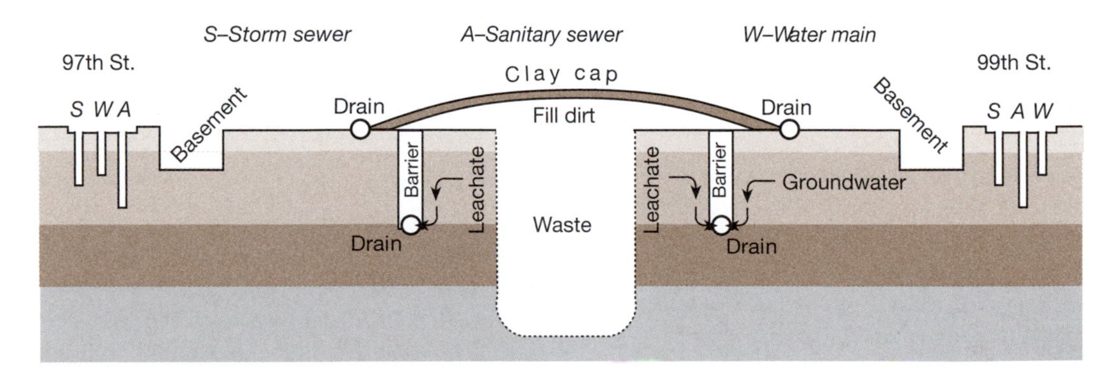

Figure 15.6 Before remediation, leachate was flowing at the level of storm sewers, sanitary sewers, and water mains. *After EPA*

See Love Canal Collection at http://library.buffalo.edu/libraries/specialcollections/lovecanal/index/html

D. Deep-well disposal

In some areas hazardous liquids can be safely injected under pressure into porous and permeable rocks containing ancient connate waters like those associated with petroleum. Such nasty waters are separated from overlying potable waters by relatively impermeable aquitards (Fig. 15.7).

The siting of a disposal well should be such that pressure is dissipated within the repository rock—thereby maintaining its integrity. In addition, care should be taken in placing well casings above the repository, like that in oil wells.

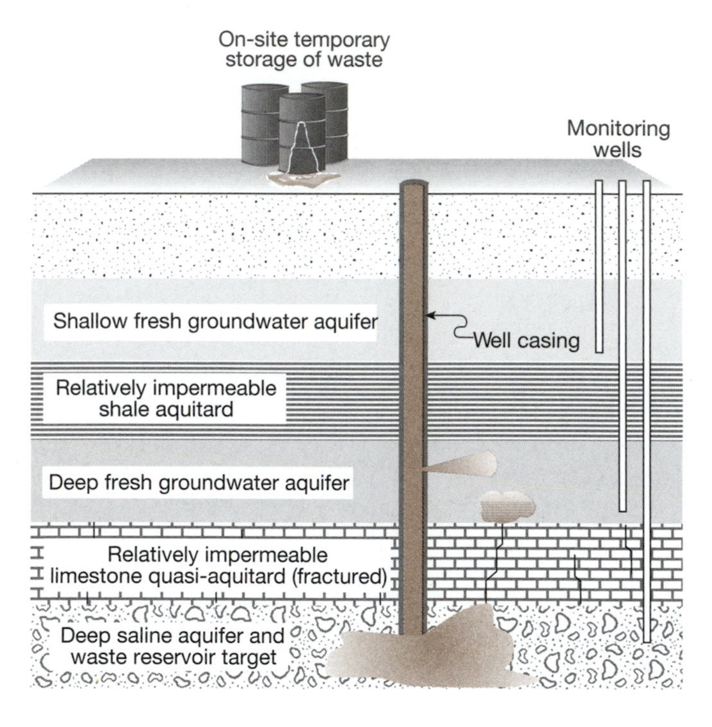

Figure 15.7 The anatomy of a deep-well disposal system.

Q15.10. Several things can go wrong with deep-well disposal of hazardous liquid. Figure 15.7 illustrates three apparent problems. Briefly describe each.

Trouble at the U.S. Army's RMA disposal well

A **troublesome development** surrounded deep-well disposal at the Rocky Mountain Arsenal (RMA) northeast of Denver, Colorado during the period 1962–1965. A well was drilled through 3,638 m of Phanerozoic sedimentary rock into a target repository of fractured Precambrian granite. The history of injection is illustrated in Figure 15.8.

With the exception of one earthquake in 1882, there was no history of seismic activity in the Denver area, but in the 1960s a swarm of earthquakes occurred (Fig. 15.9). The quakes measured 1.5–4.4 on the Richter scale of earthquake magnitude, and the depths of their foci were around 4 km.

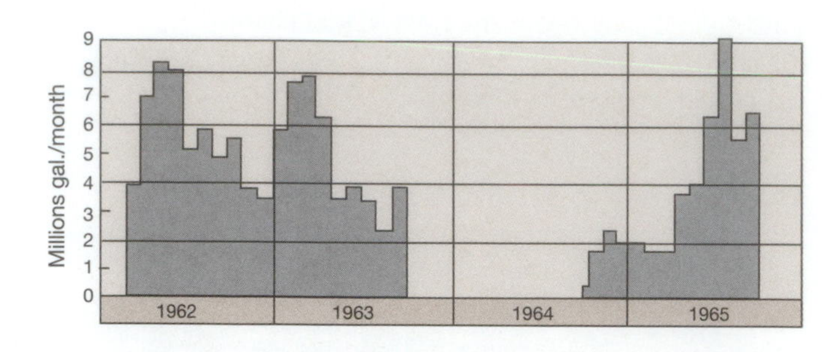

Figure 15.8 Record of injection of liquid in the Rocky Mountain Arsenal deep disposal well for the period 1962–1965.

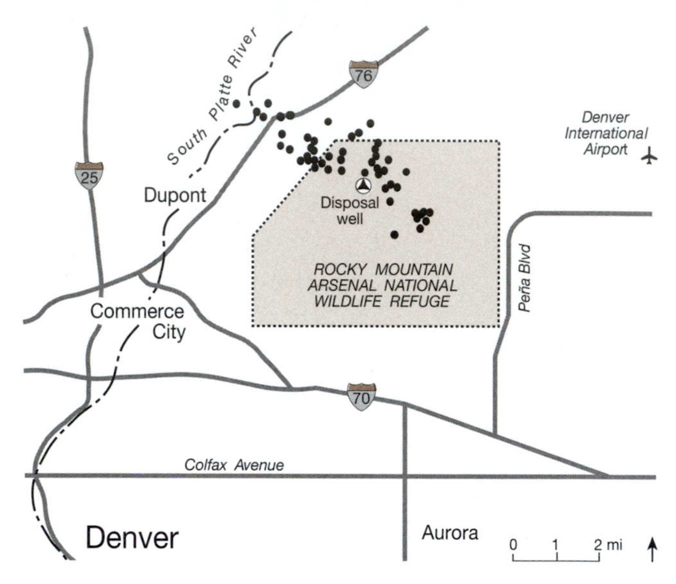

Figure 15.9 Location of the RMA deep disposal well. Black dots mark earthquakes epicenters. After J.H. Healy et al., Science, 1968.

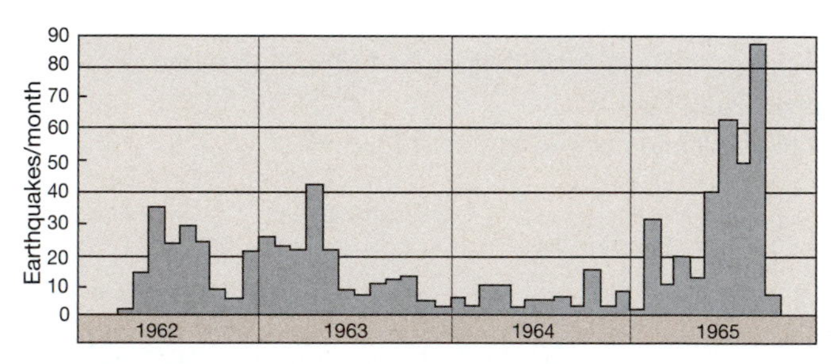

Figure 15.10 Frequency of earthquakes for the period 1962–1965.

Q15.11. (A) What is there about Figures 15.8 and 15.10 that suggests that deep-well disposal at RMA triggered earthquakes in the Denver area? (B) Are depths of foci consistent with this cause-and-effect? Explain.

The Rocky Mountain Arsenal case is not unique. Other instances of injection-induced earthquakes have been reported from such geographically disparate places as northeast Ohio, east Tennessee, south Arkansas, north Texas, Kuwait, and China.

E. Incineration

Times Beach, Missouri

The town of Times Beach, Missouri, 25 miles southwest of St. Louis, captured the attention of the nation in 1983 when *EPA* shut it down after discovering toxic levels of **dioxin**, a known human carcinogen, within its soil. *EPA* permanently relocated the 2,081 residents of Times Beach and dismantled homes and businesses at a cost of $30 million. Heating (**incineration**) of soil from Times Beach and 27 other sites in eastern Missouri cost an additional $170 million.

Times Beach dioxin was a byproduct in the manufacture of hexachlorophene in the 1970s near Springfield, Missouri. The manufacturer hired a waste oil handler to dispose of its chemical waste, which the handler mixed with used motor oil and sprayed on streets and parking lots in Times Beach as a dust control measure. *Dilution is the solution to pollution*. Or so the handler apparently thought.

Incineration, safe or unsafe? In 1983, incineration was judged by *EPA* to be a proven and cost-effective method for destroying dioxin. By the end of 1997, some 92,000 cubic yards of dioxin-contaminated soil from the 28 Missouri sites had been heat treated, and cleanup was reported by *EPA* as having been completed.

Q15.12 Given the cost of incineration of dioxin-contaminated Missouri soil, exclusive of the cost of relocating residents, what was the approximate cost per cubic yard?

Although furnace temperatures do break down dioxin, it is reconstituted in cooler parts of incinerators and exits via smokestacks in fly ash and gas (Fig. 15.11). Toxic bottom ash and fly ash captured by scrubbers must be buried in secure landfills. As concerns hazardous waste 'disposal,' incinerators are probably the worst of several bad ideas.

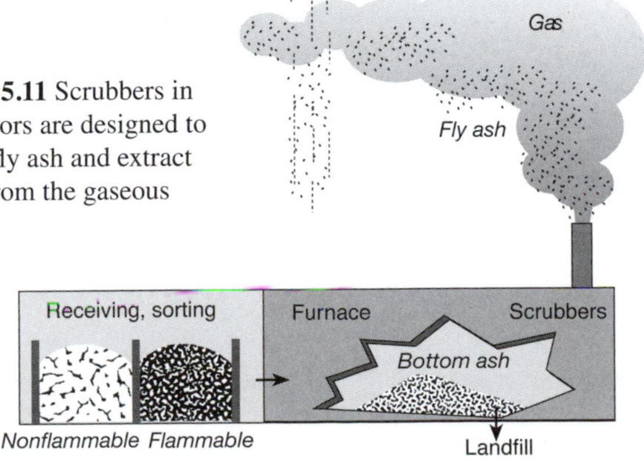

Figure 15.11 Scrubbers in incinerators are designed to capture fly ash and extract metals from the gaseous effluent.

American weapons of mass destruction

In compliance with international agreements, the Army has begun its $1 billion program to 'destroy' 2,000 tons of chemical waste stored near Anniston, Alabama.

COLUMBIA DAILY TRIBUNE, AUGUST 10, 2003

Weapons incinerations 'flawless,' first of many

ANNISTON, Ala. (AP) — Yesterday the Army fired up its first chemical weapons incinerator located near a residential area and destroyed two rockets loaded with enough sarin nerve agent to wipe out a city. Workers loaded each 6-foot-long rocket onto a conveyor belt and sent it into a sealed room, where it was drained of 1.2 gallons of the deadly chemical and chopped into eight pieces. Those pieces were fed into an 1,100-degree furnace, producing slag that will be trucked to a hazardous waste landfill in western Alabama. The sarin was directed into a holding tank, to be held until there is enough to burn in a large batch, probably in late October.

AP graphic

The nerve agent 'VS' and mustard gas are also stored at Anniston, but officials decided to begin with sarin rockets because hundreds of them are leaking. It is believed that incinerating the weapons is safer than storing them.

Sarin is so deadly that one drop on the skin can kill. Although residents have been assured by officials that incineration is safe, the Army is distributing protective hoods and other safety gear to 35,000 people who live within nine miles of the incinerator. There are also plans for outfitting some schools with ventilation equipment designed to keep out lethal fumes in case of an accident. These procedures add fuel to some environmentalists' claims that incineration is supremely invasive—both visually and materially.

Q15.13 In addition to the matter of aesthetics, which of the four earth spheres can likely be polluted by incineration: lithosphere (e.g., soil), hydrosphere (water), atmosphere (air), and biosphere (plants and animals)?

F. Radioactive waste

Beginning in the 1950s—when it was realized that the Atomic Age would result in huge quantities of commercial and military radioactive wastes—we began to brainstorm about ways in which such wastes might be managed. There has been no shortage of suggestions:

• Burying it within salt domes. Problems: Salt is the most fluid of rocks. And brines associated with salt would soon corrode any imaginable waste container.

• Dumping it into oceans. Problems: Oceans, expansive as they might appear, are already showing disturbing effects of decades of waste dumping.

• Burying it deep within ice sheets.

Q15.14. What is a notable byproduct of radioactive decay that would surely affect an ice vault?

• Shooting it into the sun.

Q15.15. What impact did the Challenger disaster of January 28, 1986 have on this proposal?

• Dumping it into a convergent plate boundary (subduction zone).

Q15.16. Northern Indian Ocean floor descended beneath Asia until India collided with Asia during the Eocene Epoch, 40–50 million years ago. What has happened—and continues to happen—since that time? *Hint:* Eocene marine fossils occur atop Mt. Everest.

• Storage at or near the point of origin in protective containers at ground level and/or within shallow-burial repositories. The principal advantage of on-site storage is not having to transport it over Amerca's highways and byways.

Q15.17. Name a few natural and human-induced disasters that can affect surface repositories.

• Storage in containers in deep-earth volcanic ash (aka tuffs). This is where our interest and efforts presently lie.

In 1971, biologist Barry Commoner, in his book, *The Closing Circle*, developed four laws of ecology:

1. *Everything is connected to everything else.*
2. *Everything must go somewhere.*
3. *Nature knows best.*
4. *There's no such thing as a free lunch.*

Q15.18. Which of Barry Commoner's four laws of ecology most applies to the debate outlined above?

Unlike chemical waste, radioactive waste cannot be mitigated with chemical neutralization or bioremediation.

Q15.19. (A) Do you suppose that radioactive waste could be disposed of through incineration, as can, say, hospital infectious waste? (B) Why, or why not?

Practically speaking, radioactive waste is *forever*. So what are we to do about the *exponential growth* in this supremely hazardous material?

Computing doubling times. The doubling time of a quantity undergoing exponential growth can be computed by dividing the number 70 by its annual rate of growth.

Q15.20. Judging from the graph in Figure 15.12, (A) what was the approximate annual rate of cumulative growth in global nuclear fuel waste stored during the period 1985–1990: 8%, 14%, or 19%? (B) Has the annual rate of cumulative growth during the period 1965–2000 decreased, remained about the same, or increased?

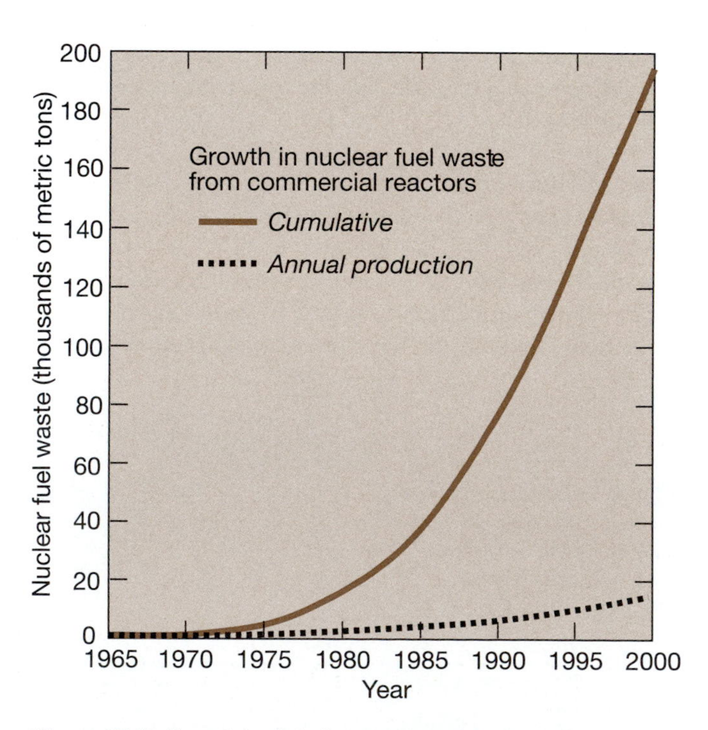

Figure 15.12 Growth in global production of nuclear fuel waste from commercial nuclear reactors, 1965–2000. (*From Worldwatch Institute, Washington, D.C., 1992)*

Three levels of radioactive waste

In the radioactive waste industry, there is a range in hazard levels and management costs. There is no one classification among nuclear-using nations, but there is a common hierarchy. (Radioactive mill tailings are excluded here.)

Low-level waste (LLW)—Materials that can be handled without special procedures. Items include implements and waste from hospitals and biological laboratories. Ocean dumping of LLW in conventional 55-gallon steel drums was a common practice during 1946–1970, but since that time shallow burial has been the preferred method. But some of the shallow-burial containers have leaked. (So what do you suppose is going on with those drums on the sea floor?)

Transuranic waste (TRW)—Nuclear fuel rods consist in large part of uranium-238, which, through the absorption of neutrons, changes to a number of heavier transuranic elements, including plutonium, fermium, nobelium, and einsteinium. The U.S. Department of Energy (DOE) began storing TRW at its Waste Isolation Pilot Plant (WIPP) within Permian salt 2,100 ft beneath the desert floor near Carlsbad, New Mexico in 1999.

High-level waste (HLW)—The most problematic HLW is spent nuclear fuel rods (SNF) from commercial and military reactors. The term "spent" does not mean that the fuel is no longer radioactive. (Don't we wish!) The transformation of uranium to transuranic elements makes the reactor less efficient, so each year one-third of fuel rods are removed and replaced with new rods. (Fuel rods can be recycled, but the process concentrates weapons-grade plutonium. For this reason, President Jimmy Carter banned further recycling.) SNF continues to be stored in water-filled pools at reactor sites and elsewhere, but the search is on for a secure permanent repository—more specifically, a stable deep-earth geologic setting.

Q15.21 (Refer to Figure 15.26.) (A) How many nuclear facilities and waste storage sites exist in your home state? (B) Which state appears to lead the pack in inventories of facilities and sites?

Q15.22 (A) What is the universal problematic substance in waste management? *Hint:* See page 262. (B) Given that fact, what is the troubling thing about the *distribution* of nuclear storage facilities in the U.S.? *Hint:* See Figures 15.13 and 15.14.

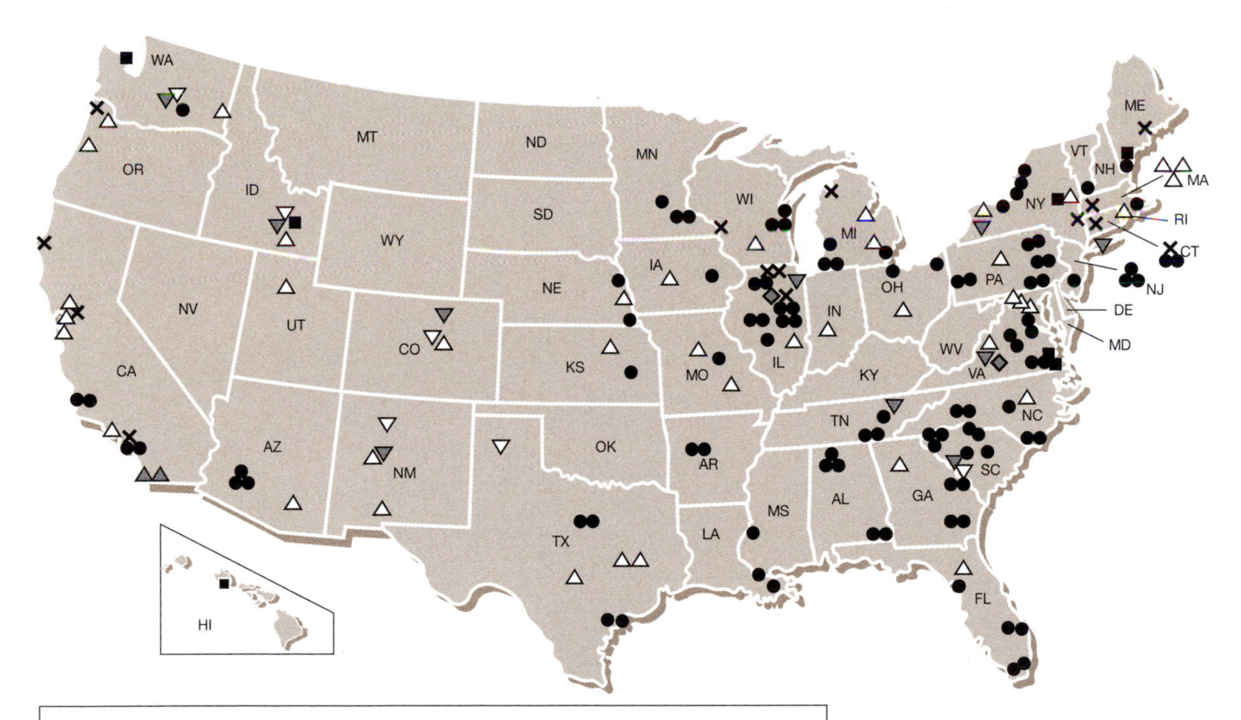

Figure 15.13 Ground-level storage continues to hold tens of thousands of metric tons of high-level radioactive waste at more than one hundred sites in 40 states. *From U.S. Department of Energy*

Yucca Mountain—a deep-earth repository?

The mandate—The Nuclear Waste Policy Act of 1982 mandated that DOE locate, build, and operate a high-level nuclear waste repository. Five sites were soon nominated by the secretary of energy (Fig. 15.14), and in February 2002 President George W. Bush announced the approval of a deep-earth site at Yucca Mountain, Nevada (Fig. 15.15).

The prospective repository is to hold 70,000 metric tons of spent nuclear fuel and high-level radioactive waste from DOE nuclear weapons facilities. Some 24,000 tons of radioactive waste presently lie in temporary storage, and spent fuel rods are accumulating at the rate of 6 tons per day.

The innumerable political, environmental, scientific, technological, engineering, and economic issues surrounding a deep radioactive waste repository fall into three categories:

(1) Transporting waste from present storage sites to the repository.

(2) Transferring waste from present containers to newly designed canisters and placing them in the repository.

(3) Isolating waste in a secure geologic environment "that will remain stable far into the future," which is taken to mean 10,000 years.

Q15.23. (A) Which of the above three categories do you suppose has/have been included in what has been called the *preclosure* period of development? (B) Which do you suppose has/have been included in the *domain of geologists and hydrologists*?

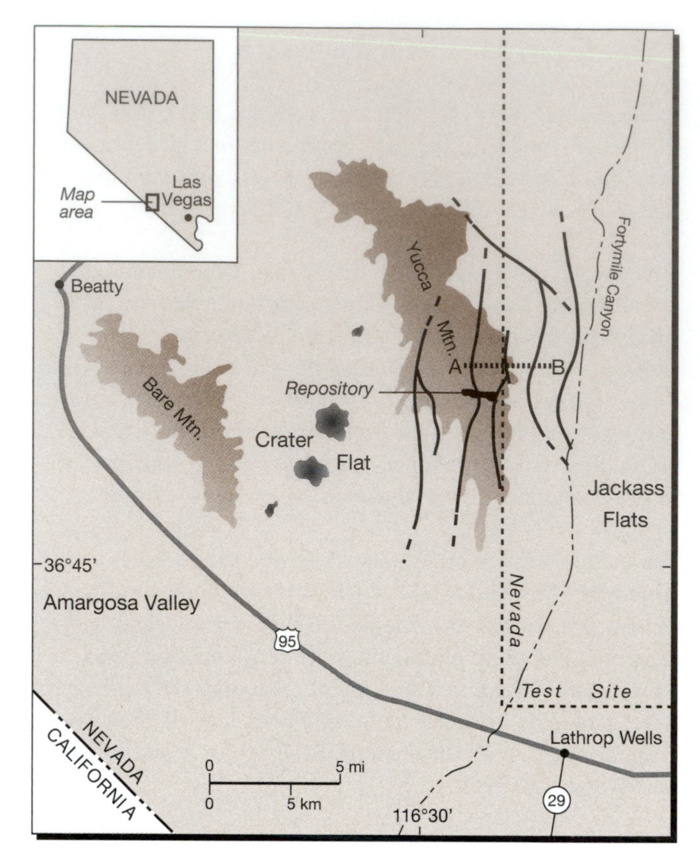

Figure 15.15 This map shows the prospective repository and cinder cones and faults in the immediate vicinity of Yucca Mountain. Line A–B is the line of cross-section in Figure 15.18 (page 272). (*After the Department of Energy*)

Figure 15.14 This map shows (a) initial potential repository sites, (b) average rainfall, and (c) people per sq. mi. in relevant states.

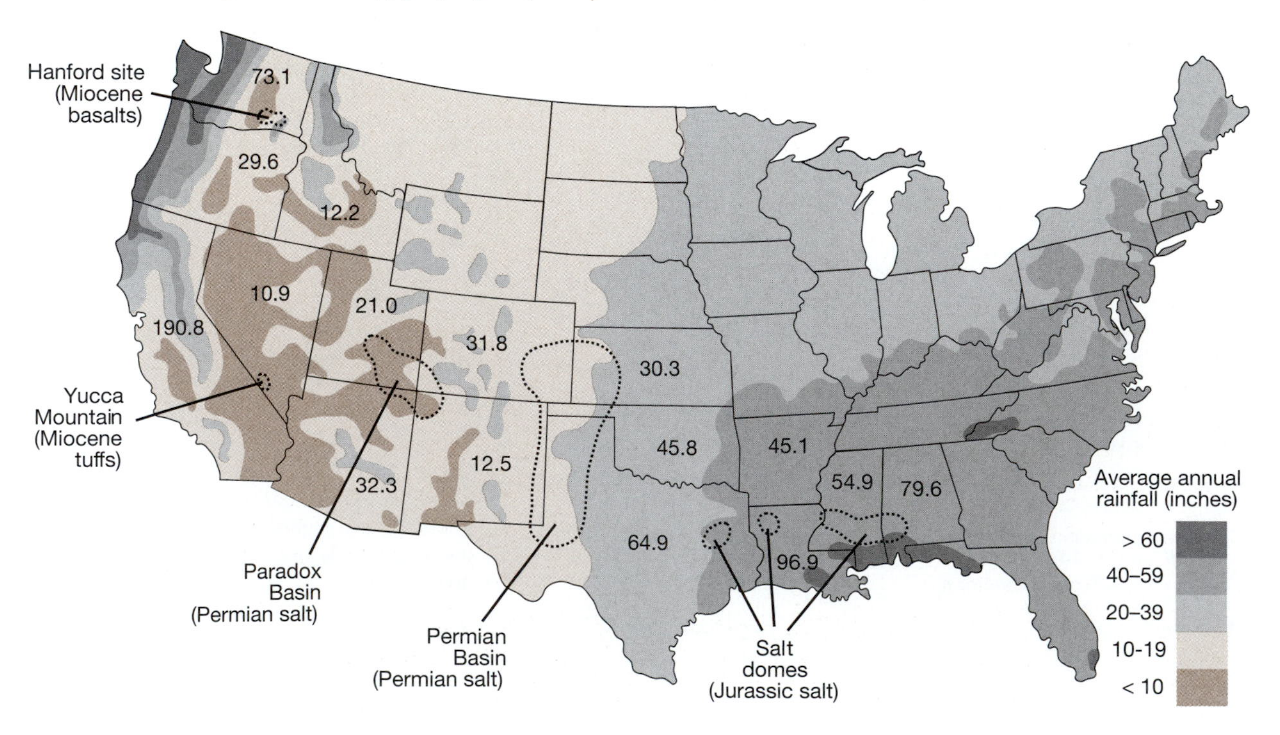

A "stable" geologic setting

Geotimes, JANUARY 1989

Geoscientists help make 10,000-year decisions

The Nuclear Waste Policy Act of 1982 gave the Department of Energy a mandate to locate, build, and operate a high-level nuclear waste repository in the U.S. The act also authorized the Nuclear Regulatory Commisssion to grant or deny DOE's application to build such a repository. And, after construction NRC may issue a license for DOE to receive and dispose of the waste. DOE's Office of Civilian Radioactive Waste Management was organized to manage the process and to begin searching for the first site. The location was to be based on objective scientific assessment of site geology, hydrology, geochemistry, geophysics, and seismic activity.

To obtain a **construction license** from the Nuclear Regulatory Commission (NRC), the Department of Energy is working to satisfy the following wish list of secure geologic features. (Again, this is looking some 10,000 years into the future.)

1. Region with negligible probability of earthquakes
2. Region with negligible probability of volcanism
3. Rocks with few fractures and low matrix permeability
4. Mineralogically, rocks that will adsorb radionuclides
5. Landscape with little erosion (as a function of rainfall)
6. Dry climate, as it affects the quantity of groundwater
7. Repository well above zone of (groundwater) saturation
8. Region with minimal resource potential
9. Low human population density

Q15.24 (Refer to the wish list above.) Which of the items in the list appear(s) to be best satisfied by the Yucca Mountain site (in comparison with other early candidate sites) as shown in Figure 15.14?

Groundwater—the universal threat in waste management—presents the potential at Yucca Mountain of transporting radionuclides to streams and springs (the aridity of the region notwithstanding). Rainfall at Yucca Mountain is on the order of 16 cm/yr. Approximately 90% of the precipitation evaporates and runs off, with some 10% soaking into the ground.

Some of the volcanic tuffs are porous, some are 'welded,' i.e., fused by residual heat following ash fall. All have been fractured and faulted to some degree by regional crustal movements. So there are two means of water circulation:

(1) Flow via macroscopic fractures
(2) Absorption via microscopic pores

To guard against moisture, spent fuel rods will be stored within double-walled canisters (Fig. 15.16) that will be placed on their sides, end-to-end, within 25-foot diameter tunnels (Fig. 15.17). To shield canisters from descending water, (a) drip shields might be placed over them, or (b) capillary liners might be fitted to tunnel walls.

Q15.25 How do you suppose a capillary liner works?

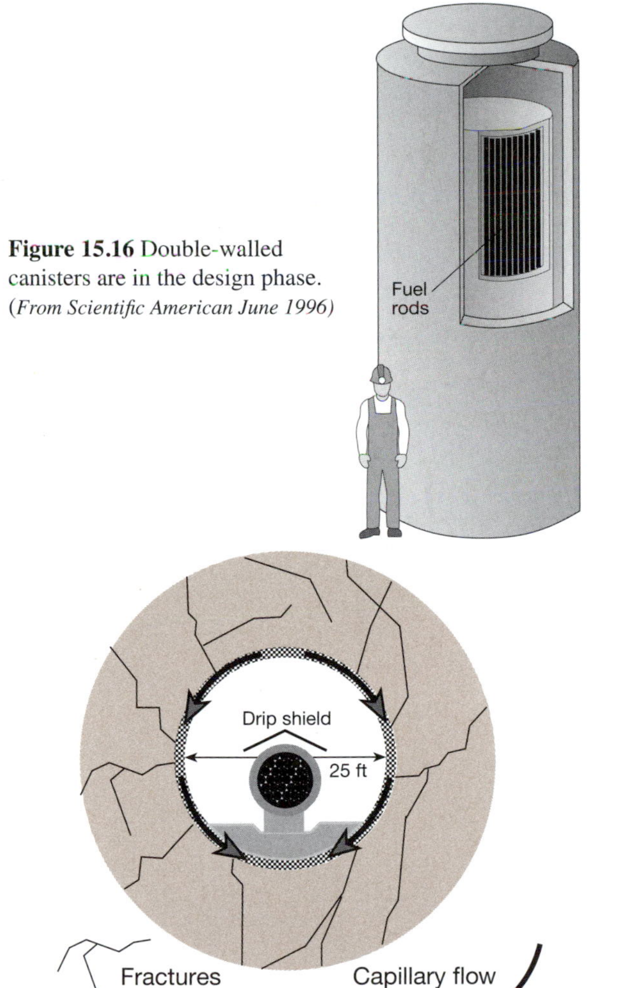

Figure 15.16 Double-walled canisters are in the design phase. (*From Scientific American June 1996*)

Figure 15.17 Cross-section of planned repository tunnel with canister in place. Arrows indicate movement of descending water through a possible capillary tunnel lining. (*After USGS Circular 1184*)

For Eureka County, Nevada Nuclear Waste Page, go to http://www.yuccamountain.org
For latest news items, including news clips, go to http://www.yuccamountain.org/new.htm

Geology and hydrology of Yucca Mountain

Synopsis of geologic events—Volcanic tuffs that comprise bedrock at Yucca Mountain (Fig. 15.18) originated from volcanoes that were active 6 to 16 million years ago. Cinder cones of Crater Flat (Fig. 15.15) resulted from renewed volcanic activity between 15,000 and 25,000 years ago. Faulting (and associated earthquakes) is judged to have been negligible for the past 10 million years. Rock layers in the immediate vicinity are slightly inclined toward the east, having been tilted in that direction by the block-faulting that is characteristic of Great Basin tectonics.

Synopsis of hydrologic setting—The site of the prospective repository at Yucca Mountain is 200 m below the surface, within the unsaturated zone. Some 300 m deeper is the water table and the zone of saturation. Flow paths of descending groundwater are guided by degrees of permeability within the tuffs, which in turn reflect degrees of lithification (welding). The tuffs are replete with fractures.

Q15.26 Notice the arrows in Figure 15.18 that indicate lateral movement of groundwater within the Paintbrush and Calico Hills Members. What is there about these members that accounts for lateral movement? ***Hint:*** **In what way do they differ from other members? (See legend.)**

Figure 15.18 This is an east-west geologic and hydrologic cross-section plotted along line A–B in Figure 15.15. (*After USGS WRI Report 84–4345, 1984*)

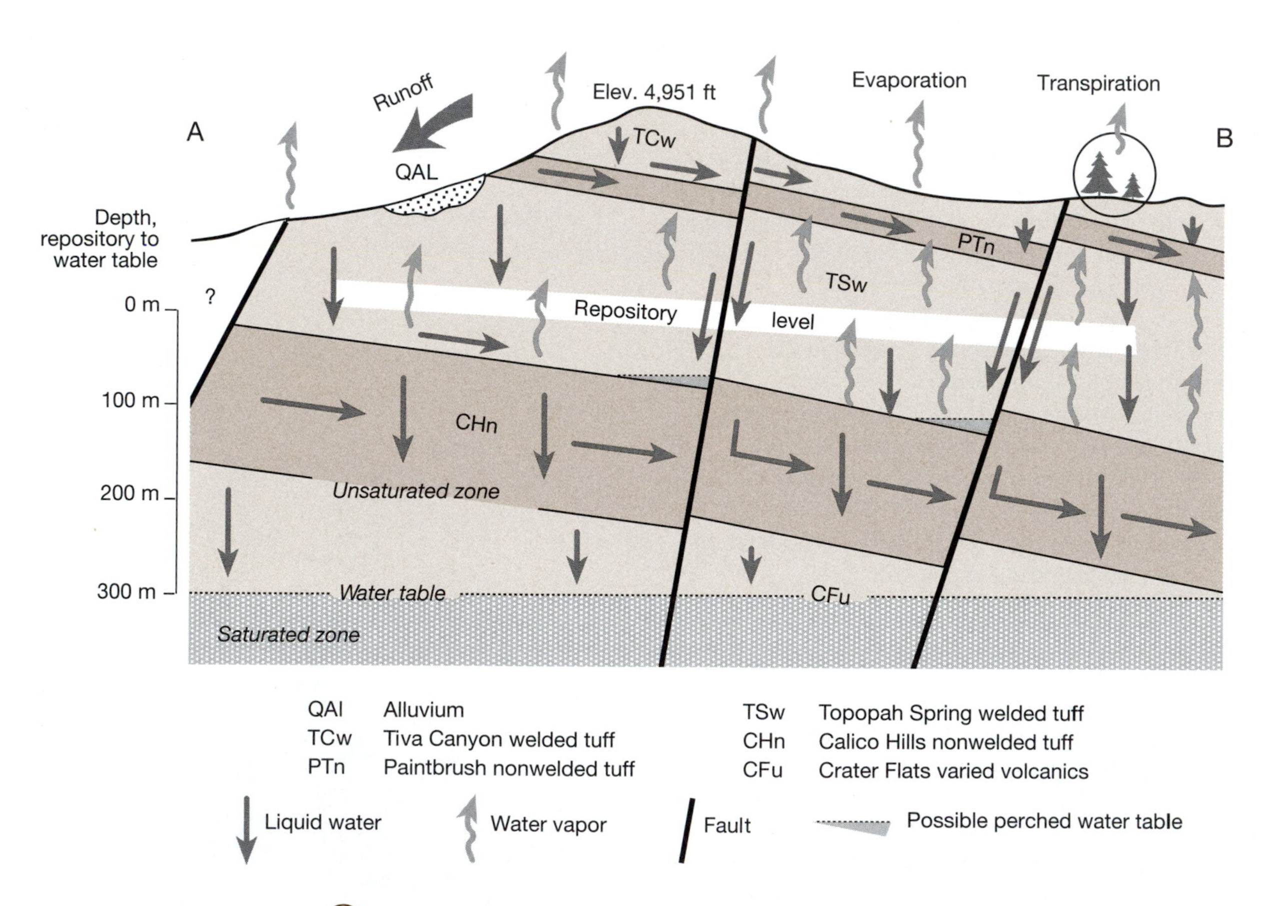

Q15.27 From what you have learned thus far about the Yucca Mountain site, on the Answer Page check each of the nine items in the 'geologic wish list of secure geologic features' with one of the following symbols:

✓ (Does satisfy the item.)
✗ (Does not satisfy the item.)
? (Inadequate information to make a judgment.)

Las Vegas Review-Journal, FEBRUARY 07, 2003

Yucca study poses 'staged' facility idea

Report suggests repository be built in modules

WASHINGTON—An influential science panel on Thursday recommended developing the Yucca Mountain Project in stages, adding small-scale test phases and raising the idea of storing more radioactive waste above ground

Timeline for development of Yucca Mountain

- 2003 DOE applies to NRC for license to construct.
- 2006 DOE designates heavy-haul routes for delivering waste to Yucca Mountain.
- 2007 NRC expected to issue license for construction.
- 2007 Receipt of waste at Yucca Mountain begins.
- 2008 DOE applies to NRC for license to operate.
- 2010 Waste emplacement begins.

Q15.28 It appears that one of the actions in the above time line is a bit rushed—relative to one or more other actions. Which action is that?

Figure 15.19 Consensus is that the development of the Yucca Mountain project will progress as per the following outline: *After Eureka County Nuclear Waste Web Page, citing information from Nuclear Energy Institute, National Safety Council, and Environmental Health Center.*

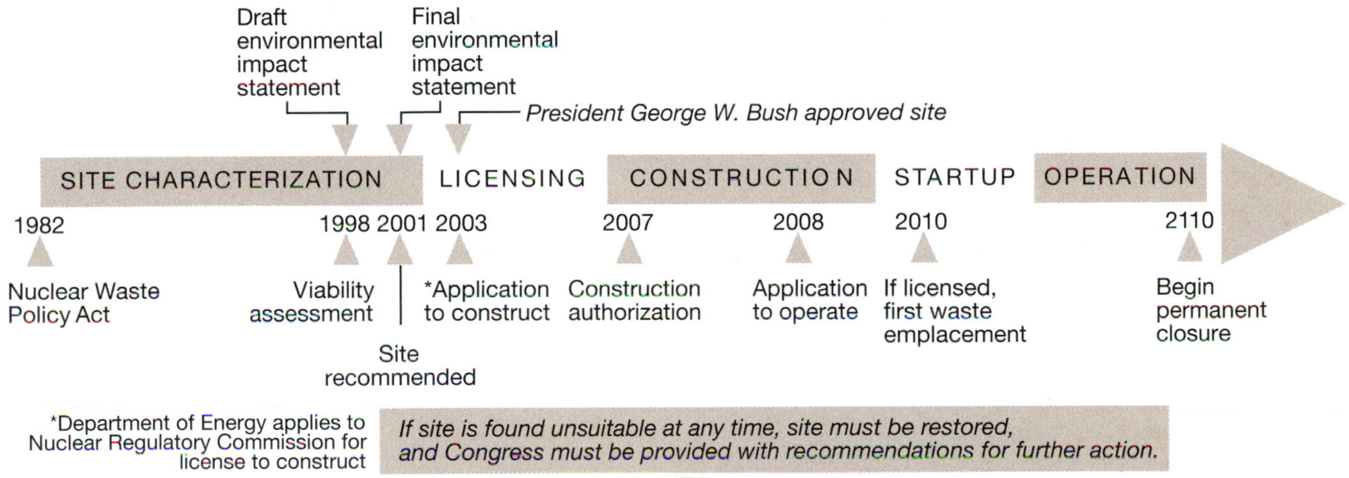

Q15.29 At a glance, it would appear that there is a problem with the grand scheme of planning and execution as outlined above. What is the problem? *Hint:* It has to do with what one might call 'financial risk assessment;' i.e., if this were a private for-profit enterprise, would you have invested your dollars in 2007? Why or why not?

In 2008 everything appeared to be going swimmingly.

But then...

COLUMBIA DAILY TRIBUNE, March 6, 2009

Obama abandons Yucca Mountain plan

WASHINGTON (AP) — For two decades, a ridge of volcanic rock in the Nevada desert 90 miles northwest of Las Vegas has been the sole focus of U.S. government plans to store highly radioactive waste.

Not anymore. Despite the $13.5 billion already spent on the aging Yucca Mountain project, the Obama administration says it is going in a different direction.

It slashed funding for Yucca Mountain in its recently announced budget.

And, yesterday, Energy Secretary Steven Chu told a Senate hearing that the Yucca Mountain site no longer was viewed as an option for storing reactor waste, brushing aside criticism from several Republican lawmakers.

Instead, Chu said the Obama administration believes the nearly 60,000 tons of used reactor fuel can remain at nuclear power plants while a new, comprehensive plan for waste disposal is developed.

Moral: Human nature is driven to believe that science will pull a technological rabbit out of a hat.

G. Biomagnification of hazardous waste

It has been said that *dilution is the solution to pollution*. The problem with this simple axiom is that toxins commonly become concentrated to hazardous levels as they move upward through food chains—a process called **biomagnification**. Pesticides consisting of **organochlorides** (organic compounds containing chlorine), such as DDT and DDD, are only slightly soluble in water, so concentrations of these compounds tend to be low in lakes and streams. However, organochlorides accumulate in plant and animal tissues. During the 1950s Clear Lake, California (Fig. 15.20) was sprayed with DDD in an effort to control pesky insects. Spraying resulted in a mere 0.02 parts per million (ppm) in the lake water, but that wasn't the end of the Clear Lake DDD story.

Figure 15.20 Clear Lake, California—the site of toxic levels of DDD within tissues of fish and aquatic birds.

Q15.30 Given the fact that one ppm contaminant in a 55-gallon drum full of water equates with approximately 4 drops of that contaminant, what fraction of one drop per 55 gallons equates with 0.02 ppm?

In the case of Clear Lake the 0.02 ppm concentration of DDD in the lake water was biomagnified upward in the food chain by the following multipliers (Fig. 11-21):

(a) Water to phytoplankton, by a factor of 250
(b) Phytoplankton to plant-eating fish, by a factor of 40
(c) Plant-eating fish to fish-eating birds, by a factor of 10

Q15.31 (A) Convert each of the above three products to ppm. (B) What principle appears to be evident in the magnitude of increase at each successive level within a food chain?

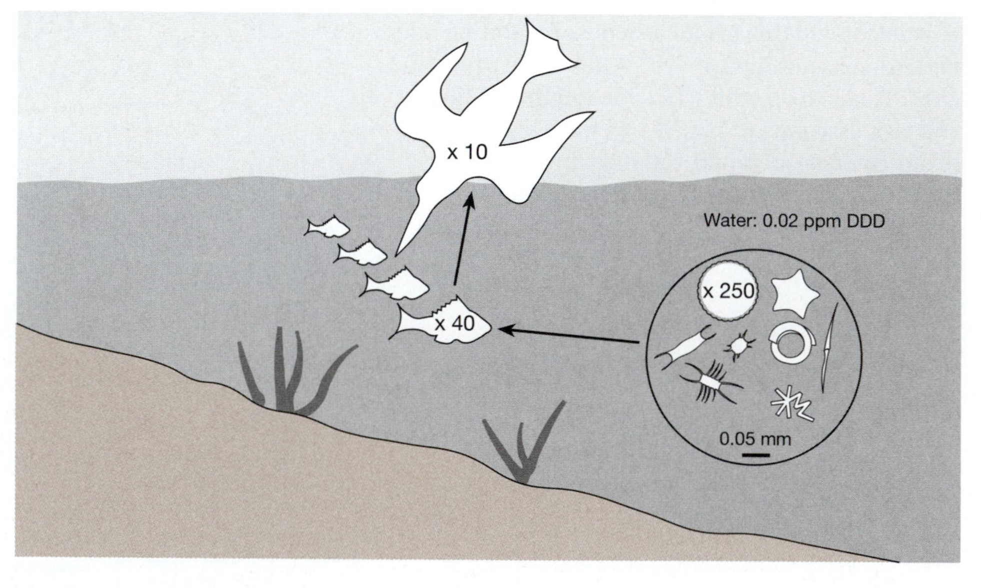

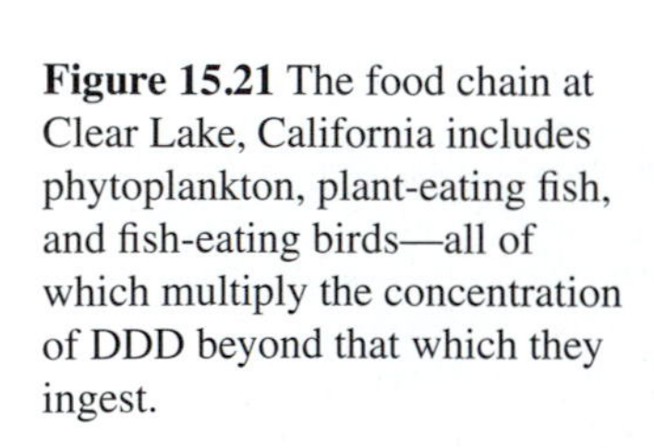

Figure 15.21 The food chain at Clear Lake, California includes phytoplankton, plant-eating fish, and fish-eating birds—all of which multiply the concentration of DDD beyond that which they ingest.

(Student's name) (Day) (Hour)

(Lab instructor's name)

ANSWER PAGE

15.1

15.2 (A)

(B)

15.3 (A) (B)

15.4

15.5 (A)

15.6

15.7

15.8

15.9

15.10

15.11 (A)

(B)

15.12

15.13

15.14

15.15

15.16

15.17

15.18

15.19 (A) (B)

15.20 (A) (B)

15.21 (A) (B)

15.22 (A)

(B)

15.23 (A) (B)

15.24

15.25

15.26

15.27

1. Negligible probability of earthquakes
2. Negligible probability of volcanism
3. Few fractures and low permeability
4. Rocks will adsorb radionuclides
5. Little erosion (i.e., dry climate)
6. Dryness minimizes groundwater
7. Repository well above water table
8. Minimal resource potential
9. Low human population density

15.28 ______

15.29 ______

15.30 ______

15.31 (A) ______

(B) ______

Intensionally blank

16 Selenium Contamination

Topics

A. In what three ways are selenium and sulfur alike? In what way are they different?
What are three ranges in amounts of selenium that are relevant to ingestion by humans?
What is locoweed, and how did it acquire its curious name?
B. Selenium contamination—What five components led to the *Kesterson effect* in California?
C. What are the three most important predictors of selenium contamination?
D. What is the purpose of the decision tree? What kinds of questions/answers comprise the tree?
E. What percent of irrigation areas investigated by *NIWQP* are selenium-contaminated?
What two things account for the cluster of eight contaminated areas in west-central U.S.?

A. Selenium—its chemistry and toxicity

Chemistry—Selenium substitutes for sulfur because of similar ionic radii and identical charges. For example, sulfur (S^{2-}) combines with iron to produce iron sulfide (FeS_2, pyrite, or fool's gold), and selenium (Se^{2-}) combines with iron to produce iron selenide ($FeSe_2$). Also, the two elements find their way into food chains with equal ease. It's only in abundances that selenium and sulfur differ markedly.

Q16.1 Within Earth's crust the weight percent of sulfur is 0.03, and the weight percent of selenium is 0.000005. So, sulfur is how many times more abundant in Earth's crust than is selenium?

Toxicity—Selenium in the range of 50–200 *µg/day is essential for the function of certain enzymes within the human body. But the *U.S. Environmental Protection Agency* views selenium as toxic when ingested in excess of 500 µg/day. In water, concentrations in excess of 50 µg/liter are considered toxic.

**One microgram (µg) is one-millionth of one gram. One part per million liquid is 4 drops per 55-gallon drum. So, a proverbial drop in the bucket is about 14 ppm.*

Selenium poisoning (*Se toxicosis, or selenosis*) was first documented in 1856 by a physician with the U.S. Cavalry who reported the splitting of horses' hooves (Fig. 16.1). Legend has it that such sad condition of their mounts prevented troops from riding to the aid of George Custer and his men at the 1876 battle of Little Bighorn.

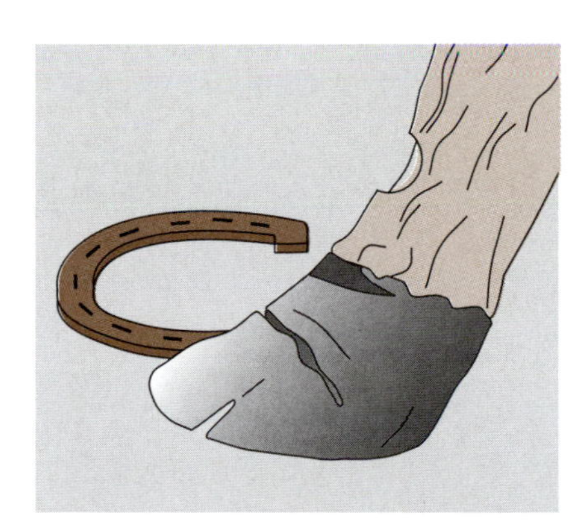

Figure 16.1 Selenosis in horses and cattle causes loss of hair and splitting of hooves.

Western ranchers have known since the 1930s that when cattle eat a particular variety of plant, they develop the blind staggers and appear to be drugged. These locoweeds (loco: *crazy* in Spanish) are species of *Astragalus* and *Oxytropis*, which accumulate toxic levels of selenium (Fig. 16.2). Sheep are affected as well (Fig. 16.3).

Figure 16.2 Locoweed, eg., *Astragalus bisulcatus*, accumulates toxic selenium.

Figure 16.3 Selenosis in sheep causes an unsteady gait and an arched back.

Selenium ingestion by adult humans in excess of five milligrams (i.e., five one-thousandths of a gram) per day can cause cancers, malformation of nails and hair, and nervous disorders.

Q16.2 Five milligrams(mg)/day is how many micrograms(µg)/day?

B. Selenium contamination and the Kesterson effect

History—Before the 1980s, concerns about irrigation in arid regions of western states focused on the accumulation of salts and pesticides…

Q16.3 Incidentally, why do you suppose that aridity promotes the accumulation of salts and pesticides on irrigated agricultural land?

…but in 1983 *U.S. Geological Survey* discovered that *Kesterson National Wildlife Refuge*—a wetlands in San Joaquin Valley, California (Fig. 16.4)—had become an environmental disaster because of **selenium contamination** ascribed in large part to irrigation.

Kesterson wetlands seemed like a good idea when it was first constructed—even though as early as 1974 *USGS* reported selenium's being concentrated by irrigation in Wyoming. Kesterson wetlands promised to provide a practical means of disposing of drain-water from Westlands irrigation area, while at the same time providing an environmentally friendly aquatic bird refuge. But the refuge proved to be anything but friendly. Ducks and other aquatic birds at Kesterson developed deformed bills and beaks and 64% mortality of embryos. To blame: Concentrations of selenium on the order of 350 μg/liter in waters within San Luis Drain.

Q16.4 How many more times greater is the concentration of 350 μg/liter than the level considered by *USEPA* to be toxic? (See the first paragraph under Toxicity on the first page of this exercise.)

Remediation: In 1987 San Luis Drain was closed, and Kesterson wetlands was drained and buried.

Sources of Kesterson selenium—Surface hydrology of the San Joaquin Valley is shown in Figure 16.4 as it appeared in 1983. In brief, (1) eastward-flowing streams drain selenium-rich Upper Cretaceous and Tertiary marine shales of the Coast Ranges, and (2) some of the selenium-rich water finds its way into the ground, where it resides until being pumped (by windmills) and used for irrigation.

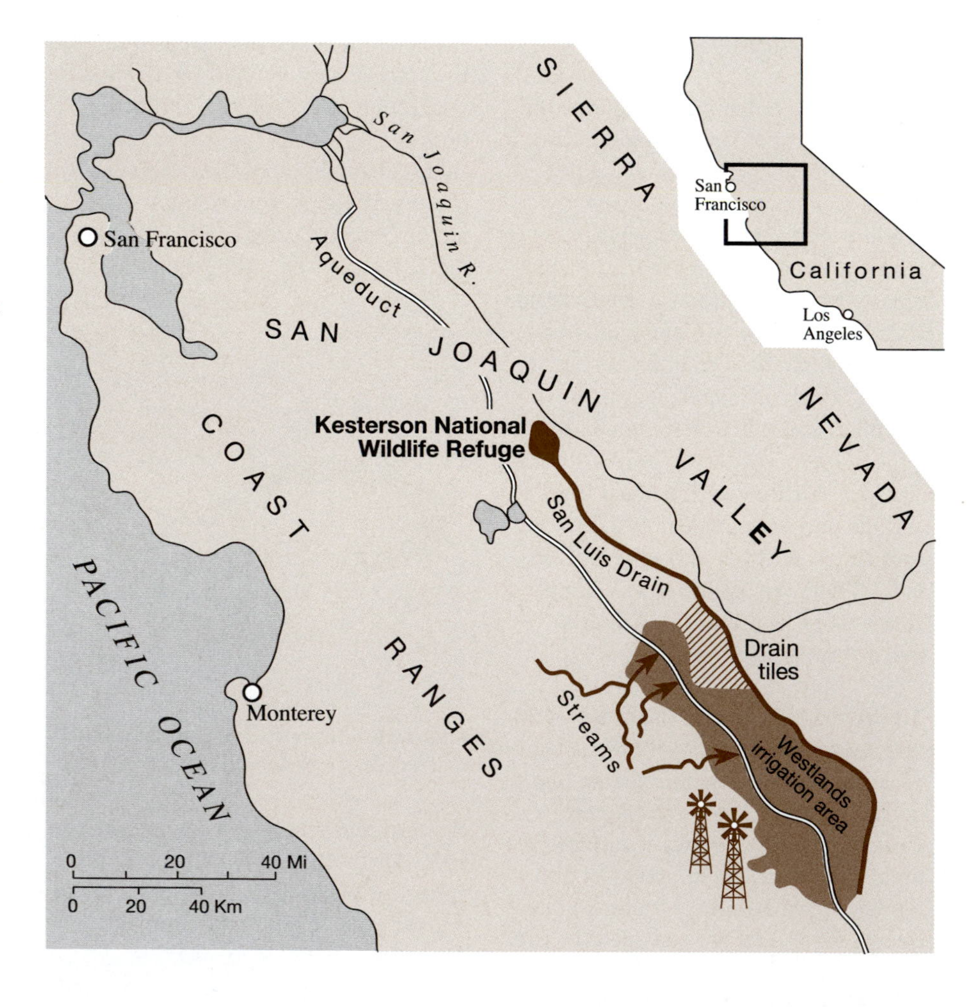

Figure 16.4 Pathways of waters rich in selenium oxides are highlighted in brown. The San Luis Drain, which once conveyed irrigation drain-water from Westlands irrigation area to Kesterson NWR wetlands, was closed in 1987. Drain water is now diverted to the San Joaquin River.

Five components of the Kesterson Effect—Theresa Presser of *USGS* traced the **biogeochemical pathway** of selenium at Kesterson NWR, which she labeled 'the **Kesterson effect**' (Fig. 16.5). Presser's model includes five components:

1. *The presence of selenium-rich source-rocks*—In this case, Upper Cretaceous and Tertiary marine shales of the Coast Ranges.

2. *Mobilization of selenium*—Oxidation of selenium—through weathering—and its dispersal into San Joaquin Valley by streams and groundwater.

3. *High rates of evaporation*—Partial evaporation of irrigation water, which concentrates selenium in drain-water.

4. *Bioaccumulation of selenium*—The uptake of selenium from water as seleno-amino acids in algae and plankton.

5. *Selenium moving upward in the food chain to aquatic birds*—The consuming of seleniferous algae and plankton by insects and fish, which are in turn consumed by aquatic birds.

And so goes Theresa Presser's **rock-to-duck** biogeochemical pathway of selenium that resulted in the environmental disaster at Kesterson National Wildlife Refuge in the San Joaquin Valley of California.

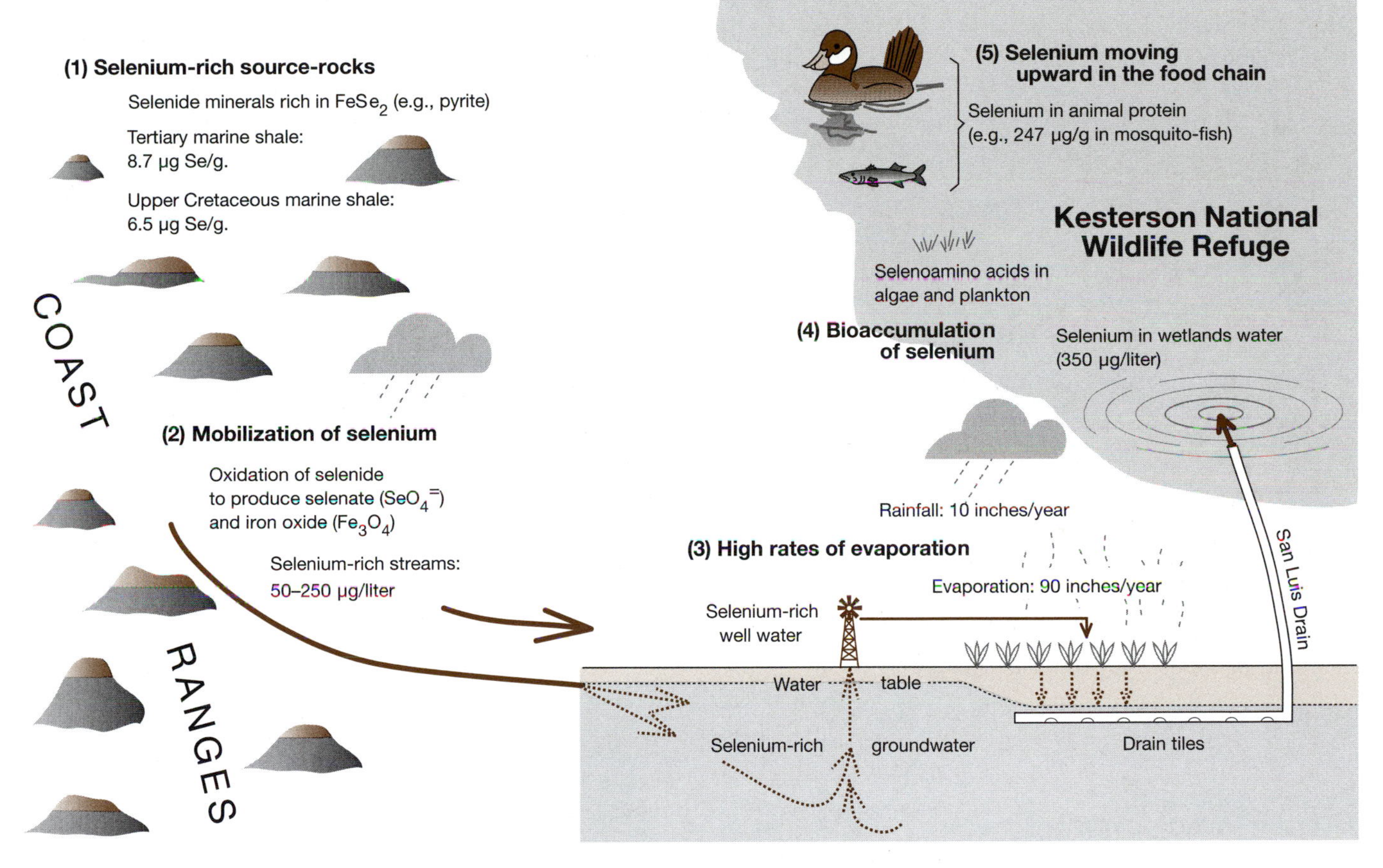

Figure 16.5 This schematic collage of maps and cross-section illustrates the processes and products that comprise the five components of the Kesterson effect in San Joaquin Valley of California.

Taking stock of processes and products illustrated in Figure 16.5—answering the following questions:

Q16.5 What label do we apply to the variety of chemical weathering that converts $FeSe_2$ to $SeO_4^{=}$?

Q16.6 By what amount is annual evaporation in excess of annual rainfall in San Joaquin Valley?

Q16.7 The water table (i.e., the boundary between water-saturated earth below and air-saturated earth above) is higher beneath the windmill than beneath the irrigated plants. A feature constructed by farmers indicated the reason for this difference in elevation of the water table. What is that feature?

Q16.8 (A) Quantitatively, how does the concentration of selenium in wetlands water compare with that in streams draining rocks of the Coast Ranges? (B) Why this difference?

Q16.9 Quantitatively, how does the concentration of selenium within protein of mosquito-fish in the wetlands compare with that in rocks of the Coast Ranges?

C. Predicting susceptibility to selenium contamination

Government agencies investigate—Several agencies* of *U.S. Department of the Interior* (*USDOI*) collectively oversee some 600 irrigation projects and associated wildlife refuges in 17 western states (Fig. 16.6). Scientists from these agencies investigated 26 such areas in efforts to determine where drain-water might have the potential to affect human health and/or wildlife. Some 12 of the 26 areas studied proved to be selenium-contaminated.

Factors that promote selenium contamination—The 12 areas that proved to be selenium-toxic are characterized by a number of factors, the most important of which are:

1. *Rocks rich in selenium*
2. *Arid to semiarid climate*
3. *Terminal ponds*

1. *Rocks rich in selenium*—Surface rocks rich in selenium in western states are those highlighted by the Kesterson NWR disaster; i.e., Upper Cretaceous and Tertiary marine shales (Fig. 16.6).

The breadth of exposures of Upper Cretaceous marine shales reflects the expansive Late Cretaceous seaway (Fig. 16.7) in which marine microorganisms rich in carbon, sulfur, and selenium accumulated in clay sediments.

Q16.10 Why is the total area of Upper Cretaceous marine shales exposed in western states (i.e., the sum of the gray patches in Figure 16.6) considerably *less* than the total area covered by the Late Cretaceous seaway shown in Figure 16.7? For example, we know that Upper Cretaceous marine sediments were deposited over the entire area of what is now the state of Texas. What *two* things account for the fact that now only a narrow band of those sediments is exposed on the Texas landscape?

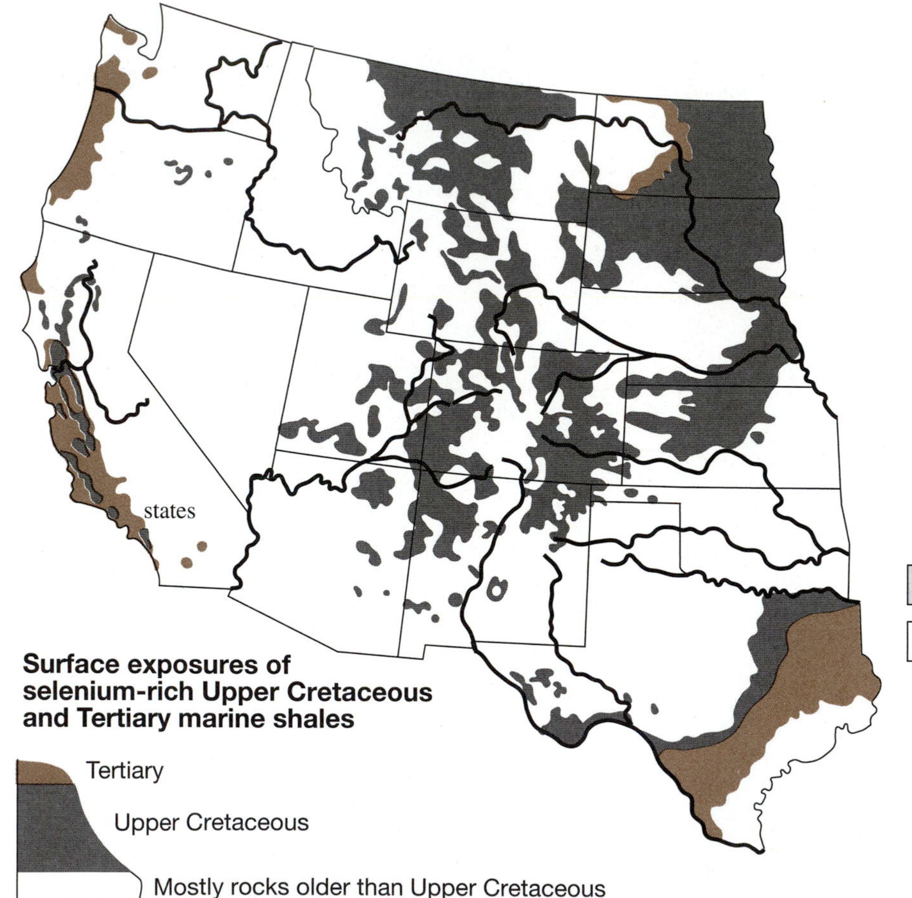

Figure 16.6 Upper Cretaceous and Tertiary rocks of western states consist of marine shales rich in organic carbon, sulfur, and selenium. Compounds containing these elements are produced through weathering and are eroded and redistributed by major river systems. (The source of this and other maps in this exercise is at the Web site shown below.)

Late Cretaceous seaway

Dry land

Figure 16.7 This map shows a part of the Late Cretaceous seaway as it appeared 75 million years ago. Volcanic ash was the source of sulfur and selenium that were metabolized by microorganisms and buried along with marine clays (now shales).

National Irrigation Water Quality Program (NIWQP), U.S. Geological Survey (USGS), U.S. Bureau of Land Management (USBLM), U.S. Bureau of Reclamation (USBOR), U.S. Fish and Wildlife Service (USFWS)

For *Methods to identify areas susceptible to irrigation-induced selenium contamination*, go to... http://water.usgs.gov/public/pubs/FS/FS-038-97/

2. *Arid to semiarid climate*—Selenium toxicity related to irrigation occurs in semiarid to arid environments where evaporation concentrates selenium in solution.

The standard measure of aridity combines the actual annual rate of precipitation and the potential rate of evaporation (as a function of temperature and humidity) into a single number called the **evaporation index**.

Evaporation index (EI) is computed from the expression…

$$EI = FWSE \div PRECIP$$

…where (in inches per year) FWSE is free-water-surface-evaporation, and PRECIP is precipitation (as rain and/or equivalent snow). In the eastern United States EI is less than 2.0, but in the American West EI is 2.5 or greater throughout more than half the region (Fig. 16.8).

Q16.11 Using the numbers for Rainfall and Evaporation in Figure 16.5 on page 281, compute the evaporation index for San Joaquin Valley.

Q16.12 As portrayed in Figure 16.8, in what four states is the largest continuous area where evaporation index is 5.0 or higher?

Q16.13 As portrayed in Figure 16.8 below, what five states most contribute to the largest continuous area where the evaporation index is 2.5–3.0?

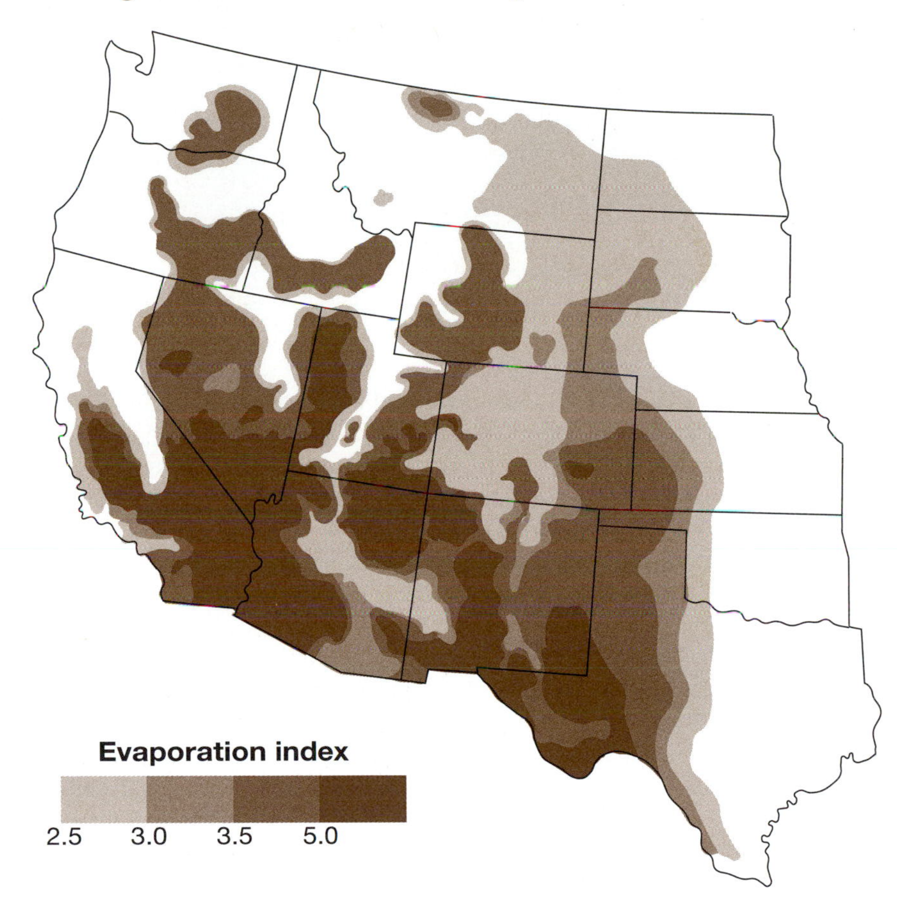

Figure 16.8 The standard measure of aridity is called the evaporation index, which is inches of potential annual evaporation divided by inches of actual annual precipitation. The Great American Desert of the western United States is characterized by values in excess of 2.5.

3. *Terminal ponds*—**Terminal ponds** of the American West typically occur within 'closed' topographic basins (i.e., valleys encircled by mountains so that water can hardly drain away). And, terminal ponds are typically ephemeral because (a) evaporation of water, plus its soaking into the ground, commonly equals or exceeds (b) the rate at which water is supplied by streams. Ephemeral terminal ponds are common in the American West for two reasons. First, mountain building in that part of our country resulted in scattered mountain ranges and closed basins; e.g., Wind River Basin in west-central Wyoming (Fig. 16.9). Second, the American West is a region of semiarid to arid climate.

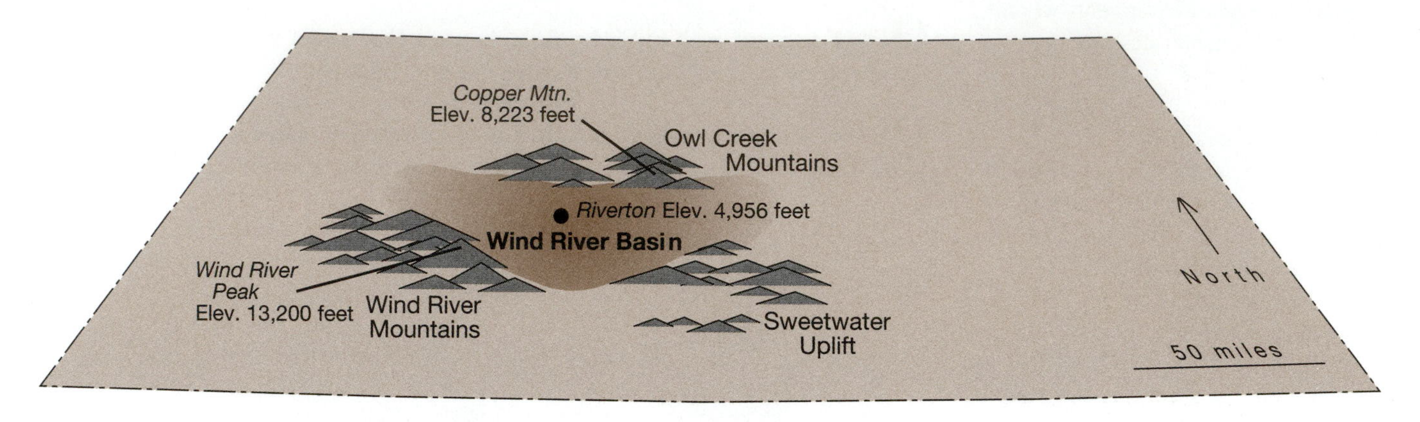

Figure 16.9 The development of Wind River Mountains, Owl Creek Mountains, and Sweetwater Uplift during the Tertiary Period resulted in the rather closed Wind River Basin. (Other mountains and basins in Wyoming are not shown.)

Q16.14 (Refer to Figure 16.9.) On the skeletal diagram on Answer Page 288, make a rough sketch—using only the three elevations given in Figure 16.9 as your guide—of a topographic profile extending from Wind River Peak, through the town of Riverton, to Copper Mountain. To make your profile more realistic, steepen it on the flanks of mountains, and flatten it within the basin.

Why do enclosed basins tend to be flat? Answer: They are partially filled (like a bowl of cereal) with sediments eroded from mountains. Why the filling? Answer: Swift mountain streams, loaded with sediments eroded from rocks that comprise mountains, diminish in velocity as they flow onto basin floors, and so tend to deposit their sediment loads.

Q16.15 Return to Answer Page 288 and the topographic profile you drew. Turn this profile into a geologic cross-section by doing the following: Beneath your profile, indicate (A) that part of your diagram occupied by *sediments* with a pattern of dots, and (B) that part of your diagram occupied by *rocks* with a pattern of horizontal lines.

In addition to sediments, enclosed basins in arid areas invariably contain mineral precipitates rich in sodium, potassium, and myriad other elements—including *selenium*. On the map in Figure 16.8 Wind River Basin lies within the tongue of high-evaporation index in the west-central part of Wyoming. (The tongue of white space indicating index of less than 2.5 marks the Wind River Mountains.)

Q16.16 What is the evaporation index for most of Wind River Basin?

Incidentally, evaporative basins were exceedingly troublesome for settlers of the American West. Not only was typical basin water unfit for drinking, but it was unfit for washing and bathing as well. It has been said that diaper rash was more of a problem for westward-bound pioneers than were marauding Indians.

Point—Terminal ponds, which are typical of enclosed basins, are—in dry regions—subject to evaporation and enrichment in *selenium*, just as are irrigation waters.

D. The decision tree—Assessing the likelihood of selenium contamination

The decision tree designed by Ralph Seiler of *USGS* (Fig. 16.10) consists of yes/no questions about the presence/absence of the three circumstances listed earlier as being most important in promoting selenium contamination in irrigated areas. Answering these questions about a specific area leads to a qualitative judgment about the likelihood of selenium contamination. A thorough assessment of these three factors requires study of the following kinds of maps and records:

Geologic maps—To assess the occurrence of rocks rich in selenium (specifically, Upper Cretaceous) on—or upstream from—the irrigated area.

Evaporation maps and precipitation records—To recognize arid or semiarid climate by computing the evaporation index at the irrigated area.

Topographic maps—To determine the extent to which upstream Upper Cretaceous rocks and the irrigated area are within the same watershed (i.e., drainage area); and, to determine the presence or absence of terminal ponds. Such studies are far beyond the scope of this exercise, so summary information from *NIWQP* studies of four specific irrigated areas is provided below, and you are asked to answer questions in the decision tree and determine the likelihood of selenium contamination in each of these four areas.

1. Maxwell National Wildlife Refuge, Colfax Co., New Mex
   Irrigation on Upper Cretaceous
   Evaporation index: 3.6
   Terminal ponds are present
2. Stillwater Wildlife Management Area, Nevada
   Not associated with Upper Cretaceous
   Evaporation index: 10.1
   Terminal ponds are present
3. West Oakes Irrigation Area, Dickey Co., North Dakota
   Irrigation is on Upper Cretaceous
   Evaporation index: 2.0
4. Riverton Reclamation Project, Wyoming
   Upstream Upper Cretaceous
   Evaporation index: 10.6
   No terminal ponds

Q16.17. What is the likelihood of selenium contamination within each of the above four irrigated areas?

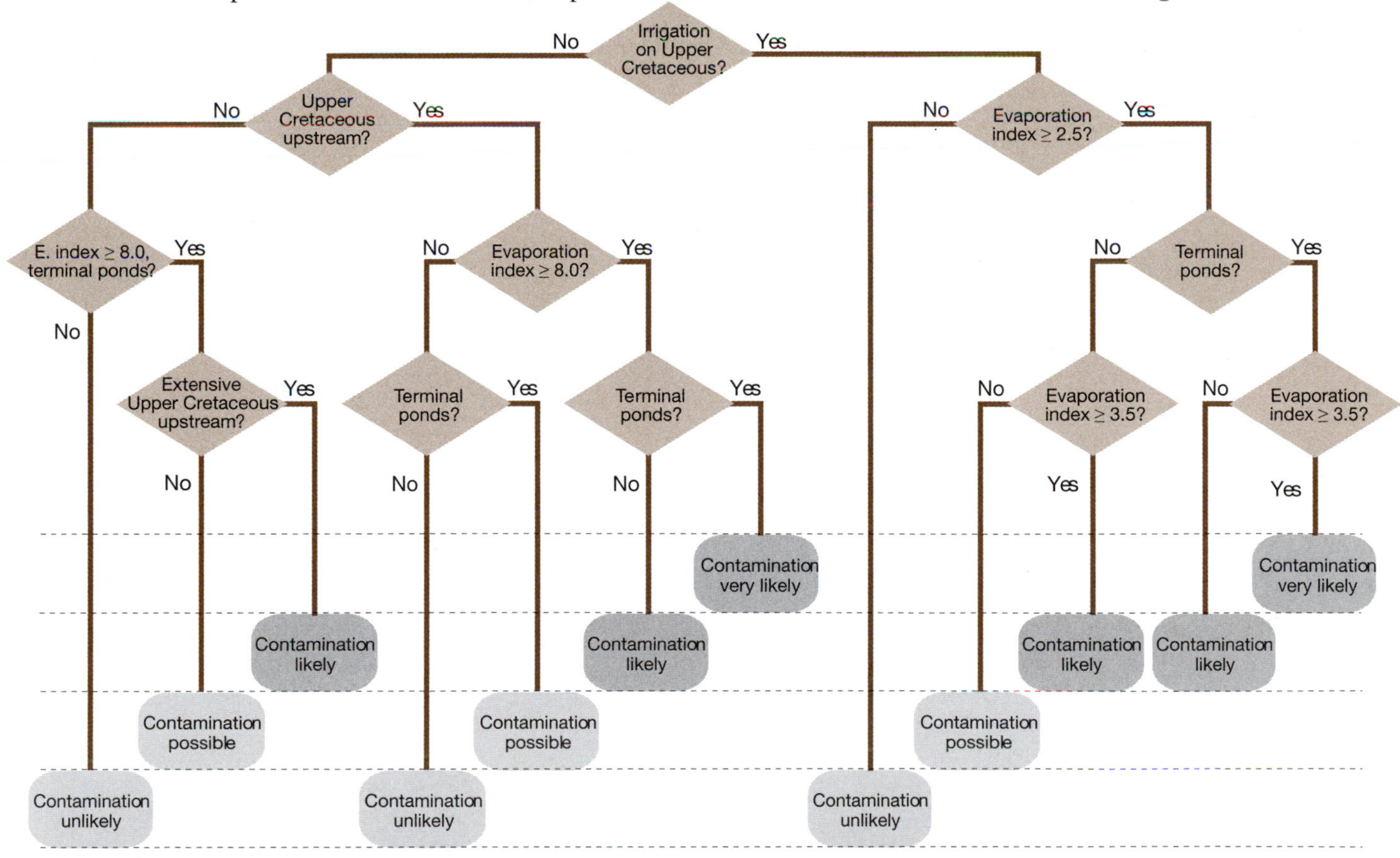

Fig. 16.10 This decision tree consists of yes/no questions about the presence/absence of the three circumstances that most promote selenium toxicity on irrigated land in western states. Answering these questions leads you to a qualitative judgment about the likelihood of selenium contamination. *(Redrawn from Ralph L. Seiler, 1995, J. Environmental Quality, v. 24, p. 973–979)*

E. Partial inventory of selenium-toxic areas in Western United States

U.S. Environmental Protection Agency (*USEPA*) specifies the range of selenium concentration safe for freshwater life as less than 5 µg/L. So an area is viewed as selenium-contaminated when 25 percent of surface-water samples contain selenium that equals or exceeds 5 µg/L.

To repeat: Some 26 areas of agricultural irrigation scattered across 17 western states were studied by *National Irrigation Water Quality Program* (*NIWQP*). Of those 26 areas, 12 were judged to be selenium-contaminated (Fig. 16.11).

Q16.18 What percent of the areas studied were judged to be selenium-contaminated?

Q16.19 What four states form a cluster that includes 8 of the 12 areas judged to be selenium-contaminated?

Q16.20 Examine Figures 16.6 (page 282) and 16.11. What two things might account for the greater concentration of selenium in basins within the cluster of states called for in Q16.19? (See again Figure 16.6 and its caption.)

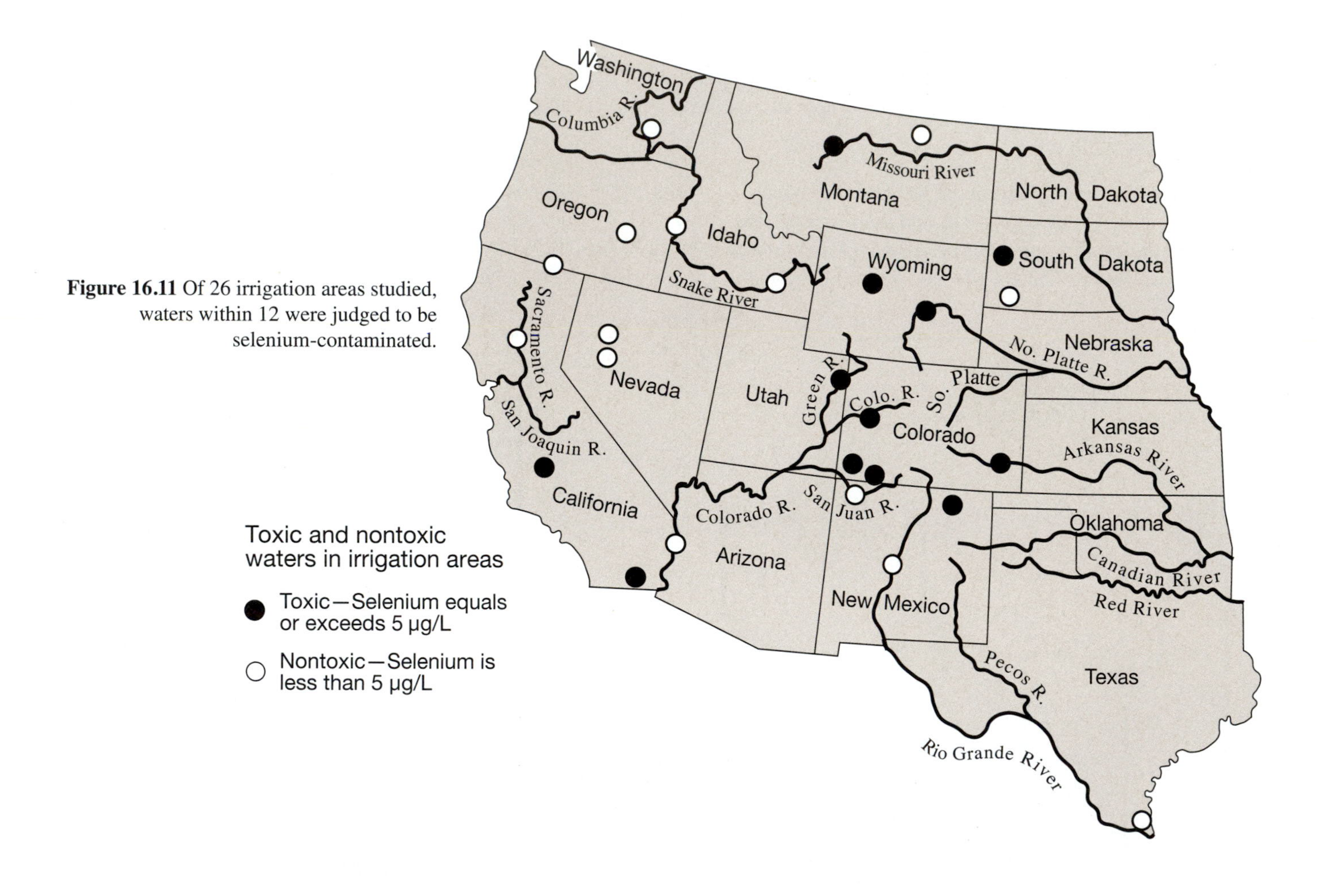

Figure 16.11 Of 26 irrigation areas studied, waters within 12 were judged to be selenium-contaminated.

(Student's name) (Weekday) (Hour)

(Lab instructor's name)

ANSWER PAGE

16.1

16.2

16.3

16.4

16.5

16.6

16.7

16.8 (A)

(B)

16.9

16.10 (Reason #1)

(Reason #2)

16.11

16.12

16.13

16.14, 16.15

Wind River Peak
Elev. 13,200 ft.

Riverton
Elev. 4,956 ft.

Copper Mtn.
Elev. 8,223 ft.

Feet
13,000
12,000
11,000
10,000
9,000
8,000
7,000
6,000
5,000
4,000
3,000

16.16 ____________________

16.17

Maxwell National Wildlife Refuge, New Mexico:

____

Stillwater Wildlife Management Area, Nevada:

____

West Oakes Irrigation Area, North Dakota:

____

Riverton Reclamation Project, Wyoming:

____

16.18 ____________________

16.19 ____________________

____

16.20 ____________________

____

____

____

____

17 Landscape Geology

Topics

A. What is the definition of a glacier? What are the two broadest categories of glaciers, and in what simple way does each move?
B. What are the various landscape features that characterize alpine glaciation? How does each develop? What is the history of development of Jenny Lake in Jackson Hole, Wyoming?
C. What are the various landscape features (and their origins) produced by continental glaciers?
D. What is the history of development of a fjord? Why are fjords commonly anoxic?
E. What are the various hazards associated with desert environments? What accounts for the curious landscape features in Monument Valley? How do desert mountains become buried? What are some of the kinds of sand dunes, and how do they indicate wind direction?
F. What is the history of Sand Hills of Nebraska, and how do they record ancient wind direction?

A. Glacial landscapes

Glacier defined—*A mass of ice that has formed through the recrystallization of snow, and which moves under the influence of gravity.* The fact that glacial ice moves is its defining feature. The work of glaciers—both erosional and depositional—derives in one way or another from the fact that glacial ice moves. The movement most easily envisioned is that of **valley glaciers**, which flow downhill like a viscous liquid (Fig. 17.1). Flow more difficult to envision is that of **continental glaciers**, which are so large that they transcend topography. A continental glacier flows outward in all directions from a center of snow and ice accumulation—sort of like mud being poured from a bucket.

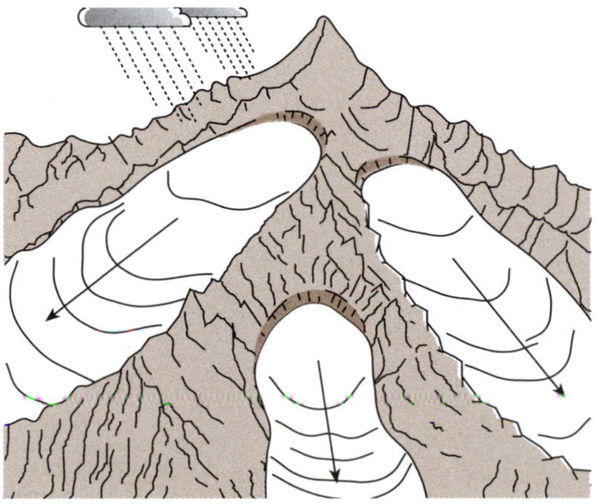

Figure 17.1 Snowfall in high country nourishes valley glaciers. The snow turns to ice and flows downhill, sculpting mountains into rugged landforms.

A present-day example of a continental glacier is that which covers 80% of Greenland (Fig. 17.2).

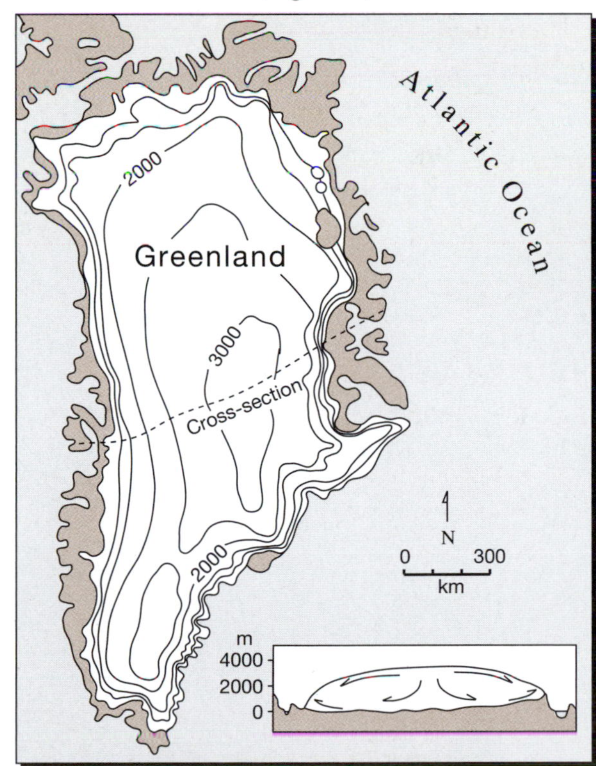

Figure 17.2 The Greenland glacier in map view. The cross-section illustrates flow lines (arrows).

Q17.1 Can you imagine why Greenland glacier doesn't quite make it to the sea along most of its coast?

B. Alpine glacial landscapes

Glaciers develop where temperatures are cold, so they occur at high elevations and at high latitudes. Those at high elevations are called **alpine glaciers**. Erosion associated with the downslope movement of alpine glaciers results in some of the most spectacular scenery in the world (Fig. 17.3).

Alpine glaciers are typically of the valley-glacier variety, carving and deepening their valleys as they flow down the flanks of mountains. The head of a glaciated valley is commonly marked by a steep arcuate wall and an amphitheater-like feature called a **cirque**. A scooped-out depression at the base of a cirque can be occupied by a pond called a **tarn**. Where several cirques surround a mountain peak, the peak is shaped into an abrupt **horn**.

U-shaped valleys, in contrast to V-shaped valleys carved by streams, are characteristic of glacial erosion because of the high viscosity of ice. And—unlike stream valleys—glaciated valleys commonly exhibit ice-scoured depressions within them. If, after the ice melts, such depressions contain water, **paternoster lakes** result. Legend has it that on one sun-lit day, a string of shimmering ponds within a glaciated valley inspired a mountaineer to borrow the term 'paternoster' from the spacing of *Our Father beads* on a rosary.

A steep spine-like ridge separating two adjacent glacial valleys is called an **arête.**

A favorite of visitors to Yosemite and Glacier National Parks is the myriad waterfalls that cascade from **hanging valleys** (Fig. 17.4). Yosemite Falls and Bridal Veil Fall in Yosemite NP plunge more than 1,400 feet to Yosemite Valley below.

Warning: A college student stepped into a stream within a hanging valley in Yosemite, lost his footing on the polished granite, and was swept over a falls to his death.

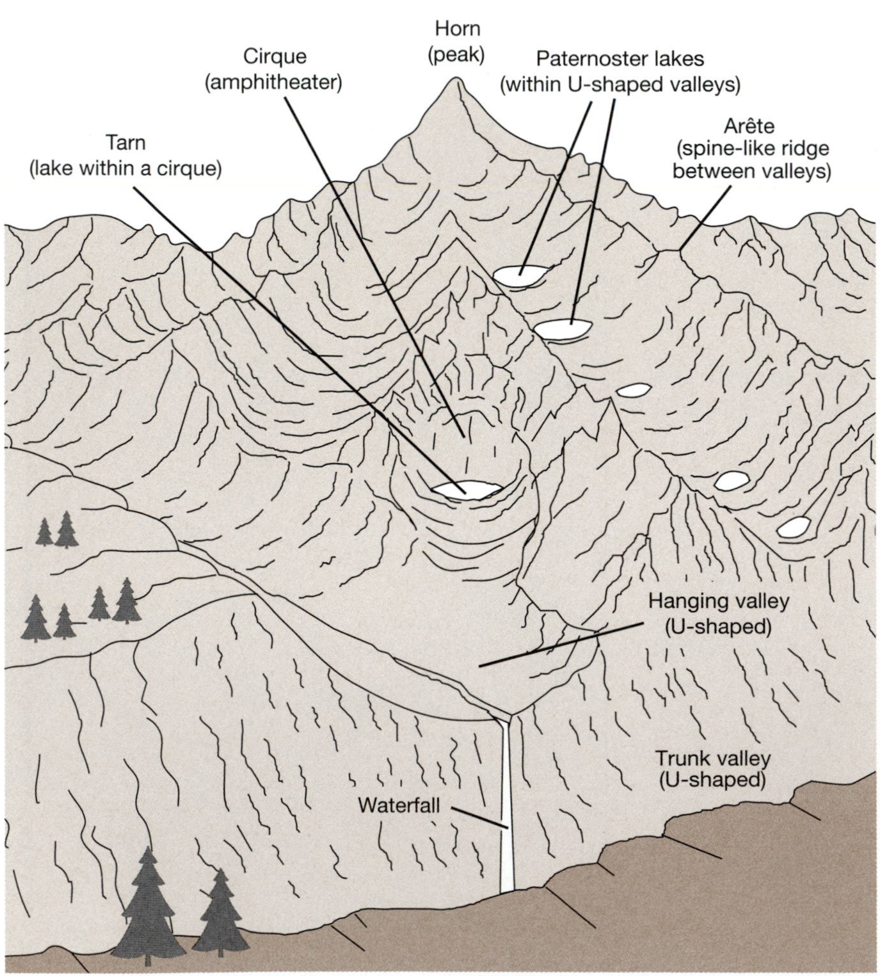

Figure 17.3. These alpine erosional features were carved by valley glaciers that have long since vanished from this scene.

Q17.2. The development of a hanging valley and associated waterfall is shown in Figure 17.4 in three stages. Try to verbalize what appears to have occurred *during glaciation* that explains the landscape *after* glaciation?

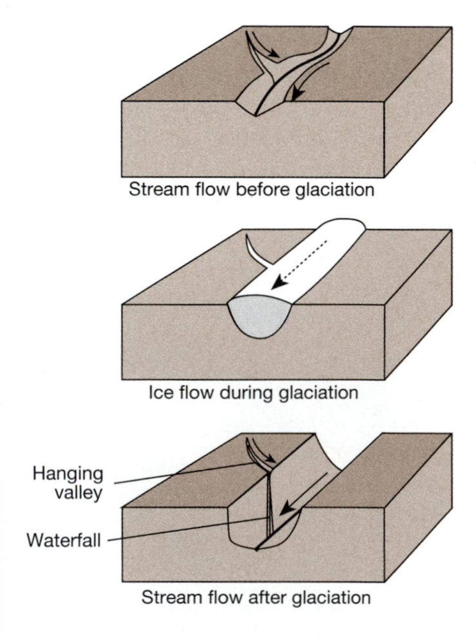

Figure 17.4. The development of a hanging valley and waterfall involves a trunk stream and its tributary before, during, and after glaciation.

Alpine glacial landscape
Glacier National Park
Montana

Mount Jackson, Montana, 7 1/2' quadrangle
N. 48° 36' 03", W. 113° 38' 33"

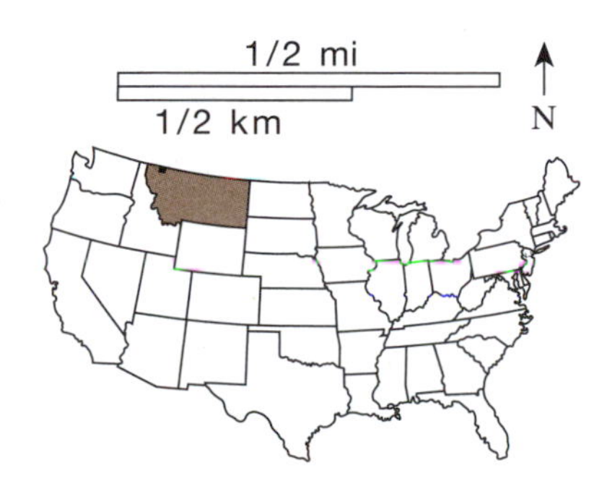

Figure 17.5 The peak in the center of the photograph is Mount Logan (elev. 9,239 ft.) in Glacier National Park. *Note*: This area is one of high relief (approximately 3,000 feet), so it is difficult to grasp the topography on a satellite image.

Q17.3 Identify the four Alpine glacial features labeled A, B, C, and D in Figure 17.5. Your choices: a hanging valley (which, in wet weather, would be marked by a waterfall at its mouth); an arête; a horn; and a U-shaped valley.

Q17.4 (A) Are these rocks in the Glacier National Park region sedimentary, or are they plutonic igneous? *Hint:* Sedimentary rocks are typically layered (i.e., stratified), whereas plutonic igneous rocks are massive by comparison.

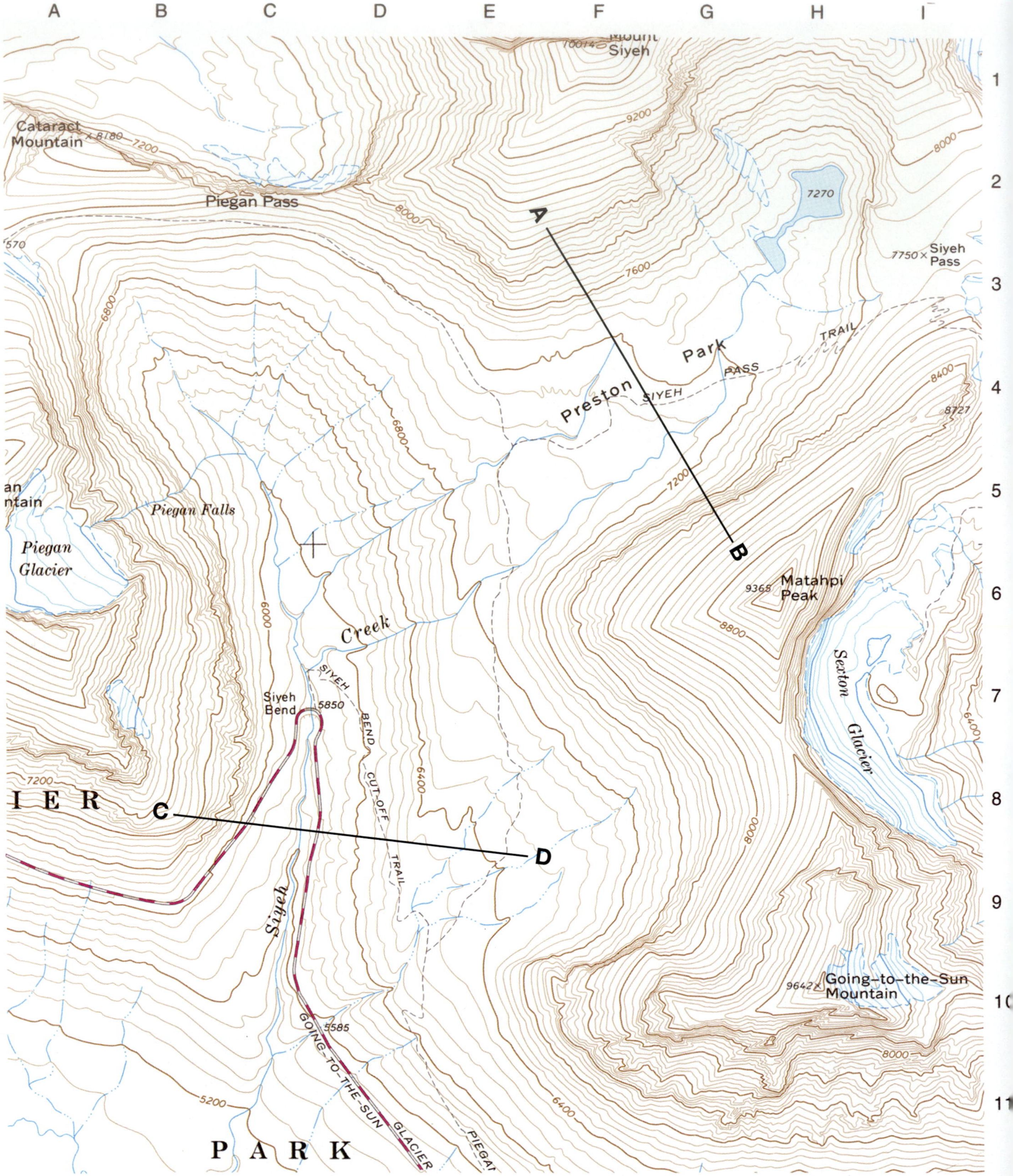

Logan Pass, Montana, 7 1/2' Quadrangle

SCALE 1:24,000

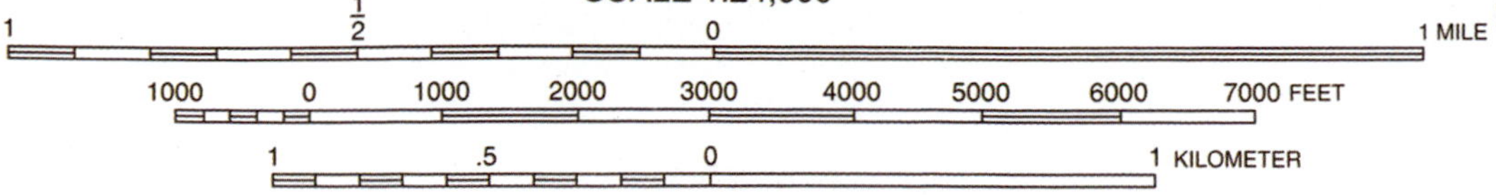

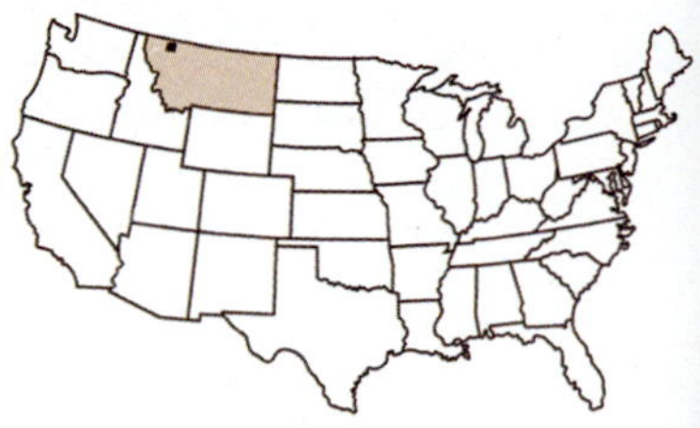

Referring to Logan Pass, Montana quadrangle on facing page 292

Q17.5 (On Answer Page 307)
(A) Draw a topographic profile connecting A and B
(B) Draw a topographic profile connecting C and D
(C) Which of these two profiles is more indicative of glacial erosion, rather than stream erosion?
(D) What is your evidence for your answer in (C)?

Q17.6 (On Answer Page 307)
For the Logan Pass quadrangle, list the map coordinates (e.g., H-6) of each of the following glacial erosional features:
(A) horn
(B) arête (*hint:* see ridge along the western border of Sexton Glacier)
(C) hanging valley with waterfall (*hint:* look for a *named* waterfall)
(D) a cirque with prominent tarn within it

Ice budget—A glacial ice budget is much like your bank account. If you deposit more dollars than you withdraw, your balance swells, but if you withdraw more dollars than you deposit, your balance shrinks. Figure 17.6 shows a longitudinal section of a valley glacier. The glacier gathers ice at its head region—the **zone of accumulation**. As it flows downhill, it encounters warmer temperatures, so it melts (and evaporates) in the **zone of ablation**.

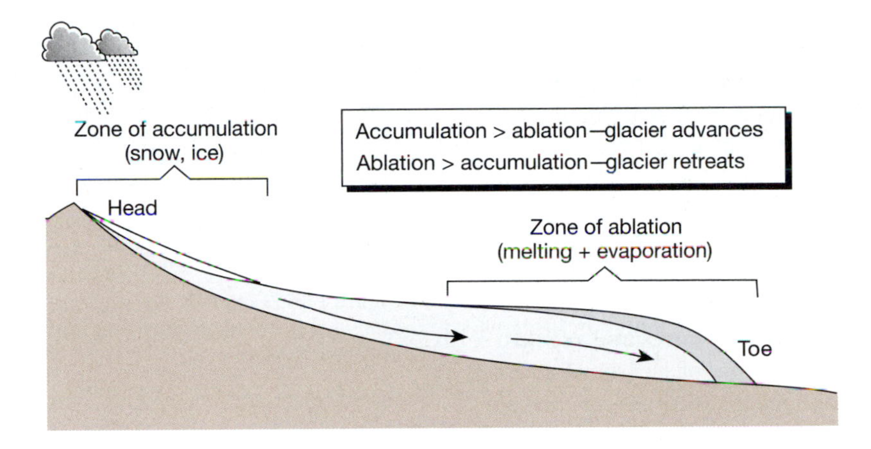

Figure 17.6 This model, although applied to a valley glacier here, applies to continental glaciers as well. For both types of glaciers, there are times when they swell, times when they shrink—depending on rates of accumulation and ablation.

Note: In the case of a 'retreating' glacier, ice does not reverse its direction and flow uphill. The *toe* of the glacier migrates toward its source, but *ice* continues to flow away from its source.

Depositional features:
Till—All rock material picked up, transported, and deposited by a glacier.

Moraine—A body of non-stratified (i.e., non-layered) till. (Till is to moraine as sand is to sand dune.)

Figure 17.6 illustrates the dynamics of advance and retreat of the toe of a glacier. When, at the farthest reach of a glacier, there is a balance between accumulation and ablation for a sufficient length of time, a **terminal moraine** develops (Fig. 17.7). In this situation, the glacier behaves like a giant conveyor belt, carrying and dumping till at its toe, forming an arcuate mass.

Outwash plain—Streams issuing from a glacier carry the finer components of till (e.g., clay, silt, sand, and pebbles). Because outwash is deposited by streams, it tends to be *stratified* (layered), forming smooth plains.

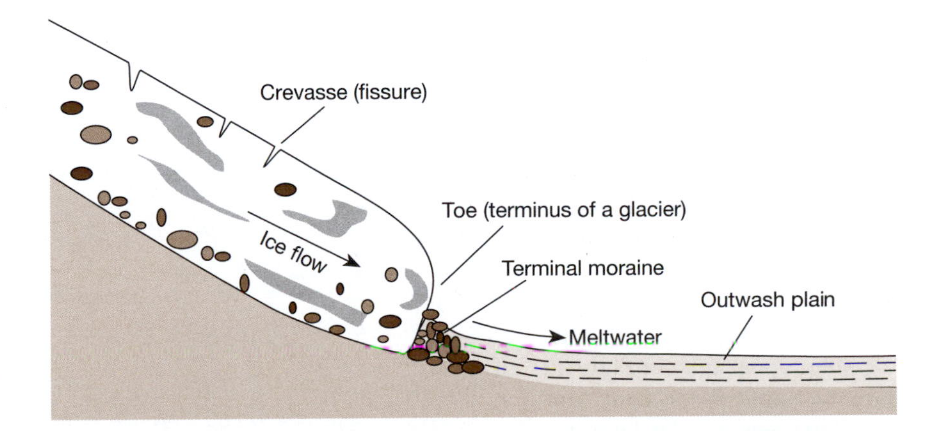

Figure 17.7 Here, a valley glacier melts where it descends onto a broad plain. Till accumulates at its toe as a terminal moraine. Beyond the moraine, sediment-rich meltwater deposits finer-grained rock debris as a stratified (layered) outwash plain.

'Little Switzerland'

U.S. National Park Service

In Figures 17.8 and 17.9, and in the satellite image on the facing page (Fig. 17.10), are illustrations of picturesque Jenny Lake in Grand Teton National Park—at the western margin of Jackson Hole, Wyoming. The Teton Range is perhaps the most photographed and painted landscape in America. It has been the setting of innumerable movies (e.g., the classic western *Shane*).

Jenny Lake is a **proglacial lake**—a lake that forms at the margin of a glacier. The lake is impounded by a terminal moraine that accumulated at the toe of Jenny Lake Glacier where it emerged from Cascade Canyon some 9,000 years ago (Figs. 17.8, 17.9). Native Americans are believed to have populated Jackson Hole at that time.

Depositional features and vegetation—Moraines of Jackson Hole consist of boulders, silt, and clay that are sufficiently porous to hold considerable water from rain and snow. This moisture, coupled with mineral nutrients, makes for fertile soil, hence the correlation between moraines and forests in Jackson Hole. In contrast, associated outwash is less porous and consists of relatively sterile quartz pebbles and quartz sand that supports only sparse grass and sagebrush.

Q17.7. On Answer Page 307—using names and terms in Figures 17.8 and 17.9—draw a topographic profile along the line on Figure 17.7 from A to B. Label your profile with the terms Cascade Canyon, Jenny Lake, terminal moraine, outwash plain, and Snake River.

Q17.8. Account for the curious stair-step topography in Figure 17.10 along the Snake River. *Hint:* You might have seen similar features along the Shoshone River (pages 160, 161) at Cody, Wyoming in Exercise 9 Streams and Rivers.

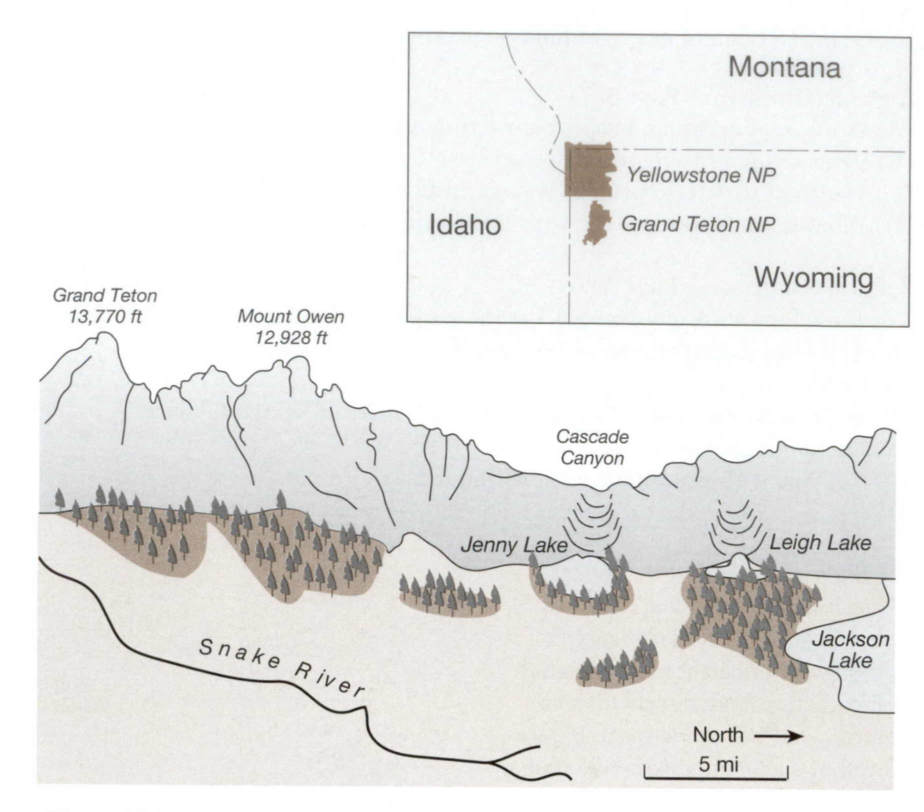

Figure 17.8 Looking westward at the Teton Range, with Jackson Hole in the foreground. Jenny Lake is at the mouth of Cascade Canyon. Trees mark moraines, sagebrush marks the outwash plain.

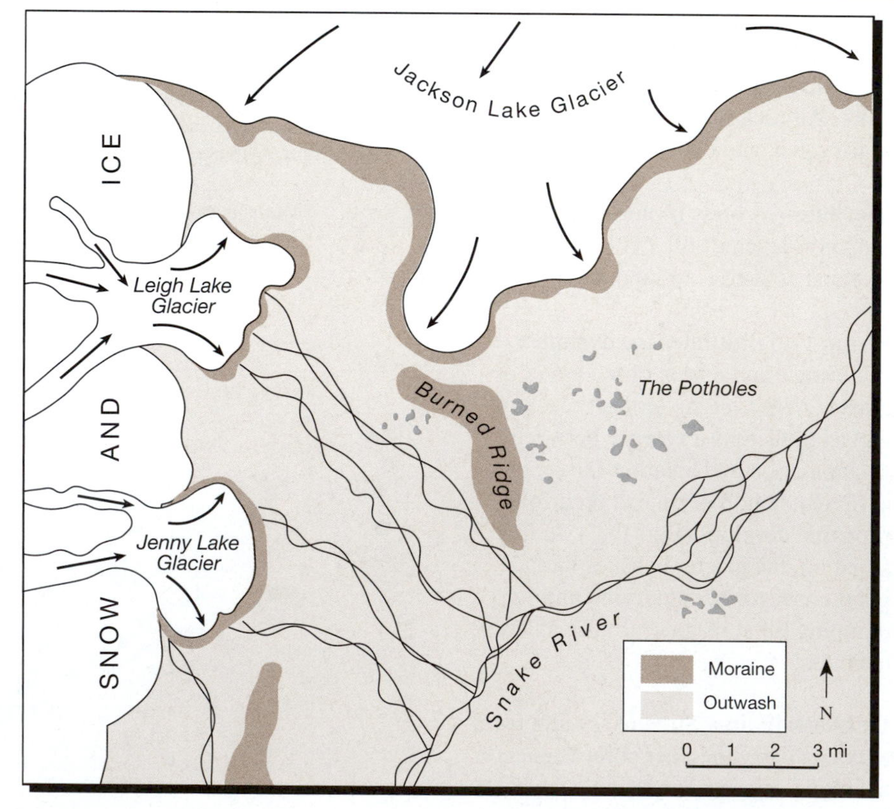

Figure 17.9 Jackson Lake glacier flowed southward out of what is now Yellowstone National Park (arrows) while Jenny Lake and Leigh Lake glaciers descended from Teton canyons.

Glacial lake impounded by a terminal moraine: Jenny Lake, Wyoming

Jenny Lake, Wyoming, 7 1/2' quadrangle
N. 43° 46' 03", W. 110° 43' 30"

Figure 17.10. A terminal moraine acts as a dam impounding Jenny Lake. The sage-covered plain beyond the dam is an outwash plain, whereas forested hills consist of moraine. Notice the terraces along the Snake River. Cascade Canyon appears to be V-shaped, rather than U-shaped, in its lower part, but lots of water has flowed through this canyon since glaciation.

C. Continental glaciers

At high latitudes (e.g., Greenland and Antarctica) glaciers presently extend across valleys and uplands alike, masking topography. These are **continental glaciers**. (Recall Greenland in Figure 17.2 on page 289).

During the Pleistocene Epoch, which ended around 10,000 years ago, depressed global temperatures sent Laurentide continental glaciers southward out of Canada into the lower 48 States (Fig. 17.11).

The Great Lakes and myriad other features in North Amreica are bold reminders of this most recent continental glacial event.

Q17.9 Underwater topographic mapping has revealed glacial deposits much like those in Figure 17.11B on the continental shelf beneath Atlantic waters. How can that possibly be?

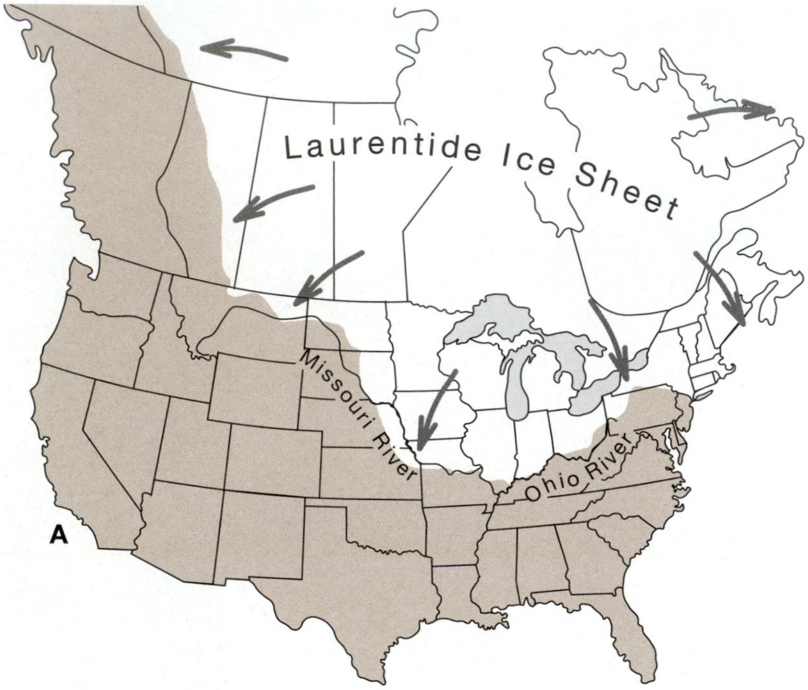

Figure 17.11 A This shows the approximate maximum extent of Pleistocene glaciation plotted on on present-day Canada and conterminous States. The southern limit of Laurentide ice was approximately along today's Missouri and Ohio Rivers. **B**. Myriad landscape features in coastal New England owe their origins to Laurentide continental glaciers.

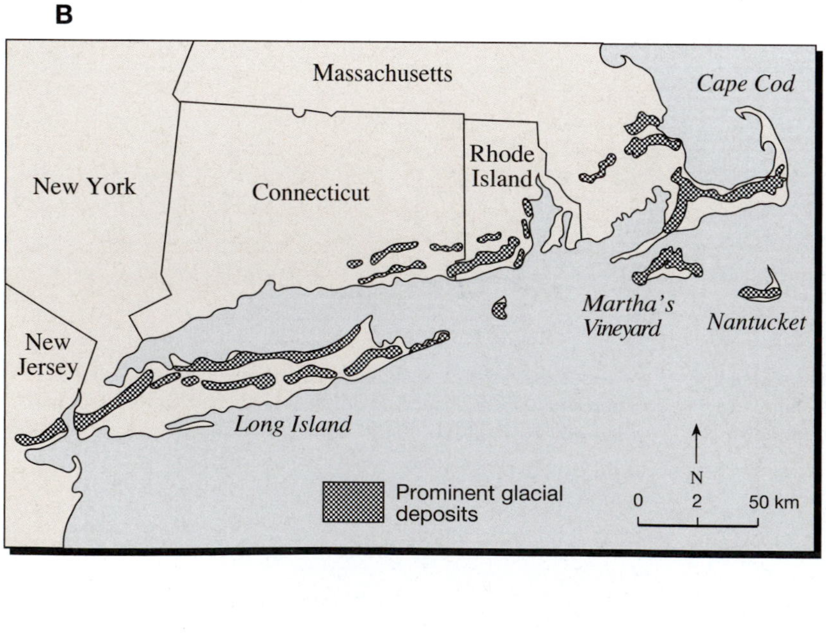

Continental glacial features

Effects of alpine glaciation are most noticeably erosional, but the record of continental glaciers is most noticeably depositional (Fig. 17.12). One or two of these features have been illustrated in our discussion of *alpine features*.

As a continental glacier retreats, it deposits rock debris (called **till**) as hummocky **ground moraine**. Where a glacier pauses in its retreat, till continues to be conveyed to its toe, where it accumulates as an arcuate ridge (much like a terminal moraine) called **recessional moraine**. When retreat resumes, a **proglacial lake** might be impounded between the recessional moraine and the glacier. (Recall Jenny Lake.) Streams issuing from the margin of a glacier carry sediments that are finer-grained than boulder-strewn till. These finer sediments are typically deposited as a stratified **outwash plain** (recall Figure 17.7 on page 293).

Material similar to that of an outwash plain can be deposited by streams flowing *within* a glacier as well. When the glacier retreats, the record of an internal stream deposit appears as a sinuous ridge called an **esker**. Residual blocks of ice become partially buried by moraine and outwash. The eventual melting of these blocks results in **kettle holes**, which commonly hold water to form ponds.

Finally, there are two kinds of hills fashioned by glacial erosion/deposition, both of which are asymmetric indicators of direction of flow. One, the **drumlin**, consists of glacial till with its steep side *upstream*. The other, the **shaped knob**, consists of bedrock with its steep side *downstream*.

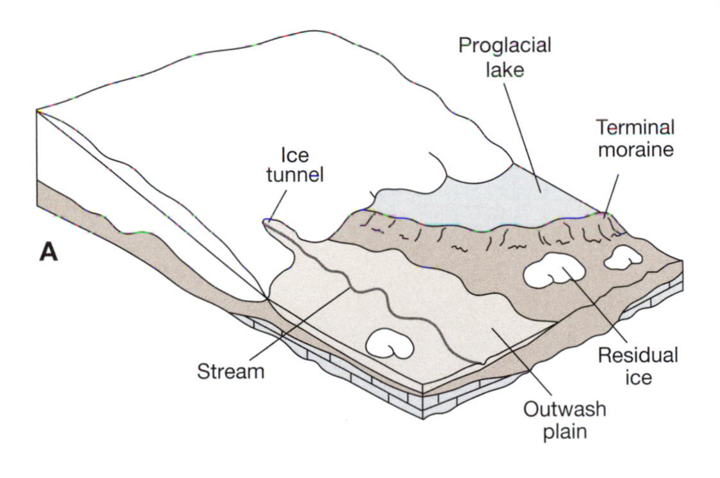

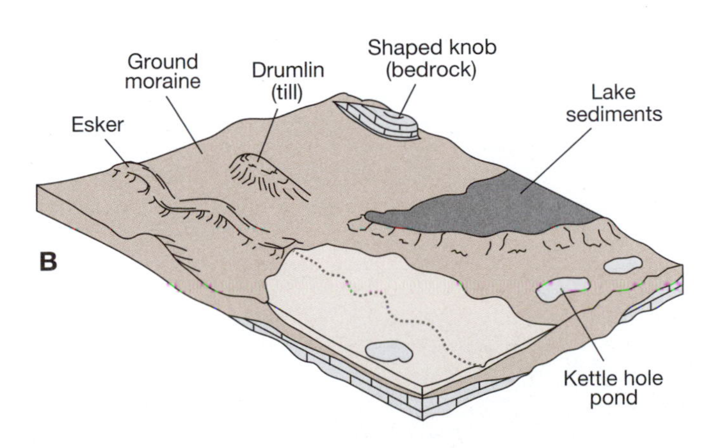

Figure 17.12 A Retreating continental glacier. **B** Ice fully retreated.

Continental glacial features Montana

Coalridge, Montana, 7 1/2' Quadrangle
N. 48° 38' 55", W. 104° 08' 38"

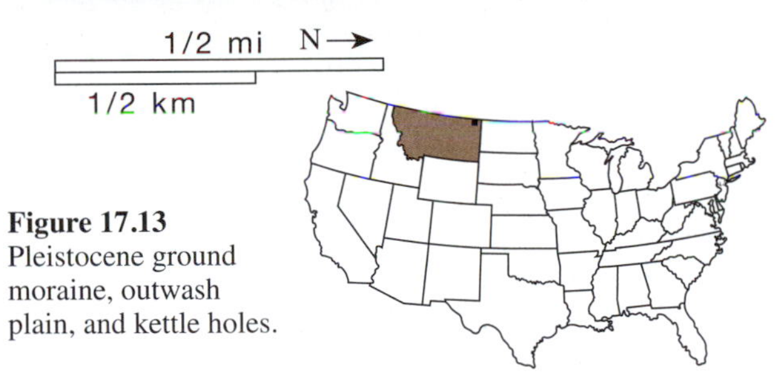

Figure 17.13
Pleistocene ground moraine, outwash plain, and kettle holes.

Q17.10 On Answer Page 308 make a map of the area shown in the top photograph of Figure 17.13, delineating ground moraine and outwash plain. Label ground moraine, outwash plain, and one or more kettle holes.

D. A glacial coastal feature—fjords

An **estuary** is a drowned coastal river valley. Present-day estuaries (e.g., the Potomac River through our nation's capital) reflect the global rise of sea level that accompanied the melting of Ice Age glaciers.

Q17.11 Why does a rise in sea level accompany the global melting of continental glaciers?

An estuary that occupies a glaciated coastal valley is called a **fjord** (or fiord)—a name borrowed from Norway where drowned glaciated valleys are numerous. Fjords also scallop the coast of Greenland (Fig. 17.2, page 289) and are common along the Alaskan coast as well.

Fjords typically contain scant marine life because of the manner in which they develop. As indicated on the map of Greenland (Fig. 17.2), glaciers commonly terminate at or near a shoreline because of the proximity of ocean water with temperatures above freezing. So terminal moraines commonly develop at the mouths of estuaries (Fig. 17.14A). With the retreat of glacial ice and the accompanying rise in sea level, a terminal moraine can act as a barrier across the mouth of a fjord (Fig. 17.14B). Organic matter that finds its way into a fjord consumes oxygen as it decays. The source of replacement oxygen is the open sea, but the terminal moraine impedes tidal exchange. So fjords are commonly **anoxic** (i.e., depleted in oxygen). In sum, a fjord is not a good place to fish.

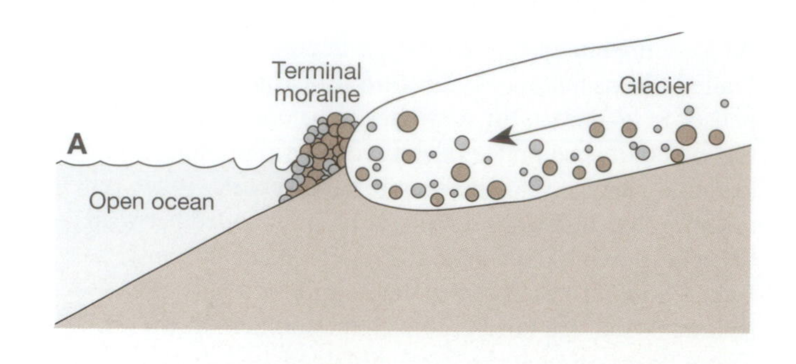

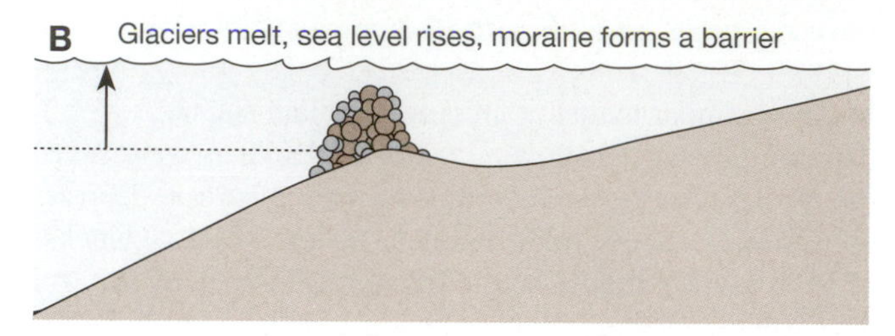

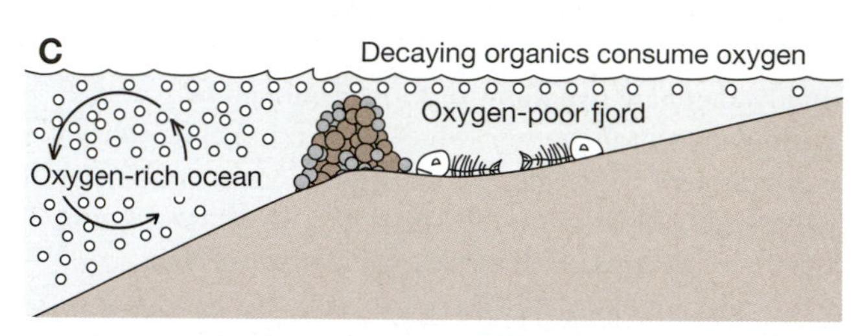

Figure 17.14 Three-step development of a barred fjord. The tiny spheres represent dissolved oxygen.

Mound Desert Island, along the coast of Maine, exhibits a number of elongate lakes. Somes Sound is similarly elongate, but it is open to the the Atlantic Ocean at its south end (Fig. 17.15).

Q17.12 What do you suppose is the origin of the lakes on Mount Desert Island? *Hint:* Being mindful of the shape and orientation of these lakes, look again at the Great Lakes in Figure 17.11A on page 296.

The Mount Desert tourist bureau represents Somes Sound as 'the only fjord in our contermi-nous 48 States.' You be the judge of that. Examine Figures 17.15 and 17.16 and then answer the related question.

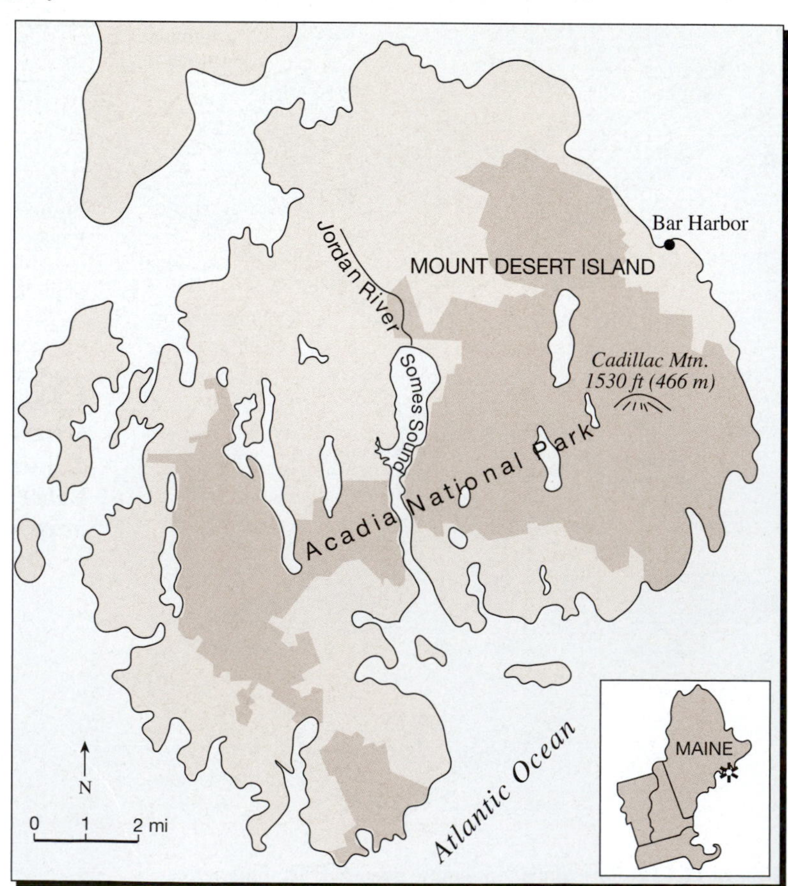

Figure 17.15 Mount Desert Island, home of Acadia National Park, boasts of Cadillac Mountain, the tallest peak on the Atlantic Seaboard. The island's curious name derives from *Isle des Monts Deserts*, (Isle of Bare Mountains), a name coined by Samuel de Champlain in 1604.

A fjord?
Somes Sound, Maine

Salsbury Cove, Maine, 7 1/2' quadrangle
N. 44° 20' 19", W. 68° 18' 46"

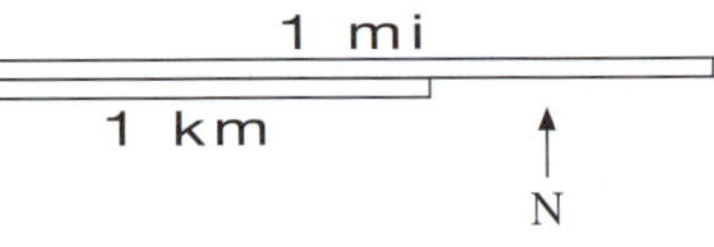

Figure 17.16 The elongate shapes of lakes on Mount Desert Island, all oriented north-south, suggest erosive action by glacial ice. The Jordan River enters Somes Sound at its north end.

Q17.13 The elongate bulbous shape of the northern one-half of Somes Sound appears to have been scooped out by a southward moving glacier, but what about its shape from there to the sea? Do you believe that the narrow valley south of the bulbous part of the sound was carved by glacial ice? Another possibility might be that a glacial lobe advanced as far as the bulbous area, with melt water creating a valley from there to the sea. Which do you favor as the eroding agent that shaped the downstream part of Somes Sound, ice or meltwater?

Incidentally, in case you're wondering about the possibility of a telltale terminal moraine at the coastline, the swift tidal exchange in and out of the sound might have removed any bay mouth moraine.

E. Desert landscapes

The Great American Desert

The definition of **aridity** is not based on precipitation alone (e.g., 10 cm or less per year). Aridity also depends on evaporation. Practically speaking, a region is viewed as dry where *potential evaporation exceeds annual precipitation,* which includes some 30% of Earth's land surface

A more sophisticated measure of dryness is the **evaporation index** (Fig. 17.17), which is inches of potential annual evaporation divided by inches of actual annual precipitation. The Great American Desert of our western States is characterized by values in excess of 2.5 on the *EI scale*. Desert landscapes occur within this region.

Evaporation index
2.5 3.0 3.5 5.0

Figure 17.17 Desert landscapes of our conterminous 48 States occur within areas where the evaporation index is in excess of 2.5.

Desert hazards

In addition to the issue of water, arid regions pose additional hazards:

Sudden dust storms—As blinding as white-out snow storms.

Surprisingly chilly nights—Arid regions typically exhibit surprising differences in daytime and nighttime temperatures.

Q17.14 Why are nights in arid lands surprisingly cold? *Hint:* Think of an insulating medium.

Maze-like corridors—One can easily become lost in the narrow passages of **badlands** (i.e., rugged arid landscapes). Ebenezer Bryce, from whom Bryce Canyon National Park takes its name, once quipped that his canyon was 'a heck of a place to lose a cow.'

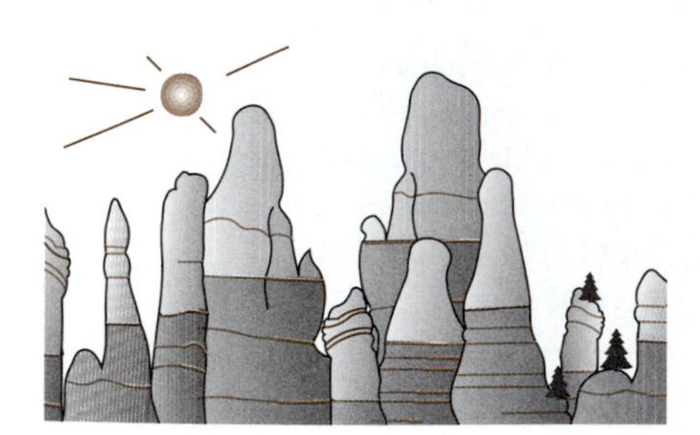

Flash floods—Certainly, humid regions experience **flash floods**, but arid regions present special threats. An arid-lands hiker strolling along a scenic **arroyo** (aka dry wash) can be totally unaware of a distant rainstorm before hearing a wall of water coursing down upon him/her (Fig. 17.18).

Figure 17.18 Arroyos are flat-floored stream channels with steep walls because the occasional water evaporates and soaks into the dry earth—depositing sediments en route.

In Almería, Spain—*perhaps the driest spot in all Europe*—an arroyo passes through the center of town en route to the nearby Mediterranean Sea (Fig. 17.19). Local geologists quip that there is a "delta of car bodies" offshore of Almería.

Q17.15 So, why all the offshore car bodies? Answer: To what municipal use do you suppose residents of Amería put the arroyo that passes through their town? *Hint:* It's a universal crowded city-center need.

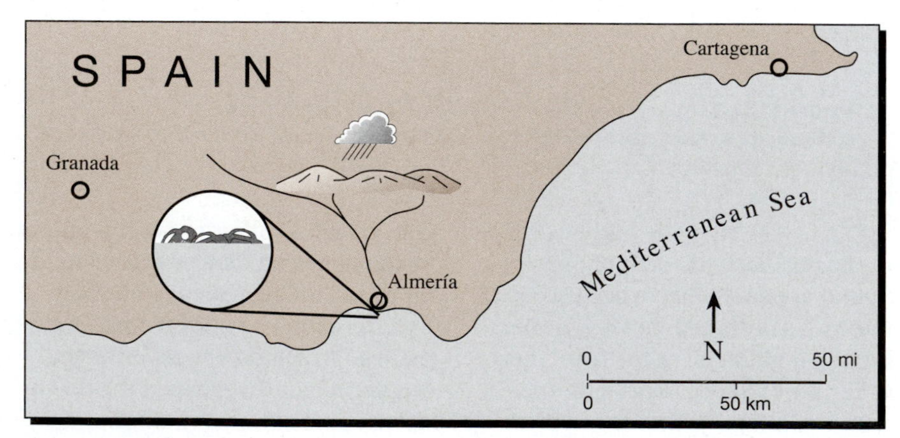

Figure 17.19 Car bodies rest in their watery grave offshore of Almería, Spain.

Desert erosion

In those deserts where sediments are effectively carried away by occasional rainwater, imposing spires and spines of rock rise above flat valley floors. An example is famous Monument Valley, Arizona (Fig. 17.20). Countless western movies have been filmed there.

Q17.16. Draw a topographic profile along the white line in Figure 17.20. Also, show, with a geologic cross-section, three horizontal rock layers. Label the layers that form vertical cliffs *sandstone*, and the layers that form less steep slopes *shale*.

Q17.17. Draw a circle around and abrupt spire standing like a 'monument,' an example of the features that give this region its name.

Q17.18. In Monument Valley erosion is along parallel fractures (called *joints*), which imparts a topographic fabric. Several peninsular spines exhibit a mutual orientation. What is their approximate azimuth in a northerly direction? (See azimuths on page 13 of Maps Exercise 1.)

Desert erosion
Monument Valley, Arizona

Mystery Valley, Arizona-Utah 7 1/2' Quadrangle
N. 36° 55' 21", W. 110° 08' 39"

Figure 17.20. Removal of sediments from this desert area by occasional rainwater has produced bold spines and spires of rock.

Desert deposition

By water—Deserts are characterized by a lack of vegetation because (a) there is insufficient water to sustain plants, and (b) there is insufficient water for chemical weathering of rock to produce nutrient-rich soil. Without plants to obstruct the surface flow of water, and without porous soil to absorb that water, the runoff of occasional rainwater is at a maximum—resulting in flash floods. However, flood waters in deserts quickly evaporate into the dry atmosphere, so streams are short-lived. As a result, sediments are deposited along stream courses, rather than being carried to the sea, and mountains become buried in their own debris (Fig. 17.21).

Q17.19 Draw a cross-section along line A-B in Figure 17.21. Show both the surface topography and your guess as to the boundary between the partially-buried mountains and the sediments. ***Hint:*** **View the sediments as a lake that has partially covered a landscape.**

Desert deposition
California

Frink NW, California, 7 1/2' quadrangle
N. 33° 25' 03", W. 115° 37' 39"

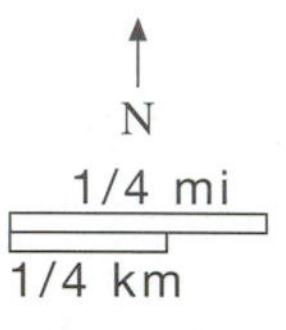

Figure 17.21 Desert mountains become buried in their own debris.

Deposition by wind—Deserts bring to mind **sand dunes**, of which there are several different types. The most common is the barchan dune (Fig. 17.22). The horns of this crescent-shaped dune point downwind.

Figure 17.23 is a Google Earth image of a part of the Salton Sea area of California. (Not the most definitive photo, but such is desert imagery.) Sparse rain reduces bedrock to sand, which is fashioned by wind into dunes.

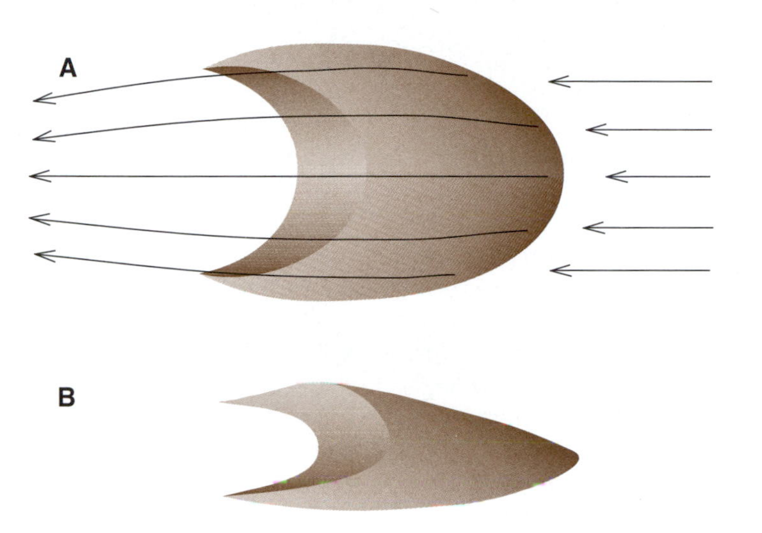

Figure 17.22 Barchan dune aerodynamics. **A** Map view with arrows indicating air flow. A constant wind direction produces a dune that is symmetrical in plan view. **B** Oblique view.

Q17.20 What is the prevailing wind direction indicated by the barchan dunes in Figure 17.23?

Q17.21. Do the dunes in Figure 17.23 reflect recent movement (i.e., within historical time), or have they been stationary for, say, a thousand years? Give two lines of evidence for your answer. *Hint:* (A) One line of evidence is indicated by the distribution of vegetation in front of a dune compared with that behind it. (Give the letter label of such a dune and explain.) (B) The other line of evidence is indicated by the relationship between one dune and a human feature. Name that dune.

Q17.22 Notice in Figure 17.23 that some of the dunes are not perfectly symmetrical like that shown in Figure 17.22. (A) Describe this asymmetry, and (B) try to explain it. *Hint:* Study the caption to Figure 17.22A.

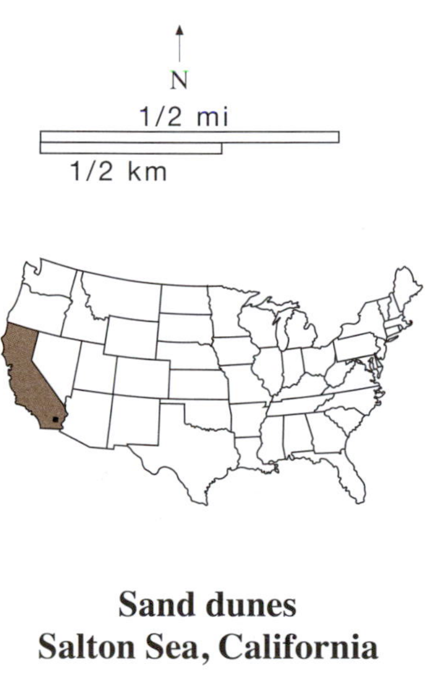

Sand dunes
Salton Sea, California

Kane Spring NE, California,
7 1/2' quadrangle
N. 33° 11' 05", W. 115° 50' 58"

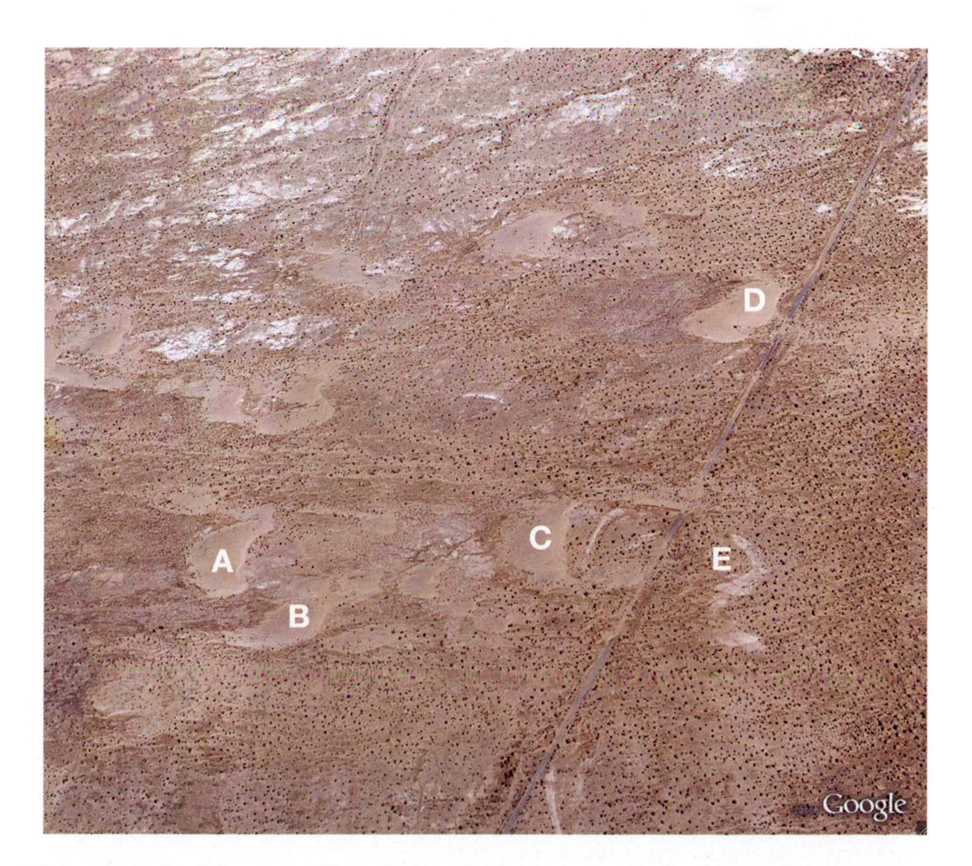

Figure 17.23 Sand dunes near Kane Spring, California, are typically of the barchan variety.

The variety of sand dunes—Sand dunes come in a variety of shapes (Fig. 17.24). Three shapes—**barchan**, **parabolic**, and **transverse**—indicate wind direction. Barchan and parabolic dunes appear in map view to have contradictory shapes, but there is one universality among the three types of dune shapes that indicate wind direction: *The windward side is the gentler slope, and the leeward side is the steeper slope.* So their cross-sections are more definitive of wind direction than are their shapes in map view. A longitudinal dune is ambiguous; wind direction can be either of two directions parallel to the dune's axis of elongation.

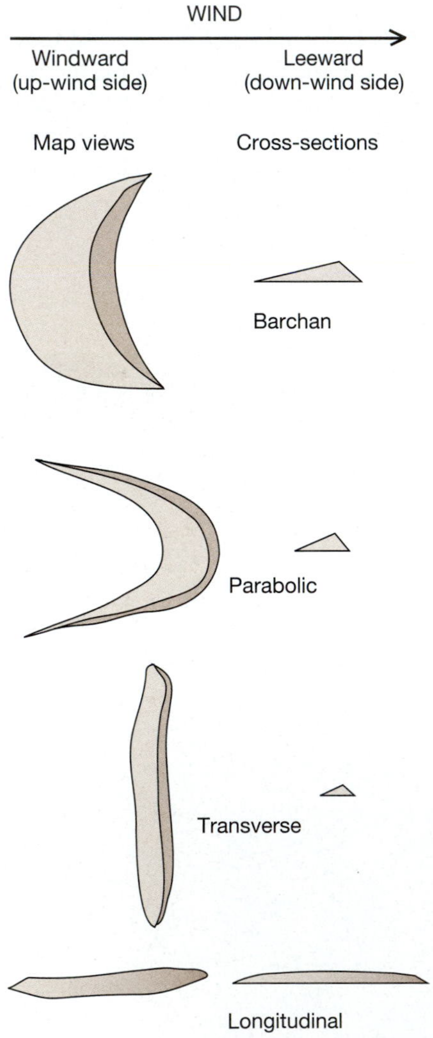

WIND

Windward (up-wind side)

Leeward (down-wind side)

Map views

Cross-sections

Barchan

Parabolic

Transverse

Longitudinal

Figure 17.24 Four types of dunes. Three of the shapes—barchan, parabolic, and transverse—indicate wind direction by their steeper slopes being on their leeward sides.

Another product of wind erosion is a depression called a **blowout**. Unlike running water, both glacial ice and wind can scoop out soil and rock creating closed depressions, which, if filled by water, become natural lakes.

F. An ancient desert: Lakeside, Nebraska

The Lakeside, Nebraska, quadrangle on the facing page covers part of an area of Pleistocene (Ice Age) sand dunes that have long since been overgrown and stabilized by grass. But these fossil dunes still exhibit much of their original shapes. As a guide to the study of sand dunes on the Lakeside quadrangle, Figure 17.25 shows an idealized contour map of a dune.

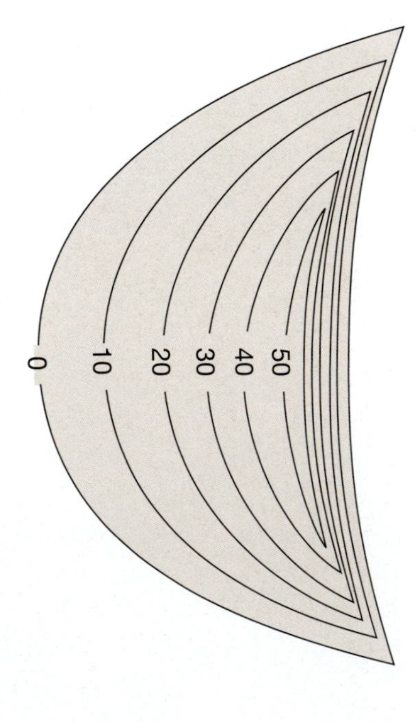

Figure 17.25 Idealized contour map of a sand dune 50+ feet high. The base of the dune is arbitrarily assigned the value of 0 feet.

Q17.23 On Answer Page 309 draw a topographic profile along A to B on the Lakeside quadrangle.

Q17.24 Which is the steeper side of this dune—the northwest side or the southeast side?

Q17.25 Which kind of dune illustrated in Figure 17.24 do you believe you traversed with your topographic profile?

Q17.26 From which direction did the wind that fashioned the dunes of the Lakeside Nebraska region blow?

Q17.27 What is there about the shape of the dune in your topographic profile that indicates that this is an ancient dune, rather than an active dune on the move?

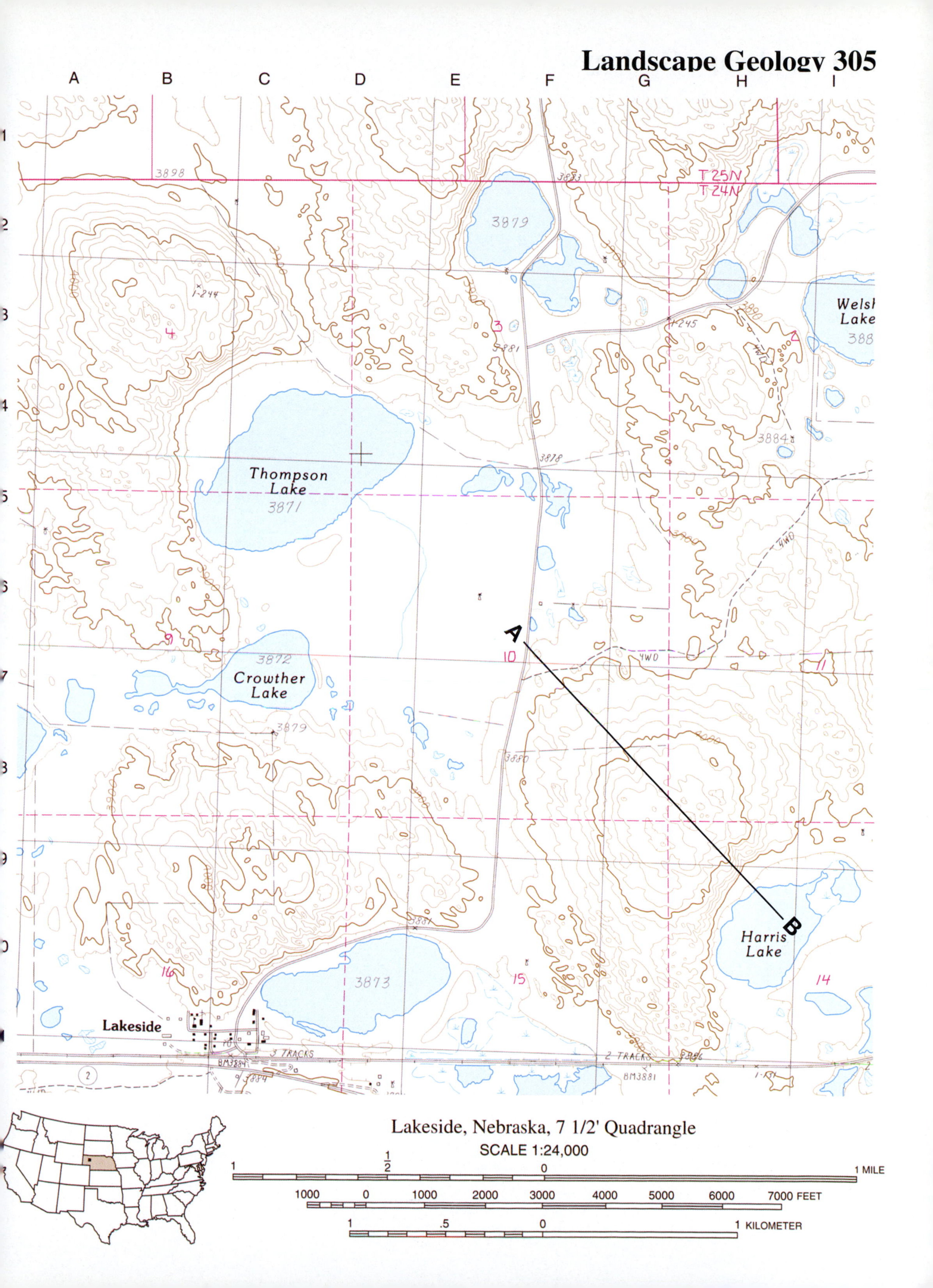

Lakeside, Nebraska, 7 1/2' Quadrangle

SCALE 1:24,000

Intentionally blank

(Student's name) (Day) (Hour)

(Lab instructor's name)

ANSWER PAGE

17.1

17.2

17.3 A

B

C

D

17.4

17.5 (A) (Logan Pass, Montana, Quadrangle)

A B

8400
8000
7600
7200
6800

(B)

C D

6400
6000
5600
5200
4800

(C)

(D)

17.6 (A)

(B)

(C)

(D)

17.7

A B

17.8

17.9

17.10

17.11

17.12

17.13

17.14

17.15

17.16

A B C D

17.17

17.18

17.19

A B

17.20

17.21 (A)

(B)

17.22 (A)

(B)

17.23 Lakeside, Nebraska, quadrangle

17.24

17.25

17.26

17.27

Intentionally blank

18 Climate Change

Topics

A. Describe the *metamorphosis* from snow to ice. Describe the internal dynamics of glacial flow.

B. Explain the stratigraphy of American mid-continent glacial deposits, one layer upon another.

C. What has been the history of the Gulf of Mexico (arm of the Atlantic) as recorded by Florida? Deltas of the world are especially at risk as concerns a rise in sea level. What is the reason, or reasons, for this special vulnerability?

D. Describe the *Milankovitch cycles* (including reference to Alley's animation of such cycles).

E. In what ways do stable isotopes differ from unstable isotopes? Cite an example of how stable isotopes of oxygen and/or carbon indicate environmental and climatic conditions of the past.

F. Describe Earth as a *greenhouse*. It has been said that CO_2 has been—*and continues to be*—the 'climate control knob' of Earth. Explain the reasoning that is the basis for that statement.

A. Climate history—most recently written in our modern landscapes

The history of climate change most obvious to the lay person is that recorded by modern landscapes—e.g., glacier-sculpted Rocky Mountains (page 295), fossil sand dunes of Nebraska (page 305), and farm lands of the American Mid-west strewn with glacial boulders (page 297). Every traveler who ventures into the high country of our American outback returns with stories about past and present glaciers.

Glacier defined—A mass of ice that forms through the recrystallization of snow (Fig. 18.1) and flows under the simple influence of gravity—like water running downhill.

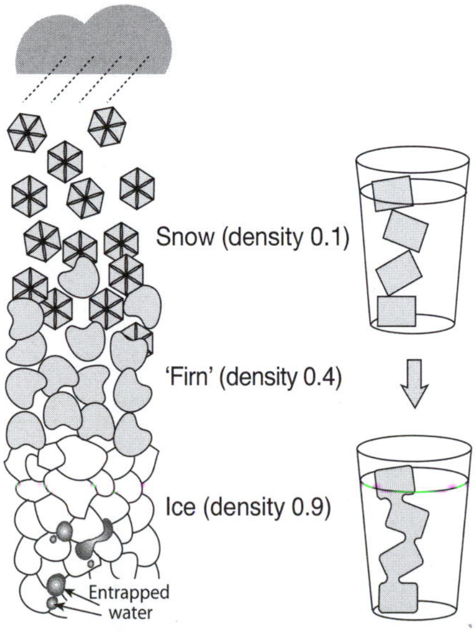

Snow (density 0.1)

'Firn' (density 0.4)

Ice (density 0.9)

Entrapped water

Figure 18.1 Metamorphosis of snow into ice brings to mind the melding of ice cubes in an ice maker. *Relevance*: Water trapped within glaciers can reveal the compositions of ancient atmospheres.

During the past 1.8 million years or so glaciers on a global scale have migrated from each of Earth's two poles into lower latitudes (Fig. 18.2). How come? Simple answer: When there is more snow in winters than melts in summers, glaciers grow and advance. And vice versa.

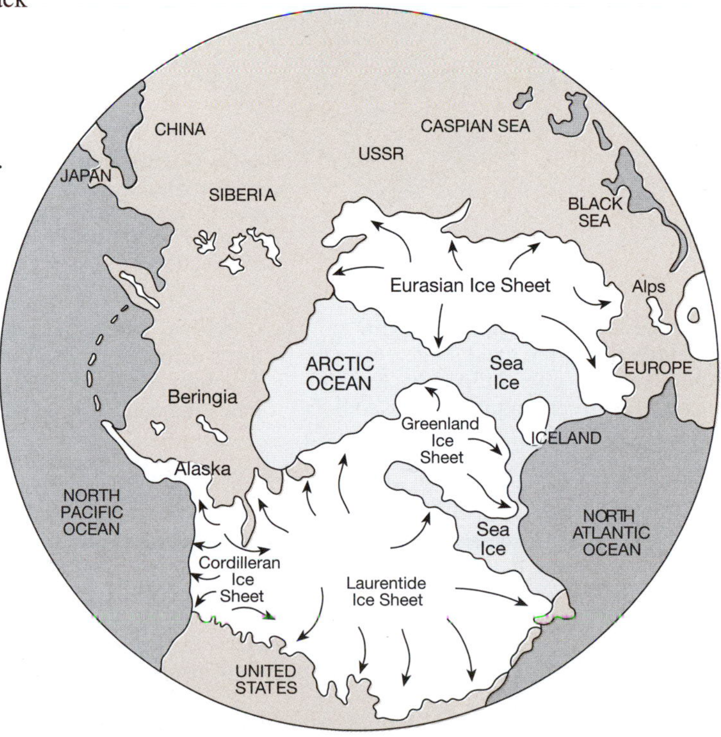

Figure 18.2 The most recent global-scale glaciation began 20,000 years ago. This ice, which coexisted with archeology's Paleolithic Age, began its retreat 10,000 years ago.

Glacier dynamics

Case study of a Glacier—Glaciers come, glaciers go, depending on the vagaries of climate. One of the earliest studies of a glacier advancing and retreating was that of the Rhone glacier in the Swiss Alps in the late 19th century (Fig. 18.3). Flags were planted in the glacier along a straight line in 1874, and then monitored for 8 years.

Q18.1 In what part of the glacier (apparent in this map view) did ice flow fastest between 1874 and 1882—along its lateral margins or down its central axis?

Q18.2 What was this fastest rate of flow within the glacier—as expressed in meters per year?

Q18.3 What was the rate at which the glacier's toe retreated (through melting and evaporating) between 1818 and 1882—as expressed in meters per year?

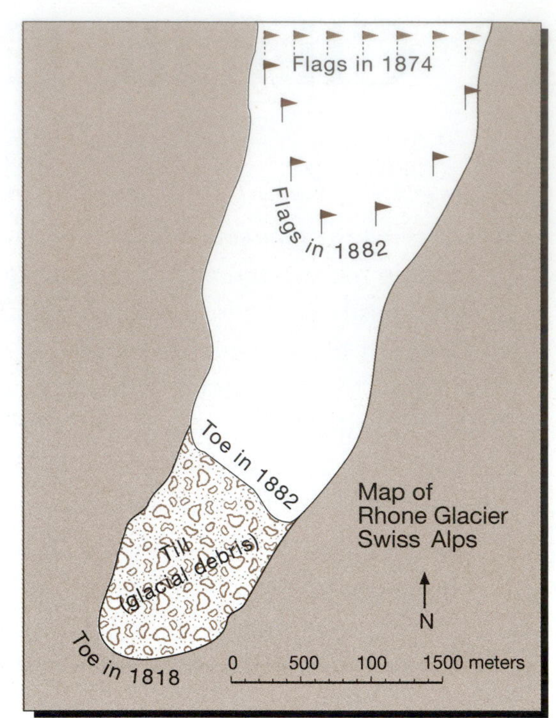

Figure 18.3 Illustrating the quantitative study of the dynamics of the Rhone glacier.

Glacier National Park (Fig. 18-4)—This part of Glacier National Park is illustrated in a satellite photograph on page 291 and on a topographic map on page 292. U.S. Geological Survey geologists predict that the few remaining glaciers in Glacier National Park will be gone by the year 1030.

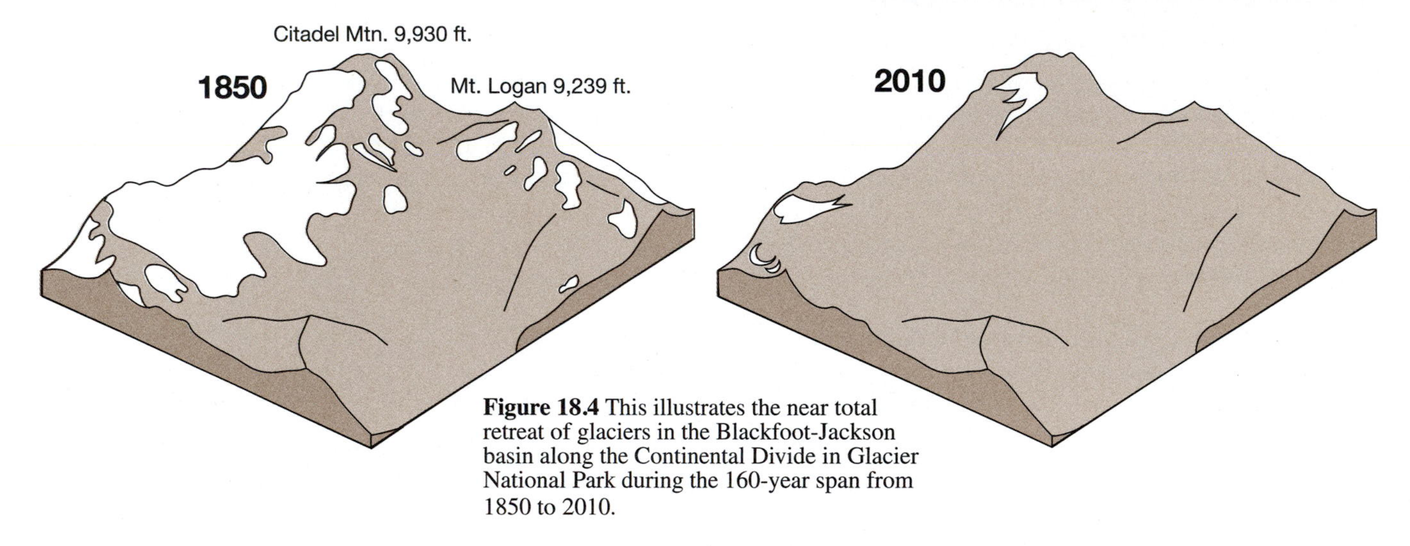

Figure 18.4 This illustrates the near total retreat of glaciers in the Blackfoot-Jackson basin along the Continental Divide in Glacier National Park during the 160-year span from 1850 to 2010.

A global view—Glaciologists now conclude that there exist some 160,000 land glaciers—plus the Arctic, Antarctic, and Greenland ice sheets—all of which, except for a mere handful, are wasting away at alarming rates. The loss of glaciers not only diminishes majestic landscape scenery, but the loss of seasonal glacial meltwater essessential for downstream farmland threatens the lives of billions of people.

Vanishing glaciers in Glacier National Park

Animated PowerPoint presentation (with 25 frames of 1 1/2-second each) at http://www.nrmsc.usgs.gov/research/glacier_model.htm

Modeled Climate-Induced Glacier Change in Glacier National Park, 1850–2100, BioScience 53: 131–140, 2003
Myrna H.P. Hall and Daniel B. Fagre

B. Climate changes of the relatively recent geologic past:' The Pleistocene Epoch

History of the concept—The Pleistocene Epoch was first subdivided in Europe into four temporal *stages* by the English geologist Charles Lyell (in 1839) on the basis of the proportions of extinct-to-living species of mollusk shells. This vertical succession of sediments was interpreted as consisting of glacial deposits and given names from their occurrences in the Alps (Fig. 18.5). European geologists later journeyed to America where they recognized geologic records of four glacial events that they correlated with European counterparts.

Years before present
10,000
WISCONSIN
(Würm)
75,000
125,000
ILLINOIAN
(Riss)
265,000
300,000
KANSAN
(Mindel)
435,000
NEBRASKAN
(Günz)
1,800,000

Figure 18.5 Idealized stratigraphic section showing the four stages of the Pleistocene Epoch (bold caps) in Midwest U.S. The names of the American stages are taken from states where the deposits bearing their names are especially plentiful and well exposed. The Kansas stage, perhaps the most extensive, is approximately bounded by the Missouri and Ohio Rivers (see the Laurentide Ice Sheet map below). The smaller names (in parentheses) are age equivalents of Pleistocene glacial sediments in the Alps.

Q18.4 There is no place in the United States where all four of the glacial stages occur together, stacked one upon the other like a 4-layer cake. So, here's a lesson in field geology: How was such a composite succession assembled? *Hint:* Recall Lyell's criteria for originally subdividing the Pleistocene.

Q18.5 Explain the relationship of the two different kinds of deposits developed in the series I through III below.

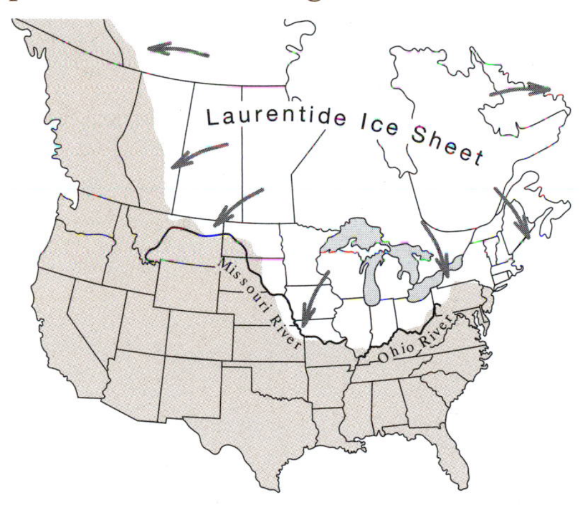

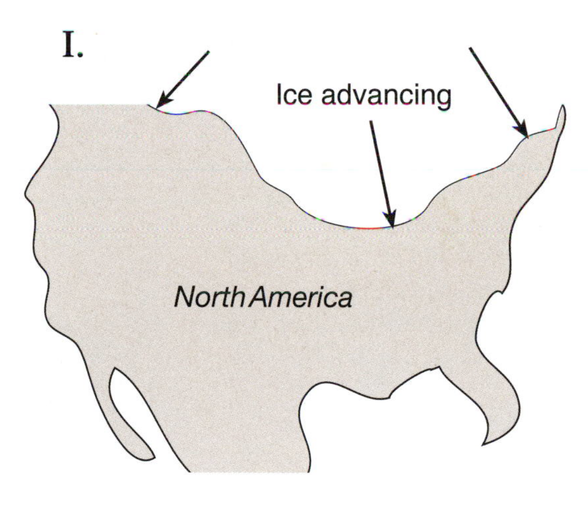

For reference—There's a slightly enlarged view of the North American Laurentide Ice Sheet on page 296. Ice reached as far south as 38° latitude, where, at present, summer temperatures of 100° F are not uncommon.

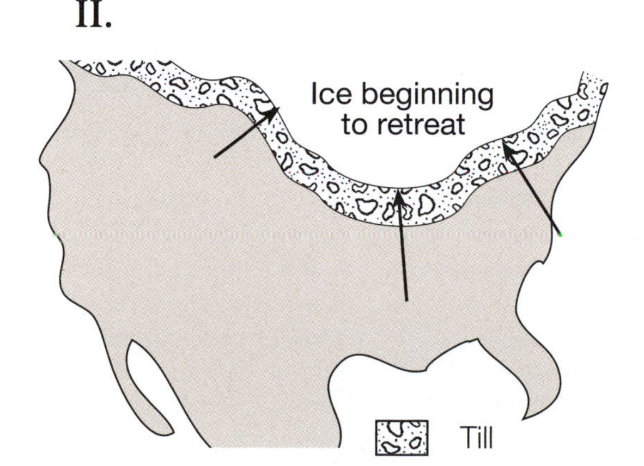

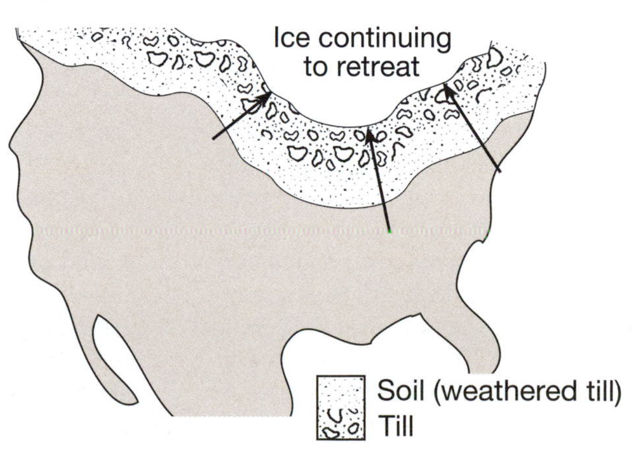

C. The Pleistocene Epoch and sea level

It's a matter of geological history. The advance and retreat of global-scale ice sheets have been accompanied by (a) a lowering of sea level, as ice sheets grew; and (b) a rise in sea level as ice sheets shrank. One of the best documented histories of the variations in sea level during the Pleistocene has been worked out for peninsular Florida (Fig. 18.6).

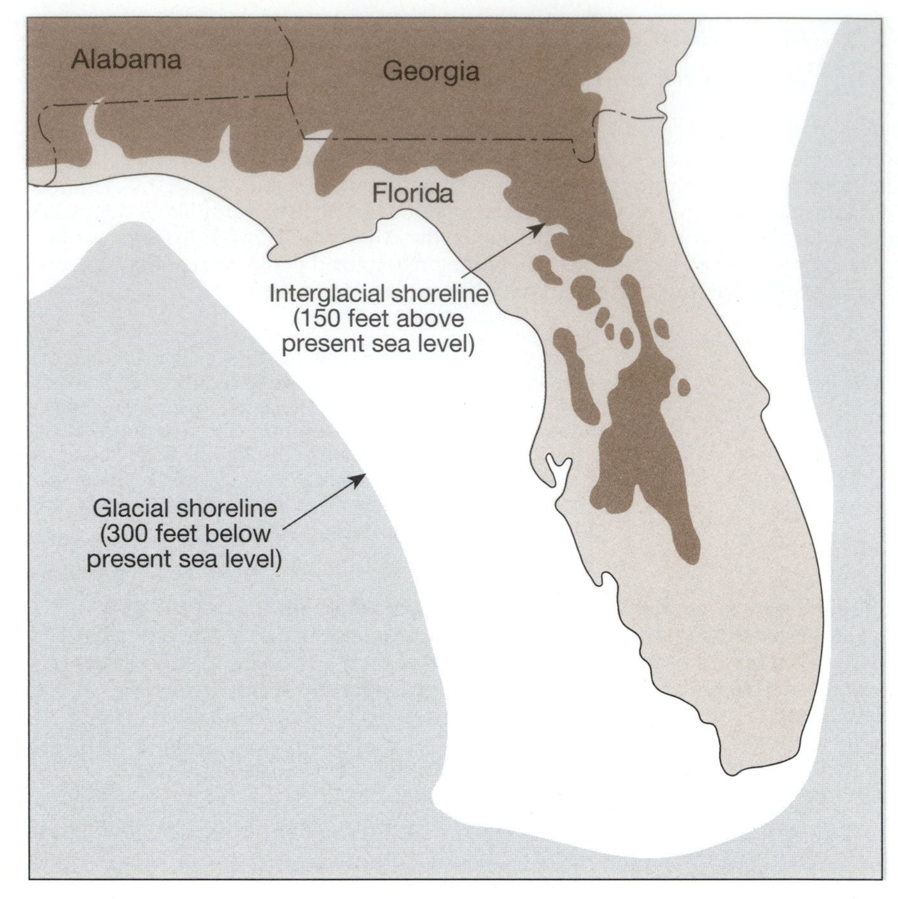

Figure 18.6 *Pleistocene shorelines in Florida.* Source: Florida's Geological History and Geological Resources, Florida Geological Survey, 1994 *Graphic by Frank R. Rupert*

Areas that are most affected by sea level changes are, of course, low-lying coastlines—best exemplified by river deltas. Deltas are viewed as bread baskets of numerous 'beans and rice' coastal populations.

Q18.6 How's your geography? Have a look at Figure 18.7. This map is from a *Time Magazine* article relating global warming to glacial melting, to related sea level rise, to innundation of vital agricultural lands. Match each letter in Figure 18.7 with a river/delta and country.

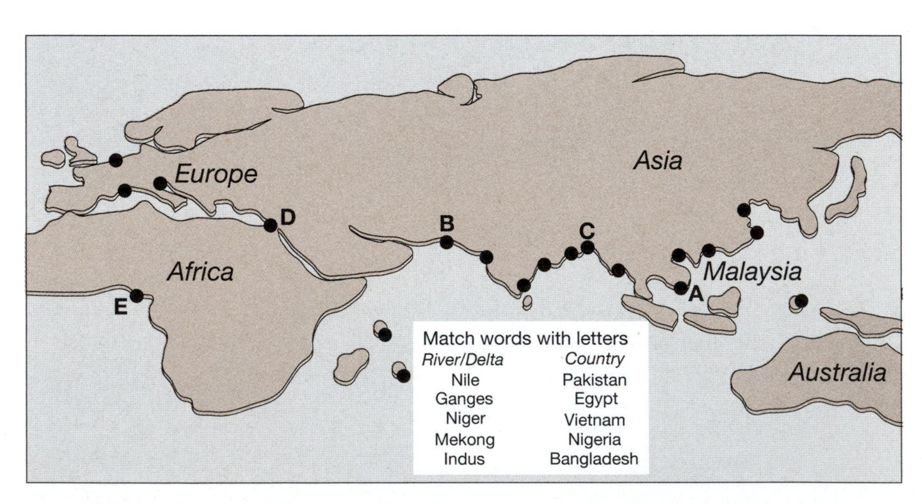

Figure 18.7 Deltas of the Eastern Hemisphere *Time Magazine, October, 02, 1995*

It's been calculated that an increase in annual mean temperature of 6°F by the year 2100 would be accompanied by a rise in sea level of 3 feet—which would displace some 100 million delta inhabitants.

Q18.7 There is yet another factor making deltas subject to ocean incursions, apart from melting ice. Our own Mississippi River delta illustrates this problem. What is it? ***Hint:*** **see page 197, Figure 11.16.**

Recent findings—Satellite measurements indicate that current sea level is 6 inches higher than it was 100 years ago, which is ascribed to the expansion of warmer sea water and runoff from melting glaciers. The present rate of rise is approximately 12 inches per century. But leading glaciologists argue that rates of sea level rise exceed such estimates because of accelerating rates of melting of mountain glaciers and changes in ice caps in Greenland and Antarctica. Studies released in 2008 and 2009 using combined satellite sensors of the sea surface and estimates of continuing ice cap and glacier melt, indicate that sea level will be three feet higher, or more, by the end of the century. This is a devastating forecast for low lying areas of Pacific islands, Bangladesh, Florida Everglades, and the lower parts of all major coastal cities.

D. Milankovitch cycles

Understanding the causes of cyclic patterns of glacial events has been a challenge to innumerable earth scientists in attempts to shape a **planetary theory** of global climate.

A working theory was proposed by Scottish geologist John Croll in the mid-nineteenth century, which was later refined by Milutin Milankovitch, a Serbian astronomer of the early 20th century. To be sure, both men *'could see further because they stood on giant shoulders'*—a remark often attributed to Isaac Newton, who, in turn, built on the work of Tycho Brahe and Johanne Kepler's *Three Laws of Planetary Motion* (both circa 1600).

Three variables comprise the Milankovitch cycles (not to be confused with Kepler's *Three Laws...*). Milankovitch cycles (Fig. 18.8) affect the degree of heating of Earth's surface by the sun, and, so, affect how and when Earth enters an ice age or a period of global warming.

Different textbooks place these three variables in different orders on the page, but no matter. It's the combined effect of these three simultaneous events that matters.

Thanks to Richard B. Alley, Evan Pugh Professor of Geosciences, Pennsylvania State University, for his animation explaining Milankovitch cycles on the National Geographic Channel (*try either of the two web sites at the bottom of page 316 to view this video*). The organization of topics on this page follows Professor Alley's script.

Figure 18.8 Milankovitch cycles—the three planetary variables most commonly attributed to the astronomer Milutin Milankovitch.

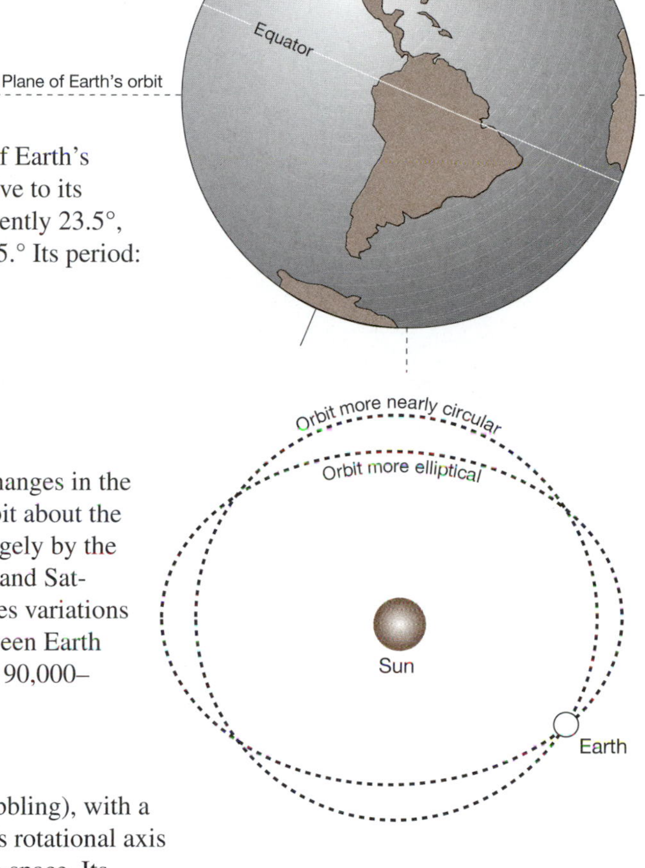

(1) **Obliquity** (tilt of Earth's rotational axis relative to its orbital plane)—currently 23.5°, its range: 21.5°–24.5.° Its period: 41,000 years.

(2) **Eccentricity** (changes in the shape of Earth's orbit about the Sun)—produced largely by the positions of Jupiter and Saturn—which produces variations in the distance between Earth and Sun. Its period: 90,000–100,000 years.

(3) **Precession** (wobbling), with a projection of Earth's rotational axis describing a cone in space. Its period: 19,000– 23,000 years.

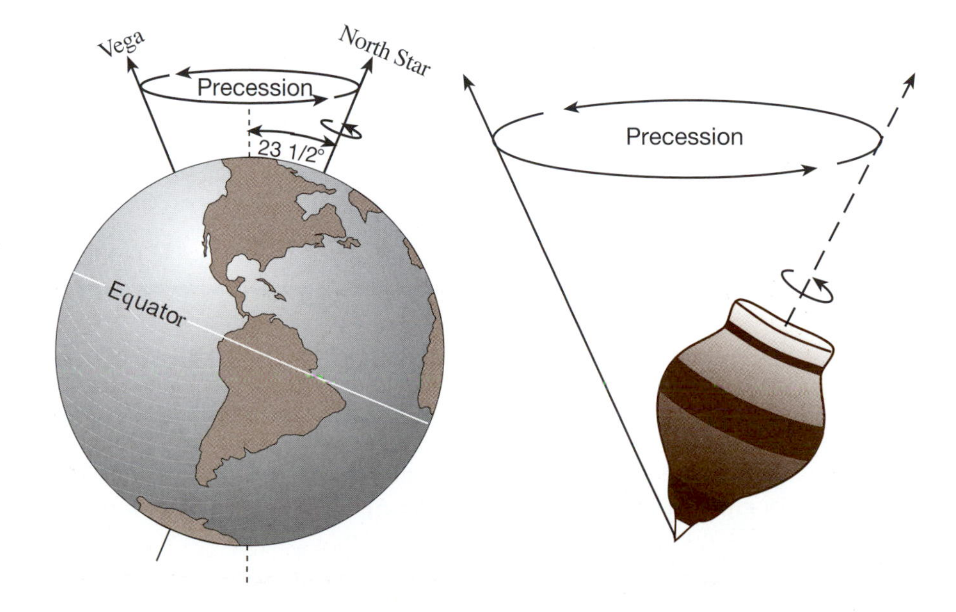

Q18.8 Which two graphs in Figure 18.9 most closely resemble each other? (The explanation comes later.)

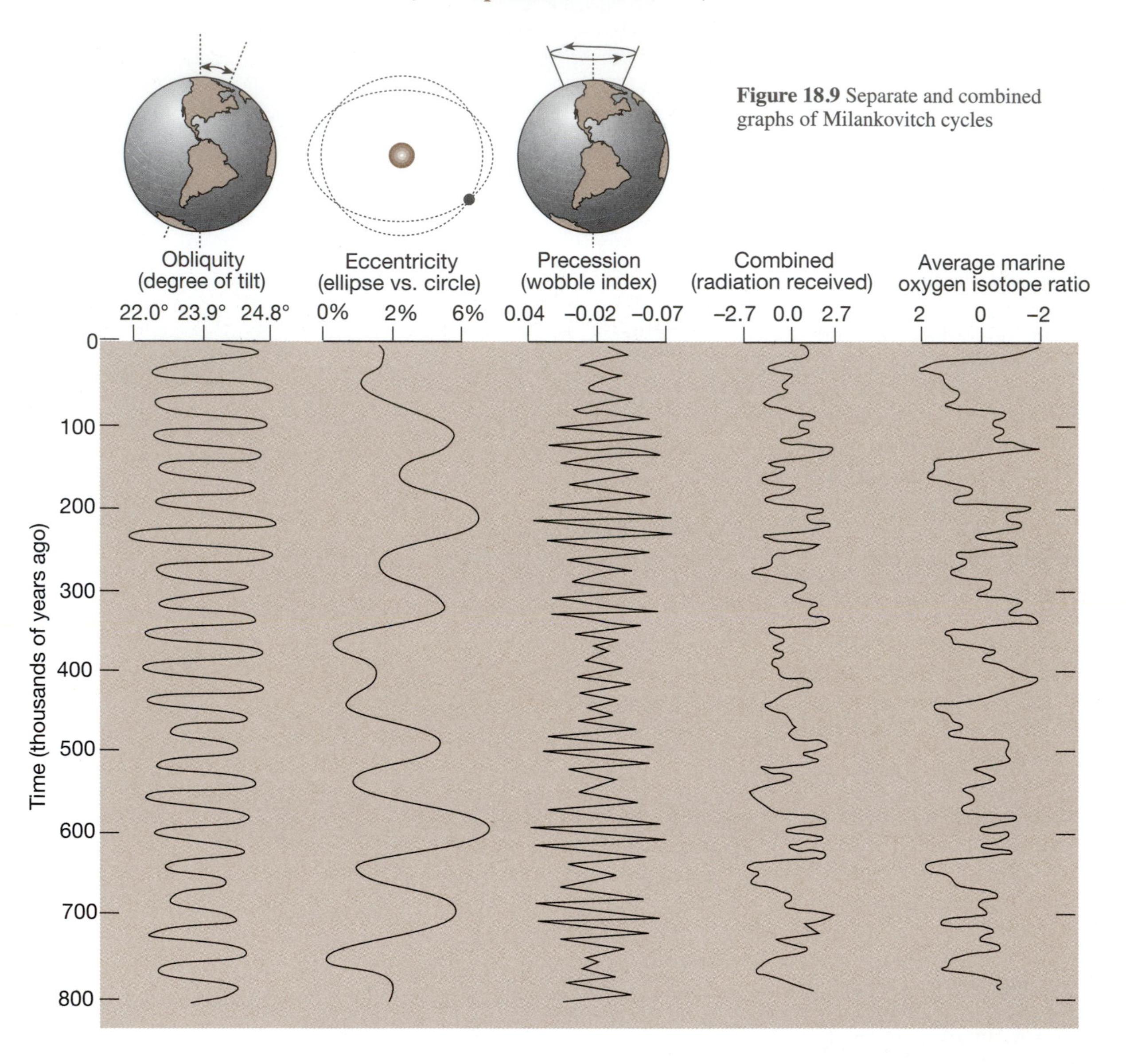

Figure 18.9 Separate and combined graphs of Milankovitch cycles

A 5-minute animated demonstration of Milankovitch cycles

by Richard B. Alley, Evan Pugh Professor of Geosciences
Pennsylvania State University

http://channel.nationalgeographic.com/series/naked-science/3491/Overview#tab-videos//06450 00, or...
Google search for: Naked Science Ice Age Meltdown Video Ice Age Cycles National Geographic Channel

But, hold it! Before going further…a question might arise: What does cause the Milankovitch precession, aka the 'wobble' of Earth about its polar rotational axis?

The answer: Earth is *not* perfectly round. Centrifugal force associated with its rotation about its own internal axis produces an equatorial bulge, but not a great one. Earth's radius at the equator is only some 11 miles greater than either of its two polar radii. But that's enough for the gravitational attraction between moon and Earth's equatorial bulge to try to pull the equatorial bulge into the plane of rotation of the moon. The result: a wobble (Fig. 18.8).

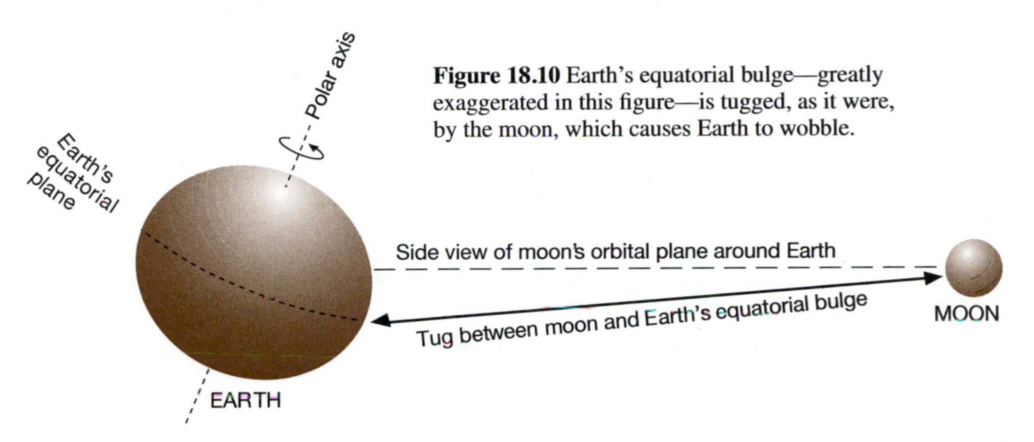

Figure 18.10 Earth's equatorial bulge—greatly exaggerated in this figure—is tugged, as it were, by the moon, which causes Earth to wobble.

Also, the Milankovitch precession of Earth's rotational axis is not to be confused with the one-year orbit of Earth about the sun (Fig. 18.11).

At the present time, in the Northern Hemisphere the date of the *winter solstice* is that of the *shortest* day of the year (December 22), whereas the date of the *summer solstice* is that of the *longest* day of the year (June 22). In the Southern Hemisphere the opposite applies.

Q18.9 The two globes in Figure 18.11 are labeled A and B. Question: Which of the two is in the position of the summer solstice, A or B?

Q18.10 Incidentally, outdoor tennis is a summer sport. So why is the Australian Open held in January, whereas the French Open is held in June?

Figure 18.11 An illustration of Earth's orbit about the sun—a visual representation of the two annual solstices. At this brief passage of one year, Earth's rotational axis is hardly wobbling. To simulate Earth's revolving about the sun while its rotational polar axis is relatively fixed, do this: Draw a complete circle with a pencil, using your hand, rather than your fingers, to describe the circle. Notice that the pencil is relatively fixed, as in the case of Earth's rotational axis over the period of one year.

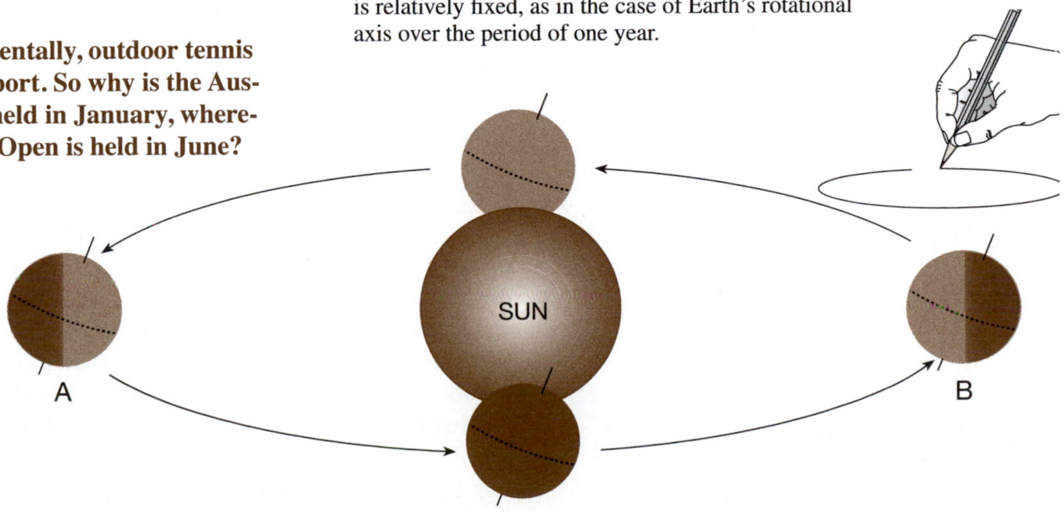

SUN

A

B

E. Stable isotopes—keys to past climates

Oxygen isotopes in reactions

Chemical elements commonly include two or more types of atoms. Members of one type have the same number of **protons** and so behave similarly in *chemical* behavior—whereas members of another type have different numbers of **neutrons** and so behave differently in *physical* behavior—such as *rates of reactions* (e.g., diffusion, evaporation, and precipitation).

Different atomic species of the same element are called **isotopes** (Gr. *isotopos*, equal place on the periodic table of chemical elements). Some isotopes are stable, whereas others are unstable, i.e., *radioactive*. Radioactive **parent isotopes** have nuclei that **decay** by emitting one or more kinds of subatomic particles to form daughter products (Fig. 18.12). Other isotopes are stable, that is they neither emit nor capture subatomic particles. Carbon occurs as the radioactive isotope ^{14}C and also as the **stable isotopes** ^{12}C and ^{13}C. **Carbon-14** serves as a 'clock' in the study of aging carbon-bearing materials, including the remains of early humans. Stable carbon isotopes, along with oxygen isotopes, have been important in the study of *past climates*.

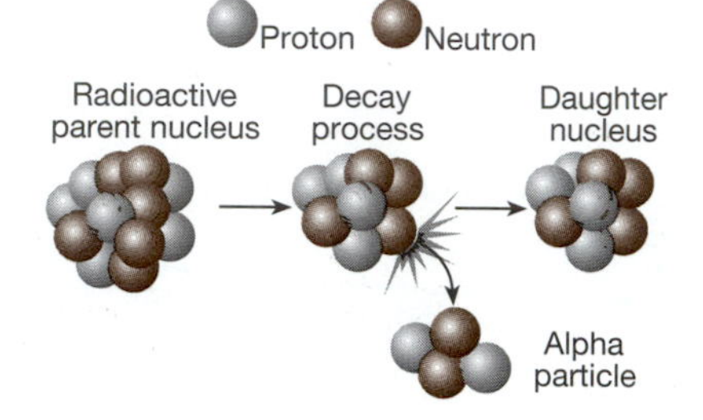

Figure 18.12 Alpha decay is one of the ways in which radioactive nuclei decay to form new isotopes. Alpha decay characterizes the decay of uranium-238, along with its family of products, ending in lead-206.

To repeat—stable isotopes of an element differ in their *rates of reactions* because—for starters—they differ in their masses (Fig. 18.13A).

Atom of ^{16}O
8 electrons
8 protons
8 neutrons
Atom of ^{18}O
8 electrons
8 protons
10 neutrons

Figure 18.13A Oxygen isotopes differ in their masses because of differences in their numbers of neutrons.

Fractionation—Lighter isotopes vibrate faster, so they tend to 'escape' in *evaporation*; whereas heavier isotopes vibrate more slowly, so they tend to be 'captured' in *precipitation* (Fig. 18.13B). This segregation, or partitioning, is known as *fractionation*.

Figure 18.13B Oxygen isotopes are selected in either evaporation or precipitation depending on their masses and vibration frequencies.

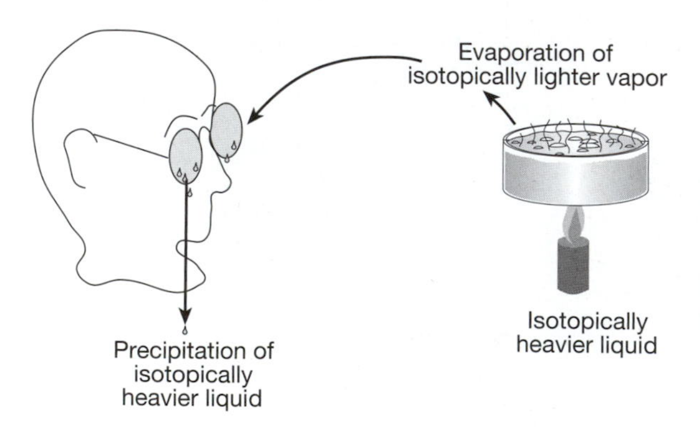

Oxygen isotopes as thermometers

The frequency of vibration of all isotopes increases with elevated temperature. However, the *difference* in vibration frequencies of heavier and lighter isotopes of an element (i.e., fractionation) decreases with increasing temperature (Fig. 18.14). Therefore, the *ratio* of lighter to heavier isotopes in a precipitate is a measure of the temperature at which that precipitate formed.

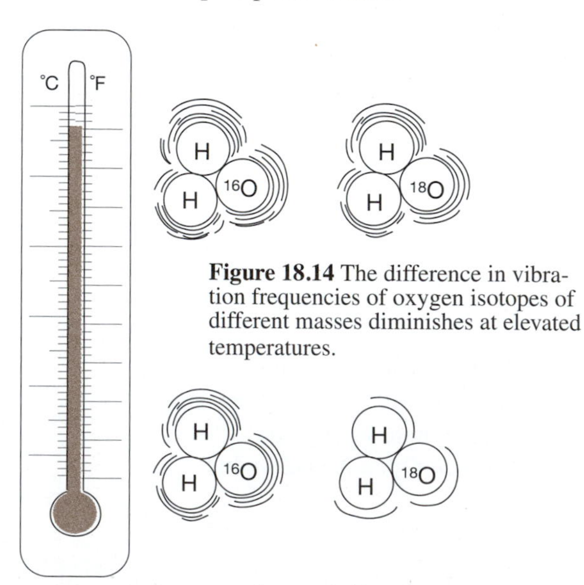

Figure 18.14 The difference in vibration frequencies of oxygen isotopes of different masses diminishes at elevated temperatures.

Q18.11 The lighter isotope carbon-12, relative to the heavier isotope carbon-13, selectively finds its way into organic materials. And so, the ratio of carbon-12 to carbon-13 can be an indicator of ancient *organic* materials. Why is this? *Hint:* Think about the probability of 'collisions' between carbon-12 (relative to that of heavier isotopes of carbon) in atmospheric CO_2 and a 'capturing' substance within plants. What is that substance?

A practical application of oxygen isotopes

Belemnites were squid-like creatures that became extinct at the end of the Cretaceous Period. The internal pin (*phragmocone*) of a belemnite grew through the development of increments analogous to tree rings—only instead of wood, pins of belemnites were made of calcium carbonate ($CaCO_3$). One of the earlier studies of oxygen isotopes as indicators of past environmental temperatures was that of belemnite pins from the Cretaceous Peedee Formation of South Carolina.

Isotopic composition of a Carolina belemnite has been adopted as the international standard zero reference point (abbreviated PDF) for carbon and oxygen ratios. Other isotopic values are known as plus (+) or minus (-) parts per thousand relative to the PDF standard.

Q18.12 In Figure 18.15, there are plotted 7 analyses of the oxygen isotopic composition of a phragmocone. Which of the following statements appears to be closer to the truth? This fossil belemnite was...
(A) Born in the spring, lived 3 years, and died in the spring.
(B) Born in the spring, lived 2 1/2 years, and died in the fall.

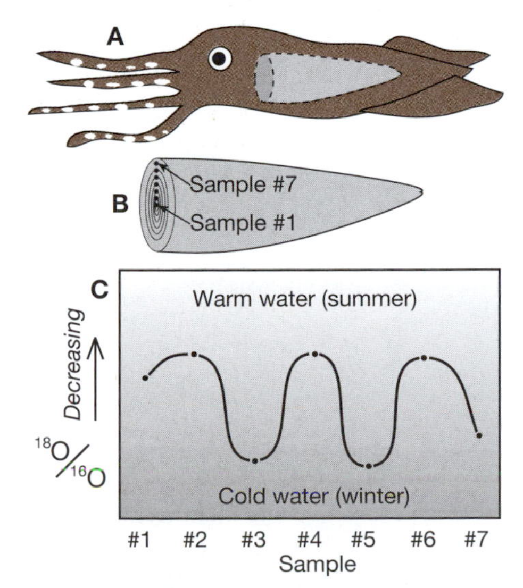

Figure 18.15 Drawings of (**A**) anatomy of a belemnite, (**B**) enlargement of its phragmocone showing growth rings, and (**C**) its life history as suggested by isotopic analyses.

Glacial studies go to sea—and to Antarctica

As we have already seen—as recently as the early twentieth century geologists believed that global glacial history consisted of only four ice advances—the basis for the Pleistocene Epoch. This narrow view was based on glacial records of the Alps, which were soon to be correlated with incomplete and scattered glacial deposits of the American Midwest. (Recall Question 18.4 on page 313.) But more recently, geologists have launched investigations of glacial deposits with much higher resolution—those of deep-sea sediment cores and ice cores from Antarctica and Greenland. Here they found virtually uninterrupted deposits of sediments and ice. And, they discovered new ways in which to apply isotopic analyses.

Q18.13 When sea water evaporates to provide for the growth of global ice sheets, ocean water 'goes heavy,' isotopically speaking. Explain. *Hint:* Review the lesson of Figure 18.14.

Q18.14 A bit of accounting from Figures 18.16, 17: How many glacial events have occurred within the past 2 million years?

Q18.15 The decade of the 1990s joins what two times (millions of years ago) as the warmest on record?

Figure 18.16 During most of the last million years, each glacial-interglacial cycle was about 100,000 years long, whereas earlier cycles were about 40,000 years long.

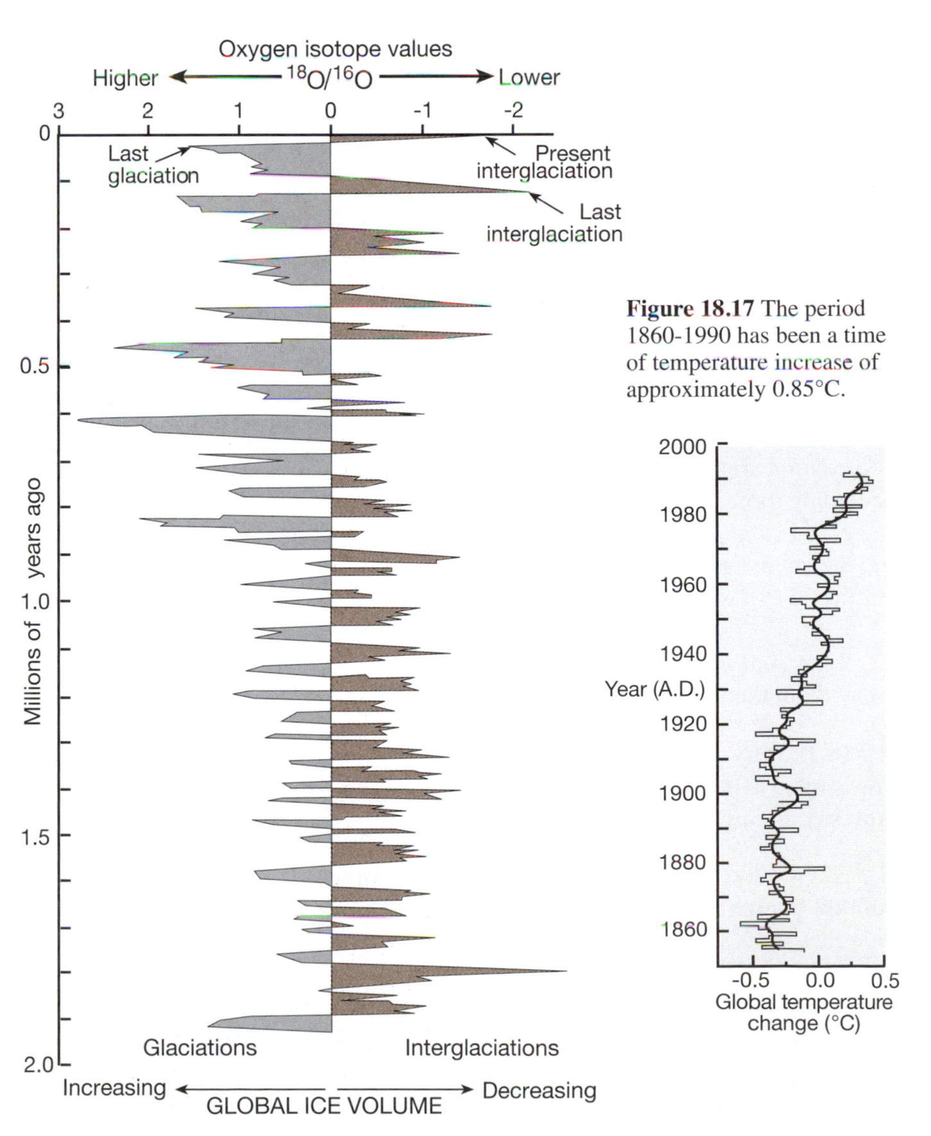

Figure 18.17 The period 1860-1990 has been a time of temperature increase of approximately 0.85°C.

F. About carbon dioxide

Earth as a greenhouse—

Q18.16 Is the *net heat flow* from Earth to the sun, or is it from the sun to Earth? List a couple of observations that support your answer. *Hint:* One observation comes from deep oil-well drilling and another comes from the worlds deepest mines.

Atmospheric greenhouse gases include *water vapor, carbon dioxide, and methane* (natural gas). As illustrated in Figure 18.18, some 30% of short-wave incoming solar energy is reflected back into space by clouds and the ocean, with 70% being absorbed by the ocean, atmosphere, land, and biosphere. Eventually, all of this heat energy is re-emitted in the form of invisible long-wave energy that is absorbed by atmospheric greenhouse gases, thereby heating the atmosphere.

Q18.17 Does this scenario sound familiar? Give a familiar example.

Figure 18.18 Some of the short-wave visible solar radiation reaching Earth is absorbed by land and seas, while some reflects back into space off of snow, ice, clouds, and atmospheric dust. Earth also radiates long-wave infrared radiation, which is entrapped by the chemistry of greenhouse gases and thereby heats the atmosphere.

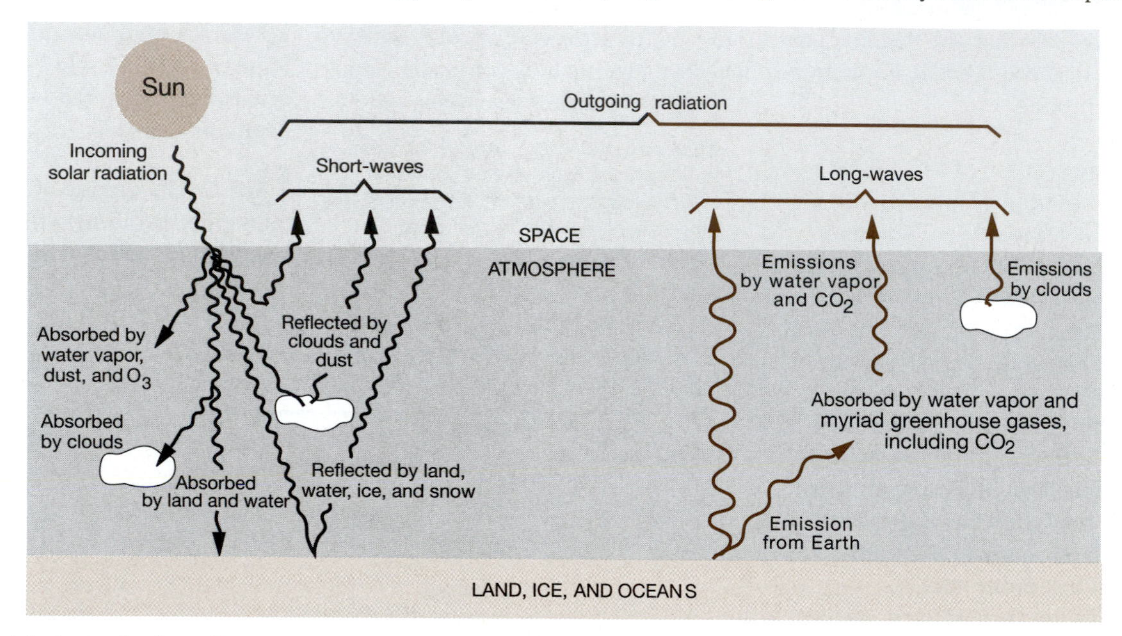

Since 1958 total atmospheric carbon dioxide has been measured at Hawaii's Mauna Loa Observatory. Two kinds of variation are evident in the graph recording those measurements:

(1) Steady increase in carbon dioxide attributed to the burning of fossil fuels.

(2) What appears to be annual fluctuation within the trend.

Q18.18 Why was Hawaii chosen for measuring atmospheric CO_2? Why not, say, in our Midwest?

Q18.19 What accounts for the annual fluctuations within the steady increase in CO_2. *Hint:* How's your knowledge of botany?

Figure 18.19 Atmospheric carbon dioxide increased from 320 parts per million volumetrically to 30 ppmv in 50 years.

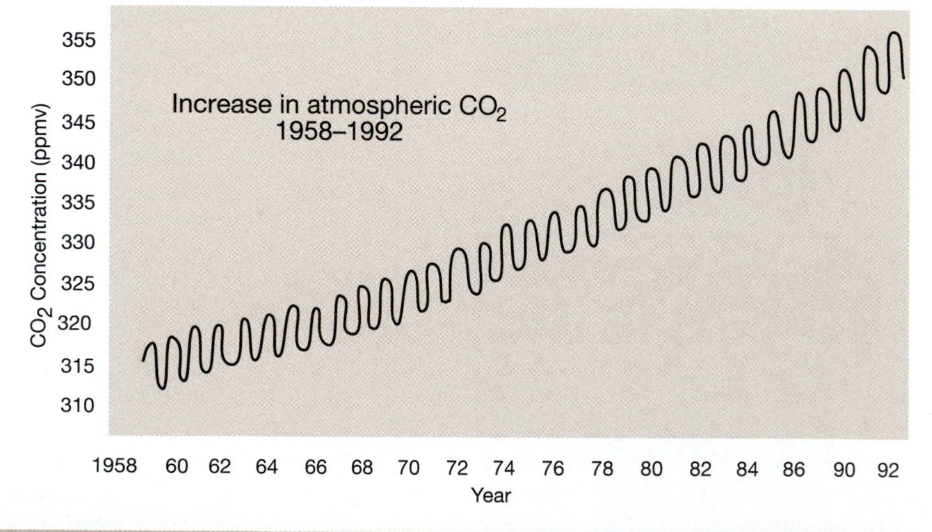

At the 2009 fall meeting of the American Geophysical Union, Robert Alley presented a 50-minute talk with the theme, "*CO_2 is, and has always been, the biggest climate control knob in Earth history.*" The video of Professor Alley's talk is at http://mind.ofdan.cal?p=2771

(Student's name) (Day) (Hour)

(Lab instructor's name)

ANSWER PAGE

18.1

18.2

18.3

18.4

18.5

18.6 (1) (2)

(3) (4) (5)

18.7

18.8

18.9

18.10

18.11

18.12

18.13

18.14

18.15

18.16

18.17

18.18

18.19